Lecture Notes in Computer Science 736

Edited by G. Goos and J. Hartmanis

Advisory Board: W. Brauer D. Gries J. Stoer

Robert L. Grossman   Anil Nerode
Anders P. Ravn   Hans Rischel (eds.)

Hybrid Systems

Springer-Verlag
Berlin Heidelberg New York
London Paris Tokyo
Hong Kong Barcelona
Budapest

Robert L. Grossman   Anil Nerode
Anders P. Ravn   Hans Rischel (Eds.)

Hybrid Systems

Springer-Verlag

Berlin Heidelberg New York
London Paris Tokyo
Hong Kong Barcelona
Budapest

Series Editors

Gerhard Goos
Universität Karlsruhe
Postfach 69 80
Vincenz-Priessnitz-Straße 1
D-76131 Karlsruhe, Germany

Juris Hartmanis
Cornell University
Department of Computer Science
4130 Upson Hall
Ithaca, NY 14853, USA

Volume Editors

Robert Lee Grossman
Department of Mathematics, Statistics & Computer Science
University of Illinois at Chicago, Chicago, IL 60680, USA

Anil Nerode
Mathematical Science Institute, Cornell University
Ithaca, NY 14853, USA

Anders Peter Ravn
Hans Rischel
Department of Computer Science,Technical University of Denmark
DK-2800 Lyngby, Denmark

CR Subject Classification (1991): C.1.m, C.3, D.2.1, F.3.1, F.1-2

ISBN 3-540-57318-6 Springer-Verlag Berlin Heidelberg New York
ISBN 0-387-57318-6 Springer-Verlag New York Berlin Heidelberg

© Springer-Verlag Berlin Heidelberg 1993
Printed in Germany

Typesetting: Camera-ready by author
Printing and binding: Druckhaus Beltz, Hemsbach/Bergstr.
45/3140-543210 - Printed on acid-free paper

PREFACE

This volume of invited refereed papers is inspired by a workshop on the Theory of Hybrid Systems, held October 19-21, 1992 at the Technical University, Lyngby, Denmark, and by a prior Hybrid Systems Workshop, June 10-12, 1991 at the Mathematical Sciences Institute, Cornell University, USA, organized by R.L. Grossman and A. Nerode. Some papers are the final versions of papers presented at these workshops. Some are invited papers from other researchers who were not able to attend the workshops.

We are very grateful to Albert Benveniste, Anil Nerode, Amir Pnueli and Willem-Paul de Roever for their help in organizing this volume. We also wish to thank the following referees: H. R. Andersen, M. Basseville, T. Gautier, P. Le Guernic, C. Jard, Y. Lakhneche, H. H. Løvengreen, E. Rutten, P. Sestoft, J. Sifakis, J. U. Skakkebæk, and A. Yakhnis.

We gratefully acknowledge the financial support of the workshop in Lyngby granted by the Technical University of Denmark under the research programme "Mathematical Modelling of Computer Based Systems".

Department of Computer Science   Anders P. Ravn
Technical University of Denmark   Hans Rischel
Lyngby, June 1993

CONTENTS

Introduction

R.L. Grossman[1], A. Nerode[2], A. Ravn[3] and H. Rischel[3]

[1] Department of Mathematics, Statistics,
& Computer Science (M/C 249)
University of Illinois at Chicago
Chicago, IL 60680, USA
[2] Mathematical Sciences Institute
Cornell University, Ithaca, New York 14850
[3] Department of Computer Science
Technical University of Denmark
DK 2800 Lyngby, Denmark

Hybrid Systems are networks of interacting digital and analog devices. Inherently unstable aircraft and computer aided manufacturing are typical control theory areas of application for hybrid systems, but due to the rapid development of processor and circuit technology modern cars, for instance, and even consumer electronics use software to control physical processes. The identifying characteristic of hybrid systems is that they incorporate both continuous components, usually called plants, which are governed by differential equations, and also digital components, i.e. digital computers, sensors and actuators controlled by programs. These programs are designed to select, control, and supervise the behaviour of the continuous components. Modelling, design, and implementation of hybrid systems has recently become an active area of research in both computer science and control engineering. Hybrid systems are in almost all cases modelled as interacting networks of automata, possibly with an infinite number of states, and input and output letters.

How are hybrid systems to be analysed? How are they to be synthesized? How can we verify that they meet performance requirements? There are many possible approaches, and more questions than answers. Issues that have been addressed for many years in computer science in concurrency, distributed computing and program development and verification have to be rethought to be extended to hybrid systems. Likewise, issues that are classical in control engineering, such as observability, controllability, and stability, have to be rethought to be useful in hybrid systems.

For sequential programs, verifying that programs satisfy specifications comprises proving by induction that (some, all, no) execution sequences satisfy conditions. For concurrent and reactive systems the notion of an execution sequence has to be generalized, but proofs go along the same inductive lines. Hybrid systems are inherently concurrent and reactive. Furthermore, a suitable formalism has to incorporate techniques for specifying real-time constraints, and relating the model of the reactive system to the differential equations which describe the plants.

A suggestion is temporal logic based systems, as in the phase transition systems by *Manna* and *Pnueli* which introduce the notions of sampling computations and important events together with an inductive proof rule for verifying properties of hybrid systems. A more radical proposal is to use interval logic, which is the basis for the extended Duration Calculus by *Zhou, Ravn* and *Hansen*, and the model used by *Henzinger, Manna* and *Pnueli* in refining hybrid systems. It is also possible, as shown by *Lamport*, to extend a notation like TLA+ with explicit variables that denote continuous states and clocks and prove properties of hybrid systems. A similar approach has been taken by *Kurki-Suonio*.

An approach based on extensions of a Hoare style proof system to real-time programs is illustrated by *Hooman*, while *Nicollin, Olivero, Sifakis* and *Yovine* base their approach on extensions of CCS style process algebras. This leads to development of a model checking algorithm for timed state graphs, as developed for integration graphs by *Kesten, Pnueli, Sifakis* and *Yovine*. This work is closely related to the Hybrid Automata of *Alur, Courcoubetis, Henzinger* and *Ho*. This work is very important because it illuminates techniques that eventually may lead to automated support for analysis and even synthesis in the above-mentioned computer science related approaches.

The SIGNAL language introduced by *Benveniste, Le Borgne* and *Le Guernic* beautifully illustrates how synthesis can be supported by a simple, yet powerful design language. It would be an interesting exercise to relate abstract, logic based specifications to such a language which would allow automatic code generation for a controller.

Analysis and synthesis are at the heart of approaches that investigate how to extend differential equation methods for continuous dynamics to cover hybrid systems. But in fact little theory has been developed, apart from solutions of differential equations with discontinuous right hand sides and solutions of non-smooth variational problems. Nevertheless, from the point of view of simulation there is progress. Dynamical systems simulators intended to explore state space, if event driven, can be extended to yield similar phase portraits for hybrid systems as shown by *Guckenheimer, Back* and *Myer*.

"Mode switching" is the most commonly used design method for digital controllers of continuous plants. The plant state space is divided into regions (modes). Changing the control currently used is typically triggered by entering such a mode. For instance, an aircraft control system may have climbing, descending, and level flight modes, in which different control laws are used. Mode switching design is ad hoc for several reasons. The main reason is that it is a very complex mathematical task to identify the possible behaviours in the plant state space of even a small continuous dynamical system. Beyond that, identifying the effects of a proposed mode switching scheme is even more daunting. However, G. Sussman of MIT and his students have had success in using his dynamical system simulator not only to analyse plant state space but also to heuristically guess control schemes which will alter the plant state trajectories to meet performance requirements. Grossman and Myer have taken the approach of precomputing the mode changes resulting from a wide collection of pairs of controls and plant states. To go from one mode to a desired mode one does a table look-up for a suitable control. For real-time applications this leads to the need for high speed database retrieval. Using a discrete event simulation point of view *Lemmon*, *Stiver* and *Antsaklis* introduce algorithms for identifying modes which are useful for control.

A fundamental problem is to find general procedures to extract digital control programs from system models and specifications. For compact convex optimization problems *Nerode* and *Kohn* use a relaxed calculus of variations to extract finite control automata, which guarantee an approximately optimal performance. In their paper on multiple agents this is done in a distributed control context. Kohn has implemented this system in his Declarative Control software. In conventional control theory, a fundamental form of stability is to insure that arbitrary small changes in control and input parameters do not lead to big changes in the resulting plant state trajectories. The usual definitions of stability are not applicable in hybrid systems because the control laws can be changed frequently. The "Models" paper by *Nerode* and *Kohn* proposes stability definitions based on continuity of system functions with respect to non-Hausdorff finite subtopologies of the usual topologies on the spaces of control theory.

Concepts familiar from control and systems theory can be carried over to hybrid systems. But the analogies of many familiar concepts from control systems have still not been worked out for hybrid systems. There are many representations to be studied, such as state space, input-output form, operators, linear representations on higher dimensional spaces, stochastic control, and Markoff process representation, etc. The paper by *Grossman* and *Larson* on hybrid flows introduces the observation space representation of hybrid systems, dual to the state space representation, and the relation to bialgebras.

The mode switching approach is illustrated by *Antsaklis*, *Stiver* and *Lemmon*. The examples of this paper and the one by *Blanke*, *Nielsen* and *Jørgensen* may hopefully be seen as a challenge by those who wish to test their theories. The same concern for application is the theme of the paper on Industrial-Scale Requirements Analysis by *Anderson*, *de Lemos*, *Fitzgerald* and *Saeed*, and the paper on Requirements Documentation by *Engel*, *Kubica*, *Madey*, *Parnas*, *Ravn* and *van Schouwen*.

Verifying Hybrid Systems *

Zohar Manna[1] and Amir Pnueli[2]

[1] Department of Computer Science, Stanford University
Stanford, CA 94305
e-mail: manna@cs.stanford.edu
[2] Department of Applied Mathematics and Computer Science
Weizmann Institute, Rehovot, Israel
e-mail: amir@wisdom.weizmann.ac.il

Abstract. Hybrid systems are modeled as *phase transition systems* with sampling semantics. By identifying a set of *important events* it is ensured that all significant state changes are observed, thus correcting previous drawbacks of the sampling computations semantics. A proof rule for verifying properties of hybrid systems is presented and illustrated on several examples.

Keywords: Temporal logic, real-time, specification, verification, hybrid systems, statecharts, proof rules, phase transition system, sampling semantics, important events.

1 Introduction

Hybrid systems are reactive systems that intermix discrete and continuous components. Typical examples are digital controllers that interact with continuously changing physical environments.

A formal model for hybrid systems was proposed in [MMP92], based on the notion of *phase transition systems* (PTS). Two types of semantics were considered in [MMP92]. The first semantics, to which we refer here as the *super dense* semantics, is based on *hybrid traces* which view computations as functions from pairs of a real number (representing time) and an index (representing discrete steps) to states. The second semantics is that of *sampling computations*, which sample the continuous behavior of a hybrid system at countably many observation points.

It is argued in [MMP92] that the super dense semantics provides a more accurate and faithful description of hybrid behaviors. In particular, it never

---

* This research was supported in part by the National Science Foundation under grant CCR-92-23226, by the Defense Advanced Research Projects Agency under contract NAG2-703, by the United States Air Force Office of Scientific Research under contract F49620-93-1-0139, by the European Community ESPRIT Basic Research Action Project 6021 (REACT) and by the France-Israel project for cooperation in Computer Science.

misses events of a variable crossing a threshold or a state predicate changing truth value. Sampling computations, on the other hand, may miss such an event if it happens to occur between two observation points.

However, when comparing the two approaches with respect to ease of verification, it is obvious that this is a wide open question under the super dense view, while verification under the sampling view can be obtained by generalizing the sequence-based methods used for reactive and real-time systems. That such a generalization is possible has been indicated in [SBM92] and [AL92].

In this paper, we improve the accuracy with which sampling computations describe hybrid behaviors. This is done by augmenting the model of phase transition systems by a finite set of *important events* that should never be missed. Subsequently, we present a rule for proving invariance properties of hybrid systems under the sampling computations semantics.

The power of this rule is illustrated by proving several invariance properties of hybrid systems.

2 Computational Model: Phase Transition System

Hybrid systems are modeled as phase transition systems [MMP92]. The definition presented here introduces an additional component: a set of important events, whose purpose is to ensure that all significant state changes are observed in every sampling computation.

As the time domain we take the nonnegative reals R^+. In some cases, we also need its extension $R^\infty = R^+ \cup \{\infty\}$.

A *phase transition system* (PTS) $\Phi = \langle V, \Theta, \mathcal{T}, \mathcal{A}, \mathcal{I} \rangle$ consists of:

- $V = \{u_1, ..., u_n\}$: A finite set of *state variables*. The set $V = V_d \cup V_c$ is partitioned into V_d the set of *discrete variables* and V_c the set of *continuous variables*. Continuous variables have always the type *real*. The discrete variables can be of any type.

  We define a *state* s to be a type-consistent interpretation of V, assigning to each variable $u \in V$ a value $s[u]$ over its domain. We denote by Σ the set of all states.

- Θ : The *initial condition*. A satisfiable assertion characterizing the initial states.

- $\mathcal{T}$: A finite set of *transitions*. Each transition $\tau \in \mathcal{T}$ is a function

$$\tau : \Sigma \mapsto 2^\Sigma,$$

  mapping each state $s \in \Sigma$ into a (possibly empty) set of *τ-successor* states $\tau(s) \subseteq \Sigma$.

  The function associated with a transition τ is represented by an assertion $\rho_\tau(V, V')$, called the *transition relation*, which relates a state $s \in \Sigma$ to its τ-successor $s' \in \tau(s)$ by referring to both unprimed and primed versions of the state variables. An unprimed version of a state variable refers to its value in s, while a primed version of the same variable refers to its value in s'. For example, the assertion $x' = x + 1$ states that the value of x in s' is greater by 1 than its value in s.

The enabledness of a transition τ can be expressed by the formula

$$En(\tau): \ (\exists V')\rho_\tau(V, V'),$$

which is true in s iff s has some τ-successor. The enabling condition of a transition τ can always be written as $\delta \wedge \kappa$, where δ is the largest subformula that does not depend on continuous variables. We call κ the *continuous part* of the enabling condition, and denote it by $En_c(\tau)$.

A transition can also change the value of a continuous variable.

- $\mathcal{A}$: A finite set of *activities*. Each activity $\alpha \in \mathcal{A}$ is a conditional differential equation of the form:

$$p \ \rightarrow \ \dot{x} = e,$$

where p is a predicate over V_d called the *activation condition* of α, $x \in V_c$ is a continuous state variable, and e is an expression over V. We say that the activity α *governs* variable x. Activity α is said to be *active* in state s if its activation condition p holds on s. If p is *true*, it may be omitted.

It is required that the activation conditions of the activities that govern any variable x are exhaustive and exclusive, i.e., exactly one of them holds on any state.

- A set of *important events* $\mathcal{I}$. This is a finite set of assertions that includes at least the assertions $En_c(\tau)$, for each $\tau \in \mathcal{T}$. These are assertions such that changes in their truth values must be observable. Usually, $\mathcal{I}$ includes, in addition to the assertions $\{En_c(\tau)\}$, all the assertions that appear in specifications of the system.

We introduce a special variable T, called the *clock variable*. At any point in an execution of a system, T has a value over R^+ representing the current time. The set of variables $V_T = V \cup \{T\}$ is called the set of *situation variables*. A type consistent interpretation of V_T is called a *situation*, and the set of all situations is denoted by Σ_T. Often, we represent a situation as a pair $\langle s, t \rangle$ where s is a state and $t \in \mathsf{R}^+$ is the interpretation of the clock T.

In [MP93], we present a similar model for hybrid systems which associates with each transition a pair of real numbers $l_\tau \leq u_\tau$ representing a lower and upper bound on the length of time a transition can be continuously enabled without being taken. The approach taken here makes no such association as part of the model. Instead, it is recommended that each transition that needs such bounds employs a special continuous *timer* variable that grows linearly with time and constrains the length of intervals in which the transition was continuously enabled.

Activity Successors

Consider a phase transition system Φ, and let $\langle s_1, t_1 \rangle$ and $\langle s_2, t_2 \rangle$ be two situations of Φ with $t_1 < t_2$. An *evolution* from $\langle s_1, t_1 \rangle$ to $\langle s_2, t_2 \rangle$ consists of a set of functions $F : \{f_x(t) \mid x \in V_c\}$ that are differentiable in the interval $[t_1, t_2]$ and satisfy the following requirements:

- $f_x(t_1) = s_1[x]$ and $f_x(t_2) = s_2[x]$. Thus, the values of $f_x(t)$ at the boundaries of the interval $[t_1, t_2]$ agree with the interpretation of x by s_1 and s_2, respectively.
- $s_1[y] = s_2[y]$ for every $y \in V_d$. That is, states s_1 and s_2 agree on the values of all discrete variables.
- For every activity $p \rightarrow \dot{x} = e$ that governs x, if p holds at s_1, then f_x satisfies the differential equation

$$\dot{f}_x(t) = e(F)$$

in the interval $[t_1, t_2]$, where the expression $e(F)$ is obtained from e by replacing each occurrence of a variable $y \in V_c$ by the function $f_y(t)$.
- For every assertion $\varphi \in \mathcal{I}$, $\varphi(t)$ has a uniform truth value for all $t \in (t_1, t_2)$, which equals either $\varphi(t_1)$ or $\varphi(t_2)$.

The last requirement ensures that the truth value of every important assertion $\varphi \in \mathcal{I}$ is uniform throughout the interior of the evolution interval, and matches its value at one of the endpoints of the interval. In particular, it disallows a change in the truth value of φ in an internal point. It also guarantees that any value assumed by φ at internal points is also represented at one of the endpoints. This implies that φ cannot be true at both endpoints but false in the middle, nor false at both endpoints but true in the middle.

If such an evolution exists, we say that the situation $\langle s_2, t_2 \rangle$ is an *activity successor* of the situation $\langle s_1, t_1 \rangle$. Assuming that the differential equations satisfy some reasonable healthiness conditions, such as the Lipschitz condition, there exists at most one evolution from $\langle s_1, t_1 \rangle$ to $\langle s_2, t_2 \rangle$. In fact, the functions F are uniquely determined by the situation $\langle s_1, t_1 \rangle$.

We denote by $\mathcal{A}(\langle s_1, t_1 \rangle)$ the set of all activity successors of $\langle s_1, t_1 \rangle$.

Consider, for example, a trivial phase transition system with a single (continuous) state variable x, no transitions, a single activity α given by $\alpha : \dot{x} = -1$, and an empty $\mathcal{I}$. Then, the following are examples of a situation and its activity successor:

$$s_1 : \langle x : \; 0\,, T : 1 \rangle \in \mathcal{A}(s_0 : \langle x : 1\,, T : 0 \rangle)$$
$$s_2 : \langle x : -1\,, T : 2 \rangle \in \mathcal{A}(s_1 : \langle x : 0\,, T : 1 \rangle)$$
$$s_2 : \langle x : -1\,, T : 2 \rangle \in \mathcal{A}(s_0 : \langle x : 1\,, T : 0 \rangle)$$

Sampling Computations

A *sampling computation* of a phase transition system $\Phi : \langle V, \Theta, \mathcal{T}, \mathcal{A}, \mathcal{I} \rangle$ is an infinite sequence of situations

$$\sigma : \langle s_0, t_0 \rangle, \; \langle s_1, t_1 \rangle, \; \langle s_2, t_2 \rangle, \; \ldots$$

satisfying:

- *Initiation:* $s_0 \models \Theta$ and $t_0 = 0$.
- *Consecution:*   For each $j = 0, 1, ...,$
  - Either $t_j = t_{j+1}$ and $s_{j+1} \in \tau(s_j)$ for some transition $\tau \in \mathcal{T}$ — transition τ is taken at j, or
  - $\langle s_{j+1}, t_{j+1} \rangle$ is an activity successor of $\langle s_j, t_j \rangle$ (implying $t_j < t_{j+1}$) — a continuous phase takes place at step j.
- *Urgency:*   No continuous step may disable an enabled transition. Thus, when some transition τ is enabled but every continuous step of positive length will cause it to become disabled, τ must be taken or disabled before any continuous step.
- *Time Divergence:*   As i increases, t_i grows beyond any bound.

Example

Consider a simple phase transition system Φ_1 given by:

State Variables $V = V_c : \{x\}$
Initial Condition Θ $: x = 1$
Transitions $\mathcal{T}$ $: \{\tau\}$, where $\rho_\tau : (x = -1) \wedge (x' = 1)$
Activities $\mathcal{A}$ $: \{\alpha\}$, where $\alpha : \dot{x} = -1$
Important events $\mathcal{I}$ $: \{x = -1\}$

Fig. 1 depicts the full behavior of system Φ_1 as a function from $T \in \mathsf{R}^+$ to the value of x. Note that this is not really a function because at $T = 2$ (and all other

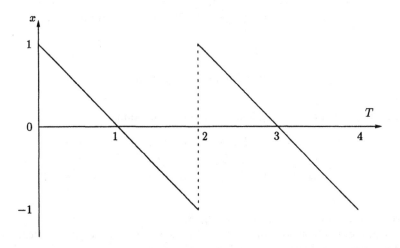

Fig. 1. Full behavior of hybrid system Φ_1.

even positive integers) x has two values: -1 which is the value it attains at the end of the continuous phase, but also $+1$ which is the result of taking transition τ at these points.

There are (uncountably) many sampling computations that correspond to this full behavior.

For example, the sampling computation

$$\sigma_1 : s_0 : \langle x : 1\,, T : 0\rangle \xrightarrow{\alpha} s_1 : \langle x : -1\,, T : 2\rangle \xrightarrow{\tau}$$
$$s_2 : \langle x : 1\,, T : 2\rangle \xrightarrow{\alpha} s_3 : \langle x : -1\,, T : 4\rangle \xrightarrow{\tau}$$
$$\cdots$$

corresponds to sampling the full behavior as shown in Fig. 2.

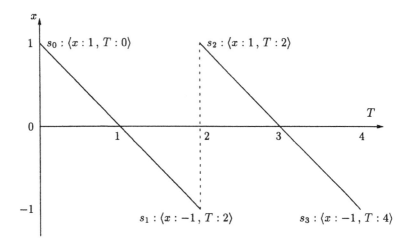

Fig. 2. Sampling computation σ_1.

A more frequent sampling leads to the computation

$$\sigma_2 : s_0 : \langle x : \quad 1\,, T : 0\rangle \xrightarrow{\alpha} s_1 : \langle x : \quad 0\,, T : 1\rangle \xrightarrow{\alpha}$$
$$s_2 : \langle x : -1\,, T : 2\rangle \xrightarrow{\tau} s_3 : \langle x : \quad 1\,, T : 2\rangle \xrightarrow{\alpha}$$
$$s_4 : \langle x : \quad 0\,, T : 3\rangle \xrightarrow{\alpha} s_5 : \langle x : -1\,, T : 4\rangle \xrightarrow{\tau}$$
$$\cdots$$

whose sampling points are shown in Fig. 3.

In comparison, consider phase transition system Φ_2 which is identical to Φ_1 in all components, except for $\mathcal{I}$, which is given by

Important events $\mathcal{I}_2 : \{x = -1,\ x = 0\}$.

Thus, system Φ_2 considers the assertion $x = 0$ to be an important event in addition to $x = -1$, which is the enabling condition of transition τ. The computation

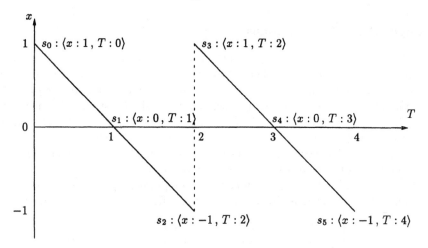

Fig. 3. Sampling computation σ_2.

σ_2 is a sampling computation of Φ_2 as well. However, the sequence σ_1 is not. Informally, this is because σ_1 fails to observe the (infinitely many) points at which x becomes 0. Formally, $s_1 : \langle x : -1, T : 2\rangle$ is no longer an activity successor of $s_0 : \langle x : 1, T : 0\rangle$ because the evolution from s_0 to s_1 did not respect the condition that important assertions do not change their truth value in the middle of a continuous step.

Zeno Computations and Progressive Systems

The requirement of time divergence excludes *Zeno computations* in which there are infinitely many state-changes within a finite time interval [AL92]. Unfortunately, not every phase transition system is guaranteed to have computations that satisfy all the requirements given above.

Consider, for example, the PTS Φ_3 given by:

State Variables V : $\{x, y\}$
Initial Condition Θ : $x = y = 0$
Transitions $\mathcal{T}$: $\{\tau\}$, where $\rho_\tau : (x = 0) \wedge (y' = y + 1)$
Activities $\mathcal{A}$: $\{\alpha\}$, where $\alpha : \dot{x} = 1$
Important events $\mathcal{I}$: $\{x = 0\}$

This PTS does not have a computation. This is because, as long as $x = 0$, transition τ is enabled. Any continuous step will disable τ. Therefore, τ must be taken infinitely many times, not allowing time to progress.

A PTS is called *progressive* if it has at least one computation. From now on, we restrict our attention to progressive transition systems.

Super-Dense Computations

In addition to sampling computations, [MMP92] presents another class of computations, to which we refer here as *super-dense computations*. Sampling computations have the signature $N \mapsto \Sigma \times R^+$, that is, each natural number $j = 0, 1, \ldots$ is mapped to a pair consisting of a state and a real-time stamp, i.e., a situation.

In contrast, super-dense computations have the signature $R^+ \times N \mapsto \Sigma$, that is, each pair $\langle t, i \rangle$, where $t \in R^+$ and $i \in N$, is mapped to a state $s \in \Sigma$. The pair $\langle t, i \rangle$ identifies a time stamp t and a step number i. The step numbers correspond to the transitions that are taken at time instant t.

For example, the (single) super-dense computation produced by phase transition system Φ_1 is given by a function $x(r, i)$ from $R^+ \times N$ to R defined as

$$
x(r, i) = \begin{cases}
1 & \text{for} \quad r = 0 \quad \text{and } i \geq 0 \\
1 - r & \text{for } 0 < r < 2 \text{ and } i \geq 0 \\
-1 & \text{for} \quad r = 2 \quad \text{and } i = 0 \\
1 & \text{for} \quad r = 2 \quad \text{and } i \geq 1 \\
3 - r & \text{for } 2 < r < 4 \text{ and } i \geq 0 \\
-1 & \text{for} \quad r = 4 \quad \text{and } i = 0 \\
1 & \text{for} \quad r = 4 \quad \text{and } i \geq 1 \\
\vdots
\end{cases}
$$

An argument, made in [MMP92], claims that the super-dense semantics provides a more precise representation of the behavior of hybrid systems than the semantics of sampling computations. The main criticism of sampling computations complains that some important events may fail to be observed, such as the event of x becoming 0, to which computation σ_1 is oblivious.

As previously seen, this difficulty has been overcome by the introduction of the important event component. Consequently, in this paper we continue to use the sampling-computation semantics. The advantages of the sampling semantics are that it is simpler than the super-dense semantics and conforms better with sequence based verification methods.

System Description by Hybrid Statecharts

Hybrid systems can be conveniently described by an extension of statecharts [Har87] called *hybrid statecharts*. The main extension is

- States may be labeled by (unconditional) differential equations. The implication is that the activity associated with the differential equation is active precisely when the state it labels is active.

We illustrate this form of description by the example of *Cat and Mouse* taken from [MMP92]. At time $T = 0$, a mouse starts running from a certain position on the floor in a straight line towards a hole in the wall, which is at a distance X_0 from the initial position. The mouse runs at a constant velocity v_m. After a delay of Δ time units, a cat is released at the same initial position and chases the mouse at velocity v_c along the same path. Will the cat catch the mouse, or will the mouse find sanctuary while the cat crashes against the wall?

The statechart in Fig. 4 describes the possible scenarios.

Initially $x_c = x_m = X_0$, $y = 0$

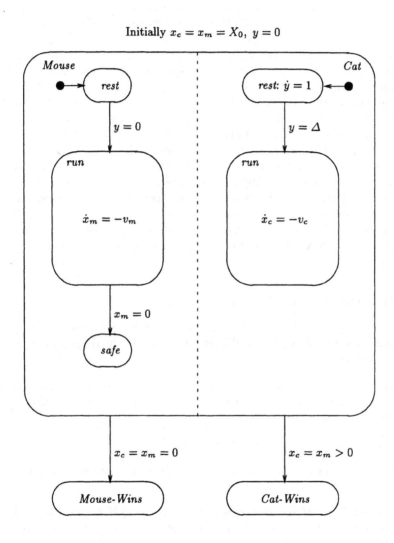

Fig. 4. Specification of Cat and Mouse.

The specification (and underlying phase transition system) uses the continuous state variables x_m and x_c, measuring the distance of the mouse and the cat, respectively, from the wall. The continuous variable y serves as a timer, measuring the waiting time of the cat before it starts running. The specification refers to the constants X_0, v_m, v_c, and Δ.

A behavior of the system starts with states $M.rest$ and $C.rest$ active, variables x_m and x_c set to the initial value X_0, and variable y set to 0. The mouse proceeds immediately to the state of running, in which its variable x_m changes continuously according to the equation $\dot{x}_m = -v_m$. The cat waits for a delay of Δ before entering its running state, using the timer variable y that increases linearly with time. There are several possible termination scenarios. If the event $x_m = 0$ happens first, the mouse reaches sanctuary and moves to state *safe*, where it waits for the cat to reach the wall. As soon as this happens, detectable by the condition $x_c = x_m = 0$ becoming true, the system moves to state *Mouse-Wins*. The other possibility is that the event $x_c = x_m > 0$ occurs first, which means that the cat overtook the mouse before the mouse reached sanctuary. In this case they both stop running and the system moves to state *Cat-Wins*. The compound conditions $x_c = x_m = 0$ and $x_c = x_m > 0$ stand for the conjunctions $x_c = x_m \wedge x_m = 0$ and $x_c = x_m \wedge x_m > 0$, respectively.

This diagram illustrates the typical interleaving between continuous activities and discrete state changes which, in this example, only involve changes of control.

The idea of using statecharts with continuous activities associated with certain states (usually basic ones) was already suggested in [Har84]. According to this suggestion, these states are associated with activities that represent physical (and therefore possibly continuous) operations and interactions with the environment.

The Underlying Phase Transition System

Following the graphical representation, we identify the phase transition system underlying the picture of Fig. 4. We refer to states in the diagram that do not enclose other states as *basic states*.

- *State Variables:* Given by $V_c = \{x_c, x_m, y\}$ and $V_d = \{\pi\}$. Variable π is a control variable whose value is a set of basic states of the statechart.
- *Initial Condition:* Given by

$$\Theta : \pi = \{M.rest, C.rest\} \ \wedge \ x_c = x_m = X_0 \ \wedge \ y = 0.$$

- *Transitions:* Listed together with the transition relations associated with them.

$$M.rest\text{-}run : M.rest \in \pi \ \wedge \ y = 0 \ \wedge \ \pi' = \pi - \{M.rest\} \cup \{M.run\}$$
$$C.rest\text{-}run : C.rest \in \pi \ \wedge \ y = \Delta \ \wedge \ \pi' = \pi - \{C.rest\} \cup \{C.run\}$$
$$M.run\text{-}safe : M.run \in \pi \ \wedge \ x_m = 0 \ \wedge \ \pi' = \pi - \{M.run\} \cup \{M.safe\}$$
$$M.win : (Active \cap \pi) \neq \phi \ \wedge \ x_c = x_m = 0 \ \wedge \ \pi' = \{Mouse\text{-}Wins\}$$
$$C.win : (Active \cap \pi) \neq \phi \ \wedge \ x_c = x_m > 0 \ \wedge \ \pi' = \{Cat\text{-}Wins\}$$

The set *Active* stands for the set of basic states

$$\{M.rest, \ M.run, \ M.safe, \ C.rest, \ C.run\}.$$

- *Activities:* Six activities represent the running activities of the two participants and the progress of the timer. Their equations are given by

$$\alpha_m^{on} : M.run \in \pi \rightarrow \dot{x}_m = -v_m$$
$$\alpha_m^{off} : M.run \notin \pi \rightarrow \dot{x}_m = 0$$
$$\alpha_c^{on} : C.run \in \pi \rightarrow \dot{x}_c = -v_c$$
$$\alpha_c^{off} : C.run \notin \pi \rightarrow \dot{x}_c = 0$$
$$\alpha_y^{on} : C.rest \in \pi \rightarrow \dot{y} = 1$$
$$\alpha_y^{off} : C.rest \notin \pi \rightarrow \dot{y} = 0$$

- *Important Events:* Given by

$$\mathcal{I} : \{x_m = 0, \ x_c = x_m, \ x_m > 0, \ y = 0, \ y = \Delta\}$$

System Description by Textual Programs

In [MP91b], we introduced a simple concurrent programming language. It is possible to extend this simple programming language to represent timed and hybrid behaviors as well. The resulting language is a subset of the language *Statext* introduced in [KP92b], and shown there to have expressive power equal to that of hybrid statecharts.

The main extensions to the language are:

- Each statement is assigned lower and upper bounds for the length of time the statement may be continuously enabled before it must be executed. Typically, these bounds may apply uniformly to all statements. For example, we may assign time bounds $[1, 5]$ to the transitions associated with all statements. These time bounds are referenced in the transition associated with the statement.
- We introduce the *preemption* statement $S_1 \ \cup \ S_2$. The intended meaning of the *preemption* statement is
  - Execute S_1 forever or until it terminates, or
  - Execute S_1 until a first step of S_2 is taken, and then continue to execute S_2.
- Differential equations are also acceptable as statements of the extended language. A differential equation statement never terminates, and the only way to get out of it is by preemption, using the $\cup$ construct. Statements consisting of differential equations are associated with activities, in contrast to all other statements which are associated with transitions. Thus, the statement

$$\ell : \dot{x} = e$$

gives rise to the activity

$$\alpha_\ell : \quad \ell \in \pi \ \rightarrow \ \dot{x} = e.$$

We refer to this activity as an *explicit activity* since it corresponds to an explicit statement in the program.

Besides the explicit activities, each continuous variable $x \in V_c$ also has an implicit *default activity* α_x^{off} which controls its continuous change when none of the explicit activities governing x is active. If $\ell_1, \ldots, \ell_m$ are all the statements giving rise to explicit activities for x, then its default activity is given by

$$\alpha_x^{off} : \quad \{\ell_1, \ldots, \ell_m\} \cap \pi = \phi \;\rightarrow\; \dot{x} = 0.$$

Instead of giving a full definition of the extended language, we will illustrate it on an example.

Example: Fig. 5 presents a simple program consisting of two processes communicating by the shared variable x, initially set to 0. Process P_1 keeps incrementing variable y as long as $x < 1$. Process P_2 represents a continuous component that lets x grow linearly from 0 until it reaches the value 1. At that point, statement m_1 intervenes and shuts off the continuous growth of x, terminating process P_2. Once process P_1 detects that $x = 1$, it terminates too. Note that, once $x = 1$, statement m_1 must be taken because, otherwise, x will continue to grow beyond 1, disabling the transition associated with m_1.

$$
\boxed{
\begin{array}{c}
y: \textbf{integer where } y = 0 \\
x: \textbf{real } \text{ where } x = 0 \\[1em]
\left[
\begin{array}{l}
\ell_0 : \textbf{while } x < 1 \textbf{ do} \\
\quad \left[\ell_1 : \; y := y + 1 \right] \\
\ell_2 :
\end{array}
\right]
\quad \| \quad
\left[
\begin{array}{l}
[m_0 : \dot{x} = 0.2] \; \cup \; [m_1 : \textbf{await } x = 1] \\
m_2 :
\end{array}
\right] \\[1em]
\quad - \quad P_1 \quad - \qquad\qquad\qquad\qquad - \quad P_2 \quad -
\end{array}
}
$$

Fig. 5. Program HYB-T: A Hybrid textual Program

With each of the statements of P_1, we associate the uniform time bounds $[1, 5]$.

We show how to define a phase transition system $\Phi_{\text{HYB-T}}$ corresponding to program HYB-T. To enforce the timing requirements for the statements of process P_1, we introduce a (continuous) timer variable z.

- *State Variables:* $V = V_c : \{x, z\} \; \cup \; V_d : \{\pi, y\}$.

  Variable π is a control variable ranging over subsets of the locations of program HYB-T: $\{\ell_0, \ell_1, \ell_2, m_0, m_1, m_2\}$. The value of π in a state denotes all the locations of the program in which control currently resides.

- *Initial Condition*

$$\Theta : \pi = \{\ell_0, m_0, m_1\} \; \wedge \; x = y = z = 0$$

- *Transitions:* $\mathcal{T}: \{\ell_0, \ell_1, m_1\}$ with transition relations:

$$\rho_{\ell_0} : \ell_0 \in \pi \ \wedge\ 1 \leq z \leq 5 \ \wedge\ \begin{pmatrix} x < 1 \ \wedge\ \pi' = \pi - \{\ell_0\} \cup \{\ell_1\} \\ \vee \\ x \geq 1 \ \wedge\ \pi' = \pi - \{\ell_0\} \cup \{\ell_2\} \end{pmatrix} \ \wedge\ z' = 0$$

$\rho_{\ell_1} : \ell_1 \in \pi \ \wedge\ 1 \leq z \leq 5 \ \wedge\ \pi' = \pi - \{\ell_1\} \cup \{\ell_0\} \ \wedge\ y' = y + 1 \ \wedge\ z' = 0$

$\rho_{m_1} : m_1 \in \pi \ \wedge\ x = 1 \ \wedge\ \pi' = \pi - \{m_0, m_1\} \cup \{m_2\}$

As usual, a transition τ preserves the value of every variable whose primed version does not appear in the transition relation for τ. For example, transition ℓ_0 preserves the values of variables x and y.

- *Activities:* $\mathcal{A}: \{\alpha_x^{on}, \alpha_x^{off}, \alpha_z\}$ given by:

$$\begin{aligned} \alpha_x^{on} &: m_0 \in \pi \rightarrow \dot{x} = 0.2 \\ \alpha_x^{off} &: m_0 \notin \pi \rightarrow \dot{x} = 0 \\ \alpha_z &: \qquad\qquad\quad \dot{z} = 1 \end{aligned}$$

- *Important Events:* $\mathcal{I}: \{x = 1, \ 1 \leq z, \ z \leq 5\}$.

To fully comprehend the behavior of this program, we present two possible computations of program HYB-T. In the presentation of these computations, we use the notation $\pi : \{\ell_i, m_{0,1}\}$ as abbreviation for $\pi : \{\ell_i, m_0, m_1\}$. The first computation attempts to maximize the value of y on termination.

$\langle \pi : \{\ell_0, m_{0,1}\}, \ x : 0.0, \ y : 0, \ z : 0, \ T : 0 \rangle \xrightarrow{cont}$
$\langle \pi : \{\ell_0, m_{0,1}\}, \ x : 0.2, \ y : 0, \ z : 1, \ T : 1 \rangle \xrightarrow{\ell_0}$
$\langle \pi : \{\ell_1, m_{0,1}\}, \ x : 0.2, \ y : 0, \ z : 0, \ T : 1 \rangle \xrightarrow{cont}$
$\langle \pi : \{\ell_1, m_{0,1}\}, \ x : 0.4, \ y : 0, \ z : 1, \ T : 2 \rangle \xrightarrow{\ell_1}$
$\langle \pi : \{\ell_0, m_{0,1}\}, \ x : 0.4, \ y : 1, \ z : 0, \ T : 2 \rangle \xrightarrow{cont}$
$\langle \pi : \{\ell_0, m_{0,1}\}, \ x : 0.6, \ y : 1, \ z : 1, \ T : 3 \rangle \xrightarrow{\ell_0}$
$\langle \pi : \{\ell_1, m_{0,1}\}, \ x : 0.6, \ y : 1, \ z : 0, \ T : 3 \rangle \xrightarrow{cont}$
$\langle \pi : \{\ell_1, m_{0,1}\}, \ x : 0.8, \ y : 1, \ z : 1, \ T : 4 \rangle \xrightarrow{\ell_1}$
$\langle \pi : \{\ell_0, m_{0,1}\}, \ x : 0.8, \ y : 2, \ z : 0, \ T : 4 \rangle \xrightarrow{cont}$
$\langle \pi : \{\ell_0, m_{0,1}\}, \ x : 1.0, \ y : 2, \ z : 1, \ T : 5 \rangle \xrightarrow{m_1}$
$\langle \pi : \{\ell_0, m_2\} \quad, \ x : 1.0, \ y : 2, \ z : 1, \ T : 5 \rangle \xrightarrow{\ell_0}$
$\langle \pi : \{\ell_2, m_2\} \quad, \ x : 1.0, \ y : 2, \ z : 0, \ T : 5 \rangle \xrightarrow{cont}$
$\langle \pi : \{\ell_2, m_2\} \quad, \ x : 1.0, \ y : 2, \ z : 1, \ T : 6 \rangle \xrightarrow{cont}$
$\ldots$

The second computation attempts to minimize the value of y on termination.

$\langle \pi : \{\ell_0, m_{0,1}\}, \ x : 0.0, \ y : 0, \ z : 0, \ T : 0 \rangle \xrightarrow{cont}$
$\langle \pi : \{\ell_0, m_{0,1}\}, \ x : 1.0, \ y : 0, \ z : 5, \ T : 5 \rangle \xrightarrow{m_1}$
$\langle \pi : \{\ell_0, m_2\} \quad, \ x : 1.0, \ y : 0, \ z : 5, \ T : 5 \rangle \xrightarrow{\ell_0}$
$\langle \pi : \{\ell_2, m_2\} \quad, \ x : 1.0, \ y : 0, \ z : 0, \ T : 5 \rangle \xrightarrow{cont}$
$\langle \pi : \{\ell_2, m_2\} \quad, \ x : 1.0, \ y : 0, \ z : 1, \ T : 6 \rangle \xrightarrow{cont}$
$\ldots$

3   A Requirement Specification Language

To specify properties of hybrid systems, we use the language of temporal logic, as presented in [MP91b], with the extension that state formulas (assertions) may refer to the variable T representing the real-time clock.

This logic has been considered under the name *Explicit Clock Temporal Logic* in [PH88], [HLP90], and [Ost90]. However, since it is always possible to introduce a continuous variable z with the activity $\dot{z} = 1$, leading to the fact that always $z = T$, one may argue that, under the hybrid context, this is no extension at all.

There have been several proposals for more radical extensions of the temporal language to express properties of real-time and hybrid systems. Some of the real-time extensions are surveyed in [AH92]. For our purposes here, classical temporal logic suffices.

We illustrate the use of temporal logic with explicit reference to the clock T to express two important timing properties of systems.

- *Bounded Response:*   Every p should be followed by an occurrence of a q, not later than d time units. This property can be specified by either of the formulas:

$$p \wedge T = t_0 \;\Rightarrow\; \Diamond\,(q \wedge T \le t_0 + d)$$
$$p \wedge T = t_0 \;\Rightarrow\; (T \le t_0 + d)\,\mathcal{W}\,q$$

- *Minimal separation:*   No q can occur earlier than d time units after an occurrence of p. This property can be specified by the formula:

$$p \wedge T = t_0 \;\Rightarrow\; \Box(T < t_0 + d \;\to\; \neg q)$$

The second formula for bounded response uses the operator $\mathcal{W}$ to state that time cannot progress beyond $t_0 + d$ before q occurs. While the first formula is a response formula (liveness), the second formula is a safety formula. The advantage of formulating properties by safety formulas is that their verification requires simple proof rules.

Temporal logic can also specify properties that refer to continuous variables. For example, to specify that the mouse will always escape the cat, for the system of Fig. 4, we can write the invariance formula

$$\Box(Cat.run \wedge (x_c = x_m) \to x_m = 0),$$

where we use names of states in a statechart as control predicates. Of course, such a property will not be valid over all computations of the Cat and Mouse system, unless some relation is established among the problem parameters: X_0, Δ, v_c, and v_m. Indeed, a sufficient condition for this property to be valid is:

$$\frac{X_0}{v_m} < \Delta + \frac{X_0}{v_c}.$$

The following axiom constrains the value of the clock T to be nonnegative.

TIME-RANGE $: 0 \le T$

We may use this axiom freely in any reasoning step.

4 Verification

In this paper, we only consider verification of *invariance formulas* of the form

$$\Box p,$$

where p is an assertion that may refer to continuous variables as well as to the clock T.

As shown, for example, in [MP91a], such invariance formulas are proven using a very simple computational induction principle, that can be summarized as:

> To establish that φ is invariant over all computations of a PTS, show that
>
> I1. φ is implied by the initial condition Θ, and
> I2. φ is preserved by any step taken in the computation.

To express the preservation of assertions over steps in a computation, we introduce the notion of a *verification condition* that claims that a computation step always leads from a state satisfying assertion p to a state satisfying assertion q.

4.1 Verification Conditions for Hybrid Systems

As indicated in the definition of a hybrid computation, there are two kinds of steps that can lead from a situation $\langle s_i, t_i \rangle$ to its successor $\langle s_{i+1}, t_{i+1} \rangle$: a discrete step, caused by taking a transition, and a continuous step in which $\langle s_{i+1}, t_{i+1} \rangle$ is an activity successor of $\langle s_i, t_i \rangle$. Consequently, we have two types of verification conditions.

- Condition $\{p\}\, \tau\, \{q\}$ states that transition τ always leads from a p-state to a q-state.
- Condition $\{p\}\, cont\, \{q\}$ states that every continuous step always leads from a p-state to a q-state.

Transformation under Transitions

The condition $\{p\}\, \tau\, \{q\}$ is given by

$$\left(\rho_\tau \wedge T' = T \wedge p \right) \to q' \tag{1}$$

where ρ_τ is the transition relation corresponding to τ. The *primed version* q' of the assertion q is obtained by replacing each variable in q by its primed version. In addition to ρ_τ, which ensures that the state variables V' are related to V according to the transition relation, the verification condition requires also that time does not progress in this step.

Consider, for example, the verification condition

$$\{at\_\ell_1 \wedge T \leq 5 + z\}\, \ell_1\, \{at\_\ell_0 \wedge T \leq 10 + z\} \tag{2}$$

for program HYB-T. Here, the expressions at_ℓ_1 and at_ℓ_0 abbreviate the formulas $\ell_1 \in \pi$ and $\ell_0 \in \pi$. Expanding the definition of the verification condition, we get

$$\underbrace{\cdots \pi' = \pi - \{\ell_1\} \cup \{\ell_0\} \cdots \wedge z \leq 5 \cdots \wedge z' = 0}_{\rho_{\ell_1}} \wedge T' = T \wedge \underbrace{\cdots \wedge T \leq 5 + z}_{p}$$

$$\rightarrow \qquad \underbrace{(at\_\ell_0)' \wedge T' \leq 10 + z'}_{q'}.$$

Clearly $\pi' = \pi - \{\ell_1\} \cup \{\ell_0\}$ implies $(at\_\ell_0)'$. Since $T' = T$, we obtain from $T \leq 5 + z$ and $z \leq 5$ the inequality $T' \leq 10$. From $z' = 0$, it follows that $T' \leq 10 + z'$.

Transformation under Continuous Steps

To formulate the verification condition over continuous steps, we consider an evolution from a situation that can be described as $\langle V, T \rangle$ to the situation $\langle V', T' \rangle$, assuming that $T' > T$.

An *activity selection* is a mapping $g : V_c \mapsto \mathcal{A}$, assigning to each continuous variable $x \in V_c$ an activity $g(x)$ in its governing set. Assume that the activity selected by g for each continuous state variable $x \in V_c$ is

$$p_x^g \rightarrow \dot{x} = e^g.$$

Let $F^g = \{f_x^g(t)\}$ be a set of functions, one for each continuous variable $x \in V_c$, such that

- $f_x^g(T) = x$,
- The equation

  $$\dot{x}_x^g(t) = e^g(F^g)$$

  is satisfied in the range $t \in [T, T']$, where $e^g(F^g)$ is obtained from e^g by replacing each occurrence of $y \in V_c$ by $f_y^g(t)$.

We assume that we know how to express the functions f_x^g in a closed form, referring to x, T, and t. For example, if g selects the activity $\dot{x} = 2$ for $x \in V_c$, then $f_x^g(t)$ is given by

$$f_x^g(t) = x + 2 \cdot (t - T).$$

The Condition

With each possible activity selection function g, we associate a verification condition $\{p\} \, g \, \{q\}$, which is given by

$$\rho_{cont}^g \wedge p \rightarrow q',$$

where the relation ρ_{cont}^g stands for

$$\bigwedge_{x \in V_d} x' = x \ \wedge \ \bigwedge_{x \in V_c} x' = f_x^g(T') \ \wedge \ \bigwedge_{x \in V_c} p_x^g \ \wedge \ T' > T \ \wedge \ \bigwedge_{\tau \in \mathcal{T}} \Big(En(\tau) \to En'(\tau) \Big)$$

$$\wedge \ \bigwedge_{\varphi \in \mathcal{I}} \left(\begin{array}{c} (\ \varphi \ \vee \ \varphi') \ \wedge \ \Big(\forall t : (T < t < T') : \ \varphi^g(t) \Big) \\ \vee \\ (\neg\varphi \ \vee \ \neg\varphi') \ \wedge \ \Big(\forall t : (T < t < T') : \neg\varphi^g(t) \Big) \end{array} \right)$$

The first conjunct of the formula states that all discrete variables are not changed in a continuous step. The next conjunct states that the value of $f_x^g(T')$ agrees with x'. The third conjunct requires that the activation condition p_x^g holds at the start of the continuous step. Since p_x^g depends only on discrete values, it must hold throughout. The fourth conjunct requires that time progresses by a positive amount. The last conjunct in the first line ensures that the progress of time cannot disable any enabled transition, corresponding to the urgency requirement.

The next line requires that every important assertion φ has the same truth value throughout the interior of the evolution interval, and that this value agrees with the value of φ at one of the interval's boundaries, i.e., with either φ or φ'. The formula $\varphi^g(t)$ is obtained from φ by replacing each occurrence of $y \in V_c$ by $f_y^g(t)$.

Consider, for example, the condition $\{-1 \leq x \leq 1\} g \{-1 \leq x \leq 1\}$ for system Φ_1. Since there is only one activity, there is only one activity selection g, i.e., the one that selects this activity. The evolution function f_x^g is given by $f_x^g(t) = x - (t - T)$. Consequently, the verification condition is given (after some simplifications) by:

$$\underbrace{\left[\begin{array}{l} x' = x - (T' - T) \ \wedge \ T' > T \ \wedge \\ (x = -1) \ \to \ (x' = -1) \qquad \wedge \\ \left[\begin{array}{l} \cdots \wedge (\forall t : (T < t < T') : x - (t - T) = -1) \\ \vee \\ \cdots \wedge (\forall t : (T < t < T') : x - (t - T) \neq -1) \end{array} \right] \end{array} \right]}_{\rho_{cont}^g} \ \wedge \ \underbrace{-1 \leq x \leq 1}_{p} \qquad \to$$

$$\underbrace{-1 \leq x' \leq 1}_{p'}$$

Clearly, since $T' > T$, it is impossible to have $x - (t - T) = -1$ for all $t \in (T, T')$. Consequently, ρ_{cont}^g implies that $x - (t - T) \neq -1$ for all $t \in (T, T')$. The case $x = -1$ is also impossible since, by the second line of ρ_{cont}^g, it would imply $x' = -1$ while, on the other hand, $x' = x - (T' - T) < x = -1$. We conclude that $x \neq -1$ which, by p, implies $x > -1$. Since $x - (t - T) \neq -1$ for all $t \in (T, T')$, $x > -1$ implies that $x - (t - T) > -1$ for all $t \in (T, T')$. Therefore, in the limit of t approaching T', $x' = x - (T' - T) \geq -1$.

For the other inequality, from $T' > T$ it follows that $x' = x - (T' - T) < x \leq 1$ and therefore $x' \leq 1$.

Finally, we define the verification condition over a continuous step to be

$$\{p\} \, cont \, \{q\} : \quad \bigwedge_g \{p\} \, g \, \{q\},$$

where the conjunction of the individual conditions $\{p\} \, g \, \{q\}$ is taken over all possible activity selection functions g.

4.2 The Invariance Rule

The following rule establishes the validity of an invariance formula $\Box p$ for an assertion p over a given PTS Φ.

Rule INV

I1. $\Theta \wedge T = 0 \rightarrow \varphi$
I2. $\{\varphi\} \, \tau \, \{\varphi\}$ for every $\tau \in \mathcal{T}$
I3. $\{\varphi\} \, g \, \{\varphi\}$ for every activity selection g
I4. $\varphi \rightarrow p$

$\Box p$

Premise I1 establishes that the auxiliary assertion φ is initially true. Premise I2 requires that φ is preserved by every transition. Premise I3 requires that φ is preserved by any continuous step. It follows that all situations in a computation of Φ satisfy φ. By premise I4, they also satisfy p.

To illustrate the use of this rule, consider system Φ_1. We will prove that the invariant $\Box(-1 \leq x \leq 1)$ is valid over all sampling computations of this system. We use rule INV with $p = \varphi : -1 \leq x \leq 1$.

Premise I1 assumes the form

$$\underbrace{x = 1}_{\Theta} \wedge T = 0 \rightarrow \underbrace{-1 \leq x \leq 1}_{\varphi},$$

which is obviously valid.

Premise I2 assumes the form

$$\underbrace{\cdots \wedge (x' = 1) \wedge \cdots}_{\rho_\tau} \wedge \underbrace{\cdots}_{\varphi} \rightarrow \underbrace{-1 \leq x' \leq 1}_{\varphi'}$$

which is also obviously valid.

Premise I3 requires showing that $-1 \leq x \leq 1$ is preserved under a continuous step. This has been verified above.

Premise I4 is trivial since $p = \varphi$.

Case Splitting

It often happens that the assertion φ appearing in rule INV naturally splits into a disjunction:

$$\varphi : \varphi_0 \vee \varphi_1 \vee \cdots \vee \varphi_m$$

In this case, we may use the following requirements to establish premises I1–I4 of rule INV.

D1. $\Theta \wedge T = 0 \rightarrow \varphi_0$.

D2. For every $i = 0, \ldots, m$ and transition τ, there exist indices $j_1, \ldots, j_k$ (depending on i and τ), such that

$$\{\varphi_i\} \tau \{\varphi_{j_1} \vee \cdots \vee \varphi_{j_k}\}.$$

We call $\varphi_{j_1}, \ldots, \varphi_{j_k}$ the τ-successors of φ_i.

D3. For every $i = 0, \ldots, m$ and activity selection g,

$$\{\varphi_i\} g \{\varphi_i\}.$$

D4. For every $i = 0, \ldots, m$,

$$\varphi_i \rightarrow p.$$

The *scheme* of an invariance proof under case splitting consists of the list of assertions $\varphi_0, \ldots, \varphi_m$ and the identification, for each $i = 0, \ldots, m$ and $\tau \in \mathcal{T}$, of the τ-successors of φ_i.

A scheme is defined to be *valid* with respect to an assertion p if requirements D1–D4 are satisfied.

It is obvious that a valid scheme identifies a set of verification conditions that can serve as the premises to rule INV. Indeed, we have the following claim.

Claim 1 *A scheme that is valid with respect to an assertion p establishes the validity of the formula*

$$\square p$$

over all sampling computations of the given PTS Φ.

In presenting an invariance proof, we rarely list the full set of verification conditions and prove them in detail. Instead, it is often sufficient to present the scheme of the proof as convincing evidence of the proof's validity.

Consider a proof of the property

$$\square(x < 1 \rightarrow at\_\ell_{0,1})$$

for program HYB-T, where $at\_\ell_{0,1}$ is an abbreviation for $at\_\ell_0 \vee at\_\ell_1$.

A scheme of the proof of this property can be given by the assertion list

$$\varphi_0 : at\_\ell_0 \wedge x \leq 1$$
$$\varphi_1 : at\_\ell_1 \wedge x \leq 1$$
$$\varphi_2 : x \geq 1$$

and the successor identification

	ℓ_0	ℓ_1	m_1
φ_0	φ_1, φ_2	φ_0	φ_2
φ_1	φ_1	φ_0	φ_2
φ_2	φ_2	φ_2	φ_2

It is not difficult to see that this scheme is valid with respect to $p : x < 1 \rightarrow at_-\ell_{0,1}$.

In some cases, we encounter an assertion φ_i and a transition τ such that φ_i implies that τ is disabled. An example of this is assertion φ_0 and transition ℓ_1, which is disabled on all states satisfying φ_0. In such a case, we can fill the τ-successors entry by an arbitrary set of assertions. This is because the conjunction $\rho_\tau \wedge \varphi_i$ appearing on the left-hand side of the implication in the verification condition is false and implies anything. In all of these cases, we consistently put φ_i as its own τ-successor.

Proof Diagrams

A convenient way of presenting a proof scheme is by proof diagrams.

A *proof diagram* representation of a proof scheme consists of

- A node corresponding to each assertion $\varphi_0, \ldots, \varphi_m$.

- The initial node is annotated by an entry arrow ●↰.
- For each node φ_i, transition τ, and the set of φ_i's τ-successors $\varphi_{j_1} \ldots, \varphi_{j_k}$, there is an edge connecting φ_i to each φ_{j_r}, $r = 1, \ldots, k$, labeled by τ. If φ_i is the only τ-successor of itself, the edge is not drawn.

In Fig. 6 we present a proof diagram corresponding to the scheme for the proof of $\Box(x < 1 \rightarrow at_-\ell_{0,1})$.

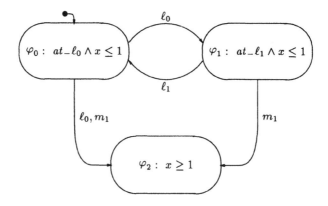

Fig. 6. A proof diagram.

A diagram is said to be *valid* with respect to assertion p if the scheme it represents is valid with respect to p.

Statechart Conventions

There are several conventions inspired by the visual language of statecharts [Har87] that improve the presentation and readability of proof diagrams. We extend the notion of a directed graph into a structured directed graph by allowing *compound nodes* that may encapsulate other nodes, and edges that may depart or arrive at compound nodes. A node that does not encapsulate other nodes is called a *basic node*.

We use the following conventions:

- Labels of compound nodes: a diagram containing a compound node n, labeled by an assertion φ and encapsulating nodes $n_1, \ldots, n_k$ with assertions $\varphi_1, \ldots, \varphi_k$, is equivalent to a diagram in which n is unlabeled and nodes $n_1, \ldots, n_k$ are labeled by $\varphi_1 \wedge \varphi, \ldots, \varphi_k \wedge \varphi$.
- Edges entering and exiting compound nodes: a diagram containing an edge e connecting node A to a compound node n encapsulating nodes $n_1, \ldots, n_k$ is equivalent to a diagram in which there is an edge connecting A to each n_i, $i = 1, \ldots, k$, with the same label as e. Similarly, an edge e connecting the compound node n to node B is the same as having a separate edge connecting each n_i, $i = 1, \ldots, k$, to B with the same label as e.

With these conventions we can redraw the proof diagram of Fig. 6 as shown in Fig. 7. Note that the common conjunct $x \leq 1$ has been factored out of nodes n_0

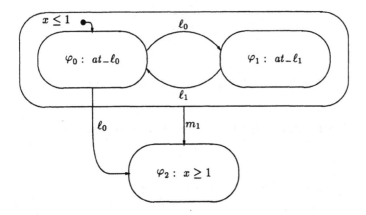

Fig. 7. A structured proof diagram.

and n_1 and now appears as the label of the compound node encapsulating them.

Termination of Program HYB-T within 15 Time Units

We will prove that program HYB-T terminates within 15 time units. This property can be expressed by the invariance

$$\Box(T \leq 15 \ \lor \ (at\_\ell_2 \ \land \ at\_m_2)).$$

This formula states that if more than 15 time units have elapsed, then the program must have terminated.

The proof diagram presented in Fig. 8 establishes the validity of this formula over program HYB-T. The notation $at\_\ell_{0,1}$ is an abbreviation for $at\_\ell_0 \lor at\_\ell_1$.

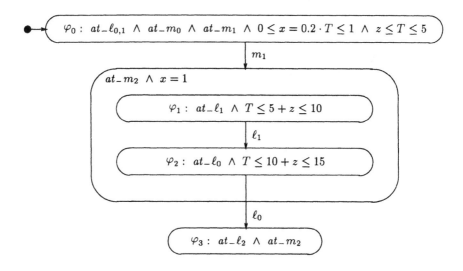

Fig. 8. A proof diagram for a timed property.

It is not difficult to show that the diagram is valid with respect to the assertion

$$T \leq 15 \ \lor \ (at\_\ell_2 \land at\_m_2).$$

For example, we show that assertion φ_0 is preserved under a continuous step. The only relevant activity selection function g is the one that picks α_x^{on} : $at\_m_0 \to \dot{x} = 0.2$ for x and $\alpha_z : \dot{z} = 1$ for z. For this choice of g, the evolution of x and z is given by $f_x^g(t) = x + 0.2 \cdot (t - T)$ and $f_z^g(t) = z + (t - T)$. Consequently, the appropriate verification condition (after some simplifications) is

$$\left.\begin{array}{l} \pi' = \pi \ \wedge \ x' = x + 0.2 \cdot (T' - T) \ \wedge \ z' = z + (T' - T) \ \wedge \ T' > T \ \wedge \ \cdots \\ \wedge \ (x = 1) \ \rightarrow \ (x' = 1) \ \wedge \ \cdots \\ \wedge \ (\forall t : T < t < T' : x + 0.2 \cdot (t - T) \neq 1) \ \wedge \ \cdots \end{array}\right\} \rho^g_{cont}$$

$$\wedge \quad \underbrace{at\_\ell_{0,1} \ \wedge \ at\_m_0 \ \wedge \ at\_m_1 \ \wedge \ 0 \leq x = 0.2 \cdot T \leq 1 \ \wedge \ z \leq T \leq 5}_{\varphi_0}$$

$$\rightarrow$$

$$\underbrace{(at\_\ell_{0,1})' \ \wedge \ (at\_m_0)' \ \wedge \ (at\_m_1)' \ \wedge \ 0 \leq x' = 0.2 \cdot T' \leq 1 \ \wedge \ z' \leq T' \leq 5}_{\varphi'_0}$$

Obviously, ρ^g_{cont} implies that $x' > x$. Therefore, $(x = 1) \rightarrow (x' = 1)$ implies that $x \neq 1$. It follows from φ_0 that $x < 1$. Since $f^g_x(t) = x + 0.2 \cdot (t - T) \neq 1$ for all $t \in (T, T')$, it follows that $f^g_x(t) < 1$ for all $t \in (T, T')$ and, in the limit of t approaching T', $f^g_x(T') = x' \leq 1$.

We use these facts to show that ρ^g_{cont} and φ_0 imply φ'_0.

- $(at\_\ell_{0,1})' \wedge (at\_m_0)' \wedge (at\_m_1)'$ follows from $\pi' = \pi$.
- $0 \leq x' = 0.2 \cdot T'$ follows from $x' = x + 0.2 \cdot (T' - T)$, $T' > T$, and $0 \leq x = 0.2 \cdot T$.
- Taking the limit over $(\forall t : T < t < T' : x + 0.2 \cdot (t - T) < 1)$ as t tends to T', yields $x' = x + 0.2 \cdot (T' - T) \leq 1$.
- $z' \leq T'$ follows from $z' = z + (T' - T)$ and $z \leq T$.
- $T' \leq 5$ follows from $x' = 0.2 \cdot T' \leq 1$.

This establishes that program HYB-T terminates within 15 time units.

Proving Untimed Properties of Timed Systems

The previous example concentrated on proving timed properties, i.e., properties in which time is explicitly mentioned. Another interesting class of properties consists of properties that do not refer to time directly but whose validity over a program P is a consequence of the timing constraints satisfied by the computations of P.

For example, the property $\square(y \leq 3)$ is valid over all computations of program HYB-T.

In Fig. 9 we present a proof diagram that is valid with respect to the assertion $y \leq 3$.

Let us present some arguments for the validity of this diagram. First, let us check that all assertions appearing in the diagram are preserved by a continuous step. The interesting cases are assertions that refer to T, x, and z. As shown above, the assertion $0 \leq x = 0.2 \cdot T \leq 1$ is preserved by any continuous step taken while at_m_1 holds. Additional assertions referring to T are

$$2 \cdot y + z \leq T \leq 5 \qquad \text{while } at\_\ell_0 \wedge at\_m_0 \wedge at\_m_1 \text{ holds,}$$
$$2 \cdot y + 1 + z \leq T \leq 5 \text{ while } at\_\ell_1 \wedge at\_m_0 \wedge at\_m_1 \text{ holds.}$$

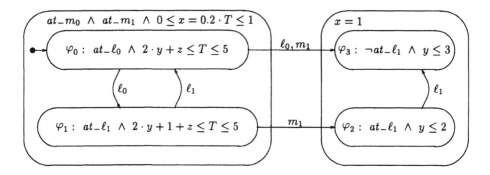

Fig. 9. A proof diagram for untimed properties.

In both cases, when time progresses from T to $T' > T$, both sides of the inequality increase by $T' - T$. The value of the clock T remains bounded by 5 as long as control is at m_1.

Next, we should check the verification conditions corresponding to the transitions labeling edges in the diagram. The verification condition corresponding to ℓ_0 is

$$\rho_{\ell_0} \wedge T' = T \wedge \underbrace{\cdots \wedge x = 0.2 \cdot T \leq 1 \wedge at\_\ell_0 \wedge 2 \cdot y + z \leq T \leq 5}_{\varphi_0} \rightarrow$$

$$\underbrace{(at\_\ell_1)' \wedge 2 \cdot y' + 1 + z' \leq T' \leq 5}_{\varphi_1'} \vee \underbrace{x' = 1 \wedge \neg(at\_\ell_1)' \wedge y' \leq 3}_{\varphi_3'}$$

The transition relation ρ_{ℓ_0} implies $y' = y$, $z \geq 1$ and $z' = 0$. According to φ_0, $x \leq 1$. We distinguish two cases: $x < 1$ and $x = 1$.

For the case $x < 1$ we show that $\rho_{\ell_0} \wedge T' = T \wedge \varphi_0$ implies φ_1'. From $T' = T \geq 2 \cdot y + z \geq 2 \cdot y + 1 = 2 \cdot y' + 1 + z'$, we obtain $2 \cdot y' + 1 + z' \leq T'$. Clauses $T' = T$ and $T \leq 5$ imply $T' \leq 5$.

For the case $x = 1$ we show that $\rho_{\ell_0} \wedge T' = T \wedge \varphi_0$ implies φ_3'. The relation ρ_{ℓ_0} implies $x' = x$ and, since we assume $x = 1$, it follows that $x' = 1$. The relation ρ_{ℓ_0}, under $x = 1$, also implies $\neg(at\_\ell_1)'$. As ρ_{ℓ_0} implies $z \geq 1$, $2 \cdot y + z \leq 5$ implies $y' = y \leq 3$. Thus, all conjuncts of φ_3' are implied by the left-hand side of the implication. This establishes the verification condition for ℓ_0.

On taking transition m_1 from either φ_0 or φ_1, we have that either $2 \cdot y \leq 5$ or $2 \cdot y + 1 \leq 5$. In both cases, this implies $y \leq 2.5 \leq 3$.

It is also straightforward to show the condition

$$\Theta \wedge T = 0 \rightarrow at\_\ell_0 \wedge at\_m_0 \wedge at\_m_1 \wedge 0 \leq x = 0.2 \cdot T \leq 1 \wedge 2 \cdot y + z \leq T \leq 5,$$

since $\Theta \wedge T = 0$ implies $at\_\ell_0 \wedge at\_m_0 \wedge at\_m_1 \wedge x = y = z = T = 0$.

This shows that the diagram is valid with respect to $y \leq 3$ and establishes that the formula

$$\Box(y \leq 3)$$

is valid over program HYB-T.

Verifying a Property of the Cat and Mouse System

Consider the property that, under the assumption

$$\frac{X_0}{v_m} < \Delta + \frac{X_0}{v_c} \tag{3}$$

all computations of the Cat and Mouse system satisfy

$$\Box(Cat.run \land (x_c = x_m) \to x_m = 0).$$

In Fig. 10, we present a proof diagram of this invariance property. In this diagram

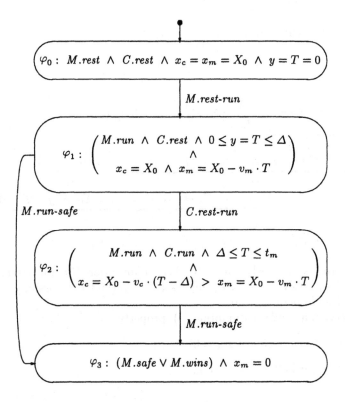

Fig. 10. A hybrid invariance proof diagram.

we use control assertions denoting that certain basic states are contained in π.

For example, $C.run$ stands for $Cat.run \in \pi$. We also use t_m for $\frac{X_0}{v_m}$, the time it takes the mouse to run the distance X_0.

It is not difficult to verify that the diagram is valid, including the preservation of all assertions under a continuous step. The only part that requires more attention is showing that the φ_2 conjunct

$$X_0 - v_c \cdot (T - \Delta) > x_m = X_0 - v_m \cdot T$$

is maintained until transition $M.run\text{-}safe$ becomes enabled, that is as long as x_m is nonnegative, which implies $T \leq t_m$. To show this, it is sufficient to show $v_c \cdot (T - \Delta) < v_m \cdot T$ which is equivalent to

$$\frac{v_m}{v_c} > 1 - \frac{\Delta}{T} \tag{4}$$

From inequality (3), we can obtain

$$\frac{v_m}{v_c} > 1 - \Delta \cdot \frac{v_m}{X_0}$$

which, using the definition of $t_m = \dfrac{X_0}{v_m}$, gives

$$\frac{v_m}{v_c} > 1 - \frac{\Delta}{t_m}. \tag{5}$$

Since $T \leq t_m$, we have $1 - \dfrac{\Delta}{t_m} \geq 1 - \dfrac{\Delta}{T}$ establishing (4).

It remains to show that

$$\underbrace{M.rest \ \wedge \ C.rest \ \wedge \ x_c = x_m = X_0 \ \wedge \ y = 0 \ \wedge T = 0}_{\Theta} \quad \rightarrow$$

$$\underbrace{M.rest \ \wedge \ C.rest \ \wedge \ x_c = x_m = X_0 \ \wedge \ y = T = 0}_{\varphi_0} \tag{6}$$

$$\varphi_0 \ \vee \ \cdots \ \vee \ \varphi_3 \rightarrow \Big(C.run \ \wedge \ x_c = x_m \ \rightarrow \ x_m = 0 \Big). \tag{7}$$

Implication (6) is obviously valid. To check implication (7), we observe that both φ_0 and φ_1 imply $\neg C.run$, φ_2 implies $x_c > x_m$ (under the assumption that $\Delta > 0$), and φ_3 implies $x_m = 0$.

This shows that under assumption (3), property

$$\Box (Cat.run \wedge (x_c = x_m) \rightarrow x_m = 0)$$

is valid for the Cat and Mouse system.

5 The Gas Burner Example

We conclude with an example of a Gas Burner system, presented in [CHR92]. Consider the hybrid statechart presented in Fig. 11. This statechart represents a

Initially $Leak = $ F, $x = 0$

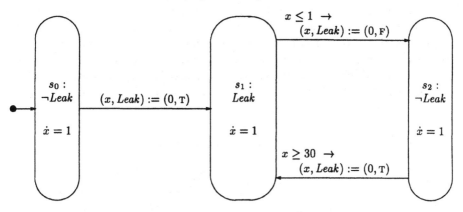

Fig. 11. GAS-BURNER: A gas burner system.

Gas Burner system that has three states: s_0, s_1, and s_2. There is a boolean state variable $Leak$ whose value represents whether the system is currently leaking. For clarity, we have labeled each state with the value of $Leak$ at the states. However, this labeling has no semantic meaning. The system employs a timer x to constrain the length of time that can be spent in each visit to states s_1 and s_2.

The verification problem posed in [CHR92] can be formulated as follows. Assuming

1. A continuous leaking period cannot extend beyond 1 time unit.
2. Two disjoint leaking periods are separated by a non-leaking period extending for at least 30 time units.

Prove:

- *Safety-Critical Requirement:* In any interval longer than 60, the *accumulated* leaking time is at most 5% of the interval length.

Obviously, the hybrid statechart of Fig. 11 satisfies assumptions 1 and 2. The only leaking state is s_1 and it is clear that the system cannot stay continuously in s_1 for more than 1 time unit and that, between two consecutive (but disjoint) visits to s_1, the system stays at the non-leaking state s_2 for at least 30 time units.

However, the property to be proved uses the notion of *accumulated time* in which some assertion, such as $Leak$, holds. This cannot be expressed directly in the temporal logic (TL) we use. The calculus of durations, introduced in [CHR92], has a special *duration* operator $\int p$ that measures the accumulated time p holds. Later, we will briefly consider an extension of TL which adopts the duration operator.

To handle this problem without extending the logic, we introduce additional auxiliary variables to the hybrid statechart representing the Gas Burner system. These variable measure the total time of an interval and the accumulated time in which variable *Leak* has been true. For simplicity, we consider the safety-critical requirement only for *initial intervals*, i.e., intervals starting at $T = 0$. The extension of the method to arbitrary intervals is straightforward.

Consider the hybrid statechart of Fig. 12. This hybrid statechart employs

Initially *Leak* = F, $x = y = z = 0$

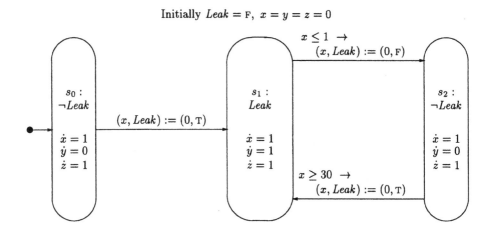

Fig. 12. H-GAS: The gas burner as a hybrid system.

three continuous variables as follows:

- As before, variable x measures the duration of time in each of the states s_0, s_1, and s_2. It is reset to 0 on entry to each of these states.
- Variable y measures the accumulated leaking time. It grows linearly in state s_2, and stays constant in any of the other states.
- Variable z measures the total elapsed time.

With these variables, we can write the requirement that the accumulated leak time does not exceed 5% of the elapsed time as $y \leq 0.05 \cdot z$ or, equivalently, as $20 \cdot y \leq z$.

Consequently, to verify that the original hybrid system of Fig. 11 maintains the safety-critical requirement for initial intervals, it is sufficient to prove that all computations of the hybrid system of Fig. 12 satisfy the invariance property

$$\Box\Big(z \geq 60 \rightarrow 20 \cdot y \leq z\Big).$$

This is the first example in which the invention of the necessary auxiliary invariants is not immediately obvious. Therefore, we will spend some time on their derivation. We try to find a relation that continuously holds between y and z and that implies the requirement

$$z \geq 60 \rightarrow 20 \cdot y \leq z. \tag{8}$$

Consider a finite prefix of a computation. Let v_1 denote the number of times the leaking state s_1 is visited in this prefix. Since on each visit variable y grows by at most 1 time unit, we have

$$y \leq v_1$$

at the end of the prefix. On the same prefix, variable z can be bounded from below by the sum of the accumulated time spent at s_1 and the accumulated time spent at s_2, ignoring the time spent at s_0 which can be arbitrarily short. The accumulated time spent at s_1 equals y. Since between two consecutive visits to s_1 the computation visits s_2, the number of visits to s_2 is at least $v_1 - 1$, and each of these visits lasts at least 30 time units. We thus obtain

$$z \geq 30 \cdot (v_1 - 1) + y \geq 31 \cdot y - 30,$$

where the last inequality is obtained by replacing v_1 by the smaller or equal value y. This leads to:

$$z \geq 31 \cdot y - 30. \tag{9}$$

We will show that this relation implies requirement (8), that is

$$z \geq 31 \cdot y - 30 \quad \rightarrow \quad (z \geq 60 \rightarrow 20 \cdot y \leq z),$$

or, equivalently,

$$z \geq 31 \cdot y - 30 \ \wedge \ z \geq 60 \quad \rightarrow \quad 20 \cdot y \leq z.$$

By $z \geq 60$, which can be written as $30 \leq \dfrac{1}{2} \cdot z$, we can replace the value 30 in $z \geq 31 \cdot y - 30$ by the bigger or equal value $\frac{1}{2} \cdot z$ and obtain

$$z \geq 31 \cdot y - \frac{1}{2} \cdot z,$$

leading to

$$\frac{3}{2} \cdot z \geq 31 \cdot y,$$
$$z \ \geq \frac{62}{3} \cdot y > 20 \cdot y.$$

We therefore start with the assumption that the inequality $z \geq 31 \cdot y - 30$ holds at all states in the computation. Working backwards, we can identify what versions of this invariant should hold on every visit to each of the states s_0, s_1, and s_2. This leads to the proof diagram presented in Fig. 13. Transitions in this diagram are identified by the names of the states in system H-GAS from which they exit. To facilitate the reading of the diagram, edges entering a node are

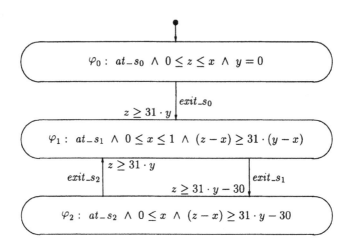

Fig. 13. An invariance proof diagram for the gas burner.

annotated by an assertion that holds whenever this node is entered. Thus, it can be shown that $z \geq 31 \cdot y$ (which is the same as $(z-x) \geq 31 \cdot (y-x)$ since $x = 0$ on entry) holds on entering the node (labeled by) φ_1 from either φ_0 or φ_2. Since x, y, and z all grow at the same linear rate within state s_1 (corresponding to node φ_1), the differences $z - x$ and $y - x$ maintain the values they had on entry. This explains why $(z - x) \geq 31 \cdot (y - x)$ is maintained within φ_1. On exit from φ_1 to φ_2, $x \leq 1$, therefore, $(z - x) \geq 31 \cdot (y - x)$ implies $z \geq 31 \cdot y - 30 \cdot x \geq 31 \cdot y - 30$ on entry to φ_2. Since, within φ_2 both x and z grow at the same rate, while y remains the same, $(z - x) \geq 31 \cdot y - 30$ is maintained.

It is not difficult to see that the initial condition implies φ_0 and that each of φ_0, φ_1, or φ_2, implies $z \geq 31 \cdot y - 30$. Consequently, $z \geq 31 \cdot y - 30$ holds over all computations, establishing the validity of

$$\Box \left(z \geq 60 \rightarrow 20 \cdot y \leq z \right).$$

To generalize this analysis to arbitrary (not necessarily initial) intervals, we can add a transition that nondeterministically resets the values of y and z to 0. This will start the measurements corresponding to an interval at an arbitrary point in time. In fact, the proof diagram of Fig. 13 is also valid for this system. It can be checked that assertions φ_0 and φ_1 in this diagram are preserved under simultaneous reset of y and z to 0.

Assertion φ_2 has to be weakened to

$$\varphi_2^* : \quad at\_s_2 \wedge 0 \leq x \wedge \begin{pmatrix} (z-x) \geq 31 \cdot y - 30 \\ \vee \\ z \geq 31 \cdot y = 0 \end{pmatrix}$$

Proof by an Extended Version of TL

The previous proof is based on a transformation that augments the original Gas Burner problem of Fig. 11 with the two additional variables y and z. We will now consider an alternative approach, which does not modify the given system but uses a stronger logic.

As is shown in [KP92a], it is possible to extend TL by adding to it the *duration function* $\int p$, which measures the accumulated time in which p has been true up to the present. We denote the extended logic by TL $_\int$.

Several extensions are needed as a result of the new operator. The first extension is axiom DURATION-RANGE which bounds the range of the duration function and also relates it to the clock T.

$$\text{DURATION-RANGE} : 0 \leq \int \psi \leq T \quad \text{for every formula } \psi$$

Since duration expressions can appear in assertions, it is also necessary to define the primed version of a duration expression $\int r$, denoted $(\int r)'$ for some assertion r. We will only consider the case that r depends on discrete variables. For this case, the value of $(\int r)'$ is given by

$$\left(\int r\right)' = \text{if } r \text{ then } \int r + T' - T \text{ else } \int r.$$

This definition states that, if r holds at s, then the value of $\int r$ at $\langle s', t' \rangle$ is its value at $\langle s, t \rangle$ plus the time difference $t' - t$. Otherwise, it retains the same value it has at $\langle s, t \rangle$. Since r depends only on discrete variables, which are not changed by a continuous step, it is sufficient to check the value of r at $\langle s, t \rangle$.

If r depends also on continuous variables, then it should be included in the set $\mathcal{I}$, and $(\int r)'$ defined as part of the relation ρ^g_{cont}.

Using the logic TL $_\int$, we can express the safety-critical requirement of system GAS-BURNER of Fig. 11 by the formula

$$\square \left(T \geq 60 \rightarrow 20 \cdot \int Leak \leq T \right)$$

We can prove this property, using rule INV. The auxiliary invariant assertion used is inspired by the proof diagram of Fig. 13 and is given by

$$at\_s_0 \ \wedge \ \int Leak = 0$$
$$\vee$$
$$at\_s_1 \ \wedge \ (T - x) \geq 31 \cdot \left(\int Leak - x\right)$$
$$\vee$$
$$at\_s_2 \ \wedge \ (T - x) \geq 31 \cdot \int Leak - 30.$$

This proof can also be presented as a proof diagram, resembling very much the diagram of Fig. 13. The main difference is that we replace y and z by $\int Leak$ and T, respectively.

Acknowledgement

We gratefully acknowledge the help of Eddie Chang, Yonit Kesten, and Oded Maler for their careful reading of the manuscript and thank them for many helpful suggestions.

References

[AH92] R. Alur and T. Henzinger. Logics and models of real time: A survey. In J.W. de Bakker, C. Huizing, W.P. de Roever, and G. Rozenberg, editors, *Proceedings of the REX Workshop "Real-Time: Theory in Practice"*, volume 600 of *Lect. Notes in Comp. Sci.*, pages 74–106. Springer-Verlag, 1992.

[AL92] M. Abadi and L. Lamport. An old-fashioned recipe for real time. In J.W. de Bakker, C. Huizing, W.P. de Roever, and G. Rozenberg, editors, *Proceedings of the REX Workshop "Real-Time: Theory in Practice"*, volume 600 of *Lect. Notes in Comp. Sci.*, pages 1–27. Springer-Verlag, 1992.

[CHR92] Z. Chaochen, C.A.R Hoare, and A.P. Ravn. A calculus of durations. *Information Processing Letters*, 40(5):269–276, 1992.

[Har84] D. Harel. Statecharts: A visual approach to complex systems. Technical report, Dept. of Applied Mathematics, Weizmann Institute of Science CS84-05, 1984.

[Har87] D. Harel. Statecharts: A visual formalism for complex systems. *Sci. Comp. Prog.*, 8:231–274, 1987.

[HLP90] E. Harel, O. Lichtenstein, and A. Pnueli. Explicit clock temporal logic. In *Proc. 5th IEEE Symp. Logic in Comp. Sci.*, pages 402–413, 1990.

[KP92a] Y. Kesten and A. Pnueli. Age before beauty. Technical report, Dept. of Applied Mathematics, Weizmann Institute of Science, 1992.

[KP92b] Y. Kesten and A. Pnueli. Timed and hybrid statecharts and their textual representation. In J. Vytopil, editor, *Formal Techniques in Real-Time and Fault-Tolerant Systems*, volume 571 of *Lect. Notes in Comp. Sci.*, pages 591–619. Springer-Verlag, 1992.

[MMP92] O. Maler, Z. Manna, and A. Pnueli. From timed to hybrid systems. In J.W. de Bakker, C. Huizing, W.P. de Roever, and G. Rozenberg, editors, *Proceedings of the REX Workshop "Real-Time: Theory in Practice"*, volume 600 of *Lect. Notes in Comp. Sci.*, pages 447–484. Springer-Verlag, 1992.

[MP91a] Z. Manna and A. Pnueli. Completing the temporal picture. *Theor. Comp. Sci.*, 83(1):97–130, 1991.

[MP91b] Z. Manna and A. Pnueli. *The Temporal Logic of Reactive and Concurrent Systems: Specification*. Springer-Verlag, New York, 1991.

[MP93] Z. Manna and A. Pnueli. Models for reactivity. *Acta Informatica*, 1993. To appear.

[Ost90] J.S. Ostroff. *Temporal Logic of Real-Time Systems*. Advanced Software Development Series. Research Studies Press (John Wiley & Sons), Taunton, England, 1990.

[PH88] A. Pnueli and E. Harel. Applications of temporal logic to the specification of real time systems. In M. Joseph, editor, *Formal Techniques in Real-Time and Fault-Tolerant Systems*, volume 331 of *Lect. Notes in Comp. Sci.*, pages 84–98. Springer-Verlag, 1988.

[SBM92] F. B. Schneider, B. Bloom, and K. Marzullo. Putting time into proof outlines. In J.W. de Bakker, C. Huizing, W.P. de Roever, and G. Rozenberg, editors, *Proceedings of the REX Workshop "Real-Time: Theory in Practice"*, volume 600 of *Lect. Notes in Comp. Sci.*, pages 618–639. Springer-Verlag, 1992.

An Extended Duration Calculus for Hybrid Real-Time Systems[*]

Zhou Chaochen[1] Anders P. Ravn[2], and Michael R. Hansen[3]

[1] UNU/IIST
Apartado 3058, Macau
E-mail: zcc@iist.unu.edu and zcc@unuiist.attmail.com
[2] Department of Computer Science
Technical University of Denmark
E-mail: apr@id.dth.dk
[3] Fachbereich Informatik
Universität Oldenburg, Germany
E-mail: michael.hansen@informatik.uni-oldenburg.de.
On leave from: Department of Computer Science
Technical University of Denmark

Abstract. Duration Calculus is a real-time interval logic which can be used to specify and reason about timing and logical constraints on discrete states in a dynamic system. It has been used to specify and verify designs for a number of real-time systems. This paper extends the Duration Calculus with notations to capture properties of piecewise continuous states. This is useful for reasoning about hybrid systems with a mixture of continuous and discrete states. The proof theory of Duration Calculus is extended such that results proven using mathematical analysis can be used freely in the logic. This provides a flexible interface to conventional control theory.

Keywords: requirements, software design, hybrid systems, real-time systems, control theory, interval logic, durations.

1 Introduction

A central step in development of software for real-time, embedded systems is to capture the requirements for a program from the wider requirements for the system of which the program is a component. The Duration Calculus [16] introduces a notation for that task. It has already been used to specify requirements and designs for a number of real-time systems, including an on-off gas-burner [1, 13], a railway crossing [14], and in an early paper [12] a simple auto pilot.

Duration Calculus is more than a specification notation, it is a calculus that allows the designer of a system to verify a design with respect to requirements through a deduction. Duration Calculus is fundamentally an interval logic for reasoning about timing and logical properties of systems modelled by discrete

---

[*] Partially funded by **ProCoS** ESPRIT BRA 7071, by the Danish Technical Research Council **RAPID**, and by the Danish Natural Science Research Council

(Boolean) states. This is not sufficient when dealing with hybrid systems, where states can be both discrete and continuous. Already the auto pilot example, mentioned above, required certain extensions to give an interface to differential equations. The purpose of this paper is to develop a fuller theory of this kind with a simple interface to mathematical analysis. Such an interface permits development of a hybrid system using both control theory and Duration Calculus within a consistent formal framework. Mathematical analysis or its specialisation: control theory, is used where it is most suitable, e.g. to analyse continuous parts of the systems, while the Duration Calculus is used to analyse critical durations and discrete progress properties of the system.

The following section gives a short introduction to the Duration Calculus (DC) by means of a simple example and motivates the need for an interface to control theory. Section 3 defines the syntax and semantics of the Extended Duration Calculus (EDC), while section 4 develops the proof theory. Section 5 illustrates the use of the calculus on larger examples. Finally in section 6 there is a summary and a comparison to other work.

2 Controlling a valve

This section introduces the concepts of embedded real-time systems and illustrates the use of DC to specify and reason about properties of such systems. The description is based on a very simple example of a valve, which controls the gas supply to a burner. We shall first take a typical programmer's view of the valve as a Boolean variable which can be in only one of two states. Later on we shall take the more detailed view of a control engineer and see it as having a state varying continuously between 0 (off) and 1 (on). The challenge is now to ensure that the two views are consistent, and to be able to deduce properties of the abstract representation from properties of the concrete representation.

2.1 Embedded real-time computing systems

An embedded computing system contains a *program* placed in one or more digital computers, which are connected through *channels* to sensors and actuators that measure or change quantities in a piece of *equipment*. The equipment is a physical system evolving over time and used in a certain *environment* for a definite purpose (Figure 1).

When it is required that the computer reacts within specific time bounds, the system is called a (hard) real-time system. The real-time constraints usually reflect that the physical process state defined by the equipment should not deviate too much from an internal state of a control program run by the computer.

The concept of an embedded, real-time system is very broad, and instead of attempting a general definition we illustrate it by the simple example of a valve, which is an actuator. It is used in a gas burner (a piece of equipment), and can influence the environment by leaking gas.

A valve can conveniently be modelled as a function of time. Since we are essentially dealing with physical processes, the time domain, *Time*, is taken to

Environment

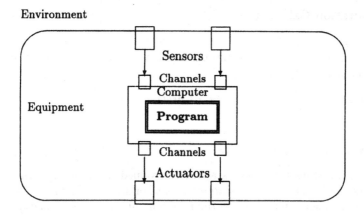

Fig. 1. Embedded software system

be the set of real numbers. Furthermore, we are mainly concerned with absence or presence of gas, so we let the function range over the Boolean values, which we encode as 1 and 0. Thus, a gas valve is modelled by a function Gas with the signature

$$Gas : Time \rightarrow \{0, 1\}$$

We shall call such a function Gas for a *boolean state* of the gas burner and we say that the gas is off at time t, when $Gas(t) = 0$, and on at time t' when $Gas(t') = 1$.

The safety of a gas burner system depends on the presence and absence of gas. Figure 2 illustrates a particular function Gas on an interval $(b, e) \subset Time$.

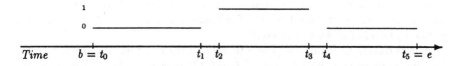

This diagram shows that the valve is definitely off from time $b = t_0$ to time t_1 and from time t_4 to $t_5 = e$. The valve is definitely on from time t_2 to t_3. Gas is not precisely determined in the transient periods from t_1 to t_2 and from t_3 to t_4. Depending on the application, it may be safer to assume either that Gas is on or it is off throughout these intervals.

Fig. 2. Timing diagram for Gas

2.2 Duration Calculus

Any timing diagram can, in principle, be defined in the conventional notation of mathematical analysis. The timing diagram in Figure 2 can for instance be described by formulas like

$$\forall t \bullet t_0 < t < t_1 \vee t_4 < t < t_5 \Rightarrow \neg Gas(t)$$

and

$$\forall t \bullet t_2 < t < t_3 \Rightarrow Gas(t)$$

Note that the state Gas denotes a Boolean valued function, thus the Boolean operators can be used to form *state assertions* or short: *assertions*. Thus, $\neg Gas$ is a state assertion which denotes the function: $\neg Gas(t) = 1 - Gas(t)$.

While such a description is possible, it is far too tedious to keep track of all the t_0, t_1 etc. that enters into a description. Temporal and interval logics, see e.g. the work by Pnueli, Manna, Moszkowski, Koymans and many others [7, 11, 9, 4] improve on this by eliding explicit references to time. We will now use Duration Calculus to describe all the "desirable" timing diagrams of the gas burner without explicit references to absolute time. To this end we assume that the gas burner has another boolean state (function) *Flame*, which models a sensor that indicates whether the gas is burning in the gas burner.

Critical durations

In informal descriptions of the required behaviours of a system, the state space can be partitioned into non-critical and critical states. Non-critical states are those which the system can maintain for an indefinite duration without any harm, e.g. in the gas burner $\neg Gas$ or $Gas \wedge Flame$ are non-critical states, where the burner is either off or the gas is burning. Critical states occur as transitions between non-critical states, e.g. $Leak = Gas \wedge \neg Flame$ in the gas burner. Critical states are unavoidable, but shall have a limited duration. One generic formulation of such a *critical duration* constraint is: "The critical state may occur at most A % of the time within a period of at least T time units". For the gas burner we could have: "*Leak* may occur at most 5 % of the time within a period of at least 1 minute".

The Duration Calculus uses the integral of a state to formalize such a critical duration constraint. The integral $\int_b^e Leak(t) \, dt$ measures the *duration* of *Leak* in the interval (b, e). (In some earlier work, the definitions have been based on closed intervals; but this is unessential, because the duration is insensitive to the state value in isolated points.) When the reference to the particular interval (b, e) is elided, the duration $\int Leak$ is a real valued term for any interval.

An atomic duration formula can now be formed by comparing reals and durations, and duration formulas are formed from the atomic ones using the connectives of predicate logic. A duration formula can be considered a boolean valued function on time intervals. An example is the formula expressing the critical duration constraint on *Leak*

Req: $(\ell \geq 60) \Rightarrow (\int Leak \leq 0.05 \cdot \ell)$

The premiss $(\ell \geq 60)$ expresses that we consider an interval for which the length (ℓ) is at least 60 time units (we assume a time unit of seconds). The conclusion $(\int Leak \leq 0.05 \cdot \ell)$ expresses the desired limitation of the duration of *Leak*. Note that the length of the interval is equal to the duration of the trivial state 1

$$\ell = \int 1$$

A design

A design is a refinement of the requirements. It is thus a more constrained formula. Consider now a simple gas burner, consisting of just a gas valve and a flame sensor. The design decisions for the burner are expressed by timing constraints on its three possible state: the *idle* state where both the gas and the flame are off, the *burning* state where both the gas and the flame are on, and the leaking state. (Notice that the state $\neg Gas(t) \wedge Flame(t)$ is impossible.)

1. The burner stays in the critical leaking state at most one second, i.e. we require that leak is detectable and stoppable within one second.
2. The burner always stays in the idle state for at least 30 seconds before risking another leak. This will (at least for small time intervals) limit the accumulation of unburned gas caused by repeated unsuccessful attempts to ignite the gas.
3. There are no time restrictions for the burning state.

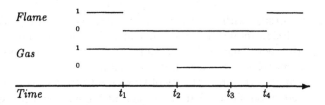

This diagram illustrates two *Leak = Gas $\wedge \neg Flame$* periods, from t_1 to t_2 and from t_3 to t_4, separated by a $\neg Leak$ period from t_2 to t_3. The non-leaking period should be sufficently long to prevent dangerous levels of unburned gas.

Fig. 3. Timing diagram for *Flame* and *Gas*

To formalize the above design decisions, we need the notion of an assertion P holding (everywhere) in a proper interval $\lceil P \rceil$. It is linked to durations by the following law

$$\lceil P \rceil \Rightarrow (\int P = \ell) \wedge (\ell > 0)$$

Note that $\lceil P \rceil$ is false for an empty interval.

If P holds everywhere in a proper interval, then this interval can be partitioned into two intervals such that P holds everywhere in both. This fact is formalized using the "chop" operator ; of interval temporal logic:

$$\lceil P \rceil \Rightarrow \lceil P \rceil; \lceil P \rceil$$

The converse does not necessarily hold.

An empty interval is characterized by $\ell = 0$, and this introduces another convenient abbreviation

$$\lceil\ \rceil \ \widehat{=}\ \ell = 0$$

The fact that *Leak* shall cease to hold within 1 second can now be written as $\lceil Leak \rceil \Rightarrow \ell \leq 1$. The fact that *Gas* shall be off for 30 seconds can be specified by the formula: $\lceil Leak \rceil; \lceil \neg Leak \rceil; \lceil Leak \rceil \Rightarrow \ell > 30$, cf. Figure 3. This left hand side reads: The interval of consideration can be partitioned into three parts where $\lceil Leak \rceil$ holds for the first and last part and $\lceil \neg Leak \rceil$ holds for the middle part.

These formulas should, however, hold for any subinterval of a behaviour. Thus we need to introduce the modality $\Diamond$ ("for some subinterval") and its dual $\Box$ ("for every subinterval") for a formula D:

$$\Diamond D \ \widehat{=}\ true; D; true$$
$$\Box D \ \widehat{=}\ \neg(\Diamond(\neg D))$$

Using the abbreviations, we can write the design decisions:

Des-1: $\Box(\lceil Leak \rceil \Rightarrow \ell \leq 1)$
Des-2: $\Box(\lceil Leak \rceil; \lceil \neg Leak \rceil; \lceil Leak \rceil \Rightarrow \ell > 30)$

Verification

Verification of a design with respect to the requirements is a deduction which shows that the design implies the requirements. This means that a design will always satisfy the requirements; but of course there may be behaviours allowed by the requirements that the design does not allow.

Verification applies the proof theory of Duration Calculus. As an illustration, the following deduction shows that the design above ensures that any interval of less than 30 seconds has a limit of 1 on the duration of leaks.

We assume that states have 'finite variability', which means that for an arbitrary assertion P, any proper interval can be divided into finitely many subintervals such that either $\lceil P \rceil$ or $\lceil \neg P \rceil$ hold in a subinterval.

$\ell \leq 30$

$\hspace{8cm} \Rightarrow \{\text{Finite Variability}\}$

$\lceil\ \rceil \vee$
$\lceil Leak \rceil; (\lceil \neg Leak \rceil \vee \lceil\ \rceil) \vee$
$\lceil \neg Leak \rceil; (\lceil Leak \rceil \vee \lceil\ \rceil) \vee$
$\lceil \neg Leak \rceil; \lceil Leak \rceil; \lceil \neg Leak \rceil \vee$
$(\ell \leq 30 \wedge \Diamond(\lceil Leak \rceil; \lceil \neg Leak \rceil; \lceil Leak \rceil))$

$\hspace{8cm} \Rightarrow \{\text{Des-2}\}$

$$\lceil\,\rceil \vee$$
$$\lceil Leak\rceil; \;(\lceil\neg Leak\rceil \vee \lceil\,\rceil) \vee$$
$$\lceil\neg Leak\rceil; \;(\lceil Leak\rceil \vee \lceil\,\rceil) \vee$$
$$\lceil\neg Leak\rceil; \;\lceil Leak\rceil; \;\lceil\neg Leak\rceil$$

$$\Rightarrow \{\text{Des-1}\}$$

$$\lceil\,\rceil \vee$$
$$(\textstyle\int Leak \le 1); \;(\lceil\neg Leak\rceil \vee \lceil\,\rceil) \vee$$
$$\lceil\neg Leak\rceil; \;((\textstyle\int Leak \le 1) \vee \lceil\,\rceil) \vee$$
$$\lceil\neg Leak\rceil; \;(\textstyle\int Leak \le 1); \;\lceil\neg Leak\rceil$$

$$\Rightarrow \{\lceil\neg P\rceil \vee \lceil\,\rceil \Rightarrow \textstyle\int P = 0\}$$

$$\textstyle\int Leak = 0 \vee$$
$$\textstyle\int Leak \le 1; \;\int Leak = 0 \vee$$
$$\textstyle\int Leak = 0; \;\int Leak \le 1 \vee$$
$$\textstyle\int Leak = 0; \;\int Leak \le 1; \;\int Leak = 0$$

$$\Rightarrow \{\textstyle\int P \le r; \;\int P \le s \Rightarrow \int P \le r + s\}$$

$$\textstyle\int Leak \le 1$$

When this fact has been established, it is easy to deduce that the requirement is satisfied by dividing any interval of length greater than 1 minute into a sequence of intervals of length equal to 30 seconds followed by a single interval of length less than 30 seconds. Induction on the number of intervals can then be used to establish the requirement.

Note that correctness relies on proper adjustment of the timing constants. In a practical design, one will often start out with symbolic constants representing limit values, and then as a side effect of the verification determine constraints on the actual values.

2.3 Control Theory

Control theory, cf. [5], is a theory about properties of the class of dynamical systems that can be described by difference or differential equations over state functions. In most applications linear or piecewise linear equations are used, because mathematical analysis of non-linear systems is difficult. The dynamic system describes all relevant states, i.e. both the environment and the equipment of the embedded system. In a more concrete model for the valve, we would be interested in describing the transient periods. In order to do so, we introduce a new environment state v, which describes the movement of the valve between being fully closed 0 and fully open 1

$$v : Time \to [0, 1]$$

The state v should obviously be continuous, and we would assume that it moves as shown in Figure 4.

Given that v in the transient periods are given by the differential equations

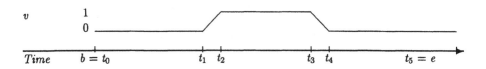

This diagram shows movement of the valve as function of time. In the transient periods, we assume that the movement is governed (approximately) by differential equations of the form $\dot{v} = k$ where k denotes some constant speed.

Fig. 4. Timing diagram for v

Eq: $\dot{v} = k_{on}$ and $\dot{v} = k_{off}$

where the constants k_{on} and k_{off} satisfies

$$k_{on} > 0 \text{ and } k_{off} < 0$$

it is easy to use elementary mathematical analysis to find that the tightest estimates for the bounds on the transient periods are given by $\delta_{off} = -1/k_{off}$ and $\delta_{on} = 1/k_{on}$.

In order to use that result, the Duration Calculus is extended with two interval functions (actually functionals) $\mathbf{b}.(\cdot)$ and $\mathbf{e}.(\cdot)$. Applied to a (piecewise) continuous function f on a non-empty interval $\mathbf{b}.f$ ($\mathbf{e}.f$) gives the value of f at the beginning (end) of the interval. (Strictly speaking they give the limit from the right and left of the respective end points. Their values on empty intervals are discussed later.)

The gas burner system can now be linked to the valve system above as follows. First the *Gas* and v states are related

$$Gas \Leftrightarrow (v > 0)$$

We have here chosen that *Gas* holds whenever the valve is open. Cf. the remarks about transient periods in the text for Figure 2.

We can now start to introduce the results of the control theory for the valve into our analysis of the system. Note in that connection that, if it can be proven in mathematical analysis that P holds for every proper interval, then $\lceil\ \rceil \vee \lceil P \rceil$ is valid in the calculus.

The following formulas constitute a possible 'theory of valves':

V0: The range of v is $[0, 1]$:

$$\Box(\lceil\ \rceil \vee \lceil 0 \leq v \leq 1\rceil)$$

V1: The valve has finite variability: It is either stable, closing or opening for some time

$$\Box\left(\lceil\ \rceil \vee (\lceil\dot{v} = 0\rceil;\ true) \vee (\lceil\dot{v} = k_{off}\rceil;\ true) \vee (\lceil\dot{v} = k_{on}\rceil;\ true)\right)$$

and

$$\Box(\lceil\,\rceil \vee (true;\ \lceil\dot{v} = 0\rceil) \vee (true;\ \lceil\dot{v} = k_{off}\rceil) \vee (true;\ \lceil\dot{v} = k_{on}\rceil))$$

V2: Once it starts to close, it will complete closing. Or in other words: It is not the case that it ceases to close before it is closed

$$\Box((\lceil v > 0\rceil \wedge (\lceil\dot{v} = k_{off}\rceil;\ true)) \Rightarrow \lceil\dot{v} = k_{off}\rceil)$$

Similarily for opening.
V3: The movement v is totally continuous

$$\Box(\lceil\,\rceil \vee \lceil Continuous(v)\rceil)$$

A consequence of this fact is that ending and beginning values of v are identical over a "chop"

$$(\lceil Continuous(v)\rceil \wedge ((\ell > 0 \wedge \mathbf{e}.v = x);\ (\mathbf{b}.v = y \wedge \ell > 0))) \Rightarrow x = y$$

Assuming that the valve in the gas burner system works as specified, we can consider a refinement of design decision **Des-1**

Des'-1: When the flame disappears, the valve should be closing within δ seconds

$$\Box(\lceil Leak\rceil \wedge \lceil\dot{v} \neq k_{off}\rceil \Rightarrow \ell \leq \delta)$$

We can now verify that this is a correct refinement of **Des-1** by the following deduction

$\lceil Leak\rceil$	$\Rightarrow \{Gas \Leftrightarrow v > 0\}$
$\lceil\neg Flame \wedge v > 0\rceil$	$\Rightarrow \{\lceil P_1 \wedge P_2\rceil \Leftrightarrow \lceil P_1\rceil \wedge \lceil P_2\rceil\}$
$\lceil\neg Flame\rceil \wedge \lceil v > 0\rceil$	$\Rightarrow \{Des'-1\}$
$\ell \leq \delta;\ (\lceil\,\rceil \vee (\lceil v > 0\rceil \wedge \lceil\dot{v} = k_{off}\rceil;\ true))$	$\Rightarrow \{V2\}$
$\ell \leq \delta;\ (\lceil\,\rceil \vee \lceil\dot{v} = k_{off}\rceil)$	$\Rightarrow \{Control\ theory,\ Eq\}$
$\ell \leq \delta;\ (\lceil\,\rceil \vee \ell \leq -1/k_{off})$	$\Rightarrow \{Analysis\ etc.\}$
$\ell \leq \delta - 1/k_{off}$	

We conclude that **Des'-1** is a correct refinement of **Des-1** whenever $\delta - 1/k_{off} \leq 1$, i.e. $\delta \leq 1 + 1/k_{off}$. Note that this tells us that if the valve is slow, $k_{off} > -1$, the refinement is infeasible. Only a miracle can then implement the system.

2.4 Summary

We have motivated the extensions to the Duration Calculus, and illustrated how it can be used to reason about a hybrid system, which is partly continuous and partly discrete.

3 Syntax and Semantics

This section defines the formal language of Extended Duration Calculus. It builds on previous work reported in [2]. The Extended Duration Calculus (EDC) is a first order logic with special duration and interval operators and corresponding special axioms. It is based on a mathematical theory of functions which is summarized in subsection 1. The syntax of EDC is presented in subsection 2, while the semantics is presented in subsection 3.

3.1 A Mathematical Theory about Functions

As the underlying model for EDC we assume a mathematical theory about state functions to be given. This theory, which is named MT, has the following constituents:

1. (a) *State names*: $f, g, h, \dot{f}, \dot{g}, \ldots$, where $\dot{f}$ is the derivative of f. Each state name denotes a real valued function on time i.e. it has the signature: $\mathbf{R} \to \mathbf{R}$.

   (b) *Boolean state names*: *Flame*,.... Each boolean state name denotes a boolean valued function on time, i.e. it has the signature: $\mathbf{R} \to \{0, 1\}$.

2. (a) Operators on real numbers, e.g.: $+, -, \ldots$ can be applied on state names to form (composite) *states*, e.g. $f + g$. A state has the signature $\mathbf{R} \to \mathbf{R}$, and it is defined pointwise from its constituent state names, e.g. $(f + g)(t) \mathrel{\widehat{=}} f(t) + g(t)$. We assume from now on that the operators $+, -, \ldots$ have their usual meaning, and we will use the word state for either a state name or a composite state.

   (b) Propositional logic operators (e.g. $\neg, \vee, \wedge$) can be applied on boolean state names to form *boolean states*. We also allow 0 and 1 as the constant boolean states which at any time have the values 0 and 1, respectively.

3. States can be compared by the relational operators of real arithmetics to form the *relations*, e.g. $f < g$. A relation A denotes a boolean valued function of time: $A : \mathbf{R} \to \{0, 1\}$. This function is also defined pointwise, e.g. $(f < g)(t) = 1$ iff $f(t) < g(t)$.

4. An *atomic assertion* A is either a relation or a boolean state. An *assertion* P is either atomic or it is constructed from atomic formulas using predicate logic. E.g. *Continuous*(s), for any state s, is an assertion defined pointwise as:

$$Continuous(s)(t) \mathrel{\widehat{=}}$$
$$\forall \varepsilon > 0 \ \exists \delta > 0 \ \forall t' \bullet |\, t - t' \,| \leq \delta \Rightarrow |\, s(t) - s(t') \,| \leq \varepsilon$$

An *interpretation* $\mathcal{I}$ is an assignment of a function $f_{\mathcal{I}} : \mathbf{R} \to \mathbf{R}$ to each state name f and an assignment of a function $b_{\mathcal{I}} : \mathbf{R} \to \{0, 1\}$ to each boolean state b.

For a given interpretation $\mathcal{I}$, any state s of MT denotes a real valued function $s_{\mathcal{I}}$ on time and any assertion P of MT denotes a boolean valued function $P_{\mathcal{I}}$ on time.

We assume from now on that an interpretation $\mathcal{I}$ obeys the following Dirichlet condition:

Condition 1 *For any state s (or assertion P), the function denoted by s (or P) in $\mathcal{I}$ must be bounded, and sectionally continuous on any bounded interval.*

A sectional (piecewise) continuous function has a finite number of discontinuities in a bounded interval.

A consequence is that an assertion has a finite variability, which is the basis for well-founded induction over assertions (see section 4.3). Other consequences are that for any state assertion P, its interpretation $P_{\mathcal{I}}$ is Riemann integrable, and that for a state s both the limit from the left

$$s_{\mathcal{I}}(t_0^-) \;\widehat{=}\; \lim_{t \to t_0, t < t_0} s_{\mathcal{I}}(t)$$

and the limit from the right

$$s_{\mathcal{I}}(t_0^+) \;\widehat{=}\; \lim_{t \to t_0, t > t_0} s_{\mathcal{I}}(t)$$

exist for any interpretation $\mathcal{I}$ and for any t_0 in **R**.

3.2 Syntax of Extended Duration Calculus

Assume in the following a mathematical theory MT to be given. I.e. states s and assertions P are already defined for this theory. Based on this mathematical theory the syntax of Extended Duration Calculus is now defined.

We allow universal quantification $\forall x$ in Extended Duration Calculus. Therefore, we introduce a syntactical category *Var* of *global variables*. Each global variable x ranges over the real numbers.

Duration terms

Duration terms, dt_i $i = 0, 1, 2, \ldots$, are real valued terms, which are built from the special symbol ℓ (length), global variables (x), and initial (**b**.s) and final values (**e**.s) for states s of MT .

$$dt ::= \ell \mid x \mid \mathbf{b}.s \mid \mathbf{e}.s \mid \oplus(dt_1, \ldots, dt_n)$$

where $\oplus$ is an arbitrary n-ary operator on reals defined in MT.

Duration formulas

Atomic duration formulas are built from duration terms and assertions P of MT by the rule:

$$D ::= \lhd(dt_1, \ldots, dt_n) \mid \lceil P \rceil$$

where $\lhd$ is an arbitrary n-ary relational operator on reals in MT.

Duration formulas are formed from the atomic ones using the propositional connectives, the "chop" connective, and universal quantification:

$$
\begin{array}{rl}
D ::= & \neg D \\
& | \ D_1 \wedge D_2 \\
& | \ D_1;\ D_2 \\
& | \ \forall x \bullet D
\end{array}
$$

where x is any global variable from *Var*.

Remark: The same symbols (logical connectives, arithmetic operators and relations) can be used in assertions and duration formulas. Since assertions occur within brackets $\lceil \cdot \rceil$, this does not give any ambiguity problems.

Abbreviations

The abbreviations ($\vee$, $\Rightarrow$, $\Leftrightarrow$) and the existential quantifier are defined as usual. The usual convention of abbreviating multiple quantifiers of the same kind, e.g. $\forall x_1 \bullet \forall x_2 \bullet D$ to $\forall x_1, x_2 \bullet D$ is also used.

Abbreviations corresponding to logical constants are:

$$
\begin{array}{ll}
\mathit{true} \ \widehat{=} \ \ell \geq 0 & \text{holds for any interval} \\
\mathit{false} \ \widehat{=} \ \neg\mathit{true} & \text{holds for no interval}
\end{array}
$$

The following syntactical abbreviation denotes an empty interval:

$$\lceil\ \rceil \ \widehat{=} \ \ell = 0 \quad \text{the empty interval}$$

Furthermore, we introduce the standard abbreviations from interval temporal logic:

$$
\begin{array}{ll}
\Diamond D \ \widehat{=} \ \mathit{true};\ D;\ \mathit{true} & \text{for some sub-interval } D \\
\Box D \ \widehat{=} \ \neg\Diamond\neg D & \text{for every sub-interval } D
\end{array}
$$

Precedence rules

In order to avoid an excessive number of parentheses the following precedence rules are used:

first:	**b., e.**
second:	real operators
third:	real predicates
fourth:	$\neg$, $\Box$, $\Diamond$,
fifth:	;
sixth:	$\vee$, $\wedge$
seventh:	$\Rightarrow$
eight:	$\forall x \bullet$, $\exists x \bullet$

3.3 Semantics

We define now the truth value of an arbitrary duration formula D. This truth value depends, in general, on the interpretation $\mathcal{I}$ to the state names and boolean state names, the value of the global variables, and the interval (c, d) of consideration. So let a *valuation* be a pair $(\mathcal{V}, (c, d))$ where $\mathcal{V} : \mathit{Var} \to \mathbf{R}$ is a *value assignment* associating a real value with every global variable. Let Val be the set of all valuations.

We assume that each interpretation $\mathcal{I}$ is equiped with a constant $k_{\mathcal{I}} \in \mathbf{R}$, which we shall use to define the semantics of **b**.s and **e**.s for empty intervals.

The *semantics of a duration term dt* in an interpretation $\mathcal{I}$ is a function:

$$\mathcal{I}(dt) : \mathit{Val} \to \mathbf{R}$$

defined inductively on the structure of duration terms by:

$$\mathcal{I}(\ell)(\mathcal{V}, (c, d)) \qquad\qquad = d - c$$

$$\mathcal{I}(x)(\mathcal{V}, (c, d)) \qquad\qquad = \mathcal{V}(x)$$

$$\mathcal{I}(\mathbf{b}.s)(\mathcal{V}, (c, d)) \qquad\qquad = \begin{cases} s_{\mathcal{I}}(c^+), & \text{if } c < d \\ k_{\mathcal{I}}, & \text{if } c = d \end{cases}$$

$$\mathcal{I}(\mathbf{e}.s)(\mathcal{V}, (c, d)) \qquad\qquad = \begin{cases} s_{\mathcal{I}}(d^-), & \text{if } c < d \\ k_{\mathcal{I}}, & \text{if } c = d \end{cases}$$

$$\mathcal{I}(\oplus(dt_1, \ldots, dt_n))(\mathcal{V}, (c, d)) = \oplus(\mathcal{I}(dt_1)(\mathcal{V}, (c, d)), \ldots, \mathcal{I}(dt_n)(\mathcal{V}, (c, d)))$$

Note that we use the meaning of the operator $\oplus$ in MT at the right hand side of the last definition.

The semantics of a duration formula D in an interpretation $\mathcal{I}$ is a function:

$$\mathcal{I}(D) : \mathit{Val} \to \{tt, f\!f\}$$

defined inductively on the structure of duration formulas below.

The following shorthand notation will be used:

$$\mathcal{I}, \mathcal{V}, (c, d) \models D \ \widehat{=} \ \mathcal{I}(D)(\mathcal{V}, (c, d)) = tt$$
$$\mathcal{I}, \mathcal{V}, (c, d) \not\models D \ \widehat{=} \ \mathcal{I}(D)(\mathcal{V}, (c, d)) = f\!f$$

The function $\mathcal{I}(D)$ is defined by:

$$\mathcal{I}, \mathcal{V}, (c, d) \models \lhd(dt_1, \ldots, dt_n) \text{ iff } \lhd(\mathcal{I}(dt_1)(\mathcal{V}, (c, d)), \ldots, \mathcal{I}(dt_n)(\mathcal{V}, (c, d)))$$
$$= tt$$

$\mathcal{I}, \mathcal{V}, (c, d) \models \lceil P \rceil$ iff $c < d$ and $P_{\mathcal{I}}(t) = 1$ for every $t \in (c, d)$

$\mathcal{I}, \mathcal{V}, (c, d) \models \neg D$ iff $\mathcal{I}, \mathcal{V}, (c, d) \not\models D$

$\mathcal{I}, \mathcal{V}, (c, d) \models D_1 \wedge D_2$ iff $\mathcal{I}, \mathcal{V}, (c, d) \models D_1$ and $\mathcal{I}, \mathcal{V}, (c, d) \models D_2$

$\mathcal{I}, \mathcal{V}, (c, d) \models D_1; \ D_2$ iff $\mathcal{I}, \mathcal{V}, (c, m) \models D_1$ and $\mathcal{I}, \mathcal{V}, (m, d) \models D_2$
                                                      for some $m \in (c, d)$

$\mathcal{I}, \mathcal{V}, (c, d) \models \forall x_i \bullet D$ iff $\mathcal{I}, \mathcal{V}', (c, d) \models D$ for all assignments $\mathcal{V}'$
                                                      for which $\mathcal{V}'(x_k) = \mathcal{V}(x_k)$ for all $k \neq i$.

Note that we use the meaning of the relational operator $\lhd$ in MT at the right hand side of the first definition.

3.4 Satisfaction and validity

A duration formula D *is true in the interpretation* $\mathcal{I}$ (written $\mathcal{I} \models D$) iff $\mathcal{I}, \mathcal{V}, (c, d) \models D$ for every valuation $(\mathcal{V}, (c, d))$. We also say that $\mathcal{I}$ *satisfies* D. Furthermore, D is *valid* (written $\models D$) iff D is true in every interpretation.

4 Proof system

In this section we give a proof system which has valid formulas as its theorems.

The language of EDC is evidently an extension of the first order theory MT that includes real analysis and thereby real arithmetic. We shall use MT to derive properties of states and assertions. The notation $\vdash_{MT} \mathcal{F}$ is used to signify that $\mathcal{F}$ is a provable formula in MT.

4.1 Interval Temporal Logic

The Duration Calculus adopts axioms and rules from interval temporal logic (ITL). E.g.

1.1 *Monotonicity* :

     If $D_1 \Rightarrow D_1'$ and $D_2 \Rightarrow D_2'$ then $D_1; \ D_2 \Rightarrow D_1'; \ D_2'$.

1.2 *Associativity* :

     $(D_1; \ D_2); \ D_3 \ \Leftrightarrow \ D_1; \ (D_2; \ D_3)$

1.3 *False − Zero* :

     $D; \ false \ \Rightarrow \ false$ and $false; \ D \ \Rightarrow \ false$

1.4 *Point − Unit* :

$$D \Leftrightarrow (D; \lceil \rceil) \Leftrightarrow (\lceil \rceil; D)$$

1.5 *Chop − Or* :

$$(D_1 \vee D_2); D_3 \Rightarrow D_1; D_3 \vee D_2; D_3$$

and

$$D_1; (D_2 \vee D_3) \Rightarrow D_1; D_2 \vee D_1; D_3$$

The reverse implication follows from monotonicity.

1.6 *Exists − Chop* :

$$(\exists x \bullet D_1); D_2 \Leftrightarrow \exists x \bullet D_1; D_2$$

provided x does not occur free in D_2.

$$D_1; (\exists x \bullet D_2) \Leftrightarrow \exists x \bullet D_1; D_2$$

provided x does not occur free in D_1.

1.7 *Continuum*

$$\ell = r + s \quad \Leftrightarrow \quad \ell = r; \ell = s \quad \text{for } r, s \geq 0$$

4.2 Axioms of $\lceil P \rceil$ and $\lceil \rceil$

2.1

$$\lceil 1 \rceil \Leftrightarrow (\ell > 0)$$

2.2

$$\lceil 0 \rceil \Leftrightarrow false$$

2.3

$$\lceil P \rceil \Rightarrow \Box(\lceil \rceil \vee \lceil P \rceil)$$

2.4

If $\quad \vdash_{MT} P \Rightarrow Q$
then $\lceil P \rceil \Rightarrow \lceil Q \rceil$

2.5

$$\lceil P \rceil \wedge \lceil Q \rceil \Rightarrow \lceil P \wedge Q \rceil$$

Note: that the implication in the opposite direction follows from 2.4.

We can now prove the following

Theorem 2.6. Let $x, y > 0$.

$$\lceil P \rceil \wedge \ell = x + y$$
$$\Rightarrow (\lceil P \rceil \wedge \ell = x); (\lceil P \rceil \wedge \ell = y)$$

Proof:

$$\lceil P \rceil \wedge \ell = x + y$$
$$\Rightarrow \lceil P \rceil \wedge (\ell = x; \ell = y) \qquad\qquad 1.7$$
$$\Rightarrow \Box(\lceil \ \rceil \vee \lceil P \rceil) \wedge (\ell = x; \ell = y) \qquad 2.3$$
$$\Rightarrow (\ell = x \wedge (\lceil \ \rceil \vee \lceil P \rceil)); (\ell = y \wedge (\lceil \ \rceil \vee \lceil P \rceil)) \text{ ITL}$$
$$\Rightarrow (\lceil P \rceil \wedge \ell = x); (\lceil P \rceil \wedge \ell = y) \qquad \text{def. } \lceil \ \rceil$$

Since $\lceil P \rceil \Rightarrow \ell > 0$ by 2.4 and 2.1 we have the following

Corollary 2.7.

$$\lceil P \rceil \Rightarrow \lceil P \rceil; \lceil P \rceil$$

4.3 Dirichlet Condition

The following two induction principles uses finite variability to cover an interval. Assume in these induction rules that

$$\vdash_{MT} P_1 \vee P_2 \vee \cdots \vee P_n$$

3.1 *Forward induction* :

Let X denote a formula letter occurring in a formula $D(X)$:
If $D(\lceil \ \rceil)$ is provable and $D(X \vee X; \lceil P_i \rceil)$ is provable from $D(X)$ for $i = 1, \ldots, n$, then $D(true)$.

3.2 *Backward induction* :

Let X denote a formula letter occurring in a formula $D(X)$:
If $D(\lceil \ \rceil)$ is provable, and $D(X \vee \lceil P_i \rceil; X)$ is provable from $D(X)$ for $i = 1, \ldots, n$, then $D(true)$.

3.3 *Sectional Continuity* :

$$\lceil \neg Continuous(f) \rceil \Leftrightarrow false$$

3.4 *Boundedness* :

$$\exists r. \lceil | f | \leq r \rceil$$

Since we have that $\vdash_{MT} Continuous(f) \vee \neg Continuous(f)$, we can by 3.1 and 3.3 derive the following induction principle:

Theorem 3.5. If $D(\lceil \ \rceil)$ is provable and $D(X \vee X; \lceil Continuous(f) \rceil)$ is provable from $D(X)$, then $D(true)$.

4.4 Interface to control theory

4.1

$$\lceil\,\rceil \Rightarrow \mathbf{b}.f = \mathbf{e}.f$$

4.2

$$\mathbf{b}.f = r_1 \wedge ((\mathbf{b}.f = r_2 \wedge \ell > 0);\ true) \Rightarrow r_1 = r_2$$

4.3

$$\mathbf{e}.f = r_1 \wedge (true;\ (\ell > 0 \wedge \mathbf{e}.f = r_2)) \Rightarrow r_1 = r_2$$

4.4

$$\lceil Continuous(f) \rceil \wedge ((\ell > 0 \wedge \mathbf{e}.f = r_1)\,;\ (\mathbf{b}.f = r_2 \wedge \ell > 0)) \Rightarrow r_1 = r_2$$

4.5

$$\lceil Continuous(f) \rceil \Rightarrow (\lceil H(f) \rceil \Leftrightarrow \neg((\ell > 0 \wedge \neg H(\mathbf{e}.f))\,;\ \ell > 0))$$

4.6

Let $R(\lceil H(f) \rceil,\ \mathbf{b}.f,\ \mathbf{e}.f,\ \ell)$ be a formula of EDC without "chop" (;).
If $\vdash_{MT} \forall c < d \bullet R(\forall t \in (c,d) \bullet H(f(t)),\ f(c^+),\ f(d^-),\ d-c)$ then

$$\lceil\,\rceil \vee R(\lceil H(f) \rceil, \mathbf{b}.f, \mathbf{e}.f, \ell)$$

Using these axioms we can prove

Theorem 4.7.

$$\lceil Continuous(f) \rceil \wedge ((\lceil H(f) \rceil \wedge H(\mathbf{e}.f))\,;\ \lceil H(f) \rceil) \Rightarrow \lceil H(f) \rceil$$

Proof: By 4.5 it is sufficient to prove:

$$\left\{ \begin{array}{l} \lceil Continuous(f) \rceil \wedge \\ (\lceil H(f) \rceil \wedge H(\mathbf{e}.f) \wedge \ell = x)\,;\ (\lceil H(f) \rceil \wedge \ell = y) \wedge \\ (\ell > 0 \wedge \neg H(\mathbf{e}.f) \wedge \ell = u)\,;\ (\ell > 0 \wedge \ell = v) \end{array} \right\} \Rightarrow false$$

Assume the premise of this implication. We consider the three cases: $x = u, x < u$, and $x > u$. The case $x = u$ is trivial, so assume now that

$x < u$: By 1.7 $u < x + y$ and we derive $\lceil H(f) \rceil \wedge \ell = y \wedge (\ell = u - x \wedge \neg H(\mathbf{e}.f)\,;\ \ell > 0)$ by 4.3. But this contradicts 4.5. (The case $x > u$ is similar.)

4.5 Duration Calculus

In the Extended Duration Calculus, there is no syntactical construct for the duration $\int P$ of an assertion P. In this section, we show that this notion can be derived, and that axioms for $\int P$ can be proven in the Extended Duration Calculus.

Throughout the section we assume that for any assertion P, there is a function f_P which is defined by:

$$(\lceil\ \rceil \vee \lceil Continuous(f_p)\rceil) \wedge \Box(\lceil P\rceil \Rightarrow (\dot{f}_p = 1)) \wedge \Box(\lceil \neg P\rceil \Rightarrow (\dot{f}_p = 0))$$

Then we define:

$$\int P \,\hat{=}\, \mathbf{e}.f_p - \mathbf{b}.f_p$$

With these definitions the axioms of the original Duration Calculus [16] are proved to be theorems in the extended calculus:

(A1) *Zero* :

$$\int 0 = 0$$

Proof: Follows from the following:

$$\ell = 0 \Rightarrow (\mathbf{e}.f_0 = \mathbf{b}.f_0)\ (4.1)$$
$$\Rightarrow \int 0 = 0 \qquad (def.\ \textstyle\int 0)$$

$$\ell > 0 \Rightarrow \lceil \neg 0\rceil \qquad (2.1)$$
$$\Rightarrow \lceil \dot{f}_0 = 0\rceil \qquad (def.\ f_0)$$
$$\Rightarrow \mathbf{e}.f_0 = \mathbf{b}.f_0 \quad (4.6)$$
$$\Rightarrow \int 0 = 0$$

(A2) *Positive*:

$$\int P \geq 0$$

Proof: We prove first

$$\lceil P\rceil;\ true\ \Rightarrow\ \textstyle\int P \geq 0$$

This proof is by induction (3.1) with P_1 and P_2 being P and $\neg P$, respectively. Let $R(X)$ be $\lceil P\rceil;\ X\ \Rightarrow \int P \geq 0$. The base case $R(\lceil\ \rceil)$ of the induction follows from:

$$\lceil P\rceil \Rightarrow \lceil \dot{f}_P = 1\rceil \qquad (def.\ f_P)$$
$$\Rightarrow \mathbf{e}.f_P \geq \mathbf{b}.f_P\ (4.6)$$

The inductive step $R(X; \lceil P\rceil)$ follows from:

$$\lceil P\rceil;\ X;\ \lceil P\rceil \Rightarrow (\mathbf{e}.f_P \geq \mathbf{b}.f_P \wedge \ell > 0);\ \lceil \dot{f}_P = 1\rceil\ (R(X),\ def\ f_P)$$
$$\Rightarrow (\mathbf{e}.f_P \geq \mathbf{b}.f_P \wedge \ell > 0);$$
$$(\mathbf{e}.f_P \geq \mathbf{b}.f_P \wedge \ell > 0) \qquad (4.6)$$
$$\Rightarrow \mathbf{e}.f_P \geq \mathbf{b}.f_P \qquad (4.4,\ 4.3,\ 4.2)$$

The other inductive case $R(X; \lceil \neg P \rceil)$ can be established similarily.

By a similar induction we can prove: $\lceil \neg P \rceil$; $true \Rightarrow \int P \geq 0$. Furthermore, by 4.1 we get $\lceil \; \rceil \Rightarrow \int P \geq 0$. I.e. we have

$$(\lceil \; \rceil \vee \lceil P \rceil; \; true \vee \lceil \neg P \rceil; \; true) \Rightarrow \int P \geq 0$$

The proof is completed by noting that $(\lceil \; \rceil \vee \lceil P \rceil; \; true \vee \lceil \neg P \rceil; \; true)$ follows from 3.1.

(A3) *Sum* :

$$(\int P = r_1); \; (\int P = r_2) \Leftrightarrow (\int P = r_1 + r_2)$$

where r_1, r_2 are non-negative reals.

The implication $\Rightarrow$ follows by a), b), and c) below:

a) $((\mathbf{e}.f_P - \mathbf{b}.f_P) = r \wedge \ell > 0); \; ((\mathbf{e}.f_P - \mathbf{b}.f_P) = s \wedge \ell > 0) \Rightarrow (\mathbf{e}.f_P - \mathbf{b}.f_P) = r + s$ by 4.4, 4.3, and 4.2.
b) $((\mathbf{e}.f_P - \mathbf{b}.f_P) = r \wedge \ell = 0); \; (\mathbf{e}.f_P - \mathbf{b}.f_P) = s \Rightarrow r = 0 \wedge (\mathbf{e}.f_P - \mathbf{b}.f_P) = s$ by 4.1.
c) $(\mathbf{e}.f_P - \mathbf{b}.f_P) = r; \; ((\mathbf{e}.f_P - \mathbf{b}.f_P) = s \wedge \ell = 0) \Rightarrow (\mathbf{e}.f_P - \mathbf{b}.f_P) = r \wedge s = 0$ by 4.1.

We use this result $(A3, \Rightarrow)$ in the proof of the other implication: $(A3, \Leftarrow)$. Furthermore, observe that

(*) $\lceil P \rceil \Rightarrow \int P = \ell$

follows by the definition of f_P and 4.6.

We prove the implication $(A3, \Leftarrow)$ by induction (3.1). Let P_1 and P_2 be P and $\neg P$, respectively, and let $R(X)$ be $X \Rightarrow (\int P = r+s \Rightarrow \int P = r; \int P = s)$.

The base case $R(\lceil \; \rceil)$ is trivial since $\lceil \; \rceil \Rightarrow r = s = 0$, by 4.1 and definition of f_P.

To establish the inductive step $R(X; \lceil P \rceil)$, we prove that

$$(X; \; (\lceil P \rceil \wedge \ell = u)) \wedge \int P = r + s \Rightarrow (\int P = r; \int P = s)$$

We consider two cases:

$s \geq u > 0$:

$$(X; \; (\lceil P \rceil \wedge \ell = u)) \wedge \int P = r + s$$
$$\Rightarrow ((X \wedge \exists x. \int P = x); \; (\lceil P \rceil \wedge \int P = u = \ell)) \wedge \int P = r + s \text{ by } (*)$$
$$\Rightarrow (X \wedge \int P = r + s - u); \; (\lceil P \rceil \wedge \int P = u) \qquad\qquad \text{by } (A3, \Rightarrow)$$
$$\Rightarrow \int P = r; \int P = s - u; \int P = u \qquad\qquad\qquad\qquad \text{by } R(X)$$
$$\Rightarrow \int P = r; \int P = s \qquad\qquad\qquad\qquad\qquad\qquad \text{by } (A3, \Rightarrow)$$

$s < u$:

$$(X; (\lceil P \rceil \wedge \ell = u)) \wedge \int P = r + s$$
$$\Rightarrow X; (\lceil P \rceil \wedge (\ell = u - s; \ell = s)) \wedge \int P = r + s \qquad \text{by (1.7)}$$
$$\Rightarrow (X; (\lceil P \rceil \wedge \ell = u - s); (\lceil P \rceil \wedge \ell = s)) \wedge \int P = r + s \text{ by 2.3, ITL}$$
$$\Rightarrow (\exists x. \int P = x; \int P = s) \wedge \int P = r + s \qquad \text{by (*)}$$
$$\Rightarrow \int P = r; \int P = s \qquad \text{by } (A3, \Rightarrow)$$

The proof is completed by establishing the other inductive step $R(X; \lceil \neg P \rceil)$. This proof, which exploits that $\lceil \neg P \rceil \Rightarrow \int P = 0$, is easier than the above. So it is left for the reader.

($A4$) *Triangle* :

$$\int P_1 + \int P_2 = \int (P_1 \vee P_2) + \int (P_1 \wedge P_2)$$

Proof: The proof is by induction 3.1. Let $P_1 \cong P \wedge Q$, $P_2 \cong P \wedge \neg Q$, $P_3 \cong \neg P \wedge Q$, and $P_4 = \neg P \wedge \neg Q$. Let $R(X)$ be $X \Rightarrow (\int P + \int Q = \int (P \vee Q) + \int (P \wedge Q))$

The base case $R(X)$ is trivial. To prove the inductive cases we exploit the following

Fact: $\lceil P_i \rceil \Rightarrow \int P + \int Q = \int (P \vee Q) + \int (P \wedge Q)$ for $i = 1, 2, 3, 4$.

We sketch how to prove the fact for the case $P_1 \cong P \wedge Q$. (The other cases are similar.) Observe first that $\lceil P \wedge Q \rceil \Rightarrow (\lceil P \rceil \wedge \lceil Q \rceil \wedge \lceil P \vee Q \rceil)$ by 2.4. Then use the definition of f_P to establish the desired.

The inductive step $R(X; \lceil P_i \rceil)$ is proved by:

$$X; \lceil P_i \rceil$$
$$\Rightarrow \int P + \int Q = \int (P \vee Q) + \int (P \wedge Q);$$
$$\int P + \int Q = \int (P \vee Q) + \int (P \wedge Q)$$
$$\Rightarrow \int P + \int Q = \int P \vee Q + \int P \wedge Q$$

where the first step follows from $R(X)$ and the above fact, and the second step follows from ($A3$) and interval temporal logic.

5 Example: Cat and Mouse

The Cat and Mouse example suggested by Maler, Manna and Pnueli [6] will now be formalized and proved in EDC.

The Problem

A cat and a mouse are distances sc and sm ($sc > sm$) from a hole in a wall. The mouse runs towards the hole with speed vm. L time units later, the cat awakens and chases the mouse with speed vc.

Does the mouse escape? or rather, under which conditions does the mouse escape?

A specification

We assume it to be a one dimensional problem. That is, the mouse runs on a straight line to the hole, and L time units later, the cat runs on the same straight line to catch the mouse. I.e. the states are the distance m from the mouse to the hole, and the distance a from the cat to the hole.

$a, m : Time \rightarrow \mathbf{R}$

Let the movement of the mouse be specified by:

$$M \mathrel{\hat{=}} \mathbf{b}.m = sm \wedge \lceil a > m > 0 \Rightarrow \dot{m} = -vm \rceil \wedge \lceil Continuous(m) \rceil$$

Similarily, the movement of the cat is specified by:

$$C \mathrel{\hat{=}} \begin{cases} \mathbf{b}.a = sc \wedge \lceil Continuous(a) \rceil \\ \wedge\, 0 < \ell \le L \Rightarrow \lceil \dot{a} = 0 \rceil \\ \wedge\, \ell > L \Rightarrow \ell = L;\ \lceil a > m > 0 \Rightarrow \dot{a} = -vc \rceil \end{cases}$$

Notice that the specifications say nothing about the movements after either the mouse reaches the hole, or is caught. (Except that both the cat and the mouse move continuously.)

The question whether the mouse is caught is given by:

$$Catch \mathrel{\hat{=}} \lceil 0 < m < a \rceil \wedge \mathbf{e}.m = \mathbf{e}.a > 0$$

where we use $\mathbf{e}.a = \mathbf{e}.m > 0$ instead of $a = m > 0$, because (1) we cannot express that $a = m > 0$ at a point, and (2) we cannot use $\lceil a = m > 0 \rceil$, as the movements are not specified after catching.

Proof

The mouse does not escape under any circumstances. So we assume that the following condition holds:

$$A \mathrel{\hat{=}} sc > sm \ge 0 \wedge vc \ge 0 \wedge vm \ge 0 \wedge sc \cdot vm \ge vc \cdot (sm - L \cdot vm)$$

We prove that the mouse is not caught under condition A, i.e.

$$M \wedge C \wedge A \Rightarrow \neg Catch$$

Proof: The proof is by contradiction, i.e. we show that

$$M \wedge C \wedge A \wedge Catch \Rightarrow false$$

Two cases are considered. The first is: $\ell \leq L$.

$C \wedge M \wedge A \wedge Catch \wedge \ell \leq L$ $\Rightarrow$

b.$a = sc \wedge \lceil Continuous(a) \rceil \wedge \lceil \dot{a} = 0 \rceil$
$\wedge \ \mathbf{b}.m = sm \wedge \lceil Continuous(m) \rceil \wedge \lceil a > m > 0 \Rightarrow \dot{m} = -vm \rceil$
$\wedge \ A \wedge \lceil a > m > 0 \rceil \wedge \mathbf{e}.a = \mathbf{e}.m > 0 \wedge \ell \leq L$ $\Rightarrow$

$false$

This case uses 4.6.

Consider the case: $\ell > L$:

$C \wedge M \wedge A \wedge Catch \wedge \ell > L$ $\Rightarrow$

b.$a = sc \wedge \lceil Continuous(a) \rceil$
$\wedge \ (\lceil \dot{a} = 0 \rceil \wedge \ell = L); \lceil a > m > 0 \Rightarrow \dot{a} = -vc \rceil$
$\wedge \ \mathbf{b}.m = sm \wedge \lceil Continuous(m) \rceil \wedge \lceil a > m > 0 \Rightarrow \dot{m} = -vm \rceil$
$\wedge \ A \wedge \lceil a > m > 0 \rceil \wedge \mathbf{e}.a = \mathbf{e}.m > 0$ $\Rightarrow$

$\lceil Continuous(a) \rceil \wedge \lceil Continuous(m) \rceil \wedge A$
$\wedge \ (\mathbf{b}.a = sc \wedge \mathbf{b}.m = sm \wedge \ell = L \wedge \lceil \dot{a} = 0 \rceil \wedge \lceil \dot{m} = -vm \rceil$
$\quad \wedge \lceil a > m > 0 \rceil);$
$\quad (\lceil \dot{a} = -vc \rceil \wedge \lceil \dot{m} = -vm \rceil \wedge \lceil a > m > 0 \rceil \wedge \mathbf{e}.a = \mathbf{e}.m > 0)$

In this step we have used theorem 2.6, 4.2, and 4.3.

To arrive at a contradiction, we proceed as follows: Apply 4.6 to the first part ($\ell = L$) to calculate **e**.m and **b**.m. Pass these values to the second part using 4.4. Thereafter, apply 4.6 to the second part and derive $false$.

We have now derived a contradiction for any interval, and the proof is completed.

6   Discussion

The Extended Duration Calculus has been developed as a tool to allow us to combine reasoning in logic with reasoning in mathematical analysis. It was initially motivated by a simple case study of an auto pilot [12], where we saw a need for reasoning about continuous states, and their values as determined by a differential equation. In that paper the initial **b**.$(\cdot)$ and final **e**.$(\cdot)$ values were introduced for totally continuous functions, where they coincide with the point values. A differential equation $\dot{x} = F(x)$ was interpreted as an integral equation **e**.$x = $ **b**.$x + \int F(x)$, where the duration was generalized to an 'integral' functional, which can be applied to any (continuous, real valued) state.

We were not quite satisfied with this solution even though it solved our immediate problems. One source of irritation was the restriction to continuous states; because point values might be useful also for sectionally continuous states. Another was that even though the differential equation was introduced in the model, we had no means in the logic to reason about such equations.

In the further development, we studied the approaches reported in [10] and in [8] where a differential equation or some other characterisation of a continuous function, governs the evolution of a continuous state, associated with a transition of an automaton. In the extension to ATP presented in [10], the automaton alternates between normal, event labelled, transitions, that takes zero time, and state transformations governed by an *evolution function*. The Mean Value Calculus, introduced in [17], is another extension of Duration Calculus which in addition to durations of states also can express events, which take zero time. The Mean Value Calculus has, for example, been used to model and to reason about timed automata. In [8], the emphasis is on modelling sensors and their failure modes. An abstract sensor is simply modelled as a sectionally continuous function from a physical state to an interval of real numbers. The reasoning is done using conventional mathematical analysis, and contains a number of theorems which would seem very artificial if developed outside this 'natural' theory of continuous functions. This paper, together with further studies of control theory, convinced us that we should have a simple, general interface to mathematical analysis, as exemplified by the lifting rule (4.6) in the proof theory.

The Cat and Mouse example comes from [6], which extends a timed transition system as defined in [3] to a phase transition system by associating conditional differential equations, called *activities* with transitions that take time. The paper includes an inference rule for proving properties of such systems, using an *age* operator over formulas. (An age is the extent of the largest interval ending at the current time, where a given formula holds.)

A final development has been the discovery that finite variability, or non-Zeno or whatever the constraint that prohibits an infinite number of discrete transitions to take place in finite time is called, is the well known Dirichlet conditions of mathematical analysis.

Acknowledgements The authors are grateful to Hans Rischel for instructive comments.

References

1. K.M. Hansen, A.P. Ravn, H. Rischel: Specifying and Verifying Requirements of Real-Time Systems, Proceedings of the ACM SIGSOFT '91 Conference on Software for Critical Systems, New Orleans, December 4-6, 1991, *ACM Software Engineering Notes*, vol. 15, no. 5, pp. 44-54, 1991.
2. M.R. Hansen, Zhou Chaochen: Semantics and Completeness of Duration Calculus, in J. W. de Bakker, C. Huizing, W.-P. de Roever and G. Rozenberg (Eds.) *Real-Time: Theory in Practice, REX Workshop*, pp. 209-225, *LNCS 600*, 1992.

3. T. A. Henzinger, Z. Manna and A. Pnueli: Timed transition systems , in J. W. de Bakker, C. Huizing, W.-P. de Roever and G. Rozenberg (Eds.) *Real-Time: Theory in Practice, REX Workshop*, pp. 226-251, *LNCS 600*, 1992.

4. R. Koymans: Specifying Real-Time Properties with Metric Temporal Logic, *Real-Time Systems*, 2, 4, pp. 255-299, Kluwer Academic Publishers, 1990.

5. David G. Luenberger: *Introduction to Dynamic Systems. Theory, Models & Applications*, Wiley, 1979.

6. O. Maler, Z. Manna and A. Pnueli: From Timed to Hybrid Systems, in J. W. de Bakker, C. Huizing, W.-P. de Roever and G. Rozenberg (Eds.) *Real-Time: Theory in Practice, REX Workshop*, pp. 447-484, *LNCS 600*, 1992.

7. Z. Manna and A. Pnueli: *The Temporal Logic of Reactive and Concurrent Systems*, Springer Verlag, 1992.

8. K. Marzullo: *Tolerating Failures of Continuous-Valued Sensors*, Technical Rep. TR90-1156, Dept. Comp. Sc. Cornell University, Itacha, NY, USA.

9. B. Moszkowski: A Temporal Logic for Multilevel Reasoning about Hardware, *IEEE Computer*, 18,2, pp. 10-19, 1985.

10. X. Nicollin, J. Sifakis and S. Yovine: From ATP to Timed Graphs and Hybrid Systems, in J. W. de Bakker, C. Huizing, W.-P. de Roever and G. Rozenberg (Eds.) *Real-Time: Theory in Practice, REX Workshop*, pp. 549-572, *LNCS 600*, 1992.

11. A. Pnueli and E. Harel: Applications of Temporal Logic to the Specification of Real-Time Systems (extended abstract), in M. Joseph (ed.) *Proceedings of a Symposium on Formal Techniques in Real-Time and Fault-Tolerant Systems*, *LNCS 331*, pp. 84-98, Springer Verlag, 1988.

12. A.P. Ravn, H. Rischel: Requirements Capture for Embedded Real-Time Systems, *Proc. IMACS-MCTS'91 Symp. Modelling and Control of Technological Systems, Vol 2*, pp. 147-152, Villeneuve d'Ascq, France, 1991.

13. A.P. Ravn, H. Rischel and K.M. Hansen: Specifying and Verifying Requirements of Real-Time Systems, *IEEE Trans. Softw. Eng.* Vol. 19, No. 1, Jan. 1993, pp. 41-55.

14. J. U. Skakkebæk, A.P. Ravn, H. Rischel and Zhou Chaochen: Specification of embedded real-time systems, *Proc. 4th Euromicro Workshop on Real-Time Systems*, IEEE Press, pp 116–121, June 1992.

15. E.V. Sørensen, A.P. Ravn, H. Rischel: *Control Program for a Gas Burner: Part 1: Informal Requirements, ProCoS Case Study 1*, ProCoS Rep. ID/DTH EVS2, March 1990.

16. Zhou Chaochen, C.A.R. Hoare, A.P. Ravn: A Calculus of Durations, *Information Processing Letters*, 40, 5, pp. 269-276, 1991.

17. Zhou Chaochen, Li Xiaoshan: *A Mean Value Calculus of Durations*, UNI/IIST Report No. 5, 1993.

Towards Refining Temporal Specifications into Hybrid Systems*

Thomas A. Henzinger[1]   Zohar Manna[2]   Amir Pnueli[3]

[1] Department of Computer Science
Cornell University, Ithaca, New York 14853
e-mail: tah@cs.cornell.edu

[2] Department of Computer Science
Stanford University, Stanford, California 94305
e-mail: zm@cs.stanford.edu

[3] Department of Applied Mathematics
The Weizmann Institute of Science, Rehovot, Israel 76100
e-mail: amir@wisdom.weizmann.ac.il

Abstract. We propose a formal framework for designing hybrid systems by stepwise refinement. Starting with a specification in hybrid temporal logic, we make successively more transitions explicit until we obtain an executable system.

1 Introduction

We present the foundations of a methodology for the systematic development of hybrid systems. As high-level specification language, we suggest *Abstract Phase Transition Systems* (APTS's). The behavior of an APTS consists of a sequence of phases, and each phase may be described implicitly by temporal constraints. For this purpose we introduce *Hybrid Temporal Logic* (HTL), a hybrid extension of interval temporal logic. The notion of one APTS *refining* (implementing) another is defined, and corresponds to inclusion between the sets of behaviors allowed by each system. We also propose a criterion for judging an APTS to be *executable*, i.e., directly implementable on available architectures. A *development sequence*, then, is envisioned to start at a high-level implicit APTS, which is refined by a sequence of steps into an executable APTS. Ultimately, each refinement step ought to be accompanied by verification.

2 Hybrid Temporal Logic

The behavior of a hybrid system is modeled by a function that assigns to each real-numbered time a system state, i.e., values for all system variables. We re-

* This research was supported in part by the National Science Foundation under grants CCR-92-00794 and CCR-92-23226, by the Defense Advanced Research Projects Agency under contract NAG2-703, by the United States Air Force Office of Scientific Research under contracts F49620-93-1-0056 and F49620-93-1-0139, and by the European Community ESPRIT Basic Research Action Project 6021 (REACT).

quire that, at each point, the behavior function has a limit from the left and a limit from the right. Discontinuities are points where the two limits differ. We assume the following *uncertainty principle*: limits of function values (defined over nonsingular intervals) are observable; individual function values (at singular points) are not observable — that is, we cannot know (and do not care) if at a discontinuity the function value coincides with the limit from the left or the limit from the right.

To specify properties of behavior functions, we use an interval temporal logic with a chop operator denoted by semicolon [Mos85]. Consistent with our interpretation of behavior functions, only limits and derivatives of the behavior function can be constrained by atomic formulas; individual function values cannot appear in specifications.

Syntax

Let V be a set of typed variables. The allowed types include *boolean*, *integer*, and *real*. We view the booleans and the integers as subsets of the reals, where *false* and *true* correspond to 0 and 1, respectively. For a variable $x \in V$, we write $\overleftarrow{x}$ and $\overrightarrow{x}$ for the limit from the right (the *right limit*) and the limit from the left (the *left limit*) of x, and $\overleftarrow{\dot{x}}$ and $\overrightarrow{\dot{x}}$ for the first derivative from the right (the *right derivative*) and the first derivative from the left (the *left derivative*) of x.

A *local formula* is an atomic formula over the right and left limits and derivatives of variables in V. The formulas ϕ of *Hybrid Temporal Logic* (HTL) are defined inductively as follows:

$$\phi ::= \psi \mid \neg\phi \mid \phi_1 \vee \phi_2 \mid \phi_1 ; \phi_2 \mid \forall x.\, \phi$$

where $x \in V$ and ψ is a local formula.

A *state formula* is an atomic formula over the variables in V. If ψ is a state formula, we write $\overleftarrow{\psi}$ (and $\overrightarrow{\psi}$) for the local formula that results from ψ by replacing each variable occurrence x in ψ with its right limit $\overleftarrow{x}$ (left limit $\overrightarrow{x}$, respectively).

Semantics

Let R be the set of real numbers. A *state* $\sigma : V \to R$ is a type-consistent interpretation of the variables in V (i.e., boolean variables may only be interpreted as 0 or 1, and a similar restriction holds for integer variables). We write Σ_V for the set of states.

Time is modeled by the nonnegative real line R^+. An *open interval* (a, b), where $a, b \in R^+$ and $a < b$, is the set of points $t \in R^+$ such that $a < t < b$; that is, we consider only open intervals that are nonempty and bounded. Let $I = (a, b)$ be an open interval. A function $f : I \to R$ is *piecewise smooth* on I if

- at a, the right limit and all right derivatives of f exist;

- at all points $t \in I$, the right and left limits and all right and left derivatives of f exist, and f is continuous either from the right or from the left;
- at b, the left limit and all left derivatives of f exist.

Two functions $f, g \colon I \to \mathbb{R}$ are *indistinguishable* on I if they agree on almost all (i.e., all but finitely many) points $t \in I$. Thus, if two piecewise smooth functions are indistinguishable on the open interval I, then they agree on all limits and derivatives throughout I, on the right limit and right derivatives at a, and on the left limit and left derivatives at b.

A *phase* $P = (b, f)$ over V is a pair consisting of

1. A positive real number $b > 0$, the *length* of P.
2. A type-consistent family $f = \{f_x \mid x \in V\}$ of functions $f_x \colon I \to \mathbb{R}$ that are piecewise smooth on the open interval $I = (0, b)$ and assign to each point $t \in I$ a value for the variable $x \in V$.

It follows that the phase P assigns to every real-valued time $t \in I$ a state $f(t) \in \Sigma_V$. Furthermore, the right limit of f at 0 and the left limit of f at b are defined. We write

$$\overleftarrow{P} \;=\; \lim_{t \to 0}\{f(t) \mid 0 < t < b\}$$

for the *left limiting state* $\overleftarrow{P} \in \Sigma_V$ of the phase P, and

$$\overrightarrow{P} \;=\; \lim_{t \to b}\{f(t) \mid 0 < t < b\}$$

for the *right limiting state* $\overrightarrow{P} \in \Sigma_V$ of P.

Let $P_1 = (b, f)$ and $P_2 = (c, g)$ be two phases. The phases P_1 and P_2 are *indistinguishable* (*equivalent*) if $b = c$ and for all variables $x \in V$, the two functions f_x and g_x are indistinguishable on the open interval $(0, b)$. The phase P_2 is a *subphase* of P_1 at position a, where $0 \le a < b$, if

- $a + c \le b$ and
- for all $0 < t < c$, $g(t) = f(a + t)$.

If $a = 0$, then P_2 is a *leftmost* subphase of P_1; if $a = b - c$, a *rightmost* subphase. The two phases P_1 and P_2 *partition* a third phase $P = (d, h)$ if

- $d = b + c$,
- P_1 is a leftmost subphase of P, and
- P_2 is a rightmost subphase of P.

Notice that since P is a phase, for all variables $x \in V$, at b the function h_x is continuous either from the right or from the left. Hence, if P_1 and P_2 partition P, then the value $h_x(b)$ is either the left limit of f_x at b or the right limit of g_x at 0. It follows that there are several phases that are partitioned by the two phases P_1 and P_2. All of these phases, however, are indistinguishable, because they disagree at most at b.

The formulas of hybrid temporal logic are interpreted over phases. A phase $P = (b, f)$ *satisfies* the hybrid temporal formula ϕ, denoted $P \models \phi$, according to the following inductive definition:

$P \models \psi$ iff the local formula ψ evaluates to *true*, where
- $\overleftarrow{x}$ is interpreted as the right limit of f_x at 0,

$$\overleftarrow{x} \;=\; \lim_{t \to 0}\{f_x(t) \mid 0 < t < b\};$$

- $\overrightarrow{x}$ is interpreted as the left limit of f_x at b,

$$\overrightarrow{x} \;=\; \lim_{t \to b}\{f_x(t) \mid 0 < t < b\};$$

- $\overleftarrow{\dot{x}}$ is interpreted as the right derivative of f_x at 0,

$$\overleftarrow{\dot{x}} \;=\; \lim_{t \to 0}\left\{ (f_x(t) - \overleftarrow{x})/t \mid 0 < t < b \right\};$$

- $\overrightarrow{\dot{x}}$ is interpreted as the left derivative of f_x at b,

$$\overrightarrow{\dot{x}} \;=\; \lim_{t \to b}\left\{ (f_x(t) - \overrightarrow{x})/(t - b) \mid 0 < t < b \right\}.$$

$P \models \neg\phi$ iff $P \not\models \phi$.
$P \models \phi_1 \vee \phi_2$ iff $P \models \phi_1$ or $P \models \phi_2$.
$P \models \phi_1 ; \phi_2$ iff there are two phases P_1 and P_2 that partition P such that
$P_1 \models \phi_1$ and $P_2 \models \phi_2$.
$P \models \forall x.\,\phi$ iff $P' \models \phi$ for all phases $P' = (b, f')$ that differ from P at most
in the interpretation f'_x of x.

Notice that right limits and right derivatives are applied at the left end of a phase, while left limits and left derivatives are applied at the right end. Also observe that if two phases P_1 and P_2 are indistinguishable, then $P_1 \models \phi$ iff $P_2 \models \phi$; that is, HTL-formulas cannot distinguish between phases that differ only at finitely many points.

From now on, we shall write x and $\dot{x}$ synonymous for the right limit $\overleftarrow{x}$ and the right derivative $\overleftarrow{\dot{x}}$, respectively. This convention allows us to read a state formula ψ as a hybrid temporal formula, namely, as the local formula $\overleftarrow{\psi}$.

Some sample formulas

It is convenient to define abbreviations for common temporal formulas. The following abbreviations express that a leftmost subphase, a rightmost subphase, or any subphase of a phase satisfies the formula ϕ:

$\lhd\phi$ stands for $\phi \vee (\phi; \mathit{true})$
$\rhd\phi$ stands for $\phi \vee (\mathit{true}; \phi)$
$\Diamond\phi$ stands for $(\lhd\phi) \vee (\rhd\phi) \vee (\mathit{true}; \phi; \mathit{true})$

Thus we can express that all subphases of a phase satisfy ϕ:

$\Box\phi$ stands for $\neg\Diamond\neg\phi$

We now define temporal *until* and *unless* operators over a phase P. The until formula $\phi_1 \mathcal{U} \phi_2$ asserts that the phase P can be partitioned into two subphases P_1 (which may be empty) and P_2 such that ϕ_1 holds throughout P_1 and ϕ_2 holds on a leftmost subphase of P_2; the unless formula $\phi_1 \mathcal{W} \phi_2$ asserts that either ϕ_1 holds throughout the phase P, or P satisfies $\phi_1 \mathcal{U} \phi_2$:

$\phi_1 \mathcal{U} \phi_2$ stands for $(\triangleleft \phi_2) \vee (\square \phi_1); (\triangleleft \phi_2)$
$\phi_1 \mathcal{W} \phi_2$ stands for $(\square \phi_1) \vee (\phi_1 \mathcal{U} \phi_2)$

The following formula asserts that the variable $u \in V$ is *rigid* on a phase; that is, the function f_u is constant throughout the phase:

$u \in Rigid$ stands for $\square(\overleftarrow{u} = \overrightarrow{u})$

Using rigid variables, we can specify that a function is continuous, and that its first derivative is continuous throughout a phase:

$x \in C^0$ stands for $\forall u, v \in Rigid. \left[(\overrightarrow{x} = u); (\overleftarrow{x} = v) \rightarrow u = v \right]$
$x \in C^1$ stands for $x \in C^0 \wedge \forall u, v \in Rigid. \left[(\overrightarrow{\dot{x}} = u); (\overleftarrow{\dot{x}} = v) \rightarrow u = v \right]$

The formula $x \in C^0$ requires that for any partition of a phase P into two subphases, the left and right limits of x at the point of partitioning coincide; the formula $x \in C^1$ adds the analogous requirement for the first derivatives of x.

Hybrid temporal logic subsumes many real-time temporal logics [AH92] and the duration calculus [CHR92]. For example, the formula

$$\square \forall x \in C^0. \left[\left(p \wedge x = 0 \wedge \square(\dot{x} = 1) \wedge \overrightarrow{x} > 5 \right) \rightarrow \Diamond q \right]$$

asserts that every p-state of a phase is followed within 5 time units either by a q-state or by the end of the phase. The variable x is a "clock" that measures the length of all subphases starting with a p-state. The formula

$$\forall x \in C^0. \left[\left(x = 0 \wedge \square(p \rightarrow \dot{x} = 1) \wedge \square(\neg p \rightarrow \dot{x} = 0) \right) \rightarrow \overrightarrow{x} \leq 10 \right]$$

asserts that the cumulative time that p is true in a phase is at most 10. Here the variable x is an "integrator" that measures the accumulated duration of p-states.

3 Phase Transition Systems

Following [MMP92] and [NSY92], we model hybrid systems as transition systems. Just as discrete transitions are usually represented as binary relations on states, hybrid transitions can be represented as binary relations on phases.

Abstract phase transition systems

An *abstract phase transition system* (APTS) $S = (V, P, P_0, P_F, T)$ consists of five components:

1. A finite set V of *state variables*.
2. A set P of *phases* over V.
3. A set $P_0 \subseteq P$ of *initial phases*.
4. A set $P_F \subseteq P$ of *final phases*.
5. A set T of *transitions*. Each transition $\tau \in T$ is a binary relation on the phases in P, i.e., $\tau \subseteq P^2$.

A *phase sequence* is a finite or infinite sequence of phases. Let $\overline{P} = P_0, P_1, P_2, \ldots P_n$ be a finite phase sequence with $P_i = (b_i, f_i)$ for all $0 \le i \le n$. The finite phase sequence $\overline{P}$ *partitions* a phase $P = (b, f)$ if

- $b = \sum_{0 \le i \le n} b_i$ and
- for all $0 \le i \le n$, P_i is a subphase of P at position $\sum_{0 \le j < i} b_j$.

The finite phase sequence $\overline{P}$ can thus be viewed as a set of indistinguishable phases, namely, those phases that are partitioned by $\overline{P}$. Consequently, we may interpret HTL-formulas over finite phase sequences. The finite phase sequence $\overline{P}$ *satisfies* the hybrid temporal formula ϕ, denoted $\overline{P} \models \phi$, if some phase that is partitioned by $\overline{P}$ satisfies ϕ.

Two finite phase sequences $\overline{P}_1$ and $\overline{P}_2$ are *equivalent* if there are two indistinguishable phases P_1 and P_2 such that $\overline{P}_1$ partitions P_1 and $\overline{P}_2$ partitions P_2. It follows that all equivalence classes of finite state sequences are *closed under stuttering*: if a phase P_i of the finite phase sequence $\overline{P}$ is split into two phases P' and P'' that partition P_i, the resulting finite phase sequence

$$P_0, \ldots P_{i-1}, P', P'', P_{i+1}, \ldots P_n$$

is equivalent to $\overline{P}$.

Let $\overline{P} = P_0, P_1, P_2, \ldots$ be an infinite phase sequence with $P_i = (b_i, f_i)$ for all $i \ge 0$. The infinite phase sequence $\overline{P}$ *diverges* if the infinite sum $\sum_{i \ge 0} b_i$ of phase lengths diverges, i.e., for all nonnegative reals $t \in \mathbb{R}^+$ there is an integer $n \ge 0$ such that $t < \sum_{0 \le i \le n} b_i$.

A finite phase sequence $\overline{P}$ is a *run fragment* of the APTS S if $\overline{P}$ is equivalent to a finite phase sequence $P_0, P_1, \ldots P_n$ that satisfies three conditions:

Initiality $P_0 \in P_0$.

Continuous activities For all $0 \le i \le n$, $P_i \in P$.

Discrete transitions For all $0 \le i < n$, there is a transition $\tau \in T$ such that $(P_i, P_{i+1}) \in \tau$.

The run fragment $\overline{P}$ is *complete* if $P_n \in P_F$.

An infinite phase sequence $\overline{P}$ is a *run (computation)* of the APTS S if

Safety All finite prefixes of $\overline{P}$ are run fragments of S.

Liveness $\overline{P}$ diverges.

The APTS S *satisfies* a hybrid temporal formula ϕ, written $S \models \phi$, if all run fragments of S satisfy ϕ.

Activity transition graphs

Both timed transition systems [HMP92] and timed safety automata [HNSY92] specify APTS's. We use activity transition graphs to specify APTS's.

An *activity transition graph* (ATG) is a directed labeled multigraph $A = (V_D, L, E, \mu_1, \mu_2, \mu_3, \kappa)$ consisting of the following components:

- A finite set V_D of *data variables*.
- A finite set L of vertices called *locations*. Each location $\ell \in L$ is labeled by
  - an *initial condition* $\mu_1(\ell)$, a state formula over the variables in V_D,
  - an *activity* $\mu_2(\ell)$, a hybrid temporal formula over V_D, and
  - a *final condition* $\mu_3(\ell)$, a state formula over V_D.
- A finite set E of *edges* between the locations in L. Each edge $e \in E$ is labeled by a guarded command $\kappa(e) = (\gamma \rightarrow \alpha)$, where γ is a state formula over the variables in V_D (the *guard* of e) and α is an assignment to some of the variables in V_D.

The ATG A defines the APTS $S_A = (V, \mathcal{P}, \mathcal{P}_0, \mathcal{P}_F, \mathcal{T})$:

1. The state variables are $V = V_D \cup \{\pi\}$, where the *control variable* π ranges over the locations L.
2. For each location $\ell \in L$, $\mathcal{P}$ contains all phases P such that

$$P \models \Box(\pi = \ell) \wedge \left(\bigwedge_{x \in V_D} x \in C^0 \right) \wedge \mu_2(\ell)$$

3. For each location $\ell \in L$, $\mathcal{P}_0$ contains all phases $P \in \mathcal{P}$ such that

$$P \models (\pi = \ell) \wedge \mu_1(\ell)$$

4. For each location $\ell \in L$, $\mathcal{P}_F$ contains all phases $P \in \mathcal{P}$ such that

$$P \models (\overleftarrow{\pi} = \ell) \wedge \overrightarrow{\psi}$$

where $\psi = \mu_3(\ell)$ is the final condition of ℓ.
5. For each edge $e \in E$, $\mathcal{T}$ contains a transition $\tau_e \subseteq \mathcal{P}^2$. Let $\ell_1, \ell_2 \in L$ be the source and target locations of the edge e, and let γ and α be the guard and assignment associated with e. Then $(P_1, P_2) \in \tau_e$ iff

$$P_1 \models (\overleftarrow{\pi} = \ell_1) \wedge \overleftarrow{\gamma}$$
$$P_2 \models (\pi = \ell_2)$$

and $\overleftarrow{P_2}$ results from $\overrightarrow{P_1}$ by executing the assignment α.

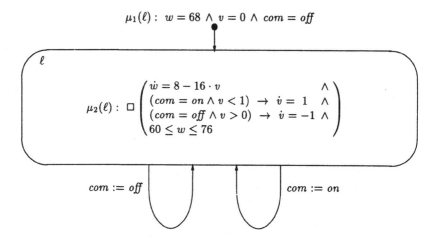

$\mu_1(\ell): \; w = 68 \wedge v = 0 \wedge com = off$

$\mu_2(\ell): \; \Box \begin{pmatrix} \dot{w} = 8 - 16 \cdot v & \wedge \\ (com = on \wedge v < 1) \;\rightarrow\; \dot{v} = 1 & \wedge \\ (com = off \wedge v > 0) \;\rightarrow\; \dot{v} = -1 & \wedge \\ 60 \leq w \leq 76 & \end{pmatrix}$

$com := off \qquad com := on$

Fig. 1. Specification $\widehat{A}$

Specifications

An ATG with a single location is called a *specification*.

Consider, for example, the specification $\widehat{A}$ presented in Figure 1. The ATG $\widehat{A}$ specifies a water level controller that opens and closes a valve regulating the outflow of water from a container. The container has an input vent through which water flows at the constant rate 8. When the valve is fully open, there is an outflow of 16, leading to an overall water level decrease of $16 - 8 = 8$ per time unit. The controller can command the valve to open and close using the switch *com*.

The specification $\widehat{A}$ refers to three variables:

- The boolean variable *com*, ranging over the values $\{on, off\}$, represents the latest command issued by the controller. The value *on* causes the opening of the valve; the value *off*, the closing of the valve.
- The real-valued variable v represents the status of the valve. It assumes values between 0 and 1, corresponding to the valve being fully closed (zero water outflow) and fully open, respectively. When $com = on$ and $v < 1$, v increases at the constant rate of 1. When $com = off$ and $v > 0$, v decreases at the same constant rate. By default, when none of these conditions hold, v maintains a constant value, satisfying the equation $\dot{v} = 0$.
- The real-valued variable w represents the water level. When the valve is fully closed, w increases at the rate 8; when the valve is fully open, w decreases at the rate 8. At intermediate values of v, the water level increases at the rate $8 - 16 \cdot v$.

The controller is supposed to keep the water level between 60 and 76 units. The two transitions represent the ability of the controller to set the variable *com* to the values *on* and *off*.

We point out that the specification describes both the actions of the controller (giving commands to open and close the valve) and the response of the controlled environment (valve closing and opening and water level rising and falling). When specifying controllers, the set of data variables $V_D = V_C \cup V_E$ of an ATG can be partitioned into a set V_C of *controlled variables*, which may be modified by the controller, and a set V_E of *environment variables*, which vary according to the laws of physics. In our example, the switch *com* is a controlled variable, while the valve v and water level w are environment variables. Note, however, that the equations for the behavior of the environment variable v are influenced by the value of the controlled variable *com*.

4 Stepwise Refinement

Let S_1 and S_2 be two APTS's over the sets V_1 and V_2 of state variables, respectively. The APTS S_1 *refines* the APTS S_2 if $V_2 \subseteq V_1$ and the projection of every run of S_1 to the variables in V_2 is a run of S_2.

Hierarchical activity transition graphs

An APTS that is given by an ATG can be refined by expanding activities — i.e., hybrid temporal formulas labeling locations — into ATG's. Thus we obtain *hierarchical* (nested) *activity transition graphs* (HATG's), which are defined inductively:

1. Every ATG is an HATG.
2. Let $B = (V_D, L, E, \mu_1, \mu_2, \mu_3, \kappa)$ and C_ℓ be two HATG's, and let A be the tuple $B[\ell := C_\ell]$ that results from B by replacing the activity $\mu_2(\ell)$ of the location $\ell \in L$ with the HATG C_ℓ. Then C is also an HATG.

Every HATG A defines an APTS S_A. Roughly speaking, a location ℓ whose activity is defined by an HATG C_ℓ contributes all phase sequences $\overline{P}$ such that $\overline{P}$ satisfies $\square(\pi = \ell)$ and some extension of $\overline{P}$ is a complete run fragment of C_ℓ. The phase sequence $\overline{P}$ needs to be extended by data variables that are local to C_ℓ and a control variable for C_ℓ.

To define the APTS S_A formally, we inductively translate the nested HATG A into a flat ATG *flat(A)*. If A is an ATG, then *flat(A)* $= A$; otherwise, A is of the form $B[\ell := C_\ell]$, for two ATG's $B = (V_D, L, E, \mu_1, \mu_2, \mu_3, \kappa)$ and $C_\ell = (V_\ell, L_\ell, E_\ell, \nu_1, \nu_2, \nu_3, \lambda)$ (we assume that the locations in L and L_ℓ are disjoint). Then the ATG *flat(A)* is defined as follows:

- The data variables of *flat(A)* are $V_D \cup V_\ell$.
- The locations of *flat(A)* are $(L - \{\ell\}) \cup L_\ell$.
  - Each location $\ell' \in (L - \{\ell\})$ is labeled by the initial condition $\mu_1(\ell')$, the activity $\mu_2(\ell')$, and the final condition $\mu_3(\ell')$.
  - Each location $\ell' \in L_\ell$ is labeled by the initial condition $\mu_1(\ell) \wedge \nu_1(\ell')$, the activity $\nu_2(\ell')$, and the final condition $\mu_3(\ell) \wedge \nu_3(\ell')$.

– The graph *flat(A)* contains the following edges.
  - Let $e \in E$ be an edge of B from ℓ_1 to ℓ_2, where $\ell_1, \ell_2 \in (L - \{\ell\})$. Then *flat(A)* contains an edge from ℓ_1 to ℓ_2 that is labeled by the guarded command $\kappa(e)$.
  - Let $e \in E_\ell$ be an edge of C_ℓ from ℓ_1 to ℓ_2. Then *flat(A)* contains an edge from ℓ_1 to ℓ_2 that is labeled by the guarded command $\lambda(e)$.
  - Let $e \in E$ be an edge of B from ℓ_1 to ℓ, where $\ell_1 \in (L - \{\ell\})$ and $\kappa(e) = (\gamma \to \alpha)$. Then *flat(A)* contains, for all locations $\ell_2 \in V_\ell$, an edge from ℓ_1 to ℓ_2 that is labeled by the guarded command

$$(\nu_1(\ell_2) \wedge \gamma) \;\to\; \alpha.$$

  - Let $e \in E$ be an edge of B from ℓ to ℓ_2, where $\ell_2 \in (L - \{\ell\})$ and $\kappa(e) = (\gamma \to \alpha)$. Then *flat(A)* contains, for all locations $\ell_1 \in V_\ell$, an edge from ℓ_1 to ℓ_2 that is labeled by the guarded command

$$(\nu_3(\ell_1) \wedge \gamma) \;\to\; \alpha.$$

A *run (fragment)* of the HATG A, then, is a run (fragment) of the ATG *flat(A)*.

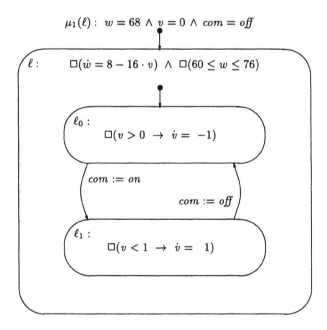

Fig. 2. Refinement $\widehat{B}$

For example, the specification $\widehat{A}$ of the water level controller can be refined into the HATG $\widehat{B}$ presented in Figure 2, which separates the phases at which

$com = on$ from those at which $com = off$. The two conjuncts $\Box(\dot{w} = 8 - 16 \cdot v)$ and $\Box(60 \leq w \leq 76)$, which label the location ℓ, are common to both locations ℓ_0 and ℓ_1 of the inner graph. Also note that the initial conditions of the inner graph are represented by special entry edges: the initial condition of ℓ_0 is *true* and the initial condition of ℓ_1 is *false*. All final conditions are, by default, *true*.

Verification conditions

Let B be an ATG over the set V_D of data variables and let π be a control variable for B. A refinement step replaces a hybrid temporal formula ϕ defining the activity of a location ℓ of B with an ATG C_ℓ. The refinement step is *correct* if the projection of every run of the resulting two-level HATG $B[\ell := C_\ell]$ to the variables in $V_D \cup \{\pi\}$ is a run of A.

Suppose that the replaced formula ϕ is an *invariant*, i.e., of the form $\Box\phi'$. Then the following two conditions suffice to establish the correctness of a refinement step:

1. For every edge of the ATG C_ℓ labeled by the guarded command $\gamma \to \alpha$, the ATG B contains a reflexive edge from ℓ to ℓ labeled by $\gamma' \to \alpha$ such that the state formula γ implies the state formula γ'.
2. There exists a state formula ψ over V_D such that

$$(i) \quad \mathcal{S}_B \models (\pi = \ell \to \psi) \wedge \Box\Big(\pi \neq \ell \to (\pi \neq \ell)\, \mathcal{W}\, (\pi = \ell \wedge \psi)\Big)$$
$$(ii) \quad \mathcal{S}_{C_\ell} \models \psi \to \phi$$

The first condition ensures that all discontinuities of the inner graph C_ℓ are admitted by the outer graph B. The proof obligation (i) asserts that the state formula ψ is true in every left limiting state of a phase in which the control of the outer graph B enters the location ℓ. Now let $\overline{P}$ be a run fragment of the inner graph C_ℓ such that ψ is true in the left limiting state of $\overline{P}$. The proof obligation (ii) guarantees that the hybrid temporal formula ϕ holds over $\overline{P}$, and since ϕ is an invariant, it also holds over any continuous segment of $\overline{P}$.

For example, to show that the water level controller $\widehat{B}$ refines the specification $\widehat{A}$, we obtain two proof obligations:

$$(i) \quad \mathcal{S}_{\widehat{A}} \models \left(\begin{array}{l} (\pi = \ell \to (w = 68 \wedge v = 0 \wedge com = off)) \\ \Box(\pi \neq \ell \to (\pi \neq \ell)\, \mathcal{W}\, (w = 68 \wedge v = 0 \wedge com = off \wedge \pi = \ell)) \end{array} \right. \left. \begin{array}{l} \wedge \\ \, \end{array} \right)$$

$(ii) \quad \mathcal{S}_{\widehat{B}_\ell} \models \underbrace{(w = 68 \wedge v = 0 \wedge com = off)}_{\psi} \to$

$$\underbrace{\Box \left(\begin{array}{l} \dot{w} = 8 - 16 \cdot v \qquad\qquad\qquad \wedge \\ (com = on \wedge v < 1) \to \dot{v} = 1 \quad \wedge \\ (com = on \wedge v = 1) \to \dot{v} = 0 \quad \wedge \\ (com = off \wedge v = 0) \to \dot{v} = 0 \quad \wedge \\ (com = off \wedge v > 0) \to \dot{v} = -1 \wedge \\ 60 \leq w \leq 76 \end{array} \right)}_{\phi}$$

where $\widehat{B}_\ell$ is the ATG labeling the location ℓ in the refinement $\widehat{B}$. In a future paper, we will present a proof system for verifying that an APTS satisfies a hybrid temporal formula.

Executability

The refinement of a specification typically proceeds in several stages until we reach an HATG that can be directly implemented, i.e., executed by stepwise simulation. Formally, an HATG A is *executable* if two conditions are met:

Effectiveness For each hybrid temporal formula ϕ defining an activity of A, the set of models of ϕ (i.e., the set of phases that satisfy ϕ) is effectively computable.

NonZenoness The APTS S_A is nonZeno: an APTS S is *nonZeno* if every run fragment of S is a finite prefix of a run of S.

The runs of an executable HATG can be generated by adding one phase at a time. The effectiveness condition ensures that, in each state, a stepwise interpreter can compute the set of possible successor phases. The nonZenoness condition ensures that the stepwise interpreter cannot make a nondeterministic choice among the possible successor phases that will result, later on, in a deadlock state from which the system cannot proceed or in a Zeno state from which time cannot diverge [Hen92, LA92].

It is not difficult to see that both systems $\widehat{A}$ and $\widehat{B}$ violate the nonZenoness condition and, therefore, are not executable. For the system $\widehat{A}$, consider the phase $P = (1, f)$, where $f(t)$ interprets the data variables as follows, for all $0 < t < 1$:

$$\begin{aligned}
f_{com}(t) &= \textit{off} \\
f_v(t) &= 0 \\
f_w(t) &= 68 + 8 \cdot t
\end{aligned}$$

Thus, as t approaches 1, the value of w approaches 76. This leads to the right limiting state

$$\overrightarrow{P} = \{com = \textit{off}, v = 0, w = 76\}.$$

This state is a deadlock state. There is no way to proceed from $\overrightarrow{P}$ without violating the constraint $w \leq 76$. The same phase can be reproduced by the system $\widehat{B}$.

In Figure 3, we present a further refinement of the system $\widehat{B}$. The refinement $\widehat{C}$ improves on the system $\widehat{B}$ in two respects. First, the phase in which $com = on$ and $v = 1$ has been separated from the phase in which $com = on$ and v is still increasing. This has been achieved by refining the location ℓ_1 of the HATG $\widehat{B}$ with a graph consisting of the two locations ℓ_1 (initial condition *true*, final condition *false*) and ℓ_2 (initial condition *false*, final condition *true*) of the HATG $\widehat{C}$. A similar separation has been carried out for the case that $com = \textit{off}$.

The more important improvement, however, is that the system $\widehat{C}$ forces the setting of com to on before w rises above 74, and the setting of com to $\textit{off}$ before

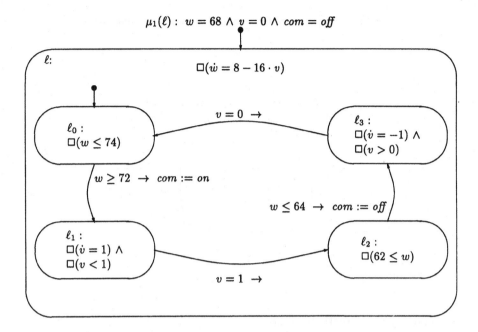

Fig. 3. Implementation $\widehat{C}$

w falls below 62. This ensures that the valve starts opening or closing in time to guarantee that w never exceeds the range $[60, 76]$. Indeed, the system $\widehat{C}$ cannot deadlock and is, therefore, executable.

We say that the HATG A *implements* the specification B if $\mathcal{S}_A$ refines $\mathcal{S}_B$ and A is executable. For example, the water level controller $\widehat{C}$ implements the specification $\widehat{A}$.

Environment versus control

In the example of the water level controller, the environment controls the environment variables v and w only through differential equations (activities); the environment does not modify any variables through guarded commands (transitions). If the environment can modify variables through transitions, we have to adopt a stricter notion of refinement. In particular, a refinement step should not introduce new constraints on the environment [AL90].

Consider, for example, the ATG $\widehat{D}$ presented in Figure 4. The specification $\widehat{D}$ is a generalization of the specification $\widehat{A}$. The container now has two valves, v_i and v_o. The valve v_o controls, as before, the rate of outflow. The valve v_i controls the inflow of water into the container and, similar to v_o, can assume a real value between 0 and 1. The two valves are regulated by the two boolean command variables com_i and com_o that can assume the values 0 (*off*) and 1 (*on*).

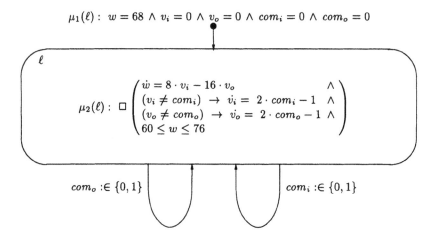

Fig. 4. Specification $\widehat{D}$

The conditional differential equation

$$v_i \neq com_i \ \rightarrow \ \dot{v}_i = 2 \cdot com_i - 1$$

should be interpreted as a compact representation of the two cases

$$((com_i = 1 \wedge v_i \neq 1) \rightarrow \dot{v}_i = 1) \quad \wedge$$
$$((com_i = 0 \wedge v_i \neq 0) \rightarrow \dot{v}_i = -1)$$

The only variable that can be directly modified by the controller is the outflow switch com_o. All other variables, including the inflow switch com_i, are environment variables. The continuous environment variables w, v_i, and v_o are governed by differential equations. The boolean environment variable com_i, however, is modified by transitions. The notation $com_i :\in \{0, 1\}$ labeling the edge in Figure 4 means that there are two transitions setting com_i nondeterministically to 0 or 1, respectively.

We will assume that all transitions of an APTS can be partitioned into *system transitions*, which modify only controlled variables, and *environment transitions*, which modify only environment variables. For example, the specification $\widehat{D}$ contains two system transitions, which update com_o, and two environment transitions, which update com_i.

Let us consider what type of refinements the specification $\widehat{D}$ can have. If we retain the notion of refinement as defined above, namely, that the APTS $\mathcal{S}_1$ refines the APTS $\mathcal{S}_2$ if every computation of $\mathcal{S}_2$ is also a computation of $\mathcal{S}_1$, we can point at the HATG $\widehat{E}$ of Figure 5 as a possible refinement of $\widehat{D}$. This refinement omits all transitions and, as a result, the command variables com_i and com_o and, consequently, the valve variables v_i and v_o all remain at 0, leading to a constant value of the water level w at 68, which is well within the prescribed range.

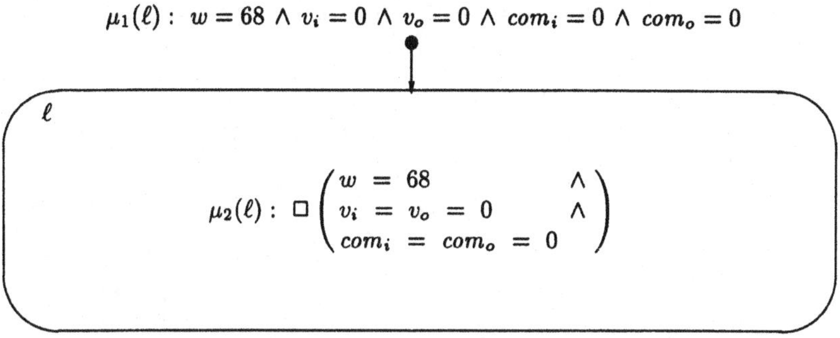

$$\mu_1(\ell): \quad w = 68 \land v_i = 0 \land v_o = 0 \land com_i = 0 \land com_o = 0$$

$$\mu_2(\ell): \quad \square \begin{pmatrix} w & = & 68 & & \land \\ v_i & = & v_o & = & 0 & \land \\ com_i & = & com_o & = & 0 & \end{pmatrix}$$

ℓ

Fig. 5. Lazy refinement $\widehat{E}$

It is quite obvious that $\widehat{E}$ cannot be accepted as a reasonable implementation of $\widehat{D}$. The easy solution that $\widehat{E}$ opted for is to prevent the environment from ever commanding the valve v_i to open. By contrast, the intended meaning of the specification $\widehat{D}$ is that, no matter how the environment commands the input valve to open or close, the controller will find a way to open and close the output valve so that the water level will remain within the prescribed limits.

Consequently, in the presence of environment transitions, we adopt the following definition for refinement between two APTS's. Let $\mathcal{S}_1$ and $\mathcal{S}_2$ be two APTS's over the sets V_1 and V_2 of state variables, respectively, such that $V_2 \subseteq V_1$ and $V_E \subseteq V_2$ is a set of environment variables. The APTS $\mathcal{S}_1$ *refines* the APTS $\mathcal{S}_2$ if

1. The projection of every run of $\mathcal{S}_1$ to the variables in V_2 is a run of $\mathcal{S}_2$.
2. Every environment transition of $\mathcal{S}_2$ is duplicated in $\mathcal{S}_1$; that is, if (P_1, P_2) is an environment transition of $\mathcal{S}_2$ and the $\mathcal{S}_1$-phases P_1' and P_2' are extensions of the $\mathcal{S}_2$-phases P_1 and P_2 to the variables in V_1, then (P_1', P_2') is a transition of $\mathcal{S}_1$.

Thus, the refinement of a specification must respect all environment transitions that appear in the specification.

With this more stringent notion of refinement, the HATG $\widehat{E}$ no longer refines the specification $\widehat{D}$, because it fails to duplicate the environment transitions of $\widehat{D}$. In Figure 6 we present an HATG $\widehat{F}$ that does implement the specification $\widehat{D}$, i.e., $\widehat{F}$ refines $\widehat{D}$ and is executable. The self-loop on the bottom represents the environment transitions that may set com_i to a new value at any point. Drawing a self-loop at an enclosing box is interpreted as if there is a similar self-loop at each of the four internal locations.

Acknowledgements. We gratefully acknowledge the help of Luca de Alfaro, Eddie Chang, Arjun Kapur, and Henny Sipma for their careful reading of the manuscript and thank them for many helpful suggestions.

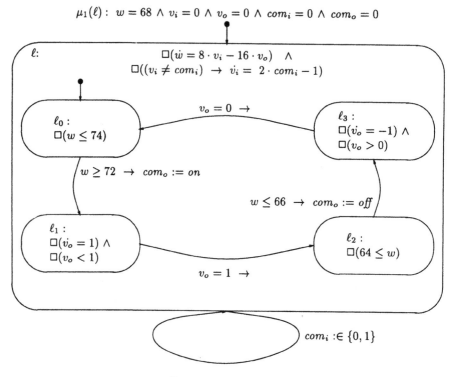

Fig. 6. Proper implementation $\widehat{F}$

References

[AL90]   M. Abadi and L. Lamport. Composing specifications. In J.W. de Bakker, W.-P. de Roever, and G. Rozenberg, editors, *Stepwise Refinement of Distributed Systems*, Lecture Notes in Computer Science 430. Springer-Verlag, 1990.

[AH92]   R. Alur and T.A. Henzinger. Logics and models of real time: a survey. In J.W. de Bakker, K. Huizing, W.-P. de Roever, and G. Rozenberg, editors, *Real Time: Theory in Practice*, Lecture Notes in Computer Science 600, pages 74–106. Springer-Verlag, 1992.

[CHR92] Z. Chaochen, C.A.R. Hoare, and A.P. Ravn. A calculus of durations. *Information Processing Letters*, 40:269–276, 1992.

[Hen92] T.A. Henzinger. Sooner is safer than later. *Information Processing Letters*, 43:135–141, 1992.

[HMP92] T.A. Henzinger, Z. Manna, and A. Pnueli. Timed transition systems. In J.W. de Bakker, K. Huizing, W.-P. de Roever, and G. Rozenberg, editors, *Real Time: Theory in Practice*, Lecture Notes in Computer Science 600, pages 226–251. Springer-Verlag, 1992.

[HNSY92] T.A. Henzinger, X. Nicollin, J. Sifakis, and S. Yovine. Symbolic model checking for real-time systems. In *Proceedings of the Seventh Annual Symposium on Logic in Computer Science*, pages 394–406. IEEE Computer Society Press, 1992.

[LA92] L. Lamport and M. Abadi. An old-fashioned recipe for real time. In J.W. de Bakker, K. Huizing, W.-P. de Roever, and G. Rozenberg, editors, *Real Time: Theory in Practice*, Lecture Notes in Computer Science 600, pages 1–27. Springer-Verlag, 1992.

[MMP92] O. Maler, Z. Manna, and A. Pnueli. From timed to hybrid systems. In J.W. de Bakker, K. Huizing, W.-P. de Roever, and G. Rozenberg, editors, *Real Time: Theory in Practice*, Lecture Notes in Computer Science 600, pages 447–484. Springer-Verlag, 1992.

[Mos85] B. Moszkowski. A temporal logic for multi-level reasoning about hardware. *IEEE Computer*, 18(2):10–19, 1985.

[NSY92] X. Nicollin, J. Sifakis, and S. Yovine. From ATP to timed graphs and hybrid systems. In J.W. de Bakker, K. Huizing, W.-P. de Roever, and G. Rozenberg, editors, *Real Time: Theory in Practice*, Lecture Notes in Computer Science 600, pages 549–572. Springer-Verlag, 1992.

Hybrid Systems in TLA$^+$

Leslie Lamport

Digital Equipment Corporation
Systems Research Center

Abstract. TLA$^+$ is a general purpose, formal specification language based on the Temporal Logic of Actions, with no built-in primitives for specifying real-time properties. Here, we use TLA$^+$ to define operators for specifying the temporal behavior of physical components obeying integral equations of evolution. These operators, together with previously defined operators for describing timing constraints, are used to specify a toy gas burner introduced by Ravn, Rischel, and Hansen. The burner is specified at three levels of abstraction, each of the two lower-level specifications implementing the next higher-level one. Correctness proofs are sketched.

1 Introduction

TLA$^+$ is a formal specification language based on TLA, the Temporal Logic of Actions [5]. We use TLA$^+$ to specify and verify a toy hybrid system—a gas burner described by Ravn, Rischel, and Hansen (RRH) [8]. The TLA$^+$ specification and proof can be compared with the one by RRH that uses the Duration Calculus.

We do not expect TLA$^+$ to permit the best possible specification of this or any other particular example. The specification of a gas burner is likely to be simpler in a formalism devised expressly for the class of hybrid systems that includes the gas burner. The specification of a Modula-3 procedure is likely to be simpler in a formalism for specifying Modula-3 procedures. But, while TLA$^+$ may not be the best method for specifying any particular system, we believe it is quite good for specifying a very wide class of systems, including gas burners and Modula-3 procedures.

There are two reasons for using TLA$^+$ instead of a language tailored to the specific problem. First, specialized languages often have limited realms of applicability. A language that permits a simple specification for one gas burner might require a very complicated one for a different kind of burner. The Duration Calculus seems to work well for real-time properties; but it cannot express simple liveness properties. A formalism like TLA$^+$ that, with no built-in primitives for real-time systems or procedures, can easily specify gas burners and Modula-3 procedures, is not likely to have difficulty with a different kind of gas burner.

The second reason for using TLA$^+$ is that formalisms are easy to invent, but practical methods are not. A practical method must have a precise language and robust tools. Building tools is not easy. It is hard to define a language that is powerful enough to handle practical problems and yet has a precise formal semantics. Such a language is a prerequisite for any sound, practical method. The advantage of not having to implement a new method for every problem domain is obvious.

An important criterion for choosing a formalism is how good it is for formal verification. A method based on a logic has an advantage over one based on an abstract programming language (such as CSP) because one does not have to translate from the specification language to a logic for reasoning. But, not all logics are equal. TLA works well in practice because most of the reasoning is in the domain of actions, which is the realm of "ordinary" mathematical reasoning. The use of temporal logic is minimal and highly ritualized. The temporal structure of the proofs are the same, regardless of whether one is reasoning about a mutual exclusion algorithm or a gas burner.

Modal logics, such as temporal logic and the Duration Calculus, are more difficult to use than ordinary logic. One reason is that the deduction principle, from which one deduces $P \Rightarrow Q$ by assuming P and proving Q, is invalid for most modal logics. In our work on the mechanical verification of TLA [4], we have found formalizing temporal logic reasoning to be much more difficult than formalizing ordinary mathematical reasoning. Temporal logic proofs that look simple when done by hand can be tedious to check mechanically. We believe that mechanical verification of TLA proofs is feasible largely because it involves very little temporal logic. One should be skeptical of claims that reasoning in a modal logic is easy in the absence of experience with mechanically checked proofs.

TLA$^+$ is a complete language, with a precise syntax and formal semantics. At the moment, only a few syntactic issues remain unresolved. Because it is completely formal, some things are a little more awkward to express in TLA$^+$ than in "semi-formal" methods. For example, most semi-formal methods, such as Unity [2] and the temporal logic of Manna and Pnueli [7], allow Boolean-valued variables; TLA$^+$ does not. In TLA$^+$, one cannot declare x to be a Boolean variable and write $\Box x$, one must instead write something like $\Box(x = \text{``T''})$. Although seemingly innocuous, Boolean variables pose the following problem. A specification may require an array $x[1]$, ..., $x[17]$ of flexible variables. Formally, such an array is a variable x whose value is a function with domain $\{1, \ldots, 17\}$. If one can declare Boolean variables, then one should also be able to declare that x is a Boolean-valued array variable, with index set $\{1, \ldots, 17\}$. One can then write the formula $\Box(x[14])$. One can also write the formula $\Box(x[i^2 - 23])$. In general, one could write the formula $\Box(x[e])$ for an arbitrary integer-valued expression e. Formalizing Boolean arrays presents the following options:

- Define the meaning of $\Box(x[e])$ for any value of e. Does one consider $x[e]$ to have some special undefined value $\bot$ if e is not in the domain of x? If so, what is the meaning of $\Box\bot$? Is the formula $\Box(\bot \Rightarrow \bot)$ valid?
- Declare $\Box(x[e])$ not to be a wff (well formed formula) if e is not in the index set of x. This leads to two possibilities:
  - The class of expressions e that can appear in the formula $x[e]$ are restricted so it is syntactically impossible for the value of e not to be in the index set of x. This leads to unnatural restrictions on formulas.

- Whether or not a formula is a wff becomes a semantic rather than syntactic property. It is a strange formalism in which it is undecidable whether a formula is a wff.

TLA$^+$ avoids this problem by not having Boolean-valued variables. Neither Manna and Pnueli [7] nor Misra and Chandy [2], in books that are hundreds of pages long, indicate how they formalize Boolean arrays.

There are dozens of similar issues that must be resolved in designing a complete language. A simple informal specification might not look so simple if it had to be written formally. Of particular concern are formalisms based on a type system. It is easy to introduce the informal notation that *Length(s)* denotes the length of the sequence *s*. But, will the type system really allow a formal language in which the user can define *Length(s)* to denote the length of *any* sequence *s*? It is easy to define such a *Length* operator in TLA$^+$. However, this ability is based on a distinction between a function and an operator—a distinction one won't find in semi-formal methods.

Despite being completely formal, TLA$^+$ is simple enough for practical applications. A general treatment of hybrid systems requires continuous mathematics. Specifying the gas burner requires defining the Riemann integral $\int_a^b f$ of the function f over the closed interval $[a, b]$. Assuming only the set of real numbers with the usual arithmetic operations and ordering relations, the entire definition takes about 15 lines. Our example is a toy one, and we do not claim to have formalized any significant fraction of the mathematics that will be needed in practical applications. We do claim that a language that can define the Riemann integral in 15 lines is powerful enough to express any mathematical concepts likely to arise in real specifications.

TLA$^+$ specifications are written in ASCII. We hope eventually to write a program that converts an ASCII specification to input for a text formater that produces a "pretty-printed" version. The user will specify how individual operators are formated—for example, declaring that the `Integral` operator should be formated so `Integral(a, b, f)` is printed as $\int_a^b f$, and that `[|s|]` should be printed as $|s|$. We have simulated such a program to produce pretty-printed TLA$^+$ specifications.

2 Representing Hybrid Systems with TLA

This section describes the generic operators that can be used to specify hybrid systems. The precise TLA$^+$ definitions of some operators is deferred to Section 3. Figure 1 contains a complete list of all the predefined TLA$^+$ constant operators— the ones that describe data structures. The only additional operators are TLA's action operators (such as *Enabled*) and temporal operators (such as □). The syntax and formal semantics of these operators, which fit on one page, can be found in [5].

Logic

true false $\wedge$ $\vee$ $\neg$ $\Rightarrow$ $\Leftrightarrow$

$\forall x : p$ $\exists x : p$ $\forall x \in S : p$ $\exists x \in S : p$

choose $x : p$ [Equals some x satisfying p]

Sets

$=$ $\neq$ $\in$ $\notin$ $\cup$ $\cap$ $\subseteq$ $\setminus$ [set difference]

$\{e_1, \ldots, e_n\}$ [Set consisting of elements e_i]

$\{x \in S : p\}$ [Set of elements x in S satisfying p]

$\{e : x \in S\}$ [Set of elements e such that x in S]

subset S [Set of subsets of S]

union S [Union of all elements of S]

Functions

domain f $f[e]$ [Function application]

$[x \in S \mapsto e]$ [Function f such that $f[x] = e$ for $x \in S$]

$[S \to T]$ [Set of functions f with $f[x] \in T$ for $x \in S$]

$[f; e_1 \mapsto e_2]$ [Function $\widehat{f}$ equal to f except $\widehat{f}[e_1] = e_2$]

$[f; e : S]$ [Set of functions $\widehat{f}$ equal to f except $\widehat{f}[e] \in S$]

Records

$e.x$ [The x-component of record e]

$[[x_1 \mapsto e_1, \ldots, x_n \mapsto e_n]]$ [The record whose x_i component is e_i]

$[[x_1 : S_1, \ldots, x_n : S_n]]$ [Set of all records with x_i component in S_i]

$[[r; x \mapsto e]]$ [Record $\widehat{r}$ equal to r except $\widehat{r}.x = e$]

$[[r; x : S]]$ [Set of records $\widehat{r}$ equal to r except $\widehat{r}.x \in S$]

Tuples

$e[i]$ [The ith component of tuple e]

$\langle e_1, \ldots, e_n \rangle$ [The n-tuple whose i^{th} component is e_i]

$S_1 \times \ldots \times S_n$ [The set of all n-tuples with i^{th} component in S_i]

Miscellaneous

"$c_1 \ldots c_n$"   [A literal string of n characters]

if p **then** e_1 **else** e_2 [Equals e_1 if p true, else e_2]

case $p_1 \to e_1, \ldots, p_n \to e_n$ [Equals e_i if p_i true]

let $x_1 \stackrel{\Delta}{=} e_1 \ldots x_n \stackrel{\Delta}{=} e_n$ **in** e [Equals e in the context of the definitions]

Fig. 1. The constant operators of TLA$^+$.

─────────────────────── **module** *RealTime* ───────────────────────

import *Reals*

───

parameters
  now : **variable**
  ∞ : **constant**

───

assumption
  InfinityUnReal $\triangleq$ $\infty \notin \mathbf{R}$
temporal
  $RT(v)$ $\triangleq$ $\land$ *now* $\in$ *Real*
        $\land$ $\Box[\land$ *now'* $\in \{r \in \mathbf{R} : r > now\}$
          $\land$ $v' = v$ $]_{now}$

  VTimer(x : **state fcn**, $\mathcal{A}$: **action**, δ, v : **state fcn**) $\triangleq$
    $\land$ $x =$ **if** *Enabled*$\langle\mathcal{A}\rangle_v$ **then** *now* $+ \delta$
             else ∞
    $\land$ $\Box[x' =$ **if** (*Enabled*$\langle\mathcal{A}\rangle_v$)' **then if** $\langle\mathcal{A}\rangle_v$ **then** *now'* $+ \delta$
                         else x
                else ∞ $]_{\langle x, v \rangle}$

  MaxTimer(x) $\triangleq$ $\Box[(x \neq \infty) \Rightarrow (now' \leq x)]_{now}$

  MinTimer(x : **state fcn**, $\mathcal{A}$: **action**, v : **state fcn**) $\triangleq$ $\Box[\langle\mathcal{A}\rangle_v \Rightarrow (now \geq x)]_v$

───

Fig. 2. The *RealTime* module.

2.1 Real Time in TLA

A method for writing real-time specifications in TLA is described in [1]. We now review this approach and introduce TLA$^+$ by defining the operators from [1] in the TLA$^+$ module *RealTime* of Figure 2.

The *RealTime* module first imports the module *Reals*, which is assumed to define the set $\mathbf{R}$ of real numbers with its usual arithmetic operators.[1] The module declares two parameters: the (flexible) variable *now* and the constant ∞. The value of *now* represents the current time, which can assume any value in $\mathbf{R}$. The constant ∞ is the usual infinity of mathematicians. (All symbols that appear in a module's formulas are either parameters or symbols that are defined in terms of parameters. The **import** statement includes the parameters of the module *Reals* as parameters of *RealTime*.)

Next comes an assumption, named *InfinityUnReal*, which asserts that ∞ is not a real number. The keyword **temporal** heads a list of definitions of temporal formulas. The first definition in the list defines $RT(v)$ to be a formula asserting

─────────────────

[1] Some people mistakenly think that, because the reals are uncountable, letting time be a real number complicates proofs. The axioms about the real numbers needed to prove such real-time properties as the correctness of Fischer's protocol [1] form a decidable theory.

that *now* is initially in **R**, and every step either (i) sets the new value of *now* to a real number greater than its current value and (ii) leaves the value of v unchanged, or else it leaves *now* unchanged. (A list of formulas bulleted with $\wedge$ denotes their conjunction. The formula $[\mathcal{A}]_w$ is defined to equal $\mathcal{A} \vee (w' = w)$; a $[\mathcal{A}]_w$ step is thus either an $\mathcal{A}$ step or a step that leaves w unchanged.) In representing a system by a TLA formula, the discrete variables of the system are considered to change instantaneously, meaning that when they change, *now* remains unchanged. This is asserted by the formula $RT(v)$ when v is the tuple whose components are the system's discrete variables.

In [1], timing constraints are expressed through the use of timer variables. The formula $VTimer(x, \mathcal{A}, \delta, v)$ asserts that x is a timer whose value is initially either (i) δ greater than the initial value of *now* or (ii) ∞, depending on whether or not action $\langle \mathcal{A} \rangle_v$ is enabled. The value of x is set to $now + \delta$ when either (i) $\langle \mathcal{A} \rangle_v$ becomes enabled or (ii) an $\langle \mathcal{A} \rangle_v$ step occurs that leaves $\langle \mathcal{A} \rangle_v$ enabled, and it is set to ∞ when $\langle \mathcal{A} \rangle_v$ becomes disabled.[2] (The action $\langle \mathcal{A} \rangle_v$ is defined to equal $\mathcal{A} \wedge (v' \neq v)$; an $\langle \mathcal{A} \rangle_v$ step is thus an $\mathcal{A}$ step that changes v.) The formula $MaxTimer(x)$ asserts that changing *now* cannot make it greater than x. Thus, the formula $VTimer(x, \mathcal{A}, \delta, v) \wedge MaxTimer(x)$ implies that an $\langle \mathcal{A} \rangle_v$ action cannot be continuously enabled for more than δ seconds without having occurred. The formula $MinTimer(x, \mathcal{A}, v)$ asserts that an $\langle \mathcal{A} \rangle_v$ action cannot occur unless $now \geq x$, so $VTimer(x, \mathcal{A}, \delta, v) \wedge MinTimer(x, \mathcal{A}, v)$ implies that an $\langle \mathcal{A} \rangle_v$ action must be enabled for at least δ seconds before it can occur.

The definitions of *VTimer* and *MinTimer* explicitly indicate the *sorts* of the parameters. When no sort is specified, the sort **state fcn**, denoting a mapping from states to values, is assumed—except in the definitions of constants, where the default sort is **constant**, denoting a constant value.

There are many ways of defining timers for expressing real-time constraints, and they are all easily expressed in TLA$^+$. The method used in [1] is probably not optimal for the gas burner example. Although we might be able to simplify the specifications in this example by defining a new kind of timer, in practice one would use a fixed set of operators defined in a standard module like *RealTime*. To make the example more realistic, we have used a pre-existing set of operators.

2.2 Hybrid Systems

To represent hybrid systems in TLA, continuous system variables are represented by variables that change when *now* does. The gas-burner specification of RRH can be expressed using only the timers introduced in [1]. However, RRH's specification is somewhat artificial, apparently chosen to avoid reasoning with continuous mathematics. Instead of the natural requirement that the concentration of unburned gas never exceeds some value, RRH require that unburned gas never be released for more than 4 seconds out of any 30-second period. Because it poses an interesting new challenge for TLA$^+$, we specify the more natural requirement and prove that it is implied by the requirement of RRH.

---

[2] Another type of timer is also defined in [1], but it is not needed here.

The gas concentration g will satisfy an integral equation of the form

$$g(t) = \int_{t_0}^{t} F(g(t)) \, dt$$

where the function F depends on the discrete variables. A more general situation is a continuous variable f that satisfies an equation of the form

$$f(t) = f_0 + \int_{t_0}^{t} G(f(t), t) \, dt \tag{1}$$

A further generalization is to let f be a function whose range is the set of n-tuples of reals. However, this generalization is straightforward and is omitted.

For specifying hybrid systems, we define a TLA formula, pretty-printed as $[x = c + \int G \,|\, \mathcal{A}, v]$, having the following meaning:

- Initially, x equals c.
- In any step that changes now, the new value x' of x equals $f(now')$, where f is the solution to $f(t) = x + \int_{now}^{t} G(f(s), s) \, ds$.
- Any step that leaves now unchanged leaves x unchanged, unless it is an $\langle \mathcal{A} \rangle_v$ step, in which case x' equals zero.

Thus, $[x \doteq c + \int G \,|\, \mathsf{false}, v]$ asserts that x represents the solution to (1) with f_0 equal to c and t_0 the initial value of now. The general formula $[x = c + \int G \,|\, \mathcal{A}, v]$ adds the requirement that x is reset to 0 by an $\langle \mathcal{A} \rangle_v$ action. The precise definition of $[x = c + \int G \,|\, \mathcal{A}, v]$ is given later, in the *HybridSystems* module.

It is often useful to describe the amount of time a predicate P has been true, which is the integral over time of a step function that equals 1 when P is true and zero when it is false. The formula $[x = 0 + \int \chi(P) \,|\, \mathsf{false}, v]$ asserts that the value of x always equals this integral, where the function $\chi(P)$ is defined by

$$\chi(P)(r, s) = \begin{cases} 1 \text{ if } P = \mathsf{true} \\ 0 \text{ if } P = \mathsf{false} \end{cases}$$

The formal definition of χ appears in the *HybridSystems* module.

3 The Gas Burner

The example system is the gas burner shown in Figure 3. This is the toy example of RRH. Our goal is to write a specification that is in the spirit of RRH's; the specification of a real gas burner might be much more complex.

The discrete state of the system consists of the states of the gas, the heat-request signal, the flame, and the ignition—each of which can be either on or off. These state components are represented by the four variables declared in the module *BurnerParameters* of Figure 4. A physical state in which the gas is turned on or off will be represented by a state in which the value of the variable *gas* is the string "on" or "off". The *on* and *off* values of the other variables are similarly denoted by the strings "on" and "off". As is customary in TLA

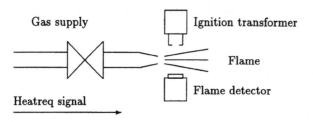

Fig. 3. The gas burner. (Figure taken from [8].)

```
┌─────────────────── module BurnerParameters ───────────────────┐
│ import HybridSystems, RealTime                                 │
├────────────────────────────────────────────────────────────────┤
│ parameters                                                     │
│   gas, heatReq, flame, ignition : variables                    │
├────────────────────────────────────────────────────────────────┤
│ state function                                                 │
│   v  ≜  ⟨gas, heatReq, flame, ignition⟩                         │
│ predicates                                                     │
│   Gas      ≜  gas = "on"                                        │
│   Flame    ≜  flame = "on"                                      │
│   Heatreq  ≜  heatReq = "on"                                    │
│   Ignition ≜  ignition = "on"                                   │
└────────────────────────────────────────────────────────────────┘
```

Fig. 4. Module *BurnerParameters*

specifications, we define a state function v that equals the tuple of all relevant variables. (In TLA$^+$, angle brackets $\langle\ \rangle$ denote tuples.) To make it easier to compare our specification with theirs, we have also defined state predicates that correspond to the Boolean variables of RRH.

For convenience, the *BurnerParameters* module imports the modules *HybridSystems* and *RealTime* that define the operators described above. Since the *RealTime* module also imports the real numbers, they are transitively imported by the *BurnerParameters* module.

We now specify the requirement that the concentration of gas remains less than some value *MaxCon*. We assume that the gas concentration g is described by the integral equation

$$g(t) = \int_{t_0}^{t} -\delta g(t) + \left\{ \begin{array}{l} \rho \text{ if gas on and flame off} \\ 0 \text{ otherwise} \end{array} \right\} dt$$

where δ is the rate at which gas diffuses away from the burner and ρ is the rate

```
┌──────────────────────── module GasConcentration ────────────────────────┐
│ import BurnerParameters                                                    │
├──────────────────────────────────────────────────────────────────────────┤
│ parameters                                                                 │
│   δ, ρ, MaxCon : {r ∈ R : r > 0} const                                     │
├──────────────────────────────────────────────────────────────────────────┤
│ state function                                                             │
│   accumRate[r, s : R]  ≜  (−δ) * r + (if  Gas ∧ ¬Flame  then  ρ  else  0)  │
│ temporal                                                                   │
│   Req0  ≜  ∀ g : [g = 0 + ∫ accumRate | false, v] ⇒ □(g < MaxCon)          │
└──────────────────────────────────────────────────────────────────────────┘
```

Fig. 5. The specification of the gas-concentration requirement

at which gas flows when it is turned on.

Using the TLA formula $[x = c + \int G \mid \mathcal{A}, v]$, it is easy to specify the requirement that the gas concentration is always less than *MaxCon*. First, the function *accumRate* that is substituted for G is defined as follows.

$$accumRate[r, s : R] \;\triangleq\; (-\delta) * r + (\text{if }\; Gas \wedge \neg Flame \;\text{ then }\; \rho \;\text{ else }\; 0)$$

This defines *accumRate* to be a function whose domain is $R \times R$ such that *accumRate*$[r, s]$ equals the expression to the right of the $\triangleq$, for every pair $\langle r, s \rangle$ in $R \times R$. (In TLA$^+$, square brackets denote function application.)

The formula $[g = 0 + \int accumRate \mid \mathsf{false}, v]$ then asserts that g describes the gas concentration. As usual, the temporal formula $\square(g < MaxCon)$ asserts that g is always less than *MaxCon*, so the formula

$$[g = 0 + \int accumRate \mid \mathsf{false}, v] \Rightarrow \square(g < MaxCon)$$

asserts that if g is the gas concentration, then g is always less than *MaxCon*. We want this formula to be true for an arbitrarily chosen, "fresh" variable g, so the desired property is obtained by universally quantifying g.[3]

The complete definition of this property, which is called *Req0*, is given in module *GasConcentration*, shown in Figure 5. This module imports the *BurnerParameters* module, which defines *Gas* and *Flame*, and declares δ, ρ, and *MaxCon* to be constant parameters whose values lie in the set of positive reals. This "type declaration" is a shorthand for the **assumption**

$$(\delta \in \{r \in R : r > 0\}) \wedge (\rho \in \{r \in R : r > 0\}) \wedge (MaxCon \in \{r \in R : r > 0\})$$

Next, we define the three requirements given by RRH. The first requirement is

---

[3] The symbols ∃ and ∀ denote quantification over flexible variables; ∃ and ∀ denote ordinary quantification over rigid variables.

For Safety, gas must never leak for more than 4 seconds in any period of at most 30 seconds.

This is a requirement for the 30-second interval beginning at time r, for every real r. It is expressed in terms of an array x of timers, where $x[r]$ is the timer used to express the requirement for the interval beginning at r. The amount of gas that has leaked during the interval $[r, r + 30]$ is obtained by "integrating" the function $G(r)$ that is defined by

$$G(r)[s, t : \mathbf{R}] \triangleq \mathbf{if}\ Gas \wedge \neg Flame \wedge (t \in [r, r + 30])\ \mathbf{then}\ 1\ \mathbf{else}\ 0$$

This defines G to be an operator such that, for any r, the expression $G(r)$ denotes a function with domain $\mathbf{R} \times \mathbf{R}$.[4]

Using $x[r]$ as a timer, the requirement for this interval is expressed by

$$[x[r] = 0 + \textstyle\int G(r)\,|\,\mathsf{false}, v] \Rightarrow \Box(x[r] \leq 4))$$

The requirement *Req1* is obtained by quantifying over all real numbers r, and then universally quantifying x. The complete definition is given in module *BurnerRequirements* in Figure 6. (Our formulas *Req1*, *Req2*, and *Req3* correspond to the formulas $\Box Req_1$, $\Box Req_2$, and $\Box Req_3$ of RRH.)

The second requirement is

Heat request off shall result in the flame being off after 60 seconds.

This condition is expressed by using a variable x that integrates the amount of time the flame has been on while the heat request was off, and is reset to zero whenever the flame goes off or the heat request comes back on.[5] Such a variable x satisfies the formula

$$[x = 0 + \textstyle\int \chi(Flame \wedge \neg HeatReq)\,|\,Heatreq' \vee \neg Flame', v]$$

Asserting that x is always less than 60 and quantifying over x yields condition *Req2* of Figure 6.

The final requirement is

---

[4] A function f has a domain; it is an ordinary value. Thus $G(7)$ and $G(7)[\sqrt{2}, .5]$ both denote values; the first is a function, the second a real number. To define a function, one must specify its domain. However, G is an operator, not a function. It is not a value, and it does not have a domain; the symbol G by itself is not a syntactically correct expression. The definition above defines $G(r)$ for any value of r, not just for real numbers. We had to define $G(r)$ to be a function because it appears as a function, without any arguments, in subsequent formulas. We could have defined G to be a function by writing $G[r : \mathbf{R}][\ldots]$, but there is no need to make G a function, and we didn't feel like writing the ": $\mathbf{R}$". The distinction between operators and functions is another example of the details that arise in defining a precise language with a formal semantics. This distinction is what allows one easily to define in TLA$^+$ an operator *Length* such that *Length*(s) is the length of *any* sequence s. One could not define *Length* to be a function, since its domain would not be a set. (Its domain would be isomorphic to the "set" of all sets.)

[5] This requirement, which comes from RRH, is satisfied if the heat request is always off and the flame is always on except for flickering out briefly every 59.9 seconds.

```
┌──────────────────── module BurnerRequirements ─────────────────────┐
│ import BurnerParameters                                             │
├────────────────────────────────────────────────────────────────────┤
```

temporal

$Req1 \;\triangleq\;$ **let** $G(r)[s, t : \mathbf{R}] \;\triangleq\;$
    if $Gas \wedge \neg Flame \wedge (t \in [r, r+30])$ **then** 1 **else** 0
    in $\forall\, x : \forall\, r \in \mathbf{R} : [x[r] = 0 + \int G(r)\,|\,\mathsf{false}, v] \Rightarrow \Box(x[r] \le 4))$

$Req2 \;\triangleq\; \forall\, x : [x = 0 + \int \chi(Flame \wedge \neg HeatReq)\,|\,Heatreq' \vee \neg Flame', v]$
        $\Rightarrow \Box(x \le 60)$

actions

$IgniteFailAct(y) \;\triangleq\; \wedge\; Gas \wedge Ignition \wedge \neg Flame$
                $\wedge\; y' \ne y$
                $\wedge\; \mathbf{unchanged}\langle v,\, now \rangle$

$Req3Reset(y) \;\triangleq\; Flame' \vee (y' \ne y) \vee \neg Heatreq'$

temporal

$IgniteFail(y) \;\triangleq\; \exists\, z : \wedge\; \Box[IgniteFailAct(y)]_y$
                $\wedge\; VTimer(z, IgniteFailAct(y), 1/2, y)$
                $\wedge\; MaxTimer(z)$
                $\wedge\; MinTimer(z, IgniteFailAct(y), y)$

$Req3 \;\triangleq\; \forall\, x, y : \wedge\; IgniteFail(y)$
            $\wedge\; [x = 0 + \int \chi(Heatreq)\,|\,Req3Reset(y), \langle v,\, y \rangle]$
            $\Rightarrow \Box(x \le 60)$

```
└────────────────────────────────────────────────────────────────────┘
```

Fig. 6. The three requirements of Ravn, Rischel, and Hansen.

Heat request shall after 60 seconds result in gas burning unless an ignite or flame failure has occurred. An ignite failure happens when gas does not ignite after 0.5 seconds. The flame fails if it disappears while gas is supplied.

A careful analysis of this condition reveals that replacing "after 60 seconds" by "within 60 seconds" makes the "or flame failure" redundant. We can therefore rewrite this requirement as

Heat request shall, within 60 seconds, result in gas burning unless an ignite failure has occurred.

The obvious approach is to define a variable x that integrates the time during which the heat request is on and is reset by the presence of a flame, an ignite failure, or the heat request being turned off. We would then define an *IgniteFail* action, define the action *Req3Reset* to be $Flame' \vee IgniteFail \vee \neg Heatreq'$ and write the requirement as

$$\forall\, x : [x = 0 + \int \chi(Heatreq)\,|\,Req3Reset, v] \Rightarrow \Box(x \le 60)$$

However, there is no actual *IgniteFail* action. An ignite failure is not a change of the discrete state variables; it is something that is caused by the passage of time. So, we must introduce a variable y that is changed when an ignite failure occurs. Since y should not be a free variable of the formula, it must be "quantified away". We want y to change when the ignition and the gas have been on for precisely .5 seconds while the flame has been off. This is accomplished with a .5-second "vtimer" for the action that changes y. We define *IgniteFailAct*(y) to be an action that changes y while leaving the other variables unchanged, and is enabled precisely when the gas and ignition are on and the flame is off. An ignition failure happens when this action has been continuously enabled for precisely .5 seconds. So, we let

$$IgniteFailAct(y) \triangleq \wedge \; Gas \wedge Ignition \wedge \neg Flame$$
$$\wedge \; (y' \neq y) \wedge \textbf{unchanged} \; \langle v, now \rangle$$

where **unchanged** f denotes $f' = f$. The TLA formula $\Box[IgniteFailAct(y)]_y$ asserts that y changes only when $Gas \wedge Ignition \wedge \neg Flame$ is true. To assert that this change occurs only when the *IgniteFailAct*(y) action has been enabled for precisely .5 seconds, we add a timer z with both a *MaxTimer* and *MinTimer* condition. The temporal formula *IgniteFail*(y) defined in Figure 6 asserts that an ignition failure has occurred iff y has changed. The definition of the formula *Req3* expressing the third requirement is now straightforward and appears in Figure 6.

We would like to prove the gas concentration condition *Req0* from the three requirements of RRH. This condition actually follows from *Req1*. The precise theorem, expressed in TLA$^+$, is that if

$$MaxCon \geq 4 * \rho * (1 + (1/(1 - \exp[(-30) * \delta])))$$

then $RT(v) \wedge Req1 \Rightarrow Req0$, where exp is the usual exponential function. Before proving this, we finish the specification by defining the operators we have been assuming.

4 The Module *HybridSystems*

The *HybridSystems* module involves the solution to an integral equation. Defining this requires first defining the definite integral. The Riemann integral of a function f on an interval is the signed area under the graph of f. It is defined as the limit of approximations obtained by breaking the interval into subintervals, where the area under the graph of f from p to q is approximated by $f(p)(q - p)$. The definition of $\int_a^b f$ appears in module *Integration* of Figure 7. The informal meanings of the defined operators are explained below.

$\mathbf{R}^+$ The set of positive reals.

$[a, b]$ The closed interval from a to b. This is an "unsigned" interval, where $[a, b]$ equals $[b, a]$. (To avoid writing explicit domains, we have defined $[-, -]$ to

```
┌──────────────────────── module Integration ────────────────────────┐
│ import Reals                                                        │
├────────────────────────────────────────────────────────────────────┤
```

constants

$\mathbf{R^+} \triangleq \{r \in \mathbf{R} : 0 < r\}$

$[a, b] \triangleq \{r \in \mathbf{R} : ((a \le r) \land (r \le b)) \lor ((a \ge r) \land (r \ge b))\}$

$|r| \triangleq$ **if** $r < 0$ **then** $-r$ **else** r

$\{m \ldots n\} \triangleq \{i \in Nat : (m \le i) \land (i \le n)\}$

$Partition(a, b, n, \delta) \triangleq$
$\quad \{ p \in [\{0 \ldots n{+}1\} \to [a, b]] : \land (p[0] = a) \land (p[n{+}1] = b)$
$\qquad\qquad\qquad\qquad\qquad \land \forall i \in \{0 \ldots n\} : \land \lor (a \le b) \land (p[i] \le p[i{+}1])$
$\qquad\qquad\qquad\qquad\qquad\qquad\qquad\qquad \lor (a \ge b) \land (p[i] \ge p[i{+}1])$
$\qquad\qquad\qquad\qquad\qquad\qquad\qquad\quad \land |p[i{+}1] - p[i]| < \delta \qquad\qquad \}$

$\Sigma(f, p)[n : Nat] \triangleq f[p[n]] * (p[n{+}1] - p[n])$
$\qquad\qquad\qquad\qquad\quad + $ **if** $n = 0$ **then** 0 **else** $\Sigma(f, p)[n - 1]$

$\int_a^b f \triangleq$ **choose** $r : \land r \in \mathbf{R}$
$\qquad\qquad\qquad\quad \land \forall \epsilon \in \mathbf{R^+} : \exists \delta \in \mathbf{R^+} : \forall n \in Nat :$
$\qquad\qquad\qquad\qquad\qquad \forall p \in Partition(a, b, n, \delta) : |r - \Sigma(f, p)[n]| < \epsilon$

$[c{+}\int_a G][r : \mathbf{R}] \triangleq$
$\quad$ **let** $f \triangleq$ **choose** $f : \land f \in [[a, r] \to \mathbf{R}]$
$\qquad\qquad\qquad\qquad\qquad\quad \land \forall t \in [a, r] : f[t] = c + \int_a^t [s \in \mathbf{R} \mapsto G[f[s], s]]$
$\quad$ **in** $f[r]$

theorem

$IntegralOfStepFcn \triangleq$
$\quad \forall G : \forall a, b, \delta \in \mathbf{R} : \forall n \in Nat : \forall p \in Partition(a, b, n, \delta) : \forall f \in [\mathbf{R} \to \mathbf{R}] :$
$\quad (\forall s, t \in \mathbf{R} : \forall i \in \{0 \ldots n\} : (p[i] < t) \land (t < p[i + 1]) \Rightarrow (G[s, t] = f(p[i])))$
$\qquad \Rightarrow ([c{+}\int_a G][b] = \Sigma(f, p)[n])$

```
└────────────────────────────────────────────────────────────────────┘
```

Fig. 7. Defining integration in TLA$^+$.

be an operator, so $[a, b]$ equals the expression to the right of the $\triangleq$ even if a and b are not real numbers. However, it has the expected meaning only when a and b are real numbers. This remark applies to all the operators in this module.)

$|r|$ The absolute value of r.

$\{m \ldots n\}$ The set of natural numbers from m through n. (We assume that the set Nat of natural numbers is imported with the $Reals$ module.)

$Partition(a, b, n, \delta)$ The set of all possible partitions of the interval $[a, b]$ into $n + 1$ subintervals each of length less than δ. Formally, a partition of an interval $[a, b]$ into $n{+}1$ subintervals is a monotonic function p from $\{0 \ldots n{+}1\}$ to $[a, b]$ such that $p[0] = a$ and $p[n + 1] = b$. ($[S \to T]$ denotes the set of all functions whose domain equals S and whose range is a subset of T.)

```
                   ──────────── module Exponentials ────────────
 import Integration
```

```
 constant
   exp  ≜  [1+∫₀[r, s ∈ R ↦ r]]
 theorems
   ExpFacts  ≜  ∧ exp ∈ [R ↦ R]
                ∧ exp[0] = 1
                ∧ ∀r ∈ R : (r > 0) ⇒ (1 − r < exp[−r]) ∧ (exp[−r] < 1)
                ∧ ∀r, s ∈ R : exp[r] ∗ exp[s] = exp[r + s]
   DiffusionSolution  ≜
     ∀p, q, a, c ∈ R : (p ≠ 0) ⇒
       [c+∫ₐ[r, s ∈ R ↦ (p ∗ r) + q]] =
       [t ∈ R ↦ c ∗ exp[p ∗ (t − a)] + (q/p) ∗ (exp[p ∗ (t − a)] − 1)]
```

Fig. 8. The exponential function.

$\Sigma(f, p)$ This is a function such that, if p is a partition of an interval into $n + 1$ subintervals, then $\Sigma(f, p)[n]$ is the approximation $\sum_{i=0}^{n} f[p[i]](p[i+1] - p[i])$ to the area under f on that interval defined by the partition. We have defined $\Sigma(f, p)$ to be a function with domain Nat to permit a recursive definition. (In TLA$^+$, only functions, not operators, can be defined recursively. A recursive function definition such as this can be expressed nonrecursively using TLA$^+$'s **choose** operator.)

$\int_a^b f$ The Riemann integral[6] of the function f is defined to be a real number r such that, for every ϵ, there is some δ such that the approximation for every partition with subintervals of length less than δ lies within ϵ of r. (The operator **choose** denotes Hilbert's ε operator [6], so **choose** $r : P(r)$ equals some value r such that $P(r)$ is true, if such an r exists; otherwise, it has an unspecified value.)

$[c+\int_a G]$ The function f such that $f[r]$ equals $c+\int_a^r G[f[t], t] \, dt$. (The expression $[s \in S \mapsto e(s)]$ denotes the function g with domain S such that $g[s] = e(s)$ for all $s \in S$.)

Finally, the module asserts the result from elementary calculus that if G is a step function on the interval $[a, b]$, then $[c+\int_a G][b]$ is equal to its approximation for the appropriate partition. This theorem is needed for the proof of property $Req0$.

Proving $Req0$ also requires introducing the exponential function exp, where $\exp[t] = e^t$. It is defined, and some theorems about exp are asserted, in the $Exponentials$ module of Figure 8. The theorem named $DiffusionSolution$ asserts

---

[6] Although this integral is commonly written $\int_a^b f(t) \, dt$, rigorous mathematicians usually write $\int_a^b f$. For example, $\int_a^b t^2 \, dt$ is just an informal way of denoting $\int_a^b f$ when f is the function defined by $f(t) = t^2$.

```
┌──────────────────── module HybridSystems ────────────────────┐
│ import Integration, RealTime                                  │
├───────────────────────────────────────────────────────────────┤
│                                                               │
│ temporal                                                      │
│    [x = c + ∫ G | A : action, v]  ≜  ∧ x = c                  │
│                                     ∧ □[x′ = if now′ = now    │
│                                           then if ⟨A⟩_v then 0 else x │
│                                           else [x + ∫_now G][now′]    ]⟨now, x, v⟩ │
│ state function                                                │
│    χ(P : predicate)  ≜  if P then 1 else 0                    │
│                                                               │
└───────────────────────────────────────────────────────────────┘
```

Fig. 9. The *HybridSystems* module.

that the solution to the equation $f(t) = c + \int_a^t (pf(t) + q)\, dt$ is

$$f(t) = ce^{p(t-a)} + \frac{q}{p}(e^{p(t-a)} - 1)$$

(Such theorems could be proved with the aid of a mathematical package such as *Maple* or *Mathematica*.) The *HybridSystems* module is now straightforward; it appears in Figure 9.

5   The Proof of Property *Req0*

We now sketch the proof of the result mentioned earlier, that *Req1* implies *Req0*. The proof requires only the theorems explicitly asserted in the modules defined above, plus the usual algebraic properties of arithmetic for the real numbers.

The proof is hierarchically structured, using the following notation. The theorem to be proved is statement ⟨0⟩1. The proof of statement ⟨i⟩j is either an ordinary paragraph-style proof or the sequence of statements ⟨i + 1⟩1, ⟨i + 1⟩2, ... and their proofs. Within a proof, ⟨k⟩l denotes the most recent statement with that number. A statement has the form

ASSUME: *Assumption*   PROVE: *Goal*

which is abbreviated to *Goal* if there is no assumption. The assertion Q.E.D. in statement number ⟨i + 1⟩k of the proof of statement ⟨i⟩j denotes the goal of statement ⟨i⟩j. The statement

CASE: *Assumption*

is an abbreviation for

ASSUME: *Assumption*   PROVE: Q.E.D.

Within the proof of statement ⟨i⟩j, assumption ⟨i⟩ denotes that statement's assumption.

We begin with the high-level proof, which essentially uses standard predicate logic to reduce the problem to proving a statement with no quantification over flexible variables. The resulting statement, ⟨1⟩2, is proved later.

ASSUME: $MaxCon \geq 4 * \rho * (1 + (1/(1 - \exp[(-30) * \delta])))$
PROVE: $RT(v) \land Req1 \Rightarrow Req0$

$\langle 1 \rangle 1.$ $\exists x : \forall r \in \mathbf{R} : [x[r] = 0 + \int G(r) \,|\, \text{false}, v]$

PROOF: Follows from a standard theorem about the validity of adding history variables [1], which asserts the validity of $\exists x : (x = f) \land \Box[x' = g]_w$ if x does not occur in f and x' does not occur in g.

$\langle 1 \rangle 2.$ $\land\ now = a$
$\quad \land\ RT(v)$
$\quad \land\ \forall r \in \mathbf{R} : [x[r] = 0 + \int G(r) \,|\, \text{false}, v]$
$\quad \land\ [g = 0 + \int accumRate \,|\, \text{false}, v]$
$\quad \land\ \forall r \in \mathbf{R} : \Box(x[r] \leq 4)$
$\quad \Rightarrow\ \Box(g < MaxCon)$

PROOF: Sketched below.

$\langle 1 \rangle 3.$ Q.E.D.

$\quad \langle 2 \rangle 1.$ $\land\ now = a$
$\qquad \land\ RT(v)$
$\qquad \land\ \forall r \in \mathbf{R} : [x[r] = 0 + \int G(r) \,|\, \text{false}, v]$
$\qquad \land\ \forall r \in \mathbf{R} : \Box(x[r] \leq 4)$
$\qquad \Rightarrow\ ([g = 0 + \int accumRate \,|\, \text{false}, v] \Rightarrow \Box(g < MaxCon))$

PROOF: From $\langle 1 \rangle 2$ by propositional logic.

$\quad \langle 2 \rangle 2.$ $\land\ now = a$
$\qquad \land\ RT(v)$
$\qquad \land\ \forall r \in \mathbf{R} : [x[r] = 0 + \int G(r) \,|\, \text{false}, v]$
$\qquad \land\ \forall r \in \mathbf{R} : \Box(x[r] \leq 4)$
$\qquad \Rightarrow\ Req0$

PROOF: $\langle 2 \rangle 1$, the definition of $Req0$, and predicate logic, since g does not occur to the left of the $\Rightarrow$.

$\quad \langle 2 \rangle 3.$ $\land\ now = a$
$\qquad \land\ RT(v) \land Req1$
$\qquad \Rightarrow\ Req0$

PROOF: By $\langle 1 \rangle 1$, $\langle 2 \rangle 2$, the definition of $Req1$, and simple predicate logic, since x does not occur free in $RT(v)$ or $Req0$.

$\quad \langle 2 \rangle 4.$ Q.E.D.

PROOF: From $\langle 2 \rangle 3$ by simple predicate logic, using the validity of $\exists a : now = a$.

Because it involves quantification over flexible variables, this part of the proof cannot be handled by TLP, the mechanical verification system for TLA based on the LP theorem prover [4].[7] The rest of the proof can, in principle, be verified using TLP. However, TLP does not yet contain the full definitional capability of TLA$^+$, so many of the definitions would have to be expanded by hand. In addition to being a possible source of error, this translation might make the

---

[7] Mechanical checking of this kind of reasoning with quantifiers is not hard. However, such reasoning is always so simple that we have not felt mechanical verification to be worth the effort.

formulas so long that verification would be impractical. We hope that future versions of TLP will permit the necessary definitions.

We now prove ⟨1⟩2. From the definitions in modules *HybridSystems* and *RealTime*, we can write

$$[x[r] = 0 + \int G(r) \,|\, false, v] \triangleq (x[r] = 0) \wedge \Box[\mathcal{N}_x(r)]_{\langle now, x[r], v \rangle}$$

$$[g = 0 + \int accumRate \,|\, false, v] \triangleq (g = 0) \wedge \Box[\mathcal{N}_g]_{\langle now, g, v \rangle}$$

$$RT(v) \triangleq (now \in \mathbf{R}) \wedge \Box[\mathcal{N}_{now}]_{\langle now \rangle}$$

for the appropriate actions $\mathcal{N}_x(r)$, $\mathcal{N}_g$, and $\mathcal{N}_{now}$. Let

$$Init \triangleq (now = a) \wedge (g = 0) \wedge (now \in \mathbf{R}) \wedge (\forall r \in \mathbf{R} : x[r] = 0)$$

$$\mathcal{N} \triangleq [\mathcal{N}_g]_{\langle now, g, v \rangle} \wedge [\mathcal{N}_{now}]_{\langle now \rangle} \wedge (\forall r \in \mathbf{R} : [\mathcal{N}_x(r)]_{\langle now, x[r], v \rangle})$$

$$J \triangleq \forall r \in \mathbf{R} : x[r] \le 4$$

Using the commutativity of conjunction and the fact that conjunction distributes over $\Box$, statement ⟨1⟩2 can be rewritten as[8]

$$Init \wedge \Box\mathcal{N} \wedge \Box J \Rightarrow \Box(g < MaxCon)$$

TLA formulas of this form are proved with the aid of an invariant. First, we define the trivial "type-correctness" part of the invariant:

$$T \triangleq \wedge (now \in \mathbf{R}) \wedge (g \in \mathbf{R})$$
$$\wedge \forall r \in \mathbf{R} : (x[r] \in \mathbf{R}) \wedge (0 \le x[r])$$

The nontrivial part I of the invariant is defined as follows, where we first define $\|r\|$ to be the largest number less than or equal to r that is some multiple of 30 seconds later than a.

$$\|r\| \triangleq \mathbf{choose}\ s \in \mathbf{R} : \wedge (s \le r) \wedge (r < s + 30)$$
$$\wedge |s - a|/30 \in Nat$$
$$k \triangleq (4 * \rho)/(1 - exp[-30 * \delta])$$
$$I \triangleq g \le (k * exp[-\delta * (now - \|now\|)] + \rho * x[\|now\|])$$

The proof of ⟨1⟩2 is a standard invariance proof, where we prove $\Box I$, using the hypothesis $\Box J$. The high-level structure of the proof is:

⟨2⟩1. *Init* $\wedge \Box\mathcal{N} \Rightarrow \Box T$
  PROOF: This is a straightforward invariance proof and is omitted.
⟨2⟩2. *Init* $\wedge \Box\mathcal{N} \wedge \Box(J \wedge T) \Rightarrow \Box(g < MaxCon)$
  ⟨3⟩1. *Init* $\Rightarrow I$
    PROOF: Trivial, because *Init* implies $g = 0$.

---

[8] This notation is rather sloppy, since $\Box\mathcal{N}$ is not a syntactically correct TLA formula. We must write $\Box[\mathcal{N}]_w$ for some state function w. However, it is easy to check that $\mathcal{N}$ equals $[\mathcal{N}]_{\langle now, g, X, v \rangle}$, where $X \triangleq [r \in \mathbf{R} \mapsto x[r]]$.

⟨3⟩2. $J \wedge J' \wedge T \wedge I \wedge \mathcal{N} \Rightarrow I'$
  PROOF: Sketched below.
⟨3⟩3. $J \wedge I \Rightarrow (g \leq MaxCon)$
  PROOF: Follows by simple algebra from the definition of k and assumption ⟨0⟩.
⟨3⟩4. Q.E.D.
  ⟨4⟩1. $\Box(J \wedge T) \wedge I \wedge \Box\mathcal{N} \Rightarrow \Box I$
    PROOF: ⟨3⟩2 and the standard TLA invariance rule.
  ⟨4⟩2. $\Box(J \wedge T) \wedge Init \wedge \Box\mathcal{N} \Rightarrow \Box I$
    PROOF: ⟨3⟩1 and ⟨4⟩1.
  ⟨4⟩3. $\Box I \Rightarrow \Box(g \leq MaxCon)$
    PROOF: ⟨3⟩3 and simple temporal logic reasoning.
  ⟨4⟩4. Q.E.D.
    PROOF: ⟨4⟩2 and ⟨4⟩3.
⟨2⟩3. Q.E.D.
  PROOF: By ⟨2⟩1 and ⟨2⟩2.

We have reduced our goal to proving ⟨3⟩2, which is an assertion of ordinary mathematics, with no temporal operators. (In the reasoning, primed and unprimed variables are considered to be separate, unrelated values.) The proof of ⟨3⟩2 is sketched below. The "algebraic calculations", which are omitted, constitute the heart of the proof. They are straightforward, but writing them out at the same level of detail as we have been using would be extremely tedious. Most mechanical verifiers would probably require that they be broken into extremely small steps, though they should not be hard to check with a mathematical package like *Mathematica* or *Maple*.

⟨3⟩2. ASSUME: $J \wedge J' \wedge T \wedge I \wedge \mathcal{N}$
    PROVE: I'
  let $mid \triangleq \lfloor now \rfloor + 30$
  ⟨4⟩1. CASE: $now' = now$
    PROOF: $[\mathcal{N}_g]_{\langle now, g, v \rangle} \wedge (now' = now)$ implies $g' = g$, and $(now' = now) \wedge \forall r : [\mathcal{N}_x(r)]_{\langle now, x[r], v \rangle}$ implies $x[\lfloor now \rfloor]' = x[\lfloor now \rfloor]$, so I' equals I.
  ⟨4⟩2. CASE: $(now' \neq now) \wedge (v' = v)$
    ⟨5⟩1. $now' > now$
      PROOF: By $[\mathcal{N}_{now}]_{\langle now \rangle}$ (assumption ⟨3⟩).
    ⟨5⟩2. CASE: $Gas \wedge \neg Flame$
      let $D(s, t) \triangleq s * \exp[-\delta * t] + (\rho/\delta)(1 - \exp[-\delta * t])$
      ⟨6⟩1. $g' = D(g, now' - now)$
        PROOF: By $[\mathcal{N}_g]_{\langle now, g, v \rangle}$ and Theorem *DiffusionSolution* of module *Exponentials*.
      ⟨6⟩2. CASE: $\lfloor now' \rfloor = \lfloor now \rfloor$
        ⟨7⟩1. $x'[\lfloor now' \rfloor] = x[\lfloor now \rfloor] + (now' - now)$
          PROOF: Case assumptions ⟨5⟩ and ⟨6⟩, $[\mathcal{N}_x(r)]_{\langle now, x[r], v \rangle}$ with $r = \lfloor now \rfloor$, and Theorem *IntegralOfStepFcn* of module *Integration*, since $now \in [\lfloor now \rfloor, \lfloor now \rfloor + 30]$.

⟨7⟩2. Q.E.D.
PROOF: Algebraic calculation, using ⟨6⟩1, ⟨7⟩1, case assumption ⟨5⟩,
and Theorem *ExpFacts* of module *Exponentials*.

⟨6⟩3. CASE: $\lVert now' \rVert \neq \lVert now \rVert$

⟨7⟩1. $g' = D(D(g, mid - now), now' - mid)$
PROOF: By ⟨6⟩1, since an algebraic calculation using Theorem *Exp-
Facts* shows that $D(D(s, t_1), t_2) = D(s, t_1 + t_2)$ for any $s, t_1, t_2 \in \mathbf{R}$.

⟨7⟩2. $x'[\lVert now \rVert] = x[\lVert now \rVert] + mid - now$
PROOF: Case assumption ⟨4⟩, $[\mathcal{N}_x(r)]_{\langle now, x[r], v \rangle}$ with $r = \lVert now \rVert$,
and Theorem *IntegralOfStepFcn*.

⟨7⟩3. $D(g, mid - now) \leq k$
PROOF: Algebraic calculation using ⟨7⟩2, $J' \wedge I$ (Assumption ⟨3⟩),
and Theorem *ExpFacts*.

⟨7⟩4. $x'[mid] = now' - mid$

⟨8⟩1. $x'[mid] = \min(now' - mid, 30)$
PROOF: Case assumption ⟨6⟩, $[\mathcal{N}_x(r)]_{\langle now, x[r], v \rangle}$ with $r = mid$, and
Theorem *IntegralOfStepFcn*.

⟨8⟩2. Q.E.D.
PROOF: ⟨8⟩1 and J' (Assumption ⟨3⟩).

⟨7⟩5. $\lVert now' \rVert = mid$

⟨8⟩1. $(now' - mid) \leq 4$
PROOF: ⟨7⟩4 and J' (assumption ⟨3⟩).

⟨8⟩2. Q.E.D.
PROOF: ⟨5⟩1, ⟨8⟩1, and the definition of *mid*.

⟨7⟩6. Q.E.D.
PROOF: Algebraic calculation using ⟨7⟩1, ⟨7⟩3, ⟨7⟩4, ⟨7⟩5, I (assumption ⟨3⟩), and Theorem *ExpFacts*.

⟨6⟩4. Q.E.D.
PROOF: ⟨6⟩2 and ⟨6⟩3.

⟨5⟩3. CASE: $\neg(Gas \wedge \neg Flame)$

⟨6⟩1. $g' = g * \exp[-\delta * (now' - now)]$
PROOF: By $[\mathcal{N}_g]_{\langle now, g, v \rangle}$ and Theorem *DiffusionSolution*.

⟨6⟩2. CASE: $\lVert now' \rVert = \lVert now \rVert$

⟨7⟩1. $x'[\lVert now \rVert] = x[\lVert now \rVert]$
PROOF: By $[\mathcal{N}_x(r)]_{\langle now, x[r], v \rangle}$ with $r = \lVert now \rVert$, case assumption ⟨6⟩,
and Theorem *IntegralOfStepFcn*.

⟨7⟩2. Q.E.D.
PROOF: Algebraic calculation using ⟨5⟩1, ⟨6⟩1, ⟨7⟩1, I, and Theorem
ExpFacts.

⟨6⟩3. CASE: $\lVert now' \rVert \neq \lVert now \rVert$

⟨7⟩1. $mid \leq \lVert now' \rVert$
PROOF: ⟨5⟩1 and case assumption ⟨6⟩.

⟨7⟩2. $g' = g * \exp[-\delta * (mid - now)] * \exp[-\delta * (now' - mid)]$
PROOF: ⟨6⟩1 and Theorem *ExpFacts*.

$\langle 7 \rangle 3. \;\; g * \exp[-\delta * (mid - now)] \leq k$

    PROOF: Algebraic calculation, using I, J (which implies $x[\![now]\!]] \leq 4$), and Theorem *ExpFacts* (since $mid > now$).

$\langle 7 \rangle 4. \;\;$ Q.E.D.

    PROOF: Algebraic calculation, using $\langle 7 \rangle 1$, $\langle 7 \rangle 2$, $\langle 7 \rangle 3$, T, and Theorem *ExpFacts*.

$\langle 6 \rangle 4. \;\;$ Q.E.D.

    PROOF: $\langle 6 \rangle 2$ and $\langle 6 \rangle 3$.

$\langle 5 \rangle 4. \;\;$ Q.E.D.

    PROOF: $\langle 5 \rangle 2$ and $\langle 5 \rangle 3$.

$\langle 4 \rangle 3. \;\;$ Q.E.D.

PROOF: $\langle 4 \rangle 1$, $\langle 4 \rangle 2$, and assumption $\langle 3 \rangle$, since $[\mathcal{N}_{now}]_{now} \wedge (now' \neq now)$ implies $v' = v$, by definition of $\mathcal{N}_{now}$.

6 An Implementation

We now specify an implementation of the gas burner inspired by RRH's "control model". RRH also specify an "architecture". However, a comparison of Figures 2 and 3 of [8] reveals that this architecture is just the same implementation expressed in a CSP-like language. A program in a toy language is no closer to a real program than the corresponding TLA$^+$ or Duration Calculus formula is; it just has a syntax that can fool some people into thinking it is closer. So, we see no reason to introduce such a toy programming language.

The control model is described by the state-transition diagram of Figure 10. In the RRH specification, the states of the control model are abstract states that are only loosely coupled with states of the physical variables. For example, the specification of RRH asserts that the ignition and gas are turned on after the Ignite1 state is reached. The easiest way to duplicate the RRH specification would be to translate the state-machine into a simple TLA$^+$ formula, and then translate the "phase requirements" of RRH into TLA$^+$ using the operators

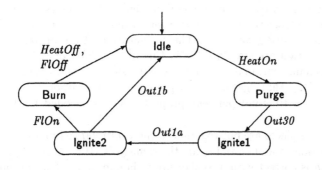

Fig. 10. The control model, adapted from Figures 2 and 3 of [8].

from the *HybridSystems* module. However, we have already presented one level of specification and verification using the *HybridSystems* module's operators, so this would be more of the same. Moreover, this is not the type of specification one would naturally write with TLA. In TLA, it is more natural to write lower level specifications purely in terms of states and transitions. One would draw a state-transition diagram for the *gas* and *ignition* variables—the transitions being enabled and disabled by changes to the control state—and write a TLA$^+$ specification of the resulting system. This representation would be the first step in refining the control model to a realistic representation of an actual implementation. However, developing this specification would take more space than seems appropriate. Instead, we present a simplified version in which the physical variables *gas* and *ignition* that are under the implementer's control change simultaneously with the control state.

The module *BurnerControl* of Figure 11 describes the time-independent behavior of the burner. It is a standard TLA specification; the next-state relation $\mathcal{N}$ is specified as the disjunction of individual actions corresponding to the transitions in Figure 10. Note that this module has a new parameter—the "internal" variable *state*. (Internal variables are ones that are existentially quantified.) The only surprising part of the specification is that the *Out1b* action can occur even if the flame is on. Timing requirements will prevent this action from occurring if the flame ignites quickly enough.

Next, module *BurnerPhaseReqs* of Figure 12 specifies the real-time requirements on the burner. It contains minimum and maximum delays on each of the transitions in Figure 10, which are expressed with the operators from the *RealTime* module. The constraints, and the new parameters ϵ_1 and ϵ_2, are taken directly from the RRH specification. RRH's condition *Ignite2Req* is split into two parts. The first, *Ignite2aReq*, asserts that an *Out1b* action cannot occur until it has been enabled (*state* = "Ignite2") for at least 1 second, and it must occur if it has been enabled for $1 + \epsilon_1$ seconds. (The formula *Control* asserts that the *Out1b* action remains enabled until an *Out1b* or *FlOn* step occurs.) The second part, *Ignite2bReq*, asserts that a *FlOn* action must be continuously enabled (*state* = "Ignite2" and flame on) for at least ϵ_2 seconds before it can occur, and it must occur if it has been continuously enabled for ϵ_1 seconds.

Next, module *BurnerEnvironment* in Figure 13 specifies the environment. The burner controls the *gas* and *ignition* variables; the environment controls the *flame* and *heatReq* variables. The specification of the initial condition and next-state relation are standard TLA. Condition ASM_1 of RRH, that "no gas results in no flame within 0.1 seconds," is expressed as usual with a timer. RRH's condition ASM_2, that "gas does not ignite when the ignition transformer is not operating," is a time-independent property that is encoded in the next-state relation $\mathcal{N}$.

Finally, the pieces are put together in the *ControlModel* module of Figure 14. The assumptions about ϵ_1 and ϵ_2 are the same as RRH's, except that our simpler specification is correct with a weaker upper bound on ϵ_1. The statement

include *BurnerEnvironment* **as** *Env*

```
┌────────────────── module BurnerControl ──────────────────┐
│ import BurnerParameters                                    │
├───────────────────────────────────────────────────────────┤
│ parameter                                                  │
│   state : variable                                         │
├───────────────────────────────────────────────────────────┤
```

predicate
$w \triangleq \langle gas, ignition, state \rangle$
$Init \triangleq w = \langle \text{"off"}, \text{"off"}, \text{"Idle"} \rangle$
actions
$HeatOn \triangleq \land HeatReq \land (state = \text{"Idle"})$
$\qquad\qquad \land (state' = \text{"Purge"}) \land \textbf{unchanged } \langle gas, ignition \rangle$

$Out30 \triangleq \land (state = \text{"Purge"}) \land (state' = \text{"Ignite1"})$
$\qquad\qquad \land \textbf{unchanged } \langle gas, ignition \rangle$

$Out1a \triangleq \land (state = \text{"Ignite1"}) \land (state' = \text{"Ignite2"})$
$\qquad\qquad \land (ignition' = \text{"on"}) \land (gas' = \text{"on"})$

$Out1b \triangleq \land (state = \text{"Ignite2"}) \land (state' = \text{"Idle"})$
$\qquad\qquad \land (ignition' = \text{"off"}) \land (gas' = \text{"off"})$

$FlOn \triangleq \land (state = \text{"Ignite2"}) \land Flame \land (state' = \text{"Burn"})$
$\qquad\qquad \land (ignition' = \text{"off"}) \land \textbf{unchanged } gas$

$GoIdle \triangleq \land (state = \text{"Burn"}) \land (\neg Flame \lor \neg HeatReq) \land (state' = \text{"Idle"})$
$\qquad\qquad \land (gas' = \text{"off"}) \land \textbf{unchanged } ignition$

$\mathcal{N} \triangleq HeatOn \lor Out30 \lor Out1a \lor Out1b \lor FlOn \lor GoIdle$
temporal
$Control \triangleq Init \land \Box[\mathcal{N}]_w$

Fig. 11. The Control Actions.

is equivalent to inserting all the definitions from the *BurnerEnvironment* module, with the names of the defined symbols preceded by "*Env.*".[9] Similarly, the statement

$$\textbf{include } BurnerPhaseReqs \textbf{ as } Burner(state)$$

is equivalent to including all the definitions of the *BurnerPhaseReqs* environment, with defined symbols renamed by prefacing them with "*Burner(state).*". This makes *state* a free parameter of all those defined symbols, which must be instantiated when the symbol is used. (Note how it is instantiated with the bound variable *s* in the definition of *Spec.*) The complete specification is the conjunction of the burner's complete specification and the environment specification,

---

[9] In contrast to an imported module, whose parameters become parameters of the importing module, parameters of an included module are instantiated. In this case, there are no explicit instantiations, so the included module's parameters are by default instantiated with parameters of the same name.

---

module *BurnerPhaseReqs* ―――――

import *BurnerControl, RealTime*

---

parameters
  $\epsilon_1, \epsilon_2 : \mathbf{R}$ **constants**

---

temporal
  $IdleReq \triangleq \exists x, y : \wedge\ VTimer(x, HeatOn, \epsilon_2, w)\ \wedge\ MinTimer(x, HeatOn, w)$
  $\qquad\qquad\qquad \wedge\ VTimer(y, HeatOn, \epsilon_1, w)\ \wedge\ MaxTimer(y)$

  $PurgeReq \triangleq \exists x, y : \wedge\ VTimer(x, Out30, 30, w)\ \wedge\ MinTimer(x, Out30, w)$
  $\qquad\qquad\qquad \wedge\ VTimer(y, Out30, 30 + \epsilon_1, w)\ \wedge\ MaxTimer(y)$

  $Ignite1Req \triangleq \exists x, y : \wedge\ VTimer(x, Out1a, 1, w)\ \wedge\ MinTimer(x, Out1a, w)$
  $\qquad\qquad\qquad \wedge\ VTimer(y, Out1a, 1 + \epsilon_1, w)\ \wedge\ MaxTimer(y)$

  $Ignite2aReq \triangleq \exists x, y : \wedge\ VTimer(x, Out1b, 1, w)\ \wedge\ MinTimer(x, Out1b, w)$
  $\qquad\qquad\qquad \wedge\ VTimer(y, Out1b, 1 + \epsilon_1, w)\ \wedge\ MaxTimer(y)$

  $Ignite2bReq \triangleq \exists x, y : \wedge\ VTimer(x, FlOn, \epsilon_2, w)\ \wedge\ MinTimer(x, FlOn, w)$
  $\qquad\qquad\qquad \wedge\ VTimer(y, FlOn, \epsilon_1, w)\ \wedge\ MaxTimer(y)$

  $BurnReq \triangleq \exists x, y : \wedge\ VTimer(x, GoIdle, \epsilon_2, w)\ \wedge\ MinTimer(x, GoIdle, w)$
  $\qquad\qquad\qquad \wedge\ VTimer(y, GoIdle, \epsilon_1, w)\ \wedge\ MaxTimer(y)$

  $PhaseReqs \triangleq IdleReq \wedge PurgeReq \wedge Ignite1Req \wedge$
  $\qquad\qquad\qquad Ignite2aReq \wedge Ignite2bReq \wedge BurnReq$

Fig. 12. The control model's timing requirements.

together with the formula $RT(v)$ that describes how *now* changes and asserts that v does not change when *now* does. The burner's specification is obtained by hiding the internal state variable in the conjunction of its time-independent specification, its timing requirements, and the RT-formula asserting that the internal state does not change when *now* does.[10]

The next step is to prove that the Control Model satisfies requirements *Req1*, *Req2*, and *Req3*. This means proving that formula *Spec* of module *ControlModel* implies *Req1* $\wedge$ *Req2* $\wedge$ *Req3*. We very briefly sketch the proof that *Spec* implies *Req1*; the other proofs are similar.

We first reduce the problem to proving an assertion without quantification over flexible variables. Let *Burner(s).VControl* be the formula that is the same as *Burner(s).Control* except with the quantifiers $\exists x, y$ removed and each x and y replaced by a unique variable. For example, let *xIgnite1* and *yIgnite1* be substituted for x and y in *Ignite1Req*. Similarly, let *Env.VSpec* be the formula obtained by removing the quantifier from *Env.Spec* and substituting *xEnv* for x. Simple predicate logic shows that, to prove *Spec* $\Rightarrow$ *Req1*, it suffices to assume

---

[10] Note that, since s does not appear in $RT(v)$, the latter formula can be moved inside the quantifier; and $RT(v) \wedge RT(s)$ is equivalent to $RT(\langle v, s \rangle)$.

```
┌──────────────────── module BurnerEnvironment ────────────────────┐
│ import BurnerParameters                                            │
├───────────────────────────────────────────────────────────────────┤
│ predicate                                                         │
│   Init  ≜  ⟨flame, heatReq⟩ = ⟨"off", "off"⟩                      │
│ actions                                                           │
│   ChangeHeatReq  ≜  ∧ heatReq′ = if HeatReq then "off" else "on"  │
│                     ∧ unchanged flame                             │
│                                                                   │
│   FlameOn  ≜  ∧ ¬Flame ∧ Gas ∧ Ignition ∧ (flame′ = "on")        │
│              ∧ unchanged heatReq                                  │
│                                                                   │
│   FlameOff  ≜  Flame ∧ (flame′ = "off") ∧ unchanged heatReq       │
│   N  ≜  ChangeHeatReq ∨ FlameOn ∨ FlameOff                        │
│ temporal                                                          │
│   Spec  ≜  ∧ Init ∧ □[N]⟨flame, heatReq⟩                          │
│           ∧ ∃x : ∧ VTimer(x, FlameOff ∧ ¬Gas, 1/10, ⟨flame, heatReq⟩) │
│                  ∧ MaxTimer(x)                                    │
└───────────────────────────────────────────────────────────────────┘
```

Fig. 13. Timing assumptions on the environment.

```
┌──────────────────── module ControlModel ────────────────────┐
│ import BurnerParameters, RealTime                            │
├──────────────────────────────────────────────────────────────┤
│ parameters                                                   │
│   ε₁, ε₂ : R constants                                       │
├──────────────────────────────────────────────────────────────┤
│ assumption                                                   │
│   EpsilonAssumption  ≜  (0 < ε₁) ∧ (ε₂ < ε₁) ∧ (ε₁ ≤ 2/3)   │
├──────────────────────────────────────────────────────────────┤
│ include BurnerPhaseReqs as Burner(state)                     │
│ include BurnerEnvironment as Env                             │
│ temporal                                                     │
│   Spec  ≜  ∧ ∃s : Burner(s).Control ∧ Burner(s).PhaseReqs ∧ RT(s) │
│           ∧ Env.Spec                                         │
│           ∧ RT(v)                                            │
└──────────────────────────────────────────────────────────────┘
```

Fig. 14. The complete control model specification.

$r \in \mathbf{R}$ and prove $\Pi \Rightarrow \Box(z \leq 4)$, where

$$\Pi \;\triangleq\; \land [z = 0 + \int G(r) \,|\, \mathsf{false}, v]$$
$$\land \; Burner(s).Control \,\land\, Burner(s).VPhaseReqs$$
$$\land \; Env.VSpec$$
$$\land \; RT(s) \,\land\, RT(v)$$

and G is defined by the **let** in the definition of *Req1*.

The proof of $\Pi \Rightarrow \Box(z \leq 4)$ has the same form as the proof of step $\langle 1 \rangle 2$ in the proof that $RT(v) \land Req1$ implies *Req0*. We first prove $\Pi \Rightarrow \Box T$ for a simple invariant T, which expresses the expected relations among the values of the variables. For example, one conjunct of T is

$$(s = \text{``Ignite1''}) \Rightarrow \land \; Gas \land Ignition$$
$$\land \; (yIgnite1 \in \mathbf{R}) \land (now \leq yIgnite1)$$

We then prove $\Pi \land \Box T \Rightarrow \Box I$ for the following invariant I.

$$\land \; z \leq 4$$
$$\land \; (z > 0) \Rightarrow (z \leq now - r)$$
$$\land \; (now \in [r, r + 30]) \Rightarrow \lor \; (s = \text{``Idle''})$$
$$\lor \; (s = \text{``Purge''}) \land ((z > 0) \Rightarrow (xPurge \geq r + 30))$$
$$\lor \; (s = \text{``Ignite1''}) \land (z + (yIgnite1 - now) \leq 1 + \epsilon_1)$$
$$\lor \; (s = \text{``Ignite2''}) \land (z + yIgnite2a - now \leq 2 + 2\epsilon_1)$$
$$\lor \land \; (s = \text{``Burn''}) \land Gas$$
$$\land \; \lor \; Flame \land (z \leq 2 + 2\epsilon_1)$$
$$\lor \; \neg Flame \land (z + yBurn - now \leq 2 + 3\epsilon_1)$$

The hard part of proving $\Pi \land \Box T \Rightarrow \Box I$ is the analog of step $\langle 3 \rangle 2$ in the proof that $RT(v) \land Req1$ implies *Req0*. As usual, this step involves ordinary mathematics, with no temporal operators.

References

1. Martín Abadi and Leslie Lamport. An old-fashioned recipe for real time. Research Report 91, Digital Equipment Corporation Systems Research Center, 1992. An earlier version, without proofs, appeared in [3, pages 1–27].
2. K. Mani Chandy and Jayadev Misra. *Parallel Program Design*. Addison-Wesley, Reading, Massachusetts, 1988.
3. J. W. de Bakker, C. Huizing, W. P. de Roever, and G. Rozenberg, editors. *Real-Time: Theory in Practice*, volume 600 of *Lecture Notes in Computer Science*. Springer-Verlag, Berlin, 1992. Proceedings of a REX Real-Time Workshop, held in The Netherlands in June, 1991.
4. Urban Engberg, Peter Grønning, and Leslie Lamport. Mechanical verification of concurrent systems with tla. In *Logics of Programs*, Lecture Notes in Computer Science, Berlin, Heidelberg, New York, June 1992. Springer-Verlag.
5. Leslie Lamport. The temporal logic of actions. Research Report 79, Digital Equipment Corporation, Systems Research Center, December 1991.

6. A. C. Leisenring. *Mathematical Logic and Hilbert's ε-Symbol.* Gordon and Breach, New York, 1969.
7. Zohar Manna and Amir Pnueli. *The Temporal Logic of Concurrent Systems.* Springer-Verlag, New York, 1991.
8. Anders P. Ravn, Hans Rischel, and Kirsten M. Hansen. Specifying and verifying requirements of real-time systems. *IEEE Transactions on Software Engineering,* January 1993. to appear.

Hybrid Models with Fairness and Distributed Clocks

Reino Kurki-Suonio

Tampere University of Technology
Software Systems Laboratory
P. O. Box 553, SF-33101 Tampere, Finland

Abstract. Explicit clocks provide a well-known possibility to introduce time into non-real-time theories of reactive systems. This technique is applied here to an approach where distributed systems are modeled with temporal logic of actions as the formal basis, and fairness as the basic force that makes events take place. The focus of the paper is on the formalization and practical proof of hybrid properties of the form "at every moment of time t, $\Phi(t)$ holds for X," where X is a set of objects with distributed clocks, and Φ is a predicate that depends both on the discrete states of $x \in X$ and on time t. The approach is illustrated by a treatment of two well-known examples from the hybrid system literature.

1 Introduction

The background of the paper is in *joint action systems* [2, 3], and in their use as language basis and in incremental modeling of reactive systems [7, 8]. As the underlying formal basis we use Lamport's *temporal logic of actions* (TLA) [13]. In this paper the joint action approach is extended to deal with hybrid systems.

The main principles of modeling that we follow are the following:

- A reactive systems is modeled as a *closed system*, which models the combined behavior of the system and its environment.
- In an operational interpretation, *fairness* is used as the underlying execution force without which nothing needs to happen.
- Time is recorded in *explicit clocks*.
- A system consists of *distributed objects* with separate clocks that are synchronized only when the objects jointly execute actions.
- Clocks are updated in ordinary state-transforming actions; no special time actions are utilized.

Although fairness is essential for operational modeling of (non-real-time) liveness properties in temporal logic [13, 16], its use has often been questioned in connection with real time (see e.g. [14]). Instead, a "real-time execution force" is usually obtained from the assumptions that time inevitably proceeds without any bound (non-Zeno postulate), and that computing resources do not stay idle when there is something that they can do (maximal parallelism principle) [6, 14].

With the explicit clock approach it is possible to preserve the role of fairness also in the presence of real time. From an operational viewpoint this can be viewed as reversing the causality between proceeding of time and execution of actions. For a detailed analysis of the relationship between fairness assumptions and real-time properties, see [1].

As shown in [1], no conceptual additions are needed when an explicit clock is used to extend the use of TLA from non-timed to real-time specifications. In [12] it is shown further, how TLA can be used to deal with hybrid systems whose properties depend on continuous functions of time. In this paper we follow similar lines of thought. Distributed clocks, and updating them in ordinary state-transforming actions (instead of special time actions) distinguish our approach from [1, 12] and from other related approaches like [15]. Here we investigate how this affects dealing with properties of hybrid systems. The main contributions are in the analysis of how to formalize and prove properties of the form "$\Phi(t)$ holds for X at all time instances t," where X is a collection of distributed objects with separate clocks, and Φ is a predicate that depends both on the discrete states of $x \in X$ and on time t.

We start Section 2 by a brief introduction to joint action systems, presenting them in a form that is suitable for the purposes of this paper. On non-timed systems we then superpose clocks in a manner that is a generalization of the construction for timed systems in [10, 11]. Section 3 addresses real-time invariance properties in hybrid systems, and these are illustrated in Sections 4 and 5 by a treatment of two problems from the literature, a gas burner [4, 5, 17, 12], and a cat-and-mouse problem [9, 15, 17]. The paper ends in Section 6 with some concluding remarks about the approach presented.

2 From Non-Timed to Timed Joint Action Systems

2.1 Non-Timed Systems

We start with an informal description of non-timed *joint action systems*, presenting them in a form that is suitable for the purposes of this paper. For their use as a basis for a specification language and an incremental design method, the reader is referred to [3, 7, 8].

Two kinds of entities are involved in joint action systems, *objects* and *actions*. An object can be understood as a collection of variables indicating the current local state of the object. The (global) state of a system consists of the local states of its objects. In the real world, an object corresponds to an entity either in the reactive system itself, or in the environment with which it interacts.

Actions describe the events that can take place in a system, and they are assumed to be atomic. In other words, once started, the execution of an action is always completed without interruption or interference by other actions. The following format will be used to describe them:

action_name[participants](parameters) : action_expression,

where

- *participants* is a list of objects that are involved,
- *parameters* is a list of parameters used in the action, and

- *action_expression* is a relation between the values that the variables of the participants have before the action (unprimed names) and after its execution (primed names).

By convention, we express an action expression A in the form $A = G \wedge B$, where G (the guard or enabling condition) is a precondition that involves only unprimed variables and parameters, and B (the body) is a collection of equations that define the new values of variables. Default equations $v' = v$ are implicitly assumed for all those variables v for which no new values are explicitly specified at the current level of abstraction.

In an operational interpretation an action is said to be executed jointly by its participants. As a closed system a joint action system specifies joint behaviors of a reactive system and its environment. Separating between the system and the environment in a model means that the responsibility for each action is assigned to either party. Our use of the term "joint action" therefore does not imply shared responsibility by the system and its environment.

An *interleaving model* is assumed, where computations are alternating sequences of states and actions, starting from an appropriate initial state. At any point of execution, the next action is selected nondeterministically from those that are then enabled. Admissible computations are also restricted by system-specific *fairness* assumptions by which such computations can be excluded where a given action is enabled infinitely often (or in the final state) but executed only finitely many times. Notice that fairness requirements not only restrict the infinite computations that are possible; in their absence a computation could also terminate at any point.

As the underlying formal basis we assume TLA [13]. Compared to this, joint action systems are closer to an execution model for a programming language. The notions of objects and action participants are intended to support distributed systems. The purpose of action parameters is to separate internal nondeterminism of actions so that deterministic bodies can be used to describe the implementable effects of actions.

Basically, TLA is a version of linear-time temporal logic where the validity of expressions is insensitive to addition and/or deletion of stuttering. Instead of using the next state operator, action expressions $\mathcal{A}$ can appear with temporal operators, but only in contexts like

$$\Box[\mathcal{A}]_X,$$

expressing that all state changes either satisfy $\mathcal{A}$ or leave all variables in X unchanged, or

$$\Diamond \langle \mathcal{A} \rangle_X,$$

expressing that there eventually is a state change that satisfies $\mathcal{A}$ and does change some variable in X. In the following we will always assume that $\mathcal{A}$ involves variables in X only.

Superposition is a transformation technique that can easily be supported for joint action systems. It allows enriching the global state in a way that, in terms of TLA, guarantees that each action in the transformed system either implies a given original action, or does not affect any variables of the original state. This means that all safety properties of the original system are preserved, which makes superposition suitable for incremental derivation of specifications.

In transforming joint action systems by superposition, we adopt here the convention of using subscripted action names to indicate the current level of refinement. For instance, an action

$$a_0[x, y] : G_0 \wedge B_0$$

could be refined in a superposition step into

$$a_1[x, y](s) : a_0[x, y] \wedge G_1 \wedge B_1,$$

where s is a new parameter, G_1 gives additional constraints for the guard, and B_1 indicates the new values for the variables that have been added to the participants. When an explicit refinement is not given for an action, an implicit refinement is assumed where $v' = v$ for all new variables in the participants.

2.2 Superposition of Distributed Clocks

Timed joint action systems are obtained by appending time to non-timed systems. We will do this in terms of distributed clocks that are updated in connection with the actions.

A universal time domain of non-negative reals is assumed, into which each execution of an action is mapped. In the presence of distributed clocks we assume a separate variable $x.clock$ in each object x, all initialized as 0. Whenever an action is executed, the local clocks $x.clock$ of its participants are synchronized, i.e., the start time of the action is taken as some value not less than the maximum of the times shown by these clocks. The clocks are subsequently updated to some values that are not less than this start time. As a result, variable $x.clock$ always shows the time when object x last participated in an action.

Such a timed version of a joint action system can be obtained from a non-timed system by simple superposition. To illustrate this, let a_0 be a non-timed action with two participants x and y:

$$a_0[x, y] : A.$$

In the timed version it gets the form

$$
\begin{aligned}
a_1[x, y](s, d_x, d_y) : \ & a_0[x, y] \\
& \wedge s \geq \max(x.clock, y.clock) \\
& \wedge d_x \geq 0 \wedge d_y \geq 0 \\
& \wedge x.clock' = s + d_x \\
& \wedge y.clock' = s + d_y,
\end{aligned}
$$

where s, d_x, and d_y are parameters standing for the start time of the action, and for its duration for the two participants. By convention we shall in the following use subscripts 0 for non-timed actions, and subscripts 1 for the level of abstraction where timings have been superposed in this non-restrictive manner.

Notice that with this superposition of distributed clocks, the mutual order of action occurrences in an interleaved computation need not conform to their real-time order. An interleaved sequence can, however, always be interpreted as a possible observation sequence of a partial-order computation, where each object determines the mutual order of those action occurrences in which it participates.

Obviously, all safety and liveness properties of the original system are preserved by this superposition. The timing and scheduling of actions can be restricted by strengthening the guards of actions with further constraints on start times and durations. This may, of course, rule out some interleaved computations of the original non-timed system, i.e., introduce new state properties.

The above superposition ensures that all clocks in the system are monotonically non-decreasing. They need not be diverging, however. As divergence is a natural property of real time, we formalize it by defining $x.clock$ to be *finitely varying* if

$$(\Box\Diamond\langle true\rangle_x) \Rightarrow (\forall t \in \mathbf{R} : \Diamond x.clock > t).$$

Since $\langle true\rangle_x$ stands for an arbitrary state change in x, this expresses the condition that the state of x cannot change infinitely often without its local clock growing unboundedly.

Compared to approaches with only instantaneous state-transforming actions, and separate time actions for advancing the clock [1], we notice that the role of fairness is here stronger in the following sense. Consider a situation where state action a would be the one to be executed next, but its start time s has a lower bound that has not yet been reached by the clock. With separate time actions it would then be possible for the clock to be incremented infinitely many times without ever reaching this bound, which means that no fairness requirement on a could force its execution. In our approach the situation is different, since a would stay continually enabled, and fairness would eventually force its execution with an appropriate start time s.

In [1] a safety specification is called *nonZeno*, if it allows to extend every finite prefix of a computation into one where time diverges. In nonZeno specifications the possibility for computations with non-diverging time causes no problems, since they can be assumed to be excluded by the reality, anyway. In other words, nonZeno systems are ones for which finite variability is an additional fairness-like condition that is not in conflict with its safety properties, and whose enforcement would not add any new safety properties, either. Hence, finite variability is not necessary for a real-time specification to be reasonable. In our approach the weaker requirement of nonZenoness would mean that finite variability is achievable by further fairness assumptions.

3   Invariance Properties in Hybrid Systems

By hybrid systems we understand timed systems whose behavior and relevant properties depend both on the discrete state of the system and on continuous time-dependent functions. In contrast to [15], for instance, "continuous variables" are not considered to be part of the system state. There may be actions in which their current values are synchronized with state variables but, as such, they "exist" only as time-dependent functions of the state. In this section we analyze how the addition of time-dependent state functions affects the formulation and proof of invariance properties.

The construction in Section 2.2 is a generalization of the timed systems in [10, 11], where the start time parameter s was always implicitly assumed to be the maximum reading of the clocks involved. For hardware/software systems such "undelayed scheduling" of actions was a natural assumption, but the situation is different for events whose enabling depends on continuous functions of time. On the other hand, the duration parameters d_x were convenient for the modeling of atomic but durational actions, which is also useful for distributed hardware/software systems. For physical events in the environment, and for other actions that depend on continuous functions, non-instantaneous atomicity seems, however, unrealistic. For this reason we restrict the following discussion to situations where all durations are always assumed to be 0 and are therefore omitted from actions.

Let Φ now be a predicate that involves both time, denoted by t, and the states of objects $x \in X$. We are interested in formalizing the property that "$\Phi(t)$ is always true for X." Informally, this property is violated by a computation at time t, if objects $x \in X$ could be synchronized at a time t when $\Phi(t)$ is false.

The nature of such real-time invariance properties in timed action systems is not entirely trivial. For instance, one should notice that "$\Phi(t)$ is always true for X" does not imply that an additional action with guard $\neg\Phi(s)$ could not be executed. In the following we will analyze such invariance properties separately for a singleton set X and for the general case. This separation in the treatment is justified by the fact that, with a singleton X, the effects of the fairness-based execution model can be seen in isolation from the effects of distributed clocks.

3.1 The Case of a Single Clock

We start with the case of one clock. That is, the set of objects involved in Φ is a singleton set $X = \{x\}$ with only one clock variable $x.clock$. In each state of x, time t may take values $t \geq x.clock$ and, for any such value of t, $\Phi(t)$ is then either true or false.

The following definition formalizes the statement "$\Phi(t)$ is always true for X" for a singleton X:

Definition 1 *For a singleton X and a time-dependent state predicate Φ,*

$$\Box_{X,t}\Phi \;\triangleq\; \Box(\forall t \geq x.clock : \neg\Phi(t) \Rightarrow \Diamond\langle x.clock' < t\rangle_X).$$

This definition expresses that Φ is always prevented from becoming false in the current state of x by another action modifying x before this could happen. In particular, $\Phi(x.clock)$ must be true for each state, and, if Φ contains no free occurrences of t, then $\Box_{X,t}\Phi$ reduces to $\Box\Phi$. Notice that, in general, $\Box_{X,t}\Phi$ is not a pure safety property, and that it does not require $x.clock$ to be finitely varying.

It is often the case that Φ is based on functions that are linear or otherwise monotonic in t. This leads to considering the special case where, in each state, the value of $\Phi(t)$ may change only once. Preparing already for non-singleton X also, we define:

Definition 2 *Let*

$$NonDecreasing(\Phi) \overset{\Delta}{=} (\max_{x \in X}(x.clock) \leq t_1 < t_2) \Rightarrow \neg\Phi(t_1) \vee \Phi(t_2),$$

$$NonIncreasing(\Phi) \overset{\Delta}{=} (\max_{x \in X}(x.clock) \leq t_1 < t_2) \Rightarrow \Phi(t_1) \vee \neg\Phi(t_2).$$

A time-dependent state predicate Φ is then called piecewise monotonic, *if*

$$\Box(NonDecreasing(\Phi) \vee NonIncreasing(\Phi)).$$

Letting Φ' denote Φ with all state variables replaced by the corresponding primed variables, we now have:

Proposition 1 *For a singleton X and a piecewise monotonic time-dependent state predicate Φ, $\Box_{X,t}\Phi$ can be expressed as*

$$\Box_{X,t}\Phi = \Box\Phi(x.clock)$$
$$\wedge \Box[\neg NonDecreasing(\Phi) \Rightarrow \Phi(x.clock')]_X$$
$$\wedge \Box(\neg NonDecreasing(\Phi) \Rightarrow \Diamond\langle true \rangle_X).$$

In other words, this requires that $\Phi(x.clock)$ is invariantly true, $\Phi(x.clock')$ holds for each action, and the system cannot halt in a state where $\Phi(t)$ would get false for some future values of t.

An important special case is the one where Φ cannot turn from false to true in any action. Intuitively this is the case when Φ depends only on some continuous functions in the environment. Preparing again for non-singleton X also, we define:

Definition 3 *A time-dependent state predicate Φ is called* regular, *if*

$$\Box[\Phi'(\max_{x \in X}(x.clock')) \Rightarrow \Phi(\max_{x \in X}(x.clock'))]_X.$$

For piecewise monotonic and regular predicates, $\Box_{X,t}\Phi$ can obviously be simplified as follows:

Proposition 2 *For a singleton X and a piecewise monotonic and regular time-dependent state predicate Φ, $\Box_{X,t}\Phi$ can be expressed as*

$$\Box_{X,t}\Phi = \Box\Phi(x.clock)$$
$$\wedge \Box(\neg NonDecreasing(\Phi) \Rightarrow \Diamond\langle true \rangle_X).$$

3.2 Distributed Clocks

With a common clock, actions in interleaved behaviors are always in a non-decreasing order of the readings of this clock. With distributed clocks and temporally independent actions the relationship between interleaved behaviors and real time is more complex. In particular, some global states that appear in interleaved behaviors may be ones that could never be observed in real time. This adds another non-trivial aspect to the real-time invariance properties that we consider.

Let $X = \{x_i\}$ be a non-empty collection of objects, and let Φ be a time-dependent state predicate involving their local states. In any state, $\Phi(t)$ makes sense only for time values $t \geq \max_{x \in X}(x.clock)$ and, for any such value of t, $\Phi(t)$ is then either true or false. The definition of $\Box_{X,t}\Phi$ now generalizes into:

Definition 4 *For a non-empty X and a time-dependent state predicate Φ,*

$$\Box_{X,t}\Phi \triangleq \Box(\forall t \geq \max_{x\in X}(x.clock) : \neg\Phi(t) \Rightarrow \exists x \in X : \Diamond\langle x.clock' < t\rangle_X).$$

Intuitively this means that a synchronized observation of objects in X is not possible at a time instance when $\Phi(t)$ would be false. If this would otherwise be possible, it is prevented by some $x \in X$ participating in another action at an earlier point of time. For states where all clocks $x.clock$, $x \in X$, have the same reading, the definition obviously implies $\Phi(x.clock)$.

If Φ contains no free occurrences of t, then $\Box\Phi$ implies $\Box_{X,t}\Phi$. The converse is, however, no longer true with distributed clocks. This is caused by the possibility for interleaving orders that do not conform to the real-time order, allowing even states that must always be followed by an action satisfying

$$\langle x.clock' < \max_{x\in X}(x.clock)\rangle_X$$

for some $x \in X$. We say that such states are *non-observable* for X. Their existence is harmless for $\Box_{X,t}\Phi$ in the sense that the invariant in Definition 4 is satisfied in such states for any Φ.

With no free occurrences of t in Φ, the distinction between $\Box\Phi$ and $\Box_{X,t}\Phi$ partitions state invariances in distributed systems into those that can be proved already in non-timed specifications and are therefore insensitive to timing changes, and those that essentially depend on timings. This seems to be a useful distinction in the design of distributed systems.

For piecewise monotonic predicates the condition that corresponds to the one in Proposition 1 is now sufficient but no longer necessary:

Proposition 3 *For a non-empty X and a piecewise monotonic Φ, the condition*

$$\Box\Phi(\max_{x\in X}(x.clock))$$
$$\wedge \Box[\neg NonDecreasing(\Phi) \wedge (s = \max_{x\in X}(x.clock') > \max_{x\in X}(x.clock)) \Rightarrow \Phi(s)]_X$$
$$\wedge \Box(\neg NonDecreasing(\Phi) \Rightarrow \Diamond\langle true\rangle_X)$$

implies $\Box_{X,t}\Phi$.

For regular Φ this leads to:

Proposition 4 *For a non-empty X and a piecewise monotonic and regular Φ, the condition*

$$\Box\Phi(\max_{x\in X}(x.clock))$$
$$\wedge \Box(\neg NonDecreasing(\Phi) \Rightarrow \Diamond\langle true\rangle_X).$$

implies $\Box_{X,t}\Phi$.

4 Gas-Burner Example

4.1 Non-Timed Specification

As an example, consider the modeling of a gas-burner [4, 5]. The interface, through which the system and the environment are supposed to interact, is defined to consist of four boolean variables, all initialized as false:

- *flow* indicates whether the system has opened the valve for gas flow,
- *flame* indicates whether a flame is sensed by the flame sensor,
- *request* indicates whether a heat request is on, and
- *ignition* indicates whether the ignition transformer has been started by the system.

All variables are assumed to belong to a single object, which will also be an implicit participant in all actions.

We say that the burner is leaking, when gas is flowing without a flame:

$$Leak \triangleq flow = true \land flame = false.$$

The critical requirement given in [4] for the design is that there is no continuous time interval of length T, $T \geq 60$, during which the accumulated leakage time would exceed $T/20$.

In accordance with our general approach, we start with non-timed modeling of the possible behaviors, distinguishing between actions that the environment does, and those that are on the responsibility of the system.

Basically, the environment can turn the heat request and the flame sensor on and off. We assume that the flame sensor cannot be turned on unless gas is flowing and the ignition transformer is on. The flame sensor can, however, go off while gas is flowing, which indicates flame failure. This leads to the following four environment actions:

$$request\_on_0 : request = false$$
$$\land \ request' = true,$$
$$request\_off_0 : request = true$$
$$\land \ request' = false,$$
$$flame\_on_0 : Leak$$
$$\land \ ignition = true$$
$$\land \ flame' = true,$$
$$flame\_off_0 : flame = true$$
$$\land \ flame' = false.$$

No fairness assumptions are assumed of these actions. This allows, for instance, situations where the flame sensor never goes off, which could indicate a fault in the sensor or in the valve.

To specify how the system should react to environment requests, we observe that a normal operation cycle should go through the following phases:

$$Idle \triangleq flow = false,$$

$$Ignite \triangleq \neg Idle \wedge ignition = true,$$

$$Burn \triangleq \neg Idle \wedge \neg Ignite \wedge flame = true.$$

If ignition does not succeed, phase *Burn* is skipped. In addition, flame failure leads from *Burn* to an additional fourth phase:

$$Fail \triangleq \neg Idle \wedge \neg Ignite \wedge flame = false.$$

This leads to the following system actions:

$$start_0 : Idle$$
$$\wedge\ request = true$$
$$\wedge\ flame = false$$
$$\wedge\ flow' = true$$
$$\wedge\ ignition' = true,$$
$$ignition\_off_0 : Ignite$$
$$\wedge\ flame = true$$
$$\wedge\ ignition' = false,$$
$$stop_0 : Ignite$$
$$\wedge\ flame = false$$
$$\wedge\ flow' = false$$
$$\wedge\ ignition' = false,$$
$$close\_normal_0 : Burn$$
$$\wedge\ request = false$$
$$\wedge\ flow' = false,$$
$$close\_failure_0 : Fail$$
$$\wedge\ flow' = false.$$

These actions are otherwise straightforward, but the conjunct *flame = false* may seem superfluous in the guard of *start*. It prevents, however, restarting when some fault prevents the flame sensor from going off. It is an advantage of non-timed systems that such properties can be analyzed and discussed in the absence of timing issues.

The operation cycle of the system is illustrated in Figure 1. Except for the transition from *Burn* to *Fail*, the transitions correspond to system actions only. A more detailed specification might avoid repeated flame failures by superposing service operations by the environment for a system that is *Idle*.

As for fairness assumptions, we assume fairness with respect to each system action. This means that the system eventually responds to stimuli from the environment (*request_on* in *Idle*, and *request_off* or *flame_off* in *Burn*), even though an

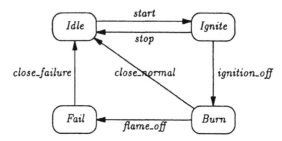

Fig. 1. Phases of a gas burner.

unbounded number of environment actions may intervene. In particular, if the heat request is repeatedly (or continually) on in *Idle*, phase *Ignite* and thereafter either *Burn* or *Idle* are eventually entered. Without further requests the system may, however, halt in either *Idle* or *Burn*.

4.2 Imposing a Real-Time Policy

In discussing the real-time properties of a gas burner we assume actions to be instantaneous, and we assume time to be superposed on the system with one clock variable *clock*.

The most obvious way to satisfy the critical requirement for gas leakage is by time bounds, a lower bound for opening the valve, and upper bounds for actions that stop a leakage. Before showing that the requirement is satisfied by certain bounds, we formulate such a policy in terms of arbitrary bounds $d_i \geq 0$.

Some auxiliary variables are needed for the policy. For simplicity we record the most recent time instances of certain events as follows: the valve was always last opened at *t_open* and closed at *t_close*, potential leakage was last stopped at *t_leak_off*, and the flame went off most recently at *t_flame_off*. All these new variables can be assumed to be initialized as 0. The general idea is to force *ignition_off* or *stop* always to take place latest at $t\_open + d_1$, *close_failure* to take place latest at $t\_flame\_off + d_2$, and *start* not to take place before $t\_leak\_off + d_3$.

For system actions this policy gives the following refinements, where actions with subscripts 0 refer to the basic timed actions with no special constraints on their start times s:

$$start_2(s) \; : \; start_1(s)$$
$$\wedge s \geq t\_leak\_off + d_3$$
$$\wedge t\_open' = s,$$
$$ignition\_off_2(s) \; : \; ignition\_off_1(s)$$
$$\wedge s \leq t\_open + d_1$$
$$\wedge t\_leak\_off' = s,$$
$$stop_2(s) \; : \; stop_1(s)$$
$$\wedge s \leq t\_open + d_1$$
$$\wedge t\_close' = t\_leak\_off' = s,$$

$$close\_normal_2(s) \ : \ close\_normal_1(s)$$
$$\wedge \ t\_close' = s,$$
$$close\_failure_2(s) \ : \ close\_failure_1(s)$$
$$\wedge \ s \leq t\_flame\_off + d_2$$
$$\wedge \ t\_close' = t\_leak\_off' = s.$$

As for environment actions, their interplay with time leads to constraints that are less intuitive. For *request_off*, for instance, one might think that no constraints are needed, as there are no restrictions for the time instance when the heat request can go off. However, if an action takes place at time s, then no subsequent action (in the interleaved sequence) can take place earlier than s. Therefore, if *request_off* takes place when there already is a reason for the system to react, then this should not prevent the proper reaction to take place within its deadline. Such reasoning leads to:

$$request\_off_2(s) \ : \ request\_off_1(s)$$
$$\wedge \ (Ignite \Rightarrow s < t\_open + d_1)$$
$$\wedge \ (Fail \Rightarrow s < t\_flame\_off + d_2),$$
$$request\_on_2(s) \ : \ request\_on_1(s)$$
$$\wedge \ (Ignite \Rightarrow s < t\_open + d_1)$$
$$\wedge \ (Fail \Rightarrow s < t\_flame\_off + d_2),$$
$$flame\_on_2(s) \ : \ flame\_on_1(s)$$
$$\wedge \ s < t\_open + d_1,$$
$$flame\_off_2(s) \ : \ flame\_off_1(s)$$
$$\wedge \ (Ignite \Rightarrow s < t\_open + d_1)$$
$$\wedge \ t\_flame\_off' = s.$$

We now check that the resulting system is finitely varying if $d_3 > 0$. First we notice that the constraints for start times do not exclude any computations of the original non-timed model; they only restrict the start times that are possible. Therefore, if an infinite number of actions is executed, fairness forces an infinite number of cycles from *Idle* to *Ignite* and back. Since *t_leak_off* is updated in each cycle, the lower bound in *start* requires each cycle to take at least d_3 time units, which proves finite variability.

4.3 Satisfying the Requirement

We now return to the critical requirement that during each continuous time interval of length T, $T \geq 60$, the accumulated leakage time L does not exceed $T/20$. In [4] it was proved in durational calculus that this requirement is satisfied if gas is never leaking continuously for longer than 1 time unit, and two consecutive leakage intervals are separated by at least 30 time units. (For flame failures this would require an instantaneous reaction by the system, i.e., $d_2 = 0$, which need not be feasible in practice.) In other words, the proof that an implementation satisfies the requirement was carried out outside a model of computations. In contrast, the corresponding proof was carried out totally within a model of hybrid computations in [17]. Here we suggest to reduce the requirement first into a piecewise monotonic and regular predicate, which simplifies the proof that needs to be carried out in a model of computations.

As formulated above, the requirement does not lead to a piecewise monotonic predicate. This is because in phase *Idle*, for instance, the predicate could first turn false by T exceeding 60, and then again true by T exceeding $20\,L$. Therefore, we reformulate the requirement as follows: During any continuous time interval of length T, if the accumulated leakage time L is at least 3, then $T \geq 20\,L$. Obviously, an interval that violates the original requirement violates also this one, and, conversely, for any interval violating this requirement there is a (possibly longer) interval that violates the original. For simplicity, we also rewrite $T \geq 20\,L$ as $N \geq 19\,L$, where N denotes the cumulative non-leakage time, $N = T - L$.

In our timed system we could express this requirement by defining two time-dependent functions $N(t)$ and $L(t)$ that would "start measuring" N and T at an arbitrary time instance, and by showing that $\Box_{X,t}\varPhi$ is true for

$$\varPhi \triangleq L(t) \geq 3 \Rightarrow N(t) - 19\,L(t) \geq 0.$$

First we notice that, independently of when the measurement is started, $\varPhi$ is piecewise monotonic and regular, and we can therefore use Proposition 2. Since $\varPhi(0)$ is trivially true in the initial state, and the system can halt only in states where $\varPhi$ is non-decreasing, it remains to show that $\varPhi(clock)$ remains true in each action. Before introducing the necessary auxiliary variables to express $\varPhi$, we will, however, introduce some further simplifications.

Since $Leak \Rightarrow Ignition \vee Fail$, it is safe to replace $L(t)$ by the accumulated time spent in phases *Ignition* and *Fail*, and $N(t)$ by the time spent in *Idle* and *Burn*. Without restriction it is also safe to start the measurement when *Ignite* is first entered, and to update the readings only at the transitions shown in Figure 1. (For any interval that violates the requirement there would be one that starts when either *Ignition* or *Fail* is entered and ends when one of them is exited. Since the system state is independent of how many times the basic cycle has been taken, it suffices to start during the first cycle. If measurements were started at an entry to *Fail*, the proof would remain essentially the same, as will be evident from the invariants that will be derived.)

It is now sufficient to prove

$$\Box(L \geq 3 \Rightarrow N - 19\,L \geq 0)$$

for the auxiliary variables i, N, and L (all initialized as 0) that are added to count the number of cycles and to measure $N(t)$ and $L(t)$ as follows:

$$
\begin{aligned}
start_3(s) \;:\; & start_2(s) \\
& \wedge N' = \text{if } i > 0 \text{ then } N + s - t\_close \text{ else } 0 \\
& \wedge i' = i + 1, \\
ignition\_off_3(s) \;:\; & ignition\_off_2(s)
\end{aligned}
$$

$$\wedge\, L' = L + s - t\_open,$$

$$stop_3(s) \;:\; stop_2(s)$$
$$\wedge\, L' = L + s - t\_open,$$

$$close\_normal_3(s) \;:\; close\_normal_2(s)$$
$$\wedge\, N' = N + s - t\_leak\_off,$$

$$flame\_off_3(s) \;:\; flame\_off_2(s)$$
$$\wedge\, N' = \dot{N} + s - t\_leak\_off,$$

$$close\_failure_3(s) \;:\; close\_failure_2(s)$$
$$\wedge\, L' = L + s - t\_flame\_off.$$

With abbreviations $l_i = i\,(d_1 + d_2)$, and $n_i = i\,d_3$, the following invariants are now straightforward:

$$Idle \Rightarrow i = N = L = 0 \vee i \geq 1 \wedge N \geq n_{i-1} + t\_close - t\_leak\_off \wedge L \leq l_i,$$
$$Ignite \Rightarrow i \geq 1 \wedge N \geq n_{i-1} \wedge L \leq l_{i-1},$$
$$Burn \vee Fail \Rightarrow i \geq 1 \wedge N \geq n_{i-1} \wedge L \leq l_i - d_2.$$

The most critical of these is the first one, with which it is easy to check whether the original requirement is satisfied for specific values of d_i. For $d_1 + d_2 \leq 1$, for instance, condition $L \geq 3$ can be satisfied in state $Idle$ only for $i \geq 3$. Then we also have

$$N - 19\,L \geq n_{i-1} - 19\,i = i\,(d_3 - 19) - d_3 \geq 3\,(d_3 - 19) - d_3 = 2\,d_3 - 57$$

(provided that $d_3 \geq 19$), which means that $N - 19\,L \geq 0$ when $d_3 \geq 28.5$.

5 Cat-and-Mouse Example

As another example we take the cat-and-mouse problem discussed in [9, 15]. At time $t = 0$ a mouse starts running with constant speed $v_m > 0$ towards a hole in the wall that is at distance x_0. At time $t = \Delta > 0$ a cat is released at the same initial position, and it starts chasing the mouse at velocity $v_c > 0$ along the same path. Either the cat catches the mouse, or the mouse escapes and the cat stops before the wall.

We describe the mouse and the cat as distinct objects m and c, with state variables $m.state \in \{rest, run, safe, caught\}$ and $c.state \in \{rest, run, win, stop\}$, initialized as $m.state = c.state = rest$. The relevant events in the system are described by non-timed actions as follows:

$$m\_start_0[m] \;:\; m.state = rest$$
$$\wedge\, m.state' = run,$$
$$c\_start_0[c] \;:\; c.state = rest$$
$$\wedge\, c.state' = run,$$
$$m\_escape_0[m] \;:\; m.state = run$$
$$\wedge\, m.state' = safe,$$
$$catch_0[c, m] \;:\; m.state = c.state = run$$

$$\land\ c.state' = win$$
$$\land\ m.state' = caught,$$
$$c\_stop_0[c]\ :\ c.state = run$$
$$\land\ c.state' = stop.$$

As for fairness assumptions, we assume fairness with respect to each individual action.

Although the system is very simple, it generates eight different interleaved computations with the following sequences of actions:

$$\langle m\_start, c\_start, catch\rangle,$$
$$\langle m\_start, c\_start, m\_escape, c\_stop\rangle,$$
$$\langle m\_start, c\_start, c\_stop, m\_escape\rangle,$$
$$\langle m\_start, m\_escape, c\_start, c\_stop\rangle,$$
$$\langle c\_start, m\_start, catch\rangle,$$
$$\langle c\_start, m\_start, m\_escape, c\_stop\rangle,$$
$$\langle c\_start, m\_start, c\_stop, m\_escape\rangle,$$
$$\langle c\_start, c\_stop, m\_start, m\_escape\rangle.$$

The situation is not really distributed in the sense that the private action c_stop of the cat, for instance, depends also on properties of the mouse. However, if all parameters of the problem (v_m, v_c, and Δ) are given as constants, and not as attributes of the two objects, then a distributed model can be accepted. Therefore we can use the example to demonstrate distributed clocks $m.clock$ and $c.clock$.

When the mouse is running, its distance from the wall can be expressed by the time-dependent function $x_m(t) = x_0 - v_m t$. Similarly, the position of the running cat is $x_c(t) = x_0 - v_c(t - \Delta)$. Assuming all actions to be instantaneous, the constraints on the basic timed model can be given as follows:

$$m\_start_2[m](s)\ :\ m\_start_1[m](s)$$
$$\land\ s = 0,$$
$$c\_start_2[c](s)\ :\ c\_start_1[c](s)$$
$$\land\ s = \Delta,$$
$$m\_escape_2[m](s)\ :\ m\_escape_1[m](s)$$
$$\land\ x_c(s) \geq x_m(s) = 0,$$
$$catch_2[c,m](s)\ :\ catch_1[c,m](s)$$
$$\land\ x_c(s) = x_m(s) \geq 0,$$
$$c\_stop_2[c](s)\ :\ c\_stop_1[c](s)$$
$$\land\ x_c(s) \geq 0 \geq x_m(s).$$

Since x_m and x_c are not part of the state but functions, their use is not restricted to the running states. The guard of m_escape, for instance, utilizes $x_c(s)$, even though the cat need not be running either in the state where m_escape appears in an interleaved sequence, or at its start time s. Notice also that with distributed time, any of the interleaved computations listed above is still possible for appropriate parameter values s.

As an example of a time-dependent state predicate in this system, consider the assertion that a running cat never passes a running mouse. Therefore, let

$$\Phi \overset{\Delta}{=} \neg((m.state = c.state = run) \wedge x_m(t) > x_c(t)),$$

with $X = \{m, c\}$. Since $x_m(t)$ and $x_c(t)$ are linear, Φ is obviously piecewise monotonic, with $\neg NonDecreasing(\Phi)$ only in state $m.state = c.state = run$, and only if $v_c > v_m$. In this state the maximum of the two clock readings is always Δ.

Obviously, Φ is also regular. Since Φ is trivially true except when both the cat and the mouse are running, it is sufficient to check that $x_m(t) \leq x_c(t)$ for $t = \Delta$, and also for those values $t = s$ with which this state could be exited, which is explicit in the guards of all such actions (m_escape, $catch$, c_stop).

To use Proposition 4 to prove $\Box_{X,t}\Phi$, $\Phi(\max(m.clock, c.clock))$ should hold invariantly, and it should not be possible for the system to halt in a state where Φ is decreasing. The former follows directly from $x_m(\Delta) < x_c(\Delta)$. For the latter we notice that no behavior can terminate with both the mouse and the cat running, since at least one of m_escape and $catch$ is then enabled, and fairness forces one of them to take place eventually.

The situation in state $m.state = c.state = run$, in which Φ is decreasing for $v_c > v_m$, is illustrated in Figure 2, with the two clock readings and the point for which $x_m(t) = x_c(t)$ shown on the time axis. If $x_m(\Delta) < 0$, action m_escape is enabled for $s < \Delta$, and the state is then non-observable.

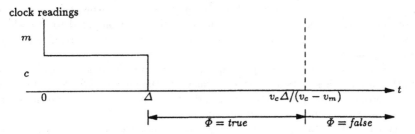

Fig. 2. Illustration of the state $m.state = c.state = run$.

6 Concluding Remarks

Having gone trough this exercise in dealing with hybrid properties in fairness-based distributed models with explicit clocks, and having seen how this affects specifications, some concluding remarks are in order.

Environment is important in any modeling of reactive systems. In general, the environment may consist of different kinds of components, including human users, other reactive systems, and the physical environment with its natural phenomena. In hybrid systems, the more demanding control aspects become dominating in dealing with the environment, emphasizing the roles of real time requirements and continuous environment variables.

Even when real-time properties are ignored, there is no consensus about the best underlying philosophy for the specification and modeling of reactive systems. With real-time and hybrid systems the situation is even more diversified. For practical design methodologies, a crucial question in this context is, whether to extend a non-real-time approach into one that can effectively deal with real-time and hybrid properties, or to look for more specialized approaches with built-in concepts for real time and other continuous variables.

The answer to this question may depend on whether the system to be designed has properties that are interesting and non-trivial already at non-timed levels of abstraction. We believe this to be the case for any system of sufficient complexity. Consequently, we feel a need for approaches where the transition between different levels of abstraction is as smooth as possible, and involves no changes in the underlying philosophy and logic. Obviously, we need more understanding of the possibilities to accomplish this. We also feel that different approaches cannot be adequately evaluated by any theoretical comparisons: their effects on specification style and design methodology are often more important than what any formal comparison can reveal. Such questions have provided the main motivation for this paper.

The starting point of the paper was in an approach that has emphasized aspects like incremental development of specifications, scalability and modularity of models, executability, and modeling of distributed systems. This has led to a fairness-based execution model with an assumption of (not necessarily instantaneous) atomic actions between distributed participants. Preparing for durational actions, which have proved convenient for other kinds of modeling situations, the proceeding of time was explicitly associated with state-transforming actions. Compared to [1], where only special time actions are allowed to advance the clock, this has an effect on specification style.

The possibility for distributed participants in actions is another major departure from other approaches to hybrid systems known to the author. Although this possibility is probably not among the first priorities in the specification and design of hybrid systems, it is important that the additional problems of distributed systems can be conveniently handled in the same general framework.

References

1. Martín Abadi and Leslie Lamport. An old-fashioned recipe for real time. In J. W. de Bakker, C. Huizing, W. P. de Roever, and G. Rozenberg, editors, *Real-Time: Theory in Practice*, volume 600 of *Lecture Notes in Computer Science*, pages 1–27. Springer-Verlag, 1992.
2. R. J. R. Back and R. Kurki-Suonio. Distributed cooperation with action systems. *ACM Trans. Programming Languages and Systems*, 10(4):513–554, October 1988.
3. R. J. R. Back and R. Kurki-Suonio. Decentralization of process nets with a centralized control. *Distributed Computing*, 3:73–87, 1989. An earlier version in *Proc. 2nd ACM SIGACT-SIGOPS Symposium on Principles of Distributed Computing*, Montreal, Canada, Aug. 1983, pages 131–142.

4. Zhou Chaochen, C. A. R. Hoare, and Anders P. Ravn. A calculus of durations. *Information Processing Letters*, 40(5):269–276, December 1991.
5. Kirsten M. Hansen, Anders P. Ravn, and Hans Rischel. Specifying and verifying requirements of real-time systems. *ACM Software Engineering Notes*, 16(5):44–54, December 1991. *Proc. ACM SIGSOFT '91, Software for Critical Systems*.
6. Jozef Hooman and Jennifer Widom. A temporal-logic based compositional proof system for real-time message passing. In E. Odijk, M. Rem, and J.-C. Syre, editors, *PARLE '89 Parallel Architectures and Languages Europe, Vol II*, volume 366 of *Lecture Notes in Computer Science*, pages 424–441. Springer-Verlag, 1989.
7. Hannu-Matti Järvinen. *The Design of a Specification Language for Reactive Systems*. Number 95 in Publications. Tampere University of Technology, Tampere, Finland, 1992. Doctoral thesis.
8. Hannu-Matti Järvinen and Reino Kurki-Suonio. DisCo specification language: marriage of actions and objects. In *Proc. 11th International Conference on Distributed Computing Systems*, pages 142–151. IEEE Computer Society Press, May 1991.
9. Y. Kesten and A. Pnueli. Timed and hybrid statecharts and their textual representation. In J. Vytopil, editor, *Formal Techniques in Real-Time and Fault-Tolerant Systems*, volume 571 of *Lecture Notes in Computer Science*, pages 591–620. Springer-Verlag, 1991.
10. Reino Kurki-Suonio. Stepwise design of real-time systems. *IEEE Transactions on Software Engineering*, January 1993. An earlier version in *Proc. ACM SIGSOFT '91, Software for Critical Systems, ACM Software Engineering Notes* 16(5):120–131, December 1991.
11. Reino Kurki-Suonio, Kari Systä, and Jüri Vain. Real-time specification and modeling with joint actions. *The Science of Computer Programming*, 1992. An earlier version in *Proc. Sixth International Workshop on Software Specification and Design*, pages 84–91. IEEE Computer Society Press, 1991.
12. Leslie Lamport. Hybrid systems in TLA$^+$. In this volume.
13. Leslie Lamport. The temporal logic of actions. Technical Report 79, Digital Systems Research Center, December 1991.
14. Leo Yuhsiang Liu and R. K. Shyamasundar. Static analysis of real-time distributed systems. *IEEE Trans. on Software Engineering*, 16(4):373–388, April 1990.
15. Oded Maler, Zohar Manna, and Amir Pnueli. From timed to hybrid systems. In J. W. de Bakker, C. Huizing, W. P. de Roever, and G. Rozenberg, editors, *Real-Time: Theory in Practice*, volume 600 of *Lecture Notes in Computer Science*, pages 447–484. Springer-Verlag, 1992.
16. Zohar Manna and Amir Pnueli. *The Temporal Logic of Reactive and Concurrent Systems*. Springer-Verlag, 1991.
17. Zohar Manna and Amir Pnueli. Verifying hybrid systems. Draft, October 1992.

A Compositional Approach to the Design of Hybrid Systems

Jozef Hooman

Dept. of Mathematics and Computing Science
Eindhoven University of Technology
P.O. Box 513, 5600 MB Eindhoven, The Netherlands
e-mail: wsinjh@win.tue.nl

Abstract. To specify and verify distributed real-time systems, classical Hoare triples are extended with timing primitives and the interpretation is modified to be able to specify non-terminating computations. For these modified triples a compositional proof system has been formulated. Compositionality supports top-down program derivation, and by using a dense time domain also hybrid systems with continuous components can be designed. This is illustrated by a process control example of a water level monitoring system. First we prove the correctness of a control strategy in terms of a continuous interface. Next, to obtain a discrete interface, a sensor and an actuator are introduced. Using their specifications only, a suitable specification of the control unit is derived. This reduces the design of the system to the conventional problem of deriving a program according to its specification. Finally the control unit is extended, in a modular way, with error detection features.

1 Introduction

An assertional method to specify and verify real-time embedded systems is presented. The method supports compositional verification, which means that we can reason with the specifications of components, describing their real-time communication interface, without knowing their implementation. Hence we can, e.g., use specifications of physical processes or hardware components. Further, when developing a computer program, compositionality makes it possible to verify design steps during the process of top-down program design. As an application of our method we consider here the design of hybrid systems, which interact with a continuously changing environment. In the remainder of this introduction, first our formalism is introduced, next we discuss our approach to an example of a water level monitoring system, and finally the structure of the paper is described.

Formal Framework

To design hybrid systems we use, as mentioned above, an assertional method which means that specifications are expressed as properties written in a certain logic, called the assertion language. Usually, the components of a hybrid system have certain real-time requirements, i.e., the timing of events is important for their correctness. Hence a logic to specify such systems should include primitives to express the real-time behaviour of processes. In this paper the timed communication interface of components is described from the viewpoint of an external observer with his own clock. Since in a hybrid system some of these components have a continuous interface, i.e., continuously changing observables, we use a dense time domain (the nonnegative reals).

During the first steps of the design process we only reason with the specifications of parallel components. Hence at this stage any sufficiently expressive real-time logic which allows the formulation of a compositional rule for parallel composition could be used. The form of our specifications is mainly motivated by the aim to design programs from their specifications and to enable compositional program verification. Observing the nice compositional rules for sequential composition and iteration of classical Hoare triples (precondition, program, postcondition) as introduced in [Hoa69], we have modified these triples to obtain a formalism for the specification and the verification of distributed real-time reactive systems. The assertion language has been extended with primitives denoting the timing of observable events. Since Hoare triples only specify properties of terminating computations (i.e. partial correctness), we have extended the framework such that we can specify progress properties of both terminating and non-terminating computations. For these modified triples a compositional proof system has been formulated. The formalism described here evolved from earlier work on Hoare-style assertional proof systems (see, e.g. [Hoo91]) and the application of these formalism to several examples such as a watchdog timer, a railway crossing, and a distributed real-time arbitration protocol.

The Water Level Monitoring System

A typical example of a hybrid system is a real-time computer system which controls physical processes. In this paper we illustrate our method by the design of a water level monitoring system, which was originally described in [vS91]. The intention is to design a control system which keeps the water level inside a vessel between certain critical values. Therefore the water level in the vessel can be measured by means of a sensor and the level can be influenced by switching a pump on or off.

Clearly, the goal can only be achieved if the water level does not change too fast and the initial level, at the start of the process, is not too close to the critical values. Hence the first step in our approach is to formalize these assumptions and to express the top-level specification. In our property-based approach this leads to a specification which expresses that the water level stays between the critical values, given assumptions about the initial level of the water and the maximal change of the water level in relation to the pump.

Next we describe a control strategy in terms of the water level and the pump. This strategy is formally expressed by a number of properties. Together with the assumptions mentioned above we can then prove that this leads to the desired properties of the water level. In general, in this step knowledge of control theory is important.

To be able to implement the control strategy, which is in terms of the continuously changing water level, we use a sensor to measure the water level and to obtain a discrete interface. There is an actuator to change the status of the pump. These hardware components are connected to a program controlled digital computer by means of channels with asynchronous message passing. The sensor and the actuator are assumed to be given, and assumptions about their behaviour are formally specified in terms of their real-time communication interface. For instance, for the sensor we express with which frequency it will send measured values to the control unit and how close this value approximates the real value of the water level. To design the control program, we first specify it in our formalism and prove that this specification and the specifications of the sensor and the actuator indeed imply the control strategy of the previous step. Note the use of compositional reasoning, since at this point we only have specifications of the three parallel components.

Once the previous step has been proved correct, we have reduced the problem to a conventional programming problem where we have to implement a program according to its specification. This, however, is not a trivial task, especially since the specification will usually express real-time properties. Here a program implementing the basic control loop is given and we show that it satisfies the required specification. Similar to program verification in untimed Hoare logic, the main point in this proof is to find suitable invariants for iterations.

Program correctness, however, is relative to large number of assumptions about, e.g., the actuator, the sensor, and the physical disturbances of the water level in the vessel. An important property of control systems is that they should be as robust as possible against errors in the model, failures of components, etc. Here we extend our specification such that certain failures are detected and appropriate alarm messages are given. To implement a program satisfying this extended specification, we again use compositionality to split up the program in two parallel components. One components monitors the sensor values and gives the appropriate alarm messages. The other component performs the standard control loop and it can be implemented by the previously developed program using channel renaming.

Structure of the Paper

The remainder of this paper is organized as follows. Section 2 contains the syntax and informal semantics of a concurrent real-time programming language. Our real-time extension of Hoare logic is introduced in section 3. For these modified Hoare triples a compositional proof system is formulated in section 4. As an application of our real-time formalism we design in section 5 a control program for a water level monitoring system. Concluding remarks can be found in section 6.

2 Programming Language

We consider a concurrent real-time programming language with asynchronous communication along unidirectional channels that connect at most two processes. In section 2.1 we give the syntax of this language and the informal meaning of statements. Our notion of time and the basic timing assumptions, which are used to reason about real-time properties of programs, are described in section 2.2.

2.1 Syntax and Informal Meaning

Let *CHAN* be a nonempty set of channel names, *CLOCK* be a nonempty set of names of local timers, *VAR* be a nonempty set of program variables, and *VAL* be a domain of values. $I\!N$ denotes the set of natural numbers (including 0). The syntax and informal meaning of our programming language is given below, with $n \in I\!N$, $n \geq 1$, $c, c_1, \ldots, c_n \in CHAN$, $clock \in CLOCK$, $x, x_1, \ldots, x_n \in VAR$, expression e ranging over VAL, and boolean expressions $b_0, b_1, \ldots, b_n$.

Atomic statements

- **skip** terminates immediately.
- Assignment $x := e$ assigns the value of expression e to the variable x.
- **delay** e suspends execution for (the value of) e time units. If e yields a negative value then **delay** e is equivalent to **skip**.
- $x := clock$ reads the value from local timer $clock$ and assigns this value to x.
- Output statement $c!!e$ is used to send the value of expression e along channel c. It does not wait for a receiver but sends immediately. There is no buffering of messages; the message is lost if there is no receiver.
- Input statement $c?x$ is used to receive a value along channel c and assign this value to the variable x. Such an input statement has to wait until a message is available.

Compound statements

- $S_1; S_2$ indicates sequential composition: first execute S_1, and continue with the execution of S_2 if and when S_1 terminates.
- Guarded command $[b_1 \rightarrow S_1 \, [] \ldots [] \, b_n \rightarrow S_n]$, abbreviated by $[[]_{i=1}^n b_i \rightarrow S_i]$, terminates after evaluation of the boolean guards b_i if none of them evaluates to true. Otherwise, non-deterministically select one of the b_i that evaluates to true and execute the corresponding statement S_i.
- Guarded command $[b_1; c_1?x_1 \rightarrow S_1 \, [] \ldots [] \, b_n; c_n?x_n \rightarrow S_n \, [] \, b_0; \mathbf{delay}\ e \rightarrow S_0]$, is abbreviated by $[[]_{i=1}^n b_i; c_i?x_i \rightarrow S_i \, [] \, b_0; \mathbf{delay}\ e \rightarrow S_0]$.
  A guard is *open* if the boolean part evaluates to true. If none of the guards is open, the guarded command terminates after evaluation of the booleans. Otherwise, wait until an input statement of the open input-guards can be executed and continue with the corresponding S_i. If the delay guard is open (b_0 evaluates to true) and no input-guard can be taken within e time units (after the evaluation of the booleans), then S_0 is executed.

  Example 1. By means of this construct we can program a *time-out*, i.e., restrict the waiting period for an input statement. Consider for instance the program
  $$[\, x \neq 0; in?x \rightarrow out!!f(x) \, [] \, y \geq 2; \mathbf{delay}\ 8 \rightarrow alarm!!y \,].$$
  If $x \neq 0$, it waits to receive a message along channel in. If $y \geq 2$ then it can wait at most 8 time units, and if no message is available within 8 time units a message is sent along channel $alarm$. $\square$

- Iteration $*G$ indicates repeated execution of guarded command G as long as at least one of the guards is open. When none of the guards is open $*G$ terminates.
- $S_1 \| S_2$ indicates parallel execution of the statements S_1 and S_2. The components S_1 and S_2 of a parallel composition are often called *processes*. We require that S_1 and S_2 do not have shared variables and channels connect at most two processes.

Henceforth we use $\equiv$ to denote syntactic equality.

A guarded command $[[]_{i=1}^{n} b_i; c_i?x_i \rightarrow S_i []b_0; \mathbf{delay}\, e \rightarrow S_0]$ with $b_0 \equiv false$ is written as $[[]_{i=1}^{n} b_i; c_i?x_i \rightarrow S_i]$. An input guard $b_i; c_i?x_i$ is abbreviated as $c_i?x_i$ if $b_i \equiv true$, and, similarly, a delay-guard $b_0; \mathbf{delay}\, e$ is written as $\mathbf{delay}\, e$ if $b_0 \equiv true$.

Let $var(S)$ be the set of program variables occurring in S. We define $obs(S)$, the set of observables of S, by induction on the structure of S. The main clauses are $obs(c!!e) = \{c!!\}$ and $obs(c?x) = \{c?, c\}$. The rest is rather straightforward:

$obs(\mathbf{skip}) = obs(x := e) = obs(\mathbf{delay}\, e) = obs(x := clock) = \emptyset$,

$obs(S_1; S_2) = obs(S_1 \| S_2) = obs(S_1) \cup obs(S_2)$, $obs([[]_{i=1}^{n} b_i \rightarrow S_i]) = \bigcup_{i=1}^{n} obs(S_i)$,

$obs([[]_{i=1}^{n} b_i; c_i?x_i \rightarrow S_i []b_0; \mathbf{delay}\, e \rightarrow S_0]) = \bigcup_{i=1}^{n}(\{c_i?, c_i\} \cup (obs(S_i)) \cup obs(S_0)$,

and $obs(*G) = obs(G)$.

2.2 Basic Timing Assumptions

In our proof system the correctness of a program with respect to a specification, which may include timing constraints, is verified relative to assumptions about:

- The execution time of atomic statements. Here we use (nonnegative) parameters representing the duration of atomic statements. We assume that
  - there exists a parameter T_a such that each assignment of the form $x := e$ takes T_a time units;
  - $\mathbf{delay}\, e$ takes exactly e time units if e is positive and 0 time units otherwise;
  - there exist a parameter $T_{asyn} > 0$ such that each asynchronous communication takes T_{asyn} time units;
  - there exists a parameter T_c such that reading a local timer by $x := clock$ takes T_c time units.
- The extra time required to execute compound programming constructs. Here we assume that there exists a parameter T_g such that the evaluation of guards in a guarded command takes T_g time units. To avoid an infinite loop in finite time, we assume that T_g is greater than a positive constant. There is no overhead for other compound statements.
- How long a process is allowed to wait with the execution of enabled statements. That is, we have to make assumptions about the progress of actions. In this paper we use the *maximal progress* model in which an enabled action will be executed as soon as possible, representing the situation that each parallel process has its own processor. Hence, a process never waits with the execution of a local, non-communication, command or with the execution of an asynchronous output. An input command can cause a process to wait, but only when no communication partner is available; as soon as a partner is available the communication must take place. Thus maximal parallelism implies minimal waiting. In [Hoo91] we show that the framework can be generalized to multiprogramming where several processes may share a single processor and scheduling is based on priorities.

3 Specifications

We express the timing behaviour of an embedded real-time system from the viewpoint of an external observer with his own clock. Thus, although parallel components of a system might have their own, physical, local clock, the observable behaviour of a system is described in terms of a single, conceptual, global clock. Since this global notion of time is not incorporated in the distributed system itself, it does not impose any synchronization upon processes. Here we use a time domain $TIME$ which equals the nonnegative real numbers.

As motivated in section 1, our specifications are based on classical Hoare triples. To distinguish our modified triples from classical Hoare logic, a slightly different notation will be introduced and we use the words "assumption" and "commitment" instead of, resp., "precondition" and "postcondition". This leads to formulas of the form $\langle\!\langle p \rangle\!\rangle \, S \, \langle\!\langle q \rangle\!\rangle$, where S is a program, as described in section 2, and p and q are assertions called, respectively, *assumption* and *commitment*.

Assertion p expresses assumptions about

- the initial state of S, i.e., the values of the program variables at the start of the execution of S,
- the starting time of S, and
- the timed occurrence of observable events.

Given assumption p, assertion q expresses a commitment of S,

- if S terminates, about the final state of S, i.e., the values of the program variables at termination,
- about the termination time (∞ if S does not terminate), and
- about the timed occurrences of observable events.

Note that, in contrast with classical Hoare triples, our commitment expresses properties of terminating as well as non-terminating computations. Together with the addition of time this means that our formalism is not restricted to partial correctness, but also deals with progress properties.

3.1 Assertion Language

The assertions p and q in a correctness formula $\langle\!\langle p \rangle\!\rangle \, S \, \langle\!\langle q \rangle\!\rangle$ are expressed in a first-order logic with the following primitives.

- Program variables, such as $x, y \in VAR$, ranging over VAL.
- Logical variables that are not affected by program execution. We have logical variables ranging over VAL, such as $v, v_0, v_1, \ldots$, and logical variables ranging over $TIME \cup \{\infty\}$, such as $t, t_0, t_1, \ldots$.
- $exp \in I\!N$ to express that the value of expression exp is a natural number.
- $clock(exp)$ to denote the value of local timer $clock$ at time exp.
- A special variable $time$, ranging over $TIME \cup \{\infty\}$, which refers to our global notion of time. An occurrence of $time$ in assumption p represents the starting time of the statement S whereas in commitment q it denotes the termination time (using $time = \infty$ for non-terminating computations).

Since our aim is to specify and verify timing properties of open systems that communicate with their environment by message passing along asynchronous channels, the logic contains the following primitives to express this communication behaviour.

- $(c!!, exp_1)$ **at** exp_2 to denote the start of an asynchronous output along channel c with value exp_1 at time exp_2.
- $c?$ **at** exp to express that a process is waiting to receive a message along channel c at time exp.
- (c, exp_1) **at** exp_2 to denote that at time exp_2 a process starts to receive value exp_1 along channel c.

Let $var(p)$ be the set of program variables occurring in assertion p. Let $obs(p)$ denote the set of observables occurring in p, which is defined by structural induction. E.g., $obs((c!!, exp_1)$ **at** $exp_2) = \{c!!\}$, $obs((c, exp_1)$ **at** $exp_2) = \{c\}$, $obs(c?$ **at** $exp) = \{c?\}$, $obs(x) = obs(time) = obs(clock(exp)) = \emptyset$, $obs(p_1 \vee p_2) = obs(p_1) \cup obs(p_2)$, etc.

We frequently use intervals of time points such as $[t_0, t_1) = \{t \in TIME \mid t_0 \leq t < t_1\}$, $\langle t_0, t_1 \rangle = \{t \in TIME \mid t_0 < t < t_1\}$, etc.

Abbreviations Based on these primitives, we define a few abbreviations.

- For a predicate P at t and an interval $I \subseteq TIME$, we use

  P **at** $I \equiv \forall t \in I : P$ **at** t,
  P **in** $I \equiv \exists t \in I : P$ **at** t.

  Note that $(\neg P)$ **at** $I \leftrightarrow \neg(P$ **in** $I)$.
- We often abstract from the value that is transmitted, using

  c **at** $t \equiv \exists v : (c, v)$ **at** t,
  $c!!$ **at** $t \equiv \exists v : (c!!, v)$ **at** t.
- $await\ c?$ **at** $t \equiv c?$ **at** $[t, \infty) \vee (\exists t_1 \in [t, \infty) : c?$ **at** $[t, t_1) \wedge c$ **at** $t_1)$

  This expresses that a process starts waiting to receive input along c at time t until it receives input, allowing the possibility that it has to wait forever.
- $await_{\geq d} c?$ **at** $t \equiv \exists t_1 \geq t : c?$ **at** $[t, t_1) \wedge (c$ **at** $t_1 \vee t_1 \geq t + d)$

  This abbreviation expresses that a process starts waiting to receive input along c at time t until it either receives input or at least d time units have elapsed.

Similar abbreviations can be defined with general expressions instead of v, t, etc. Further we will sometimes use $(P_1 \wedge P_2)$ **at** t instead of P_1 **at** $t \wedge P_2$ **at** t, etc.

3.2   Properties of Clocks and Communications

In this section we axiomatize the properties of local clocks and the communication mechanism.

Local Clock Property In this paper we assume that we can approximate durations by means of local timers. That is, the difference of two clock readings is a reasonable approximation of the elapsed real-time. Suppose local timers range over a discrete domain $CVAL$. Let u be the smallest unit of local clocks, that is, for any $T \in CVAL$, $\exists n \in \mathbb{N} : T = n \cdot u$. Further, for $T \in CVAL$, we define $tick_{clock}(T)$ as the point in real-time where local timer $clock$ changes its value to T, that is, $tick_{clock}(T) = inf\{t \in TIME \mid clock(t) = T\}$. The duration between ticks of a local timer is usually called the granularity g. Then we have

$g \approx tick_{clock}(T + u) - tick_{clock}(T)$. In general, this is not an equality since local clocks might have a certain drift r, and we have
$$g(1 - r) \leq tick_{clock}(T + u) - tick_{clock}(T) \leq g(1 + r).$$
Then for $T_1, T_2 \in CVAL$, $T_1 \leq T_2$ implies

$$g(1 - r)(T_2 - T_1) \leq (tick_{clock}(T_2) - tick_{clock}(T_1))u \leq g(1 + r)(T_2 - T_1).$$

Note that $tick_{clock}(clock(t)) \leq t \leq tick_{clock}(clock(t) + u)$. Further assume that local timers are monotonic, i.e., $t_1 \leq t_2 \to clock(t_1) \leq clock(t_2)$. Then observe that for $t_1 \leq t_2$, $(t_2 - t_1)u \leq (tick_{clock}(clock(t_2) + u) - tick_{clock}(clock(t_1)))u \leq g(1 + r)(clock(t_2) + u - clock(t_1))$. Similarly, $(t_2 - t_1)u \geq g(1 - r)(clock(t_2) - clock(t_1) - u)$. Thus $t_1 \leq t_2$ implies

$$\frac{(t_2 - t_1)u}{g(1 + r)} - u \leq clock(t_2) - clock(t_1) \leq \frac{(t_2 - t_1)u}{g(1 - r)} + u.$$

Discussion For other applications one might need other assumptions about local timers. For instance, often local clocks are internally synchronized, that is, there exists a δ such that for any two timers $clock_1$ and $clock_2$,
$$\forall t < \infty : |clock_1(t) - clock_2(t)| < \delta.$$
Another possibility, which implies the previous one, is to assume that local clocks are externally synchronized, that is, they deviate less than ε from our global notion of time. Formally,
$$\forall t < \infty : |clock(t) - t| < \varepsilon.$$
End Discussion

Communication Properties For simplicity, we assume that when a process starts sending a message it is immediately available for the receiver. Together with our maximal progress assumption this leads to the following properties.
For an asynchronous channel c we have, for all $t < \infty$,

$\neg(c!!$ at $t \wedge c?$ at $t)$
(Minimal waiting: the receiver should not be waiting when a message is transmitted.)

(c, v) at $t \to (c!!, v)$ at t
(When a process starts receiving a message, it must have been sent simultaneously.)

$(c!!, v_1)$ at $t \wedge (c!!, v_2)$ at $t \to v_1 = v_2$
(At any point of time at most one message is transmitted.)

Discussion It might be more realistic to assume that it takes Δ time units before a message transmitted by a sender is available for a receiver. Then we should change the first property into
$\neg(c!!$ at $t \wedge c?$ at $(t + \Delta))$
and the second property becomes
(c, v) at $t \to t \geq \Delta \wedge (c!!, v)$ at $(t - \Delta)$.
End Discussion

By means of these properties, a number of useful lemmas can be proved. The first lemma expresses that if the receiver is always ready to receive input before the next asynchronous output, then each output corresponds to a communication.

Lemma 1 (No Output Lost). For all $\Delta_1, \Delta_2 \in TIME$, if $await\, c?$ in $[0, \Delta_1)$, $\forall t < \infty : await\, c?$ in $[t, t + \Delta_2)$, and $\forall t < \infty : c!!$ at $t \rightarrow t \geq \Delta_1 \wedge (\neg c!!)$ at $[t - \Delta_2, t)$, then $\forall t < \infty : c!!$ at $t \leftrightarrow c$ at t.

Proof: By the communication properties, $\forall t < \infty : c$ at $t \rightarrow c!!$ at t.
Suppose $c!!$ at t. Then $t \geq \Delta_1$ and $(\neg c!!)$ at $[t - \Delta_2, t)$. Thus $(\neg c)$ at $[t - \Delta_2, t)$. Together with $await\, c?$ in $[t - \Delta_2, t)$ this implies $await\, c?$ at t. Hence with $c!!$ at t and the communication properties we obtain c at t. $\qquad\square$

If $\Delta_1 = \Delta_2$, this implies the following lemma.

Lemma 2. For any $\Delta \in TIME$, if $\forall t < \infty : await\, c?$ in $[t, t + \Delta)$ and $\forall t < \infty : c!!$ at $t \rightarrow t \geq \Delta \wedge (\neg c!!)$ at $[t - \Delta, t)$ then $\forall t < \infty : c!!$ at $t \leftrightarrow c$ at t.

The next lemma expresses that if the sender sends regularly, e.g. at least once every Δ_s time units, and the receiver is ready to receive at least once every Δ_r time units during at least Δ_s time units, then a message is received at least once every $\Delta_s + \Delta_r$ time units.

Lemma 3 (Periodic Input). For all $\Delta_s, \Delta_r \in TIME$, if $\forall t < \infty : c!!$ in $[t, t + \Delta_s)$ and $\forall t < \infty : await_{\geq \Delta_s} c?$ in $[t, t + \Delta_r)$ then $\forall t < \infty : c$ in $[t, t + \Delta_s + \Delta_r)$.

Proof: Consider $t < \infty$. Then there exists a $t_0 \in [t, t + \Delta_r)$ such that $await_{\geq \Delta_s} c?$ at t_0, i.e., there exists a $t_1 \geq t_0$ such that $c?$ at $[t_0, t_1) \wedge (c$ at $t_1 \vee t_1 \geq t_0 + \Delta_s)$. Further $c!!$ in $[t_0, t_0 + \Delta_s)$, thus $(\neg c?)$ in $[t_0, t_0 + \Delta_s)$, i.e., $\neg(c?$ at $[t_0, t_0 + \Delta_s))$. Hence $t_1 < t_0 + \Delta_s$, and thus c at t_1. Observe that $t_1 \geq t_0 \geq t$ and $t_1 < t_0 + \Delta_s < t + \Delta_r + \Delta_s$. Hence c in $[t, t + \Delta_s + \Delta_r)$. $\qquad\square$

3.3 Examples of Specifications

We give a number of small examples of specifications to illustrate our formalism. In the first triple we consider an assignment $x := x + 7$ which starts at time 6 in a state where x has the value 5 and assuming that the value 0 has been received along channel c starting at 3. Then this assignment terminates at time $6 + T_a$ in a state where x has the value 12. Further, given the assumption, we have the commitment that the value 0 has been received along channel c starting at 3.

$$\langle\!\langle x = 5 \wedge time = 6 \wedge (c, 0) \text{ at } 3 \rangle\!\rangle$$
$$x := x + 7$$
$$\langle\!\langle x = 12 \wedge time = 6 + T_a \wedge (c, 0) \text{ at } 3 \rangle\!\rangle.$$

A similar triple can be given for an output statement.

$$\langle\!\langle x = 4 \wedge time = 5 \wedge (a, 2) \text{ at } 3 \rangle\!\rangle$$
$$c!!(x + 2)$$
$$\langle\!\langle x = 4 \wedge time = 5 + T_{asyn} \wedge (c!!, 6) \text{ at } 5 \wedge (a, 2) \text{ at } 3 \rangle\!\rangle.$$

An input statement either waits forever or eventually receives a message.

$$\langle\!\langle time = 5 \rangle\!\rangle$$
$$c?x$$
$$\langle\!\langle\!\langle (c? \text{ at } [5, \infty) \wedge time = \infty) \vee$$
$$(\exists t \in [5, \infty) : c? \text{ at } [5, t) \wedge (c, x) \text{ at } t \wedge time = t + T_{asyn}) \rangle\!\rangle\!\rangle.$$

We can generalize specifications by using logical variables to represent the initial values of program variables and the starting time. For instance,

$$\langle\!\langle y = v \wedge time = t < \infty \wedge (c, v) \text{ at } (t - 5)\rangle\!\rangle$$
$$\quad \textbf{delay } 2 \,; \, d!!(y + 1) \,; \, y := y + 2$$
$$\langle\!\langle y = v + 2 \wedge time = t + 2 + T_{asyn} + T_a < \infty \wedge$$
$$\quad (c, v) \text{ at } (t - 5) \wedge (d!!, v + 1) \text{ at } (t + 2)\rangle\!\rangle.$$

Note that logical variables, such as v and t, are implicitly universally quantified.

By the next triple we specify a program FUN which should compute $f(x)$ within certain time bounds, for a particular function f, leaving x invariant:

$$\langle\!\langle x = v \wedge time = t < \infty\rangle\!\rangle \, FUN \, \langle\!\langle y = f(v) \wedge x = v \wedge t + 5 < time < t + 13\rangle\!\rangle.$$

Next consider an example of a guarded command.

$$\langle\!\langle time = 0\rangle\!\rangle$$
$$\quad [x > 5 \rightarrow c!!x \,[\!] \, y < 3 \rightarrow x := 0]$$
$$\langle\!\langle (x \leq 5 \wedge y \geq 3 \wedge time = T_g) \vee$$
$$\quad (x > 5 \wedge (c!!, x) \text{ at } T_g \wedge time = T_g + T_{asyn}) \vee$$
$$\quad (y < 3 \wedge x = 0 \wedge time = T_g + T_a)\rangle\!\rangle.$$

The following program never terminates, but we can still specify its real-time communication interface.

$$\langle\!\langle time = 0\rangle\!\rangle$$
$$\quad * [in?x \rightarrow out!!f(x)]$$
$$\langle\!\langle time = \infty \wedge \forall v \, \forall t < \infty : (in, v) \text{ at } t \rightarrow (out!!, f(v)) \text{ at } (t + T_{asyn})\rangle\!\rangle.$$

Finally note that a classical Hoare triple $\{p\} \, S \, \{q\}$, denoting partial correctness, can be expressed in our framework as

$$\langle\!\langle p \wedge time < \infty\rangle\!\rangle \, S \, \langle\!\langle time < \infty \rightarrow q\rangle\!\rangle.$$

Total correctness of S with respect to p and q can be denoted by

$$\langle\!\langle p \wedge time < \infty\rangle\!\rangle \, S \, \langle\!\langle time < \infty \wedge q\rangle\!\rangle.$$

4 Proof System

We formulate a compositional proof system for our modified Hoare triples. First we give in section 4.1 rules and axioms that are generally applicable to any statement. Next in section 4.2 we axiomatize the programming language by giving rules and axioms for the atomic statements and the compound programming constructs.

We assume that all logical variables which are introduced in the rules are fresh. Further $p[exp/var]$ is used to denote the substitution of expression exp for each free occurrence of variable var in assertion p.

4.1 General Rules and Axioms

We give three axioms to deduce invariance properties. The first two axioms expresses that an assumption which satisfies certain restrictions remains valid during the execution of any program.

Axiom 1 (Initial Invariance) $\quad \langle\!\langle p\rangle\!\rangle \, S \, \langle\!\langle p\rangle\!\rangle$

provided p does not refer to $time$ and $var(p) = \emptyset$.

In the next invariance axiom we express that the value of a program variable which does not occur in a program S is not affected by a terminating computation of S.

Axiom 2 (Variable Invariance) $\quad \langle\!\langle p \rangle\!\rangle \; S \; \langle\!\langle time < \infty \rightarrow p \rangle\!\rangle$

provided $time$ does not occur in p and $var(p) \cap var(S) = \emptyset$.

The channel invariance axiom below expresses that a program S never performs an action which does not syntactically occur in S.
For a finite set $oset$ of observables, we define

$$NoAct(oset) \text{ at } exp \equiv \bigwedge\nolimits_{O \in oset}(\neg O) \text{ at } exp.$$

Axiom 3 (Channel Invariance) $\quad \langle\!\langle time = t_0 \rangle\!\rangle \; S \; \langle\!\langle NoAct(oset) \text{ at } [t_0, time) \rangle\!\rangle$

provided $oset \cap obs(S) = \emptyset$.

The next axiom expresses that a program does not affect a non-terminating computation. By this property we obtain the classical rule for sequential composition.

Axiom 4 (Non-termination) $\quad \langle\!\langle p \wedge time = \infty \rangle\!\rangle \; S \; \langle\!\langle p \wedge time = \infty \rangle\!\rangle$

The substitution rule allows us to replace a logical variable in the assumption by any arbitrary expression if this variable does not occur in the commitment.

Rule 1 (Substitution) $\quad \dfrac{\langle\!\langle p \rangle\!\rangle \; S \; \langle\!\langle q \rangle\!\rangle}{\langle\!\langle p[exp/t] \rangle\!\rangle \; S \; \langle\!\langle q \rangle\!\rangle}$

provided t does not occur free in q.

The proof system contains a consequence rule, a conjunction rule, and a disjunction rule, which are identical to the classical rules for Hoare triples.

Rule 2 (Consequence) $\quad \dfrac{\langle\!\langle p_0 \rangle\!\rangle \; S \; \langle\!\langle q_0 \rangle\!\rangle, \; p \rightarrow p_0, \; q_0 \rightarrow q}{\langle\!\langle p \rangle\!\rangle \; S \; \langle\!\langle q \rangle\!\rangle}$

Rule 3 (Conjunction) $\quad \dfrac{\langle\!\langle p_1 \rangle\!\rangle \; S \; \langle\!\langle q_1 \rangle\!\rangle, \; \langle\!\langle p_2 \rangle\!\rangle \; S \; \langle\!\langle q_2 \rangle\!\rangle}{\langle\!\langle p_1 \wedge p_2 \rangle\!\rangle \; S \; \langle\!\langle q_1 \wedge q_2 \rangle\!\rangle}$

Rule 4 (Disjunction) $\quad \dfrac{\langle\!\langle p_1 \rangle\!\rangle \; S \; \langle\!\langle q_1 \rangle\!\rangle, \; \langle\!\langle p_2 \rangle\!\rangle \; S \; \langle\!\langle q_2 \rangle\!\rangle}{\langle\!\langle p_1 \vee p_2 \rangle\!\rangle \; S \; \langle\!\langle q_1 \vee q_2 \rangle\!\rangle}$

In the axioms and rules for the programming constructs we will often use assumption $time < \infty$. By means of the following derived rule we can discard this assumption.

Rule 5 (Split Precondition) $\quad \dfrac{p \wedge time = \infty \rightarrow q}{\dfrac{\langle\!\langle p \wedge time < \infty \rangle\!\rangle \; S \; \langle\!\langle q \rangle\!\rangle}{\langle\!\langle p \rangle\!\rangle \; S \; \langle\!\langle q \rangle\!\rangle}}$

4.2 Axiomatization of Programming Constructs

Next we give rules and axioms for the atomic statements and the compound constructs of the programming language. A skip statement terminates immediately and has no effect, so every assertion is invariant under this statement.

Axiom 5 (Skip) $\langle\!\langle p \rangle\!\rangle$ skip $\langle\!\langle p \rangle\!\rangle$

In the rule for an assignment $x := e$ we express that to obtain commitment q the assumption $q[e/x, time + T_a/time] \wedge time < \infty$ is required. Note that, in addition to the classical rule, we also update variable $time$ to express that the termination time equals the initial time plus T_a time units.

Axiom 6 (Assignment) $\langle\!\langle q[e/x, time + T_a/time] \wedge time < \infty \rangle\!\rangle \; x := e \; \langle\!\langle q \rangle\!\rangle$

Example 2. We derive
$\langle\!\langle x = 5 \wedge time = 6 \wedge (c, 0) \text{ at } 3 \rangle\!\rangle \; x := x + 7 \; \langle\!\langle x = 12 \wedge time = 6 + T_a \wedge (c, 0) \text{ at } 3 \rangle\!\rangle$.
The assignment axiom leads to
$\quad \langle\!\langle x + 7 = 12 \wedge time + T_a = 6 + T_a \wedge (c, 0) \text{ at } 3 \wedge time < \infty \rangle\!\rangle$
$\quad x := x + 7$
$\quad \langle\!\langle x = 12 \wedge time = 6 + T_a \wedge (c, 0) \text{ at } 3 \rangle\!\rangle$.
Then the consequence rule yields the required triple, since
$x = 5 \wedge time = 6 \wedge (c, 0) \text{ at } 3 \rightarrow$
$x + 7 = 12 \wedge time + T_a = 6 + T_a \wedge (c, 0) \text{ at } 3 \wedge time < \infty$.   □

Similarly, we axiomatize local clock readings and delay actions.

Axiom 7 (Clock) $\langle\!\langle q[clock(time)/x, time+T_c/time] \wedge time < \infty \rangle\!\rangle \; x := clock \; \langle\!\langle q \rangle\!\rangle$

Axiom 8 (Delay) $\langle\!\langle q[time + max(0, e)/time] \wedge time < \infty \rangle\!\rangle$ delay e $\langle\!\langle q \rangle\!\rangle$

In the rule for an output action $c!!e$ first $time$ in assumption $p \wedge time < \infty$ is replaced by t_0, which thus represents the starting time. Then we express that it starts sending the value of e at time t_0 and, for completeness, we also assert that no transmission is started after t_0 until it terminates. Further the termination time equals $t_0 + T_{asyn}$.

Rule 6 (Output)

$$\frac{(p \wedge time < \infty)[t_0/time] \wedge (c!!, e) \text{ at } t_0 \wedge (\neg c!!) \text{ at } \langle t_0, time \rangle \wedge time = t_0 + T_{asyn} \rightarrow q}{\langle\!\langle p \wedge time < \infty \rangle\!\rangle \; c!!e \; \langle\!\langle q \rangle\!\rangle}$$

To obtain a compositional proof system we do not make any assumption about the environment of a statement. Thus in the rule for an input statement no assumption is made about when a message is available, and hence this rule includes any arbitrary waiting period (including an infinite one). Further we allow any arbitrary input value. In the rule below the commitment is split up in q_{nt}, representing a non-terminating computation in which the input statement waits forever to receive a message, and q to express properties of terminating computations.

Rule 7 (Input)
$$(p \wedge time < \infty)[t_0/time] \wedge c? \text{ at } [t_0, \infty) \wedge time = \infty \rightarrow q_{nt}$$
$$(p \wedge time < \infty)[t_0/time] \wedge \exists t \in [t_0, \infty) : c? \text{ at } [t_0, t) \wedge (c, v) \text{ at } t \wedge$$
$$\frac{(\neg c? \wedge \neg c) \text{ at } \langle t, time \rangle \wedge time = t + T_{asyn} \rightarrow q[v/x]}{\langle\!\langle p \wedge time < \infty \rangle\!\rangle \; c?x \; \langle\!\langle q_{nt} \vee q \rangle\!\rangle}$$

provided $var(q_{nt}) = \emptyset$.

The rule for sequential composition is similar to the classical rule for Hoare triples.

Rule 8 (Sequential Composition)
$$\frac{\langle\!\langle p \rangle\!\rangle \; S_1 \; \langle\!\langle r \rangle\!\rangle, \quad \langle\!\langle r \rangle\!\rangle \; S_2 \; \langle\!\langle q \rangle\!\rangle}{\langle\!\langle p \rangle\!\rangle \; S_1; S_2 \; \langle\!\langle q \rangle\!\rangle}$$

Furthermore the proof system contains rules for guarded commands and iteration. For a guarded command G, first define

$$b_G = \begin{cases} \bigvee_{i=1}^{n} b_i & \text{if } G \equiv [\![]_{i=1}^{n} b_i \rightarrow S_i] \\ \bigvee_{i=0}^{n} b_i & \text{if } G \equiv [\![]_{i=1}^{n} b_i; c_i?x_i \rightarrow S_i [\!] b_0; \text{delay } e \rightarrow S_0] \end{cases}$$

We start with a simple rule for the case that none of the guards is open.

Rule 9 (Guarded Command Termination)
$$\frac{\langle\!\langle p \wedge \neg b_G \wedge time < \infty \rangle\!\rangle \; \text{delay } T_g \; \langle\!\langle q \rangle\!\rangle}{\langle\!\langle p \wedge \neg b_G \wedge time < \infty \rangle\!\rangle \; G \; \langle\!\langle q \rangle\!\rangle}$$

When at least one of the guards evaluates to true, we have, for $G \equiv [\![]_{i=1}^{n} b_i \rightarrow S_i]$,

Rule 10 (Guarded Command with Purely Boolean Guards)
$$\frac{\langle\!\langle p \wedge b_i \rangle\!\rangle \; \text{delay } T_g \; ; \; S_i \; \langle\!\langle q_i \rangle\!\rangle, \text{ for all } i \in \{1, \ldots, n\}}{\langle\!\langle p \wedge b_G \rangle\!\rangle \; [\![]_{i=1}^{n} b_i \rightarrow S_i] \; \langle\!\langle \bigvee_{i=1}^{n} q_i \rangle\!\rangle}$$

For $G \equiv [\![]_{i=1}^{n} b_i; c_i?x_i \rightarrow S_i [\!] b_0; \text{delay } e \rightarrow S_0]$ we define

- *wait in G at* $[t_0, t_1) \equiv \bigwedge_{i=1}^{n} (\forall t_2 \in [t_0, t_1) : (c_i? \text{ at } t_2 \leftrightarrow b_i) \wedge \neg c_i \text{ at } [t_0, t_1)) \wedge$
$NoAct((obs(G) - \{c_1?, c_1, \ldots, c_n?, c_n\})) \text{ at } [t_0, t_1)$
- *comm (c_i, v) in G from* $t \equiv (c_i, v) \text{ at } t \wedge time = t + T_{asyn} \wedge (\neg c_i? \wedge \neg c_i) \text{ at } \langle t_1, time \rangle \wedge$
$NoAct((obs(G) - \{c_i?, c_i\})) \text{ at } [t, t + T_{asyn})$

Rule 11 (Guarded Command with IO Guards)

$\langle\!\langle p \wedge b_G \wedge time < \infty \rangle\!\rangle \; \text{delay } T_g \; \langle\!\langle \hat{p} \rangle\!\rangle$

$\hat{p}[t_0/time] \wedge \neg b_0 \wedge \; wait \; in \; G \; at \; [t_0, \infty) \wedge \; time = \infty \rightarrow q$

$\hat{p}[t_0/time] \wedge \exists t \in [t_0, \infty) : wait \; in \; G \; at \; [t_0, t) \wedge (b_0 \rightarrow t < t_0 + max(0, e)) \wedge$
$\qquad\qquad b_i \wedge comm \; (c_i, v) \; in \; G \; from \; t \rightarrow p_i[v/x_i], \text{ for } i = 1, \ldots, n$

$\hat{p}[t_0/time] \wedge b_0 \wedge \; wait \; in \; G \; at \; [t_0, t_0 + max(0, e)) \wedge \; time = t_0 + max(0, e) \rightarrow p_0$

$\dfrac{\langle\!\langle p_i \rangle\!\rangle \; S_i \; \langle\!\langle q_i \rangle\!\rangle, \text{ for } i = 0, 1, \ldots, n}{\langle\!\langle p \wedge b_G \wedge time < \infty \rangle\!\rangle \; [\![]_{i=1}^{n} b_i; c_i?x_i \rightarrow S_i [\!] b_0; \text{delay } e \rightarrow S_0] \; \langle\!\langle q \vee \bigvee_{i=0}^{n} q_i \rangle\!\rangle}$

provided $var(q) = \emptyset$.

The rule for the iteration construct is as follows.

Rule 12 (Iteration)
$$\langle\langle I \wedge b_G \rangle\rangle \; G \; \langle\langle I \rangle\rangle$$
$$\langle\langle I \wedge \neg b_G \rangle\rangle \; G \; \langle\langle q \rangle\rangle$$
$$I \rightarrow \hat{I}$$
$$(\forall t_1 < \infty \; \exists t_2 > t_1 : \hat{I}[t_2/time]) \rightarrow q_{nt}$$
$$\overline{\langle\langle I \rangle\rangle \; * \; G \; \langle\langle (q_{nt} \wedge time = \infty) \vee q \rangle\rangle}$$

provided no program variables occur in $\hat{I}$, i.e., $var(\hat{I}) = \emptyset$.

The last hypothesis of the rule deals with the case that the iteration does not terminate. This can either be caused by a non-terminating computation inside G or by an infinite sequence of terminating computations of G. In both cases I, and thus $\hat{I}$, allows computations with an arbitrary large value for $time$.

Example 3. We apply the iteration rule to an example which shows that our proof system is not restricted to partial correctness but, e.g., also includes termination. We derive

$$\langle\langle x = v \in I\!N \wedge time = 0 \rangle\rangle$$
$$* [x \neq 0 \rightarrow x := x - 1]$$
$$\langle\langle (v < 0 \wedge time = \infty) \vee (v \geq 0 \wedge time = v(T_g + T_a) + T_g \wedge x = 0) \rangle\rangle.$$

Therefore the iteration rule is applied, using
$I \equiv (v < 0 \wedge x < 0) \vee (v \geq 0 \wedge time = (v - x)(T_g + T_a) \wedge x \in I\!N)$
$q \equiv v \geq 0 \wedge time = v(T_g + T_a) + T_g \wedge x = 0$, and
$q_{nt} \equiv v < 0$.
By the rules for guarded commands we can derive

$$\langle\langle I \wedge x \neq 0 \rangle\rangle \; [x \neq 0 \rightarrow x := x - 1] \; \langle\langle I \rangle\rangle$$

(informally this holds since x is decreased by 1, but $time$ is incremented by $T_g + T_a$ and hence $time = (v - x)(T_g + T_a)$ is invariant) and

$$\langle\langle I \wedge x = 0 \rangle\rangle \; [x \neq 0 \rightarrow x := x - 1] \; \langle\langle q \rangle\rangle.$$

Define $\hat{I} \equiv (v < 0) \vee (time \leq v(T_g + T_a))$.
Then $I \rightarrow \hat{I}$, since $x \in I\!N$ implies $(v-x)(T_g+T_a) \leq v(T_g+T_a)$. Note that $var(\hat{I}) = \emptyset$.
Further observe that $\forall t_1 < \infty \; \exists t_2 > t_1 : \hat{I}[t_2/time]$ implies
$\forall t_1 < \infty \; \exists t_2 > t_1 : (v < 0) \vee (t_2 \leq v(T_g + T_a))$.
Since $\forall t_1 < \infty \; \exists t_2 > t_1 : t_2 \leq v(T_g + T_a)$ is false, this implies $v < 0$, i.e., q_{nt}.
Then the iteration rule leads to

$$\langle\langle I \rangle\rangle \; * \; G \; \langle\langle (q_{nt} \wedge time = \infty) \vee q \rangle\rangle.$$

Since $x = v \wedge time = 0 \rightarrow I$ and $(q_{nt} \wedge time = \infty) \vee q$ implies
$(v < 0 \wedge time = \infty) \vee (v \geq 0 \wedge time = v(T_g + T_a) + T_g \wedge x = 0)$, the consequence rule yields the required triple. □

Example 4. Let $q_0 \equiv (in, v)$ at $t \rightarrow (out!!, f(v))$ at $(t + T_{asyn})$. We prove

$$\langle\langle time = 0 \rangle\rangle \; * [in?x \rightarrow out!!f(x)] \; \langle\langle time = \infty \wedge \forall v \forall t < \infty : q_0 \rangle\rangle.$$

Define $\hat{I} \equiv I \equiv \forall v \, \forall t < time : q_0$. Note that

$$\langle\!\langle I \wedge true \rangle\!\rangle \ [in?x \rightarrow out!!f(x)] \ \langle\!\langle I \rangle\!\rangle \text{ and}$$

$$\langle\!\langle I \wedge false \rangle\!\rangle \ [in?x \rightarrow out!!f(x)] \ \langle\!\langle false \rangle\!\rangle.$$

Further, $\forall t_1 < \infty \, \exists t_2 > t_1 : \hat{I}[t_2/time] \equiv \forall t_1 < \infty \, \exists t_2 > t_1 : \forall v \, \forall t < t_2 : q_0$ implies $\forall v \, \forall t < \infty : q_0$. Hence the iteration rule leads to

$$\langle\!\langle I \rangle\!\rangle \ * \ [in?x \rightarrow out!!f(x)] \ \langle\!\langle (\forall v \, \forall t < \infty : q_0 \wedge time = \infty) \vee false \rangle\!\rangle.$$

Note that $time = 0 \rightarrow I$ (since t ranges over nonnegative reals), and thus the consequence rule leads to the required triple. $\qquad\qquad\qquad\qquad\qquad\qquad$ □

Next we consider parallel composition. If $time$ does not occur in the commitments of the components of a parallel composition, then we have the following simple rule.

Rule 13 (Simple Parallel Composition)

$$\frac{\langle\!\langle p_1 \rangle\!\rangle \ S_1 \ \langle\!\langle q_1 \rangle\!\rangle, \quad \langle\!\langle p_2 \rangle\!\rangle \ S_2 \ \langle\!\langle q_2 \rangle\!\rangle}{\langle\!\langle p_1 \wedge p_2 \rangle\!\rangle \ S_1 \| S_2 \ \langle\!\langle q_1 \wedge q_2 \rangle\!\rangle}$$

provided, for $i, j \in \{1, 2\}$, $i \neq j$, $var(q_i) \cap var(S_j) = \emptyset$, $obs(p_i, q_i) \cap obs(S_j) = \emptyset$, and $time$ does not occur in q_i.

To obtain a sound rule we require that the commitment of one process does not refer to program variables of the other. Further, the assertions of one process should not refer to observable events of the other. The problem with $time$ in the commitments is that, in general, the termination times of S_1 and S_2 will be different. Therefore in the rule below we substitute logical variables t_1 and t_2 for $time$ in, resp., q_1 and q_2. Then the termination time of $S_1 \| S_2$ is the maximum of t_1 and t_2. Further, for completeness, we add predicates to express that process S_i does not perform any action between t_i and $time$, for $i = 1, 2$.

Rule 14 (Parallel Composition)

$$\frac{\begin{array}{c} \langle\!\langle p_1 \rangle\!\rangle \ S_1 \ \langle\!\langle q_1 \rangle\!\rangle, \quad \langle\!\langle p_2 \rangle\!\rangle \ S_2 \ \langle\!\langle q_2 \rangle\!\rangle \\ q_1[t_1/time] \wedge NoAct(obs(S_1)) \textbf{ at } [t_1, time) \wedge \\ q_2[t_2/time] \wedge NoAct(obs(S_2)) \textbf{ at } [t_2, time) \wedge \\ time = max(t_1, t_2) \rightarrow q \end{array}}{\langle\!\langle p_1 \wedge p_2 \rangle\!\rangle \ S_1 \| S_2 \ \langle\!\langle q \rangle\!\rangle}$$

provided, for $i, j \in \{1, 2\}$, $i \neq j$, we have $var(q_i) \cap var(S_j) = \emptyset$, and $obs(p_i, q_i) \cap obs(S_j) = \emptyset$.

Example 5. To prove a property of $c!!5 \parallel c?x$, observe that we can derive $\{time = 0\} \ c!!5 \ \{q_1\}$ with $q_1 \equiv (c!!, 5)$ at $0 \wedge time = T_{asyn}$, and $\{time = 0\} \ c?x \ \{q_2\}$ with $q_2 \equiv \exists t : c?$ at $[0, t) \wedge (c, x)$ at $t \wedge time = t + T_{asyn}$. Note that the restrictions of rule 14 are fulfilled; for instance, $obs(q_1) \cap obs(c?x) = \{c!!\} \cap \{c, c?\} = \emptyset$ and $obs(q_2) \cap obs(c!!5) = \{c?, c\} \cap \{c!!\} = \emptyset$. Further note that $q_1[t_1/time] \wedge q_2[t_2/time] \wedge time = max(t_1, t_2)$ implies $(c!!, 5)$ at $0 \wedge \exists t : c?$ at $[0, t) \wedge (c, x)$ at $t \wedge time = max(T_{asyn}, t + T_{asyn})$. Since, by the communication properties, $(c!!, 5)$ at 0 implies $(\neg c?)$ at 0, we obtain

$t = 0$ and thus $(c!!, 5)$ at $0 \wedge (c, x)$ at $0 \wedge time = max(T_{asyn}, T_{asyn})$. Again using the communication properties this leads to $(c!!, 5)$ at $0 \wedge x = 5 \wedge time = T_{asyn}$. Hence the parallel composition rule leads to
$\{time = 0\}\ c!!5\ \|\ c?x\ \{(c!!, 5)\ \text{at}\ 0 \wedge x = 5 \wedge time = T_{asyn}\}.$ $\qquad\qquad\square$

5 Example Water Level

We illustrate our formalism by a process control example of a water level monitoring system, which is inspired by [vS91]. The aim is to control the water level in a vessel. This level is influenced by certain disturbances, and water can be added by means of a pump. The monitoring system should be such that the water level stays between certain critical values, say between μ_l and μ_h. Of course this can only be achieved using a number of assumptions, e.g., about the disturbances and the pump. The system should also monitor these assumptions and give an appropriate message on channel *alarm* if it detects deviations from the assumptions.

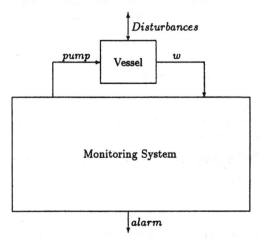

To specify this system we add a few primitives to the assertion language.

- $w(exp)$ representing the water level at time exp
- $(pump, on)$ at exp holds if the pump is on at time exp
- $(pump, off)$ at exp holds if the pump is off at time exp

In section 5.1 we design the basic control program which keeps the water level between the critical values under certain assumptions, without considering channel *alarm*. Requirements on this channel are specified in section 5.2, leading to a modified implementation.

5.1 Basic Control Program

To design the controller, first the basic assumptions about the physical model are formalized, and the top-level specification of the monitoring system is formulated.

Then a control strategy is defined in terms of the water level and the pump. Using the assumptions, we prove that this strategy implies the top-level specification. Next the control strategy is implemented. Using specifications of a sensor and an actuator we give a specification of a control unit which leads to this control strategy. Finally we give a program which satisfies the specification of the control unit.

Top-level Specification In this section we give the specification of the monitoring system and we formalize the assumptions about the effect of the pumps and the physical disturbances on the water level in the vessel. These assumptions are captured by the following properties.

$$\varphi_0^V \equiv (pump, off) \text{ at } t \lor (pump, on) \text{ at } t$$

$$\varphi_1^V \equiv (pump, off) \text{ at } [t_1, t_2) \land t_1 \le t_2 \ \rightarrow \ w(t_1) - (t_2 - t_1)\lambda \le w(t_2) \le w(t_1)$$

$$\varphi_2^V \equiv (pump, on) \text{ at } [t_1, t_2) \land t_1 \le t_2 \ \rightarrow \ w(t_1) \le w(t_2) \le w(t_1) + (t_2 - t_1)\lambda]$$

We assume that λ is a nonnegative constant. Thus when the pump is off the level decreases and when the pump is on it increases, with a rate which is at most λ.

Define $\varphi^V \equiv \forall t_1 < \infty, t_2 < \infty : \varphi_0^V \land \varphi_1^V \land \varphi_2^V$.

Now the aim is to design a monitoring system M which keeps the level inside a certain range. Formally, this is represented by the assertion

$$\varphi \equiv \forall t < \infty : \mu_l \le w(t) \le \mu_h.$$

Further we assume that system M starts at time 0, the water level has some initial value v_0, and the pump is off. Let

$$p \equiv w(0) = v_0 \land (pump, off) \text{ at } 0.$$

Then the system M is specified by

$$\langle\!\langle time = 0 \land p \land \varphi^V \rangle\!\rangle \ M \ \langle\!\langle\varphi\rangle\!\rangle.$$

Designing a Control Strategy The first design step concerns the formulation of a strategy, in terms of the water level and the pump, which leads to the top-level specification. The strategy is described by a number of properties about the relation between the water level and the status of the pump. Our intuition is that the pump should be on within D time units (for some $D \ge 0$) if the water is below a certain safety level W_l. The pump should be off within D time units if the water is above a certain safety level W_h. Thus, for $t < \infty$.

$$P_1 \equiv w(t) \le W_l \rightarrow (pump, on) \text{ in } [t, t + D)$$

$$P_2 \equiv w(t) \ge W_h \rightarrow (pump, off) \text{ in } [t, t + D)$$

Further, if the pump is on (resp. off) we have detected that the water level was below a certain level $W_l + W_d$ (resp. above $W_h - W_d$), for some parameter W_d.

$$P_3 \equiv (pump, on) \text{ at } t \rightarrow \exists t_0 \le t : w(t_0) \le W_l + W_d \land (pump, on) \text{ at } [t_0 + D, t]$$

$$P_4 \equiv (pump, off) \text{ at } t \rightarrow (pump, off) \text{ at } [0, t] \lor$$
$$\exists t_0 \le t : w(t_0) \ge W_h - W_d \land (pump, off) \text{ at } [t_0 + D, t]$$

Assuming φ^V and p, we prove the top-level specification φ from these properties. First we give a few useful lemmas.

Lemma 4. φ^V implies, for all $t_1 < \infty$, $t_2 < \infty$,

$$t_1 \leq t_2 \rightarrow w(t_1) - (t_2 - t_1)\lambda \leq w(t_2) \leq w(t_1) + (t_2 - t_1)\lambda$$

Further, using φ^V, it is easy to prove that w is a continuous function. Then standard mathematical analysis yields the following lemma.

Lemma 5 Intermediate Value Property. Consider t_1 and t_2 with $t_1 < t_2$. Then for any μ with $w(t_1) \leq \mu \leq w(t_2)$ or $w(t_2) \leq \mu \leq w(t_1)$ there exists a $t_3 \in [t_1, t_2]$ such that $w(t_3) = \mu$.

Then, as proved in Appendix A, we obtain the main theorem.

Theorem 6. If $W_l \leq v_0$, $W_l - D\lambda \geq \mu_l$, $W_h + D\lambda \leq \mu_h$, and $W_h \geq W_l + D\lambda + W_d$, then

$$\varphi^V \wedge p \wedge P_1 \wedge P_2 \wedge P_3 \wedge P_4 \rightarrow \varphi.$$

Note that $W_l - D\lambda \geq \mu_l$ expresses that our safety level W_l minus $D\lambda$, which is the maximal decrease of the water level between the point at which W_l has been reached and the point where the pump is on, is greater or equal than the critical limit μ_l.

We show that, given the theorem above, it is sufficient to implement M according to the specification

$$\langle\!\langle time = 0 \wedge p \wedge \varphi^V \rangle\!\rangle\ M\ \langle\!\langle P_1 \wedge P_2 \wedge P_3 \wedge P_4 \rangle\!\rangle.$$

Since p and φ^V do not refer to program variables or *time*, the initial invariance axiom allows us to derive

$$\langle\!\langle time = 0 \wedge p \wedge \varphi^V \rangle\!\rangle\ M\ \langle\!\langle p \wedge \varphi^V \rangle\!\rangle$$

and thus, using the new specification above, the consequence rule leads to

$$\langle\!\langle time = 0 \wedge p \wedge \varphi^V \rangle\!\rangle\ M\ \langle\!\langle p \wedge \varphi^V \wedge P_1 \wedge P_2 \wedge P_3 \wedge P_4 \rangle\!\rangle.$$

Hence, by theorem 6 and the consequence rule, we obtain

$$\langle\!\langle time = 0 \wedge p \wedge \varphi^V \rangle\!\rangle\ M\ \langle\!\langle \varphi \rangle\!\rangle,$$

provided $W_l \leq v_0$, $W_l - D\lambda \geq \mu_l$ $W_h + D\lambda \leq \mu_h$, and $W_h \geq D\lambda + W_l + W_d$.

Implementing the Control Strategy To implement the strategy given by $P_1 \wedge P_2 \wedge P_3 \wedge P_4$, we have to measure the water level and to control the pump. Therefore system M is split up into $S\|C\|A$, where sensor S measures the water level, control unit C implements the strategy, and actuator A controls the pump. We assume given certain specifications for the sensor and the actuator, and try to find a specification of the control unit such that we can prove the required properties P_1 through P_4.

For the sensor S assume given constants Δ_s and ε_s, and let

$$\varphi_1^S \equiv L!!\ \text{in}\ [t, t + \Delta_s)$$

$\varphi_2^S \equiv (L!!, v)$ at $t \to v - \varepsilon_s \leq w(t) \leq v + \varepsilon_s$

Define $\varphi^S \equiv \forall t < \infty : \varphi_1^S \wedge \varphi_2^S$. Then assume S satisfies the specification

$$\langle\!\langle time = 0 \rangle\!\rangle \ S \ \langle\!\langle \varphi^S \rangle\!\rangle.$$

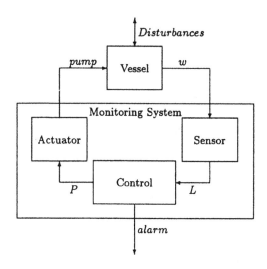

For the actuator A define, given constants Δ_a and δ_a,

$\varphi_1^A \equiv await \ P?$ in $[t_1, t_1 + \Delta_a)$

$\varphi_2^A \equiv (P, 0)$ at $t_1 \wedge (\neg P)$ at $\langle t_1, t_2 \rangle \to (pump, off)$ at $\langle t_1 + \delta_a, t_2 \rangle$

$\varphi_3^A \equiv (P, 1)$ at $t_1 \wedge (\neg P)$ at $\langle t_1, t_2 \rangle \to (pump, on)$ at $\langle t_1 + \delta_a, t_2 \rangle$

$\varphi_4^A \equiv (pump, on)$ at $t_1 \to \exists t_0 \leq t_1 : (P, 1)$ at $t_0 \wedge (pump, on)$ at $\langle t_0 + \delta_a, t_1]$

$\varphi_5^A \equiv (pump, off)$ at $t_1 \to (pump, off)$ at $[0, t_1] \vee$
$$\exists t_0 \leq t_1 : (P, 0) \ \text{at} \ t_0 \wedge (pump, off) \ \text{at} \ \langle t_0 + \delta_a, t_1]$$

Define $\varphi^A \equiv \forall t_1 < \infty, t_2 < \infty : \bigwedge_{i=1}^{5} \varphi_i^A$. Then assume A satisfies

$$\langle\!\langle time = 0 \rangle\!\rangle \ A \ \langle\!\langle \varphi^A \rangle\!\rangle.$$

Our first aim is to find a specification φ^C for the control unit C. To investigate the requirements on φ^C, assume

$$\langle\!\langle time = 0 \wedge p \wedge \varphi^V \rangle\!\rangle \ C \ \langle\!\langle \varphi^C \rangle\!\rangle$$

where $obs(\varphi^C) \subseteq \{L?, L, P!!\}$ and $var(\varphi^C) = \emptyset$.

To apply the parallel composition rule, observe that $var(\varphi^S) = var(\varphi^A) = \emptyset$. Further φ^S and φ^A only refer to the observables of the process itself, i.e., $obs(\varphi^S) = \{L!!\}$, $obs(\varphi^A) = \{P?, P\}$. Thus if we assume that $obs(S) \cap \{L?, L, P!!, P?, P\} = \emptyset$, $obs(A) \cap \{L!!, L?, L, P!!\} = \emptyset$, and $obs(C) \cap \{L!!, P?, P\} = \emptyset$, then the simple parallel composition rule leads to

$$\langle\!\langle time = 0 \wedge p \wedge \varphi^V \rangle\!\rangle \ S\|C\|A \ \langle\!\langle \varphi^S \wedge \varphi^C \wedge \varphi^A \rangle\!\rangle,$$

with $p \equiv w(0) = v_0 \wedge (pump, off)$ at 0.

By the initial invariance axiom we obtain

$$\langle\!\langle p \wedge \varphi^V \rangle\!\rangle \ S\|C\|A \ \langle\!\langle p \wedge \varphi^V \rangle\!\rangle.$$

Hence the conjunction rule leads to

$$\langle\!\langle time = 0 \wedge p \wedge \varphi^V \rangle\!\rangle \ S\|C\|A \ \langle\!\langle \varphi^V \wedge \varphi^S \wedge \varphi^C \wedge \varphi^A \wedge p \rangle\!\rangle$$

By the consequence rule, we obtain that $S\|C\|A$ is a correct implementation of M if
$$\varphi^V \wedge \varphi^S \wedge \varphi^C \wedge \varphi^A \wedge p \to P_1 \wedge P_2 \wedge P_3 \wedge P_4.$$
Hence the aim is to find φ^C should that this implication holds.

Therefore, we give a specification φ^C which expresses that if the sensor sends a value (along channel L) which is below a certain threshold V_l, a signal (the value 1) should be send to the actuator (along channel P) to switch the pump on. Similarly, a sensor value above a threshold V_h should lead to the value 0 along P to indicate that the pump should be switched off. Further these signals to the actuator are only transmitted if there is indeed a sensor value outside the safety range. To guarantee the properties P_1 through P_4, the control unit should be ready to receive a value along L during at least Δ_s in every period of (say) X_c time units. This requirement also avoids a trivial implementation which never communicates along L and P. Finally, to be able to conclude that messages along channel P are not lost, messages on channel P are transmitted with a distance of at least Δ_a time units. Formally, let $\varphi^C \equiv \forall t < \infty : \bigwedge_{i=1}^{6} \varphi_i^C$, where

$\varphi_1^C \equiv (L, v)$ at $t \wedge v \leq V_l \to (P!!, 1)$ in $[t, t + D_c]$

$\varphi_2^C \equiv (L, v)$ at $t \wedge v \geq V_h \to (P!!, 0)$ in $[t, t + D_c]$

$\varphi_3^C \equiv (P!!, 1)$ at $t \to \exists v \leq V_l : (L, v)$ in $[t - D_c, t]$

$\varphi_4^C \equiv (P!!, 0)$ at $t \to \exists v \geq V_h : (L, v)$ in $[t - D_c, t]$

$\varphi_5^C \equiv await_{\geq \Delta_s} L?$ in $[t, t + X_c)$

$\varphi_6^C \equiv P!!$ at $t \to t \geq \Delta_a \wedge (\neg P!!)$ at $[t - \Delta_a, t)$

Note that this already implies certain design decisions. For instance, to reduce the number of messages along channel P, we could have given an alternative specification where the value 1 is only transmitted along P if the previous value was 0, etc. Our specification φ^C, however, gives a strategy which is more robust against temporal failures of the actuator or the link P.

To show that $S\|C\|A$ is a correct implementation of M it remains to prove

Theorem 7. If $W_l + \varepsilon_s \leq V_l \leq W_l + W_d - \varepsilon_s$, $W_h - W_d + \varepsilon_s \leq V_h \leq W_h - \varepsilon_s$, $\delta_a < \Delta_a$, and $D \geq \Delta_s + X_c + D_c + \delta_a$, then

$$\varphi^V \wedge \varphi^S \wedge \varphi^C \wedge \varphi^A \wedge p \to P_1 \wedge P_2 \wedge P_3 \wedge P_4.$$

Proof: Observe that by $\forall t < \infty : \varphi_1^S$ and $\forall t < \infty : \varphi_5^C$ we obtain from lemma 3
$$\forall t < \infty : L \text{ in } [t, t + \Delta_s + X_c) \tag{1}$$
Further by $\forall t < \infty : \varphi_6^C$ and $\forall t < \infty : \varphi_1^A$ we obtain by lemma 2 that
$$\forall t < \infty : P!! \text{ at } t \leftrightarrow P \text{ at } t. \tag{2}$$

Proof of P_1 Assume $w(t) \leq W_l$. By (1) there exists a $t_1 \in [t, t + \Delta_s + X_c)$ and a v such that (L, v) at t_1. Hence $(L!!, v)$ at t_1, and φ_2^S leads to $v - \varepsilon_s \leq w(t_1) \leq v + \varepsilon_s$.

- If $\neg((pump, off)$ at $[t, t_1])$ then $(pump, on)$ in $[t, t_1]$, and thus $(pump, on)$ in $[t, t+D)$, since $t_1 < t + \Delta_s + X_c$ and $\Delta_s + X_c < D$ imply $t_1 < t + D$.
- Otherwise, $(pump, off)$ at $[t, t_1]$. Thus, by φ_1^V, $w(t_1) \leq w(t) \leq W_l$.
  Using $w(t_1) \geq v - \varepsilon_s$ we obtain $v \leq w(t_1) + \varepsilon_s \leq W_l + \varepsilon_s$.
  Since $W_l + \varepsilon_s \leq V_l$ we have (L, v) at $t_1 \wedge v \leq V_l$, thus by φ_1^C
  there exists a $t_2 \in [t_1, t_1 + D_c]$ with $(P!!, 1)$ at t_2. By (2) we obtain $(P, 1)$ at t_2.
  By $P!!$ at t_2 and φ_6^C we can derive $(\neg P!!)$ at $\langle t_2, t_2 + \Delta_a \rangle$, thus
  $(\neg P)$ at $\langle t_2, t_2 + \Delta_a \rangle$. Then φ_3^A leads to $(pump, on)$ at $\langle t_2 + \delta_a, t_2 + \Delta_a \rangle$.
  We show that this implies $(pump, on)$ in $[t, t + D)$. First note that the interval
  $\langle t_2 + \delta_a, t_2 + \Delta_a \rangle$ is not empty, since $\delta_a < \Delta_a$. Second, since $\delta_a \geq 0$, assumption
  $\delta_a < \Delta_a$ implies $\Delta_a > 0$ and thus $t_2 + \Delta_a \geq t_1 + \Delta_a \geq t + \Delta_a > t$. Third,
  $t_2 + \delta_a \leq t_1 + D_c + \delta_a < t + \Delta_s + X_c + D_c + \delta_a$, thus $t_2 + \delta_a < t + D$ since
  $D \geq \Delta_s + X_c + D_c + \delta_a$.

Proof of P_2 Similar to the proof of P_1, using e.g. $V_h \leq W_h - \varepsilon_s$.

Proof of P_3 Assume $(pump, on)$ at t. We have to prove
$\exists t_0 \leq t : w(t_0) \leq W_l + W_d \wedge (pump, on)$ at $[t_0 + D, t]$.
By φ_4^A there exists a $t_1 \leq t$ such that $(P, 1)$ at t_1 and $(pump, on)$ at $\langle t_1 + \delta_a, t]$.
Hence $(P!!, 1)$ at t_1 and thus by φ_3^C there exist a $t_0 \in [t_1 - D_c, t_1]$ and a $v \leq V_l$ such
that (L, v) at t_0. Thus $(L!!, v)$ at t_0 and hence, by φ_2^S, $w(t_0) \in [v - \varepsilon_s, v + \varepsilon_s]$. Then
$w(t_0) \leq v + \varepsilon_s \leq V_l + \varepsilon_s$ and thus, using $V_l \leq W_l + W_d - \varepsilon_s$, we have $w(t_0) \leq W_l + W_d$.
Since $(pump, on)$ at $\langle t_1 + \delta_a, t]$, we obtain $(pump, on)$ at $[t_0 + D, t]$ if $t_1 + \delta_a \leq t_0 + D$.
Note that $t_0 \geq t_1 - D_c$, thus $t_1 \leq t_0 + D_c$. Hence $t_1 + \delta_a \leq t_0 + D_c + \delta_a$ and thus,
using $D_c + \delta_a \leq D$, we obtain $t_1 + \delta_a \leq t_0 + D$.

Proof of P_4 Similar to the proof of P_3, using e.g. $W_h - W_d + \varepsilon_s \leq V_h$. $\qquad\square$

This leads to a proof of $P_1 \wedge P_2 \wedge P_3 \wedge P_4$, provided $W_l + \varepsilon_s \leq V_l \leq W_l + W_d - \varepsilon_s$,
$W_h - W_d + \varepsilon_s \leq V_h \leq W_h - \varepsilon_s$, $\delta_a < \Delta_a$, and $D \geq \Delta_s + X_c + D_c + \delta_a$.
Previously, we have assumed the following timing requirements
$W_l \leq v_0$, $W_l - D\lambda \geq \mu_l$ $W_h + D\lambda \leq \mu_h$, and $W_h \geq D\lambda + W_l + W_d$.
Eliminating the parameters W_l, W_h, and W_d, which have been used to formulate
the intermediate specification $P_1 \wedge \ldots \wedge P_4$, we obtain
$\mu_l + D\lambda + \varepsilon_s \leq V_l \leq \mu_h - 2D\lambda - \varepsilon_s$ and $\mu_l + 2D\lambda + \varepsilon_s \leq V_h \leq \mu_h - D\lambda - \varepsilon_s$.
Thus, for the existence of V_l and V_h, we should have
$\mu_l + D\lambda + \varepsilon_s \leq \mu_h - 2D\lambda - \varepsilon_s$ and $\mu_l + 2D\lambda + \varepsilon_s \leq \mu_h - D\lambda - \varepsilon_s$. Hence we require
$$\mu_l + 3D\lambda + 2\varepsilon_s \leq \mu_h \qquad (R1)$$
Since we need $W_l \leq v_0$ and $W_l - D\lambda \geq \mu_l$, we require that
$$v_0 \geq \mu_l + D\lambda \qquad (R2)$$
Assuming that these requirements are fulfilled, we take V_l minimal and V_h maximal
to have a minimal number of pump switches:
$$V_l = \mu_l + D\lambda + \varepsilon_s \quad \text{and} \quad V_h = \mu_h - D\lambda - \varepsilon_s \qquad (*)$$
Hence we have derived φ provided we have the requirements (R1), (R2) and
$$D \geq \Delta_s + X_c + D_c + \delta_a \qquad (R3)$$
Finally, we add $\delta_a < \Delta_a$ to the specification of the actuator. Note that this expresses

that the actuator changes the status of the pump according to a certain message before it is ready to receive a new message.

The Basic Control Program Now we can concentrate on the implementation of C according to its specification. In this case no further design steps are needed, and we directly give a program CU for the control unit:

$$CU \equiv *[L?x \rightarrow [x \leq V_l \rightarrow P!!1 \,[]\, x \geq V_h \rightarrow P!!0]\,]$$

Note that the syntactic constraint $obs(C) \cap \{L!!, P?, P\} = \emptyset$, which we have used before to apply the parallel composition rule, is satisfied. It remains to prove that this program satisfies the specification

$$\langle\!\langle time = 0 \wedge p \wedge \varphi^V \rangle\!\rangle \; CU \; \langle\!\langle \varphi^C \rangle\!\rangle.$$

This is done by means of the proof system of section 4. Here we only give the main points of this proof, namely the invariant for the iteration and the required conditions on the parameters.

- To prove $\forall t < \infty : \varphi_1^C \wedge \varphi_2^C \wedge \varphi_3^C \wedge \varphi_4^C$, we use the invariant
  $\forall t < time : \varphi_1^C \wedge \varphi_2^C \wedge \varphi_3^C \wedge \varphi_4^C$ and require $D_c \geq T_{asyn} + T_g$.
- The proof of $\forall t < \infty : \varphi_5^C$ is given by means of the invariant
  $\forall t < time - 2T_{asyn} - T_g : \varphi_5^C$ provided $X_c \geq 2T_{asyn} + 2T_g$.
- Finally, $\forall t < \infty : \varphi_6^C$ is proved assuming $Y_c \leq 2T_g + T_{asyn}$ and using the invariant
  $\forall t < time : \varphi_6^C \wedge time \geq 0 \wedge (\neg P!!)$ at $\langle time - T_{asyn}, time \rangle$.

This leads to a proof that program CU satisfies φ^C, with
$$X_c = 2T_{asyn} + 2T_g,$$
$$\Delta_a = T_{asyn} + 2T_g, \text{ and}$$
$$D_c = T_{asyn} + T_g.$$
Finally, consider the timing requirements (R1), (R2), and (R3) on D derived earlier. Observe that (R1) and (R2) represent, for a given D, constraints on the given parameters. These constraints are as weak as possible if D is minimal. Hence, considering (R3), we take $D = \Delta_s + X_c + D_c + \delta_a = \Delta_s + \delta_a + 3T_{asyn} + 3T_g$. Note that then V_l and V_h are determined by (*). Hence we have obtained a correct implementation, provided the inequalities (R1) and (R2) hold for the given parameters.

5.2 Requirements on the Alarm Channel

An important task of control programs is to detect errors in, e.g., sensors, actuators, and the modelling of physical processes. In this paper we extend the specification with a number of requirement on channel *alarm*. Here we directly give a specification for C. First we assert that there should be an alarm message within δ_e time units if C receives a sensor value outside the required critical range. Further such a message should only be send if the sensor value is indeed outside this range.

$$\varphi_1^E \equiv (L, v) \text{ at } t \wedge v > \mu_h \rightarrow (alarm!!, 0) \text{ in } [t, t + \delta_e]$$

$$\varphi_2^E \equiv (L, v) \text{ at } t \wedge v < \mu_l \rightarrow (alarm!!, 1) \text{ in } [t, t + \delta_e]$$

$$\varphi_3^E \equiv (alarm!!, 0) \text{ at } t \rightarrow \exists v > \mu_h : (L, v) \text{ in } [t - \delta_e, t]$$

$\varphi_4^E \equiv (alarm!!, 1)$ at $t \to \exists v < \mu_l : (L, v)$ in $[t - \delta_e, t]$

Note that if we obtain a value greater than μ_h from the sensor then the water need not be above μ_h, since the sensor allows a deviation of ε_s. If we want to be sure that an alarm message is only send if the water level is outside the range then we should replace μ_h and μ_l by, respectively, $\mu_h + \varepsilon_s$ and $\mu_l - \varepsilon_s$. If one prefers to send an alarm message if there is a possibility that the water level is outside the range then one should use $\mu_h - \varepsilon_s$ and $\mu_l + \varepsilon_s$, respectively.

To detect a failure of the sensor, an alarm message should be send if the sensor does not meet its promptness specification, i.e., if it does not send a message within Δ_s.

$\varphi_5^E \equiv t \geq \Delta_s \wedge L?$ at $[t - \Delta_s, t) \to (alarm!!, 2)$ in $[t, t + \delta_e]$

$\varphi_6^E \equiv (alarm!!, 2)$ at $t \to \exists t_0 \in [t - \Delta_s - \delta_e, t - \Delta_s] : L?$ at $[t_0, t_0 + \Delta_s)$

Further, if the value received from the sensor changes too fast, i.e., not according to assumption φ^V of the vessel, an alarm message should be given within δ_e.

$\varphi_7^E \equiv (L, v_1)$ at $t_1 \wedge (L, v_2)$ at $t_2 \wedge (\neg L)$ at $\langle t_1, t_2 \rangle \wedge t_1 < t_2 \wedge$
$$v_2 > v_1 + (t_2 - t_1)\lambda + \delta_d \to (alarm!!, 3) \text{ in } [t_2, t_2 + \delta_e]$$

$\varphi_8^E \equiv (L, v_1)$ at $t_1 \wedge (L, v_2)$ at $t_2 \wedge (\neg L)$ at $\langle t_1, t_2 \rangle \wedge t_1 < t_2 \wedge$
$$v_2 < v_1 - (t_2 - t_1)\lambda - \delta_d \to (alarm!!, 4) \text{ in } [t_2, t_2 + \delta_e]$$

$\varphi_9^E \equiv (alarm!!, 3)$ at $t \to \exists t_1, t_2, v_1, v_2 : (L, v_1)$ at $t_1 \wedge (L, v_2)$ at $t_2 \wedge t_1 < t_2 \wedge$
$$t_2 \in [t - \delta_e, t] \wedge v_2 > v_1 + (t_2 - t_1)\lambda - \delta_d$$

$\varphi_{10}^E \equiv (alarm!!, 4)$ at $t \to \exists t_1, t_2, v_1, v_2 : (L, v_1)$ at $t_1 \wedge (L, v_2)$ at $t_2 \wedge t_1 < t_2 \wedge$
$$t_2 \in [t - \delta_e, t] \wedge v_2 < v_1 - (t_2 - t_1)\lambda + \delta_d$$

The parameter δ_d has been added to represent a correction factor, e.g., caused by the imprecision of local timers. Define

$$\varphi^E \equiv \forall t, t_1, t_2 < \infty : \bigwedge_{i=1}^{10} \varphi_i^E.$$

We strengthen the specification of C by adding φ^E:

$$\langle\!\langle time = 0 \wedge p \wedge \varphi^V \rangle\!\rangle \; C \; \langle\!\langle \varphi^C \wedge \varphi^E \rangle\!\rangle.$$

Since we already have a program CU which satisfies φ^C, the idea is to add a parallel component SC (Sensor Control) which checks the values from the sensor and sends the appropriate alarm messages. Further SC transmits the sensor value along an internal channel B to CU.

Since program CU expects to receive this value along channel L we have to rename channels. For a program S we use $S[d/c]$ to denote the program which is obtained from S by replacing each occurrence of channel c by d. Here we use a process PC (Pump Control) which is implemented as $CU[B/L]$. In order to obtain a specification for PC, we introduce a rule for channel renaming. Let $p[d/c]$ denote the assertion obtained from p by replacing each occurrence of channel c by d. Then we have

Rule 15 (Channel Renaming) $\quad \dfrac{\langle\!\langle p \rangle\!\rangle \; S \; \langle\!\langle q \rangle\!\rangle}{\langle\!\langle p[d/c] \rangle\!\rangle \; S[d/c] \; \langle\!\langle q[d/c] \rangle\!\rangle}$

provided d does not occur in p, q and S, i.e., $d \notin obs(p) \cup obs(S) \cup obs(q)$.

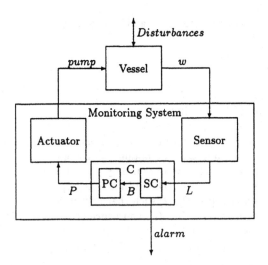

Note that we can strengthen property φ_5^C of CU to

$$\hat{\varphi}_5^C \equiv await\ L?\ \text{in}\ [0, X_c^1\rangle \wedge await\ L?\ \text{in}\ [t, t + X_c^2\rangle,$$

with $X_c^1 = T_g$ and $X_c^2 = 2T_{asyn} + 2T_g$. Further parameter D_c of φ^C is replaced by D_{pc}. Let $\hat{\varphi}^C \equiv \forall t < \infty : (\varphi_1^C \wedge \varphi_2^C \wedge \varphi_3^C \wedge \varphi_4^C)[D_{pc}/D_c] \wedge \hat{\varphi}_5^C \wedge \varphi_6^C$. Then we have

$$\langle\!\langle time = 0 \wedge p \wedge \varphi^V \rangle\!\rangle\ CU\ \langle\!\langle \hat{\varphi}^C \rangle\!\rangle$$

with $\Delta_a = T_{asyn} + 2T_g$, and $D_{pc} = T_{asyn} + T_g$. By the channel renaming rule,

$$\langle\!\langle time = 0 \wedge p \wedge \varphi^V \rangle\!\rangle\ PC\ \langle\!\langle \varphi^{PC} \rangle\!\rangle$$

where $\varphi^{PC} \equiv \hat{\varphi}^C[B/L]$. Process SC is specified by

$$\langle\!\langle time = 0 \rangle\!\rangle\ SC\ \langle\!\langle \varphi^E \wedge \varphi^{SC} \rangle\!\rangle$$

where $\varphi^{SC} \equiv \forall t < \infty : \bigwedge_{i=1}^4 \varphi_i^{SC}$ with, for some parameter D_{sc},

$\varphi_1^{SC} \equiv (L, v)\ \text{at}\ t \to (B!!, v)\ \text{in}\ [t, t + D_{sc}]$

$\varphi_2^{SC} \equiv (B!!, v)\ \text{at}\ t \to (L, v)\ \text{in}\ [t - D_{sc}, t]$

$\varphi_3^{SC} \equiv B!!\ \text{at}\ t \to t \geq X_c^1 \wedge (\neg B!!)\ \text{at}\ [t - X_c^2, t\rangle$

$\varphi_4^{SC} \equiv await_{\geq \Delta_a} L?\ \text{in}\ [t, t + X_c)$

Note that $var(\varphi^E \wedge \varphi^{SC}) = var(\varphi^{PC}) = \emptyset$. Further $obs(\varphi^E \wedge \varphi^{SC}) \cap obs(PC) = \{L?, L, alarm!!, B!!\} \cap \{B?, B, P!!\} = \emptyset$ and, assuming $obs(SC) = \{L?, L, alarm!!, B!!\}$, we have $obs(\varphi^{PC}) \cap obs(SC) = \{B?, B, P!!\} \cap \{L?, L, alarm!!, B!!\} = \emptyset$. Hence we can apply the parallel composition rule to obtain

$$\langle\!\langle time = 0 \wedge p \wedge \varphi^V \rangle\!\rangle\ SC\|PC\ \langle\!\langle \varphi^E \wedge \varphi^{SC} \wedge \varphi^{PC} \rangle\!\rangle$$

To show that $SC\|PC$ satisfies $\varphi^C \wedge \varphi^E$ of C, we prove the following theorem.

Theorem 8. If $D_c \geq D_{sc} + D_{pc}$ then $\varphi^{SC} \wedge \varphi^{PC} \to \varphi^C$.

Proof: Note that $\varphi_5^C \equiv \varphi_4^{SC}$ and $\varphi_6^C \equiv \varphi_6^{PC}$.

Further observe that from $\forall t < \infty : \varphi_5^{PC}$ and $\forall t < \infty : \varphi_3^{SC}$ we obtain by lemma 1

$$\forall t < \infty : B!! \text{ at } t \leftrightarrow B \text{ at } t \qquad (1)$$

We prove $\varphi_1^C, \ldots, \varphi_4^C$, assuming

$$D_c \geq D_{sc} + D_{pc}. \qquad (2)$$

φ_1^C Assume (L, v) at $t \wedge v \leq V_l$.
   By φ_1^{SC} we obtain $(B!!, v)$ in $[t, t + D_{sc}] \wedge v \leq V_l$.
   Using (1) this yields (B, v) in $[t, t + D_{sc}] \wedge v \leq V_l$.
   Then φ_1^{PC} leads to $(P!!, 1)$ in $[t, t+D_{sc}+D_{pc}]$. Hence, by (2), $(P!!, 1)$ in $[t, t+D_c]$.

φ_2^C Similarly, by φ_1^{SC}, (1), and φ_2^{PC} we obtain φ_2^C, using (2).

φ_3^C Assume $(P!!, 1)$ at t. By φ_3^{PC} we obtain $\exists v \leq V_l : (B, v)$ in $[t - D_{pc}, t]$.
   Using (1) this leads to $\exists v \leq V_l : (B!!, v)$ in $[t - D_{pc}, t]$.
   Thus there exists a $t_1 \in [t - D_{pc}, t]$ such that $(B!!, v)$ at t_1.
   Then, by φ_2^{SC}, (L, v) in $[t_1 - D_{sc}, t_1]$, and hence (L, v) in $[t - D_{pc} - D_{sc}, t]$.
   Hence, using (2), we have $\exists v \leq V_l : (L, v)$ in $[t - D_c, t]$.

φ_4^C Similarly, we obtain φ_4^C from φ_4^{PC}, (1), φ_2^{SC}, and (2).   □

Finally, process SC has to be implemented according to its specification. Using parameter V_{sc}, we propose the following program.

$$first := false\,;$$

$$* \; [\neg first; L?x_{old} \rightarrow B!!x_{old}\,;\, y_{old} := clock\,;$$

$$[x_{old} < \mu_l \rightarrow alarm!!1 \;[]\; x_{old} > \mu_h \rightarrow alarm!!0]\,;$$

$$first := true$$

$$[]\neg first; \mathbf{delay}\; \Delta_s \rightarrow alarm!!2$$

$$]\,;$$

$$* \; [L?x_{new} \rightarrow B!!x_{new}\,;\, y_{new} := clock\,;$$

$$[x_{new} < \mu_l \rightarrow alarm!!1 \;[]\; x_{new} > \mu_h \rightarrow alarm!!0]\,;$$

$$[x_{new} > x_{old} + (y_{new} - y_{old})V_{sc} \rightarrow alarm!!3$$

$$[]x_{new} < x_{old} - (y_{new} - y_{old})V_{sc} \rightarrow alarm!!4]\,;$$

$$x_{old} := x_{new}\,;\, y_{old} := y_{new}$$

$$[]\mathbf{delay}\; \Delta_s \rightarrow alarm!!2$$

$$]$$

Note that the assumption $obs(SC) = \{L?, L, alarm!!, B!!\}$, to justify the application of the parallel composition rule, is satisfied. To prove that this program satisfies the specification $\langle\!\langle time = 0 \rangle\!\rangle\, SC\, \langle\!\langle \varphi^E \wedge \varphi^{SC} \rangle\!\rangle$, again the main point is to find suitable invariants for the iterations. Here we only give the invariant for the second iteration.

- To prove $\forall t < \infty : \bigwedge_{i=1}^{6} \varphi_i^E$ we use the invariant $\forall t < time : \bigwedge_{i=1}^{6} \varphi_i^E$.
- For $\forall t, t_1, t_2 < \infty : \bigwedge_{i=7}^{10} \varphi_i^E$ we have the invariant $\forall t, t_1, t_2 < time : \bigwedge_{i=7}^{10} \varphi_i^E \wedge$
  $\exists t_3 < time : (L, x_{old})$ at $t_3 \wedge (\neg L)$ at $\langle t_3, time \rangle \wedge y_{old} = clock(t_3 + 2T_{asyn})$.
- We prove $\forall t < time : \varphi_1^{SC} \wedge \varphi_2^{SC}$ by the invariant $\forall t < time : \varphi_1^{SC} \wedge \varphi_2^{SC}$.

- For $\forall t < \infty : \varphi_3^{SC}$ we use $\forall t < time : \varphi_3^{SC} \wedge time \geq T_a \wedge$
  $(\neg B!!)$ at $\langle time - T_{asyn} - T_c - T_g - T_a, time \rangle$.
- Finally $\forall t < \infty : \varphi_4^{SC}$ is proved by means of $\forall t \leq time - X_c + T_g : \varphi_4^{SC}$.

Then SC satisfies the required specification, using $V_{sc} = \frac{\lambda}{g(1-r^2)u}$, and with

$\delta_d = uV_{sc}$,
$D_{sc} = T_{asyn}$,
$X_c = T_c + 4T_g + 4T_{asyn} + 2T_a$, and
$\delta_e = 3T_{asyn} + +2T_g + T_c$.

(Recall that u, g, and r are the parameters of our assumption about local timers.)
Then according to theorem 8 and the consequence rule we obtain

$$\langle\!\langle time = 0 \wedge p \wedge \varphi^V \rangle\!\rangle \; SC\|PC \; \langle\!\langle \varphi^E \wedge \varphi^C \rangle\!\rangle.$$

provided $D_c \geq D_{sc} + D_{pc} = T_{asyn} + T_{asyn} + T_g = 2T_{asyn} + T_g$, and with
$\Delta_a = T_{asyn} + 2T_g$, and X_c and δ_e as above. Finally, consider again the timing
requirements (R1), (R2), and (R3) on D derived earlier. Considering (R1) and (R2)
as requirements on the given parameters, they are as weak as possible if D is minimal.
Since D is minimal if D_c is minimal, let

$D_c = 2T_{asyn} + T_g$.

Then $D = \Delta_s + X_c + D_c + \delta_a = \Delta_s + \delta_a + 5T_g + 6T_{asyn} + 2T_a + T_c$.
Using this value of D, V_l and V_h are determined by (*)). Finally note that the
requirements (R1) and (R2) on the given parameters of the system are now stronger
than those in section 5.1, since D has a greater value for the final program. Informally
speaking, this is the price we have to pay for the additional monitoring.

6 Concluding Remarks

In this paper we have extended Hoare logic to specify and verify distributed real-time
systems. The framework has been applied to a water level monitoring system, as a
typical example of a hybrid system. Characteristic for the method is the fact that
we have a compositional proof system, thus supporting modularity and a separation
of concerns. As illustrated by the water level example, we can separate the control
theory part, designing a control strategy in terms of a continuous interface, from the
program development phase. Further, to establish a relation between the continuous
variables of the control strategy and the discrete interface of the program we have
only used the specifications of a sensor and an actuator. In this way, the design problem has been reduced to the independent development of a small control program
according to its specification. Finally, to include error detection in the program, we
could reuse the original control program and its correctness proof, by only referring
to its specification.

Note that during the first stages of the design, when reasoning about concurrent modules, the syntax of the logic and the assumption/commitment structure of
our specifications is not very important and might have been replaced by a more
concise notation. The main point at these early stages of the design is to have a
compositional rule for parallel composition which allows us to reason with the specifications of components and, together with an axiomatization of the communication

mechanism, derive properties of a compound system. Therefore the logic should be sufficiently expressive to specify the real-time communication interface of components. Similar to untimed compositional methods [Zwi89], at parallel composition we require syntactically that the specification of a component only refers to its own interface. Here we have used a simple first-order assertion language with primitives to express the timed occurrence of observables. Instead of our explicit time formalism, the specifications could have been expressed in an implicit time logic such as Metric Temporal Logic [Koy90] or the Duration Calculus [CHR91].

The structure of our specifications has been motivated by the aim to design systems down to the program code. By using the principles of Hoare triples we have obtained a convenient compositional axiomatization of the sequential constructs of the programming language. Although we have added real-time and extended the interpretation of triples to non-terminating computations, it turns out that the formalism is still convenient for program derivation. Typically, the main point in such a proof is to find suitable invariants for the iterations. Once these have been determined, most of the proof obligations are easy to check.

Obviously, for complicated programs suitable tools are required to list proof obligations, prove trivial steps automatically, collect timing assumptions on the parameters, etc. This will be a topic of future research. Another option is to investigate whether some of the properties, such as bounded readiness to receive input and minimal separation of outputs, can be verified by means of model checking algorithms. Recently these methods been extended to real-time systems using dense time [ACD90, HNSY92].

Acknowledgements

I am indebted to Anders Ravn of the Technical University of Denmark for his suggestion of the parallel implementation of the error detection component, as described in section 5.2. Many thanks goes to Marcin Engel and Marcin Kubica of the Warsaw University for their accurate comments on a draft version of this paper.

A   Proof of Theorem 6

Proof of Theorem 6
Suppose $W_l \leq v_0$, $W_l - D\lambda \geq \mu_l$, $W_h + D\lambda \leq \mu_h$, and $W_h \geq W_l + D\lambda + W_d$.
Assume $\varphi^V \wedge p \wedge P_1 \wedge P_2 \wedge P_3 \wedge P_4$. We have to prove φ.
Consider $t < \infty$. We show $\mu_l \leq w(t) \leq \mu_h$ by contradiction.
First assume $w(t) < \mu_l$. By φ_0^V, we have either $(pump, off)$ at t or $(pump, on)$ at t.

1. If $(pump, off)$ at t then by P_4 we have $(pump, off)$ at $[0, t]$ or
   $\exists t_0 \leq t : w(t_0) \geq W_h - W_d \wedge (pump, off)$ at $[t_0 + D, t]$.
   (a) Suppose $(pump, off)$ at $[0, t]$. Since $D \geq 0$, we have $W_l \geq \mu_l$ and thus, using p, $w(t) < \mu_l \leq W_l \leq v_0 = w(0)$. Hence, by the Intermediate Value Property, there exists a $t_0 \in [0, t]$ such that $w(t_0) = W_l$. By P_1 we obtain $(pump, on)$ in $[t_0, t_0 + D)$. By $(pump, off)$ at $[0, t]$, also $(pump, off)$ at $[t_0, t]$, hence $t_0 + D > t$, and thus $t - t_0 < D$. From $(pump, off)$ at $[t_0, t]$ and $t_0 \leq t$,

we obtain by φ_1^V that $w(t) \geq w(t_0) - (t - t_0)\lambda$ and thus $w(t) > W_l - D\lambda$.
Since $W_l - D\lambda \geq \mu_l$, we obtain a contradiction with $w(t) < \mu_l$.

(b) Now suppose there exists a $t_0 \leq t$ such that $w(t_0) \geq W_h - W_d$ and
$(pump, off)$ at $[t_0 + D, t]$.

   – If $t_0 + D > t$ then, using $t_0 \leq t$, lemma 4 leads to $w(t) \geq w(t_0) - (t - t_0)\lambda$.
Since $t - t_0 < D$, we obtain $w(t) \geq W_h - W_d - D\lambda$. By $W_h \geq W_l + D\lambda + W_d$
this leads to $w(t) \geq W_l$. Since $D \geq 0$ we have $W_l \geq \mu_l$, thus $w(t) \geq \mu_l$,
and hence a contradiction with $w(t) < \mu_l$.

   – Assume $t_0 + D \leq t$. By lemma 4 we obtain $w(t_0 + D) \geq w(t_0) - D\lambda$
and thus $w(t_0 + D) \geq W_h - W_d - D\lambda$. Since $W_h \geq W_l + D\lambda + W_d$
this leads to $w(t_0 + D) \geq W_l$. From $w(t) < \mu_l$ and $\mu_l \leq W_l$, we obtain
$w(t) < W_l$. Thus $w(t) \leq W_l \leq w(t_0 + D)$. Since $t \leq t_0 + D$, by the
Intermediate Value Property, there exists a $\hat{t} \in [t_0 + D, t]$ such that
$w(\hat{t}) = W_l$. Since $(pump, off)$ at $[\hat{t}, t]$ we can proceed as in case (a) to
obtain a contradiction.

2. If $(pump, on)$ at t then, since $(pump, off)$ at 0, there exists a $t' \leq t$ such that
$(pump, off)$ at t'.

Suppose that for all $t' \leq t$ with $(pump, off)$ at t' we have $w(t') \geq \mu_l$. Since the
level increases at points where the pump is on, this would imply that $w(t_0) \geq \mu_l$,
for all $t_0 \leq t$, which leads to a contradiction with $w(t) < \mu_l$.

Hence there exists a $t' \leq t$ such that $(pump, off)$ at t' and $w(t') < \mu_l$. Then for
t' we obtain a contradiction by case 1.

Thus assumption $w(t) < \mu_l$ leads in all cases to a contradiction.
Similarly, $w(t) > \mu_h$ leads to a contradiction. $\qquad\qquad\qquad\qquad\qquad\qquad\qquad$ □

References

[ACD90] R. Alur, C. Courcoubetis, and D.L. Dill. Model-checking for real-time systems.
In *Proceedings Symposium on Logic in Computer Science*, pages 414–425, 1990.

[CHR91] Zhou Chaochen, C.A.R. Hoare, and A.P. Ravn. A calculus of durations. *Information Processing Letters*, 40:269–276, 1991.

[HNSY92] T. Henzinger, X. Nicollin, J. Sifakis, and S. Yovine. Symbolic model checking
for real-time systems. In *Proceedings Symposium on Logic in Computer Science*,
pages 394–406, 1992.

[Hoa69] C.A.R. Hoare. An axiomatic basis for computer programming. *Communications
of the ACM*, 12(10):576–580,583, 1969.

[Hoo91] J. Hooman. *Specification and Compositional Verification of Real-Time Systems*.
LNCS 558, Springer-Verlag, 1991.

[Koy90] R. Koymans. Specifying real-time properties with metric temporal logic. *Real-Time Systems*, 2(4):255–299, 1990.

[vS91] A.J. van Schouwen. The A-7 requirements model: Re-examination for real-time
systems and an application to monitoring systems. Technical Report 90-276,
Telecommunications Research Institute of Ontario, Queens University, 1991.

[Zwi89] J. Zwiers. *Compositionality, Concurrency and Partial Correctness*. LNCS 321,
Springer-Verlag, 1989.

An Approach to the Description and Analysis of Hybrid Systems *

X. Nicollin   A. Olivero   J. Sifakis   S. Yovine

Laboratoire de Génie Informatique
IMAG Campus, B.P. 53X
F-38041 Grenoble cedex, France
e-mail: {nicollin|olivero|sifakis|yovine}@imag.fr

1 Introduction

The paper presents a model for *hybrid systems*, that is, systems that combine *discrete* and *continuous* components. Such systems are usually reactive real-time systems used to control an environment evolving over time.

A main assumption is that a run of a hybrid system is a sequence of two-phase steps. The first phase of a step corresponds to a continuous state transformation usually described in terms of some parameter representing the time elapsed during this phase. In the second phase the state is submitted to a discrete change taking zero time.

To illustrate this assumption, consider a temperature regulator commanding a heater so as to maintain the temperature θ of a room between two given bounds θ_{min} and θ_{max}. A run of such a system is a sequence of steps determined by the alternating state changes of the heater from ON to OFF and conversely. These transitions occur when, respectively, the conditions $\theta = \theta_{max}$ and $\theta = \theta_{min}$ are satisfied.

Between two successive transitions the temperature changes according to laws that depend on the current state (heater is ON or OFF) and that describe its evolution as a function of its initial value and the time spent in the state.

Fig. 1 shows the evolution of the system as an alternation of ON and OFF steps, each step composed of a continuous state transformation followed by a transition.

The model presented is an extended state-transition graph with a set X of variables. The vertices of the graph represent control locations. A global state of the model consists of a vertex and a valuation of the variables.

- Discrete state changes are described by associating guarded commands on X with edges, that is, pairs of the form $\langle b(X), X := f(X) \rangle$, where the *guard* $b(X)$ is a boolean expression on X, and the *command* is as assignment where $f(X)$ is a tuple of functions on X.
  From a given global state (v_0, X_0) determined by a vertex v_0 and a valuation X_0, a transition to a vertex v_1 can be executed if there exists an edge from v_0 to v_1 labeled by a guarded command $\langle b(X), X := f(X) \rangle$ such that $b(X_0)$ evaluates to true. The resulting global state is (v_1, X_1) where $X_1 = f(X_0)$.

---

* Supported by the ESPRIT project N°6021 REACT-P

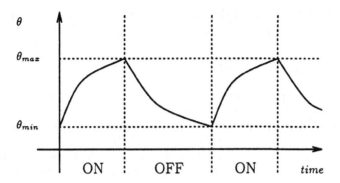

Fig. 1. Evolution of the temperature of the room

- Continuous state transformations are described by associating with each vertex
  v a function φ_v representing transformations of variables in terms of the time
  t elapsed and their initial values when the vertex v was entered. That is, the
  values of the state variables after sojourn time t at vertex v is $\varphi_v(X_0, t)$. The
  transformation stops when the vertex is left by executing a transition. Moreover,
  a predicate $\mathtt{tcp}_v$ ("time can progress") specifies the maximum sojourn time at
  v, which may depend on the values of the variables when entering the vertex.

The paper proposes a semantics of the model in terms of labeled transition systems
with two types of labels: (1) *actions*, which are names of discrete state changes,
and (2) *non-negative real numbers* representing the time elapsed.

These transition systems have features that take into account the assumptions
made about the nature of time. They are infinitely branching due to the density of
the time domain. If from some state q a state q'' is reached in a time t then for any
time $t' \leq t$ a state q' can be reached from q. Furthermore, from state q' the state q''
can be reached within time $t - t'$.

An important idea of the model is the two-phase step functioning where continuous evolutions and discrete state changes are clearly separated. This allows to
"discretize" the behavior of a system by considering a functioning by steps which
can be, of course, of arbitrary length.

The paper defines a notion of symbolic simulation allowing to compute from a
given symbolic state its successors or predecessors by a given step. A *symbolic state*
is a pair (v, b) where v is a vertex of the system and b is a boolean expression on X.
This leads to a characterization of the set of forward and backward reachable states
from some initial state as the least fixpoint of monotonic functionals.

Finally, the paper shows how the verification method by model-checking of TCTL
formulas on timed graphs presented in [HNSY92] can be adapted to hybrid systems.
In fact, it is shown that hybrid systems and TCTL admit the same class of models
and, therefore, the same fixpoint characterization of modalities holds. However, the
iterative computation of the meaning of the formulas may not converge due to the
undecidability of the properties for hybrid systems.

We consider the class of hybrid systems with linear constraints for which the application of the results on simulation and verification seems to be tractable.

Outline The paper is organized as follows.

- In Section 2 we present hybrid systems and define their semantics in terms of transition systems. We discuss also the role of the tcp predicate, which determines a *scheduling policy* for the discrete transitions. We give some examples to illustrate the use of the model for describing hybrid systems.
- Section 3 is devoted to the definition of forward and backward symbolic simulation, and to the adaptation of the symbolic model-checking method.
- In Section 4, we introduce hybrid systems with linear constraints and we present the results obtained using the tool KRONOS [NSY92] for verifying some properties of the examples presented in section 2.

2 A model for hybrid systems

2.1 Definition

A *hybrid system* is a structure $\langle V, X, \Rightarrow, \texttt{tcp}, \varphi \rangle$ where:

- $V = \{v_1, \ldots, v_m\}$ is a finite set of *vertices* (or *locations*);
- $X = (x_1, \ldots, x_n)$ is a tuple of *state variables*, ranging over some domain D_X;
- $\Rightarrow \subseteq V \times A \times B(X) \times F(X) \times V$ is a set of *edges*, where:
  - A is a vocabulary of *action names*,
  - $B(X)$ is the set of predicates on X, that is, if $b \in B(X)$ then b is a mapping from D_X to $\{\texttt{true}, \texttt{false}\}$,
  - $F(X)$ is the set of internal mappings of D_X, that is, if $f \in F(X)$ and $X_0 \in D_X$ then $f(X_0) \in D_X$.

  We write $v \xrightarrow{a, b, f} v'$ for denoting the edge $(v, a, b, f, v') \in \Rightarrow$;
- $\texttt{tcp} : V \to [(D_X \times \mathsf{R}^{\geq 0}) \to \{\texttt{true}, \texttt{false}\}]$ is the *time can progress predicate*. For $v \in V$, $X_0 \in D_X$ and $t \in \mathsf{R}^{\geq 0}$, we write $\texttt{tcp}_v[X_0](t)$ instead of $\texttt{tcp}(v)(X_0, t)$. $\texttt{tcp}_v$ is called the *time can progress predicate at v*. We require that for all $v \in V$, $X_0 \in X$ and $t \in \mathsf{R}^{\geq 0}$,
  1. $\texttt{tcp}_v[X_0](0) = \texttt{true}$;
  2. $\texttt{tcp}_v[X_0](t) \Rightarrow \forall t' \leq t \; \texttt{tcp}_v[X_0](t')$.
- $\varphi : V \to [(D_X \times \mathsf{R}^{\geq 0}) \to D_X]$ is the *evolution function*. As for tcp, we write $\varphi_v[X_0](t)$ instead of $\varphi(v)(X_0, t)$. φ_v is called the *evolution function at v*. We require that for all $v \in V$, $X_0 \in X$ and $t, t' \in \mathsf{R}^{\geq 0}$,
  3. $\varphi_v[X_0](0) = X_0$;
  4. $\varphi_v[X_0](t + t') = \varphi_v[\varphi_v[X_0](t)](t')$.

The set of edges $\Rightarrow$ defines discrete state changes exactly as in a state-transition graph with guarded commands. The vocabulary of actions used to name events can be used for synchronization purposes. The pair (b, f) defines a guarded command $\langle b, X := f(X) \rangle$. If the system is at location v, the current value of X is X_0, and

$b(X_0)$ is true, then the edge $v \xmapsto{a, b, f} v'$ is said to be *enabled*. It this case, it may be crossed: the action a is performed, the system moves to location v', and the value of X becomes $f(X_0)$.

The predicate tcp_v defines the maximum sojourn time at vertex v, and φ_v specifies how the values of variables evolve at v, as follows.

- $\text{tcp}_v[X_0](t) = \text{true}$ means that if the value of X is X_0 when entering vertex v, then it is possible to stay at v during time t. The requirements 1 and 2 above guarantee the soundness of this definition.
- $\varphi_v[X_0](t)$ gives the value of X after evolution during time t at location v, entered with initial value X_0. Since φ_v is a function, the evolution of variables at v is deterministic, and it depends only on the values of the variables when v was entered and on the time elapsed since then. Requirement 4 is not essential, as shown in section 2.3, but it is very convenient, since it allows to consider a *global state* of the system as a pair $(v, X_0) \in V \times D_X$, where v is the current location, and X_0 is the current value of X. This means that the current value of X determines the possible behaviors of the system at some location. That is, the evolution of the value of X after some instant depends only on its value $X_1 = \varphi_v[X_0](t)$ at this instant, independently of the value X_0 of X when v was entered and of the time t elapsed since then. We have, for all $t' \in R^{\geq 0}$, $\varphi_v[X_0](t + t') = \varphi_v[X_1](t')$.

2.2 Examples

We illustrate the use of the model with two examples. In order to simplify, the evolution functions associated with locations are given as a set of equations, where each equation defines the evolution law of a single variable.

The cat and the mouse This is a variation of the example presented in [MMP91]. Briefly the description is the following: at time $t = 0$, a mouse starts running from a certain position on the floor in a straight line towards a hole in the wall, which is at a distance D from the initial position. The mouse runs at a constant velocity v_m. When the mouse has run a distance $x_\delta < D$, a cat is released at the same initial position and chases the mouse at velocity v_c along the same path. Fig. 2 shows the hybrid system for this example. The equations inside a vertex define the evolution function at this vertex. For instance, for the vertex 1, we have:

$$\varphi_1[x_0, y_0](t) = (x_0 + v_m t, y_0 + v_c t)$$

The tcp_i predicates are defined by:

$$\text{tcp}_0[x_0, y_0](t) = x_0 + v_m t \leq x_\delta \ \lor \ x_0 > x_\delta$$
$$\text{tcp}_1[x_0, y_0](t) = (x_0 + v_m t \leq D \ \land \ y_0 + v_c t \leq D) \ \lor \ x_0 > D \ \lor \ y_0 > D$$
$$\text{tcp}_2[x_0, y_0](t) = \text{true}$$
$$\text{tcp}_3[x_0, y_0](t) = \text{true}$$

The conditions $x_0 > x_\delta$, $x_0 > D$ and $y_0 > D$ are only present to satisfy requirement 1 on tcp predicates. They are in fact irrelevant here, since they define values of x and y that cannot be reached from the initial state ($x = y = 0$).

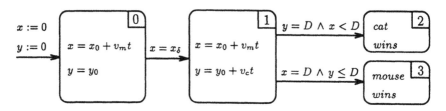

Fig. 2. Hybrid system of the cat and mouse

A temperature control system This example appears in [JLHM91]. The system controls the coolant temperature in a reactor tank by moving two independent control rods. The goal is to maintain the coolant between the temperatures θ_m and θ_M. When the temperature reaches its maximum value θ_M, the tank must be refrigerated with one of the rods. A rod can be moved again only if T time units have elapsed since the end of its previous movement. If the temperature of the coolant cannot decrease because there is no available rod, a complete shutdown is required. Fig. 3 shows the hybrid system of this example. The values of the timers x_1 and x_2 represent the times elapsed since the last use of rod 1 and rod 2, respectively. As for

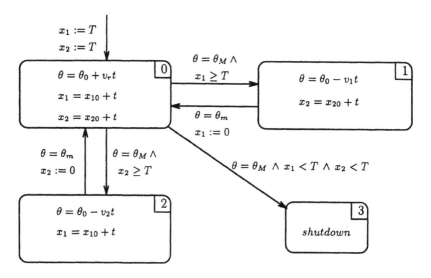

Fig. 3. Hybrid system for the temperature control system

the previous example, the evolution functions are written inside the vertices. The tcp predicates are defined by:

$$\texttt{tcp}_0[\theta_0, x_{10}, x_{20}](t) = \theta_0 + t \leq \theta_M \ \vee \ \theta_0 > \theta_M$$
$$\texttt{tcp}_1[\theta_0, x_{10}, x_{20}](t) = \theta_0 + t \geq \theta_m \ \vee \ \theta_0 < \theta_m$$
$$\texttt{tcp}_2[\theta_0, x_{10}, x_{20}](t) = \theta_0 + t \geq \theta_m \ \vee \ \theta_0 < \theta_m$$
$$\texttt{tcp}_3[\theta_0, x_{10}, x_{20}](t) = \textbf{true}$$

2.3 Semantics in terms of labeled transition systems

The behavior of a hybrid system is defined by a labeled transition system, i.e., a relation $\rightarrow$ on global states $Q = V \times D_X$ with labels in $A \cup R^{>0}$ ($\rightarrow \subseteq Q \times (A \cup R^{>0}) \times Q$). Following standard conventions we write $q \xrightarrow{l} q'$ for $(q, l, q') \in \rightarrow$.

The transitions $q \xrightarrow{a} q'$ are generated from edges of the form $v \xRightarrow{a, b, f} v'$ according to the following rule:

$$\frac{v \xRightarrow{a, b, f} v', \ b(X_0)}{(v, X_0) \xrightarrow{a} (v', f(X_0))}$$

This rule simply translates the intuitive meaning of edges labeled with guarded commands as explained above.

Transitions of the form $q \xrightarrow{t} q'$ represent state changes at some vertex v from some value X_0 and for some lapse of time t. They are specified by the rule:

$$\frac{\texttt{tcp}_v[X_0](t)}{(v, X_0) \xrightarrow{t} (v, \varphi_v[X_0](t))}$$

The conclusion expresses the effect of the evolution law on the state, its application being restrained to states from which time can progress.

The transition systems obtained are (1) *time deterministic* and (2) *time additive*, that is:

(1) $\forall q, q', q'' \in Q, \forall t \in R^{>0} \ q \xrightarrow{t} q' \ \wedge \ q \xrightarrow{t} q'' \ \Rightarrow \ q' = q''$
(2) $\forall q, q', q'' \in Q, \forall t, t' \in R^{>0} \ q \xrightarrow{t+t'} q' \ \Longleftrightarrow \ \exists q'' \ q \xrightarrow{t} q'' \ \wedge \ q'' \xrightarrow{t'} q'$

Remark As already mentioned, the requirement 4:

$$\varphi_v[X_0](t + t') = \varphi_v[\varphi_v[X_0](t)](t')$$

is useful, in the sense that it allows to consider a state of the system as a pair (v, X_0). If it is not satisfied, then the possible behaviors of the system depend on the current vertex v, the value X_0 of X when v was entered, and the time t_0 elapsed since then. That is, a state should be the triple (v, X_0, t_0), and the semantic rules should be written as follows:

$$\frac{v \xRightarrow{a, b, f} v', \ b(\varphi_v[X_0](t_0))}{(v, X_0, t_0) \xrightarrow{a} (v', f(\varphi_v[X_0](t_0)), 0)} \qquad \frac{\texttt{tcp}_v[X_0](t_0 + t)}{(v, X_0, t_0) \xrightarrow{t} (v, X_0, t_0 + t)}$$

In fact, requirement 4 does not go against generality, since it is always possible to transform a hybrid system $\langle V, X, \Rightarrow, \texttt{tcp}, \varphi \rangle$ which does not satisfy it into another one, $\langle V', X', \Rightarrow', \texttt{tcp}', \varphi' \rangle$, which does, as follows.

- $V' = V$;
- The domain of X' is $D_{X'} = D_X \times \mathsf{R}^{\geq 0}$. A value X_0' of X' is a pair $\langle X_0, t_0 \rangle$;
- If $(v_0, a, b, f, v_1) \in \Rightarrow$, then $(v_0, a, b', f', v_1) \in \Rightarrow'$, where
  - $b'(\langle X_0, t_0 \rangle) \equiv b(\varphi_{v_0}[X_0](t_0))$
  - $f'(\langle X_0, t_0 \rangle) \equiv \langle f(\varphi_{v_0}[X_0](t_0)), 0 \rangle$
- For every v, the predicate tcp_v' is defined by

$$\mathrm{tcp}_v'[\langle X_0, t_0 \rangle](t) \equiv \mathrm{tcp}_v[X_0](t_0 + t)$$

- For every v, the evolution function φ_v' is specified by

$$\varphi_v'[\langle X_0, t_0 \rangle](t) \equiv \langle X_0, t_0 + t \rangle$$

The hybrid system obtained satisfies trivially the four requirements on tcp' and φ', and the translation is correct since the transition systems associated with both systems are identical (modulo renaming of states).

2.4 Pragmatics

An important question concerning the pragmatics of our model is the choice of the tcp_v predicate. In this section, we consider a fixed vertex v, and we write φ and tcp instead of φ_v and tcp_v.

The tcp predicate may be arbitrary, provided it satisfies requirements 1 and 2 above. However, in practice, we often want to consider features such as time divergence or urgency of transitions. In this case, the choice of tcp depends on the guards of the edges outgoing from v.

The aim of this section is to show that an appropriate choice of tcp in terms of φ and the guards of the outgoing edges can express different *scheduling policies* for the execution of the transitions. We present here three such policies, emphasizing the fact that one should not restrict a priori the semantics of hybrid systems to a particular one, and that the choice should be left to the designer, depending on the kind of behavior he wants to define. The policies are illustrated by considering their effects on a very simple example.

Asynchronous scheduling policy In this policy, the tcp predicate is chosen so that time cannot progress in the vertex beyond a point where some edge is enabled and after which all edges are disabled forever. In other words, if an edge is enabled at some instant, then the vertex must be left immediately if and only if no edge will be enabled in the future.

If $P = \bigvee_i b_i$ is the disjunction of the predicates guarding the edges outgoing from v, then the asynchronous policy is realized by taking

$$\mathrm{tcp}[X_0](t) \equiv (\exists t' \geq t \ P(\varphi[X_0](t'))) \ \vee \ (\forall t' \ \neg P(\varphi[X_0](t')))$$

The second term has been added to prevent time progress from being blocked in states from which all transitions are disabled forever.

We call α the tcp predicate corresponding to the asynchronous policy. This policy has been adopted for action timed graphs in [NSY91].

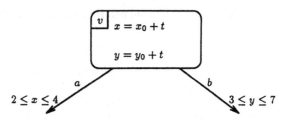

Fig. 4. A single vertex

As an example, consider the vertex v presented in Fig. 4, where the variables are x and y. Applying the definition above, we obtain:

$$\alpha[x_0, y_0](t) = x_0 + t \leq 4 \ \vee\ y_0 + t \leq 7 \ \vee\ (x_0 > 4 \ \wedge\ y_0 > 7)$$

If, for instance, v is entered with values $x = y = 0$, then the maximum sojourn time at the vertex is 7, and it may be left by crossing edge a between instants 2 and 4 or edge b between instants 3 and 7.

Synchronous scheduling policy In this policy, the vertex must be left as soon as possible, that is, as soon as an edge becomes enabled.

Let again $P = \bigvee_i b_i$. This policy is realized by taking

$$\mathtt{tcp}[X_0](t) \equiv \forall t' < t \ \neg P(\varphi[X_0](t'))$$

The $\mathtt{tcp}$ predicate for synchronous policy is represented by σ.

For the example of Fig. 4, we obtain

$$\sigma[x_0, y_0](t) = (x_0 + t \leq 2 \ \vee\ (2 \leq x_0 \leq 4 \ \wedge\ t = 0) \ \vee\ x_0 > 4) \ \wedge$$
$$(y_0 + t \leq 3 \ \vee\ (3 \leq y_0 \leq 7 \ \wedge\ t = 0) \ \vee\ y_0 > 7)$$

If the vertex v is entered with $x = y = 0$, then the sojourn time is exactly 2, and v must be left at time 2, by crossing edge a.

An intermediate scheduling policy An intermediate policy is adopted for timed transition systems in [HMP91], where time cannot progress beyond a point which disables an already enabled edge. This policy is realized by taking

$$\mathtt{tcp}[X_0](t) \equiv \bigwedge_i \ \forall t' \leq t \ [b_i(\varphi[X_0](t')) \ \vee\ \forall t'' \leq t' \ \neg b_i(\varphi[X_0](t''))]$$

The formula simply says that time can progress by t from X_0 if the disabling of an edge at some time $t' \leq t$ implies that this edge has not been enabled before t'.

Let β be the $\mathtt{tcp}$ for this policy. Then, considering again the example of Fig. 4, we get

$$\beta[x_0, y_0](t) = (x_0 + t \leq 4 \ \vee\ x_0 > 4) \ \wedge\ (y_0 + t \leq 7 \ \vee\ x_0 > 7)$$

With this policy, when entering the vertex v with $x = y = 0$, the maximum sojourn time is 4. Edge a may be crossed between instants 2 and 4, and edge b between instants 3 and 4.

Towards a comparison of the policies It is not difficult to prove that for all X_0 and t,

$$\sigma[X_0](t) \Rightarrow \beta[X_0](t) \Rightarrow \alpha[X_0](t)$$

This means that if time passing is blocked with the asynchronous (resp. intermediate) policy, then it is also blocked for the intermediate (resp. synchronous) one.

A consequence of this property is that the labeled transition system modeling the behavior of a hybrid system with a global synchronous (resp. intermediate) policy is included (in the sense of the inclusion of the transition relations) in the labeled transition system corresponding to the intermediate (resp. asynchronous) policy.

Another important property is that there exist hybrid systems whose models with the asynchronous (resp. intermediate) policy are not models of hybrid systems with the intermediate (resp. synchronous) policy.

Consider for instance the vertex v of Fig. 5, with only one variable x behaving like a timer. It is intended to describe a *strong timeout*, in the sense that if action a

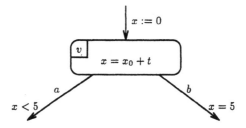

Fig. 5. A strong timeout

is not performed before time 5, then action b is performed at time 5.

If we adopt an asynchronous policy, then the tcp predicate is

$$\alpha[x_0](t) = x_0 + t \leq 5 \ \vee \ x_0 > 5$$

Thus, starting with $x = 0$, it is possible to stay in v until time 5 and the actions a and b may be performed *before time 5* and *at time 5* respectively. Since time cannot go beyond 5 in the vertex, then at time 5 the action b must be performed. So, the obtained behavior is the one expected.

It is impossible to define a hybrid system with the same behavior if we adopt the intermediate (and a fortiori the synchronous) policy. Indeed, the enabling conditions of actions a and b are disjoint; thus, in both cases, b will never be performed, because time 5 is not reachable.

Consider now Fig. 6, intended to describe a *weak timeout*, that is, if action a is not performed before or at time 5, then b is performed at time 5.

With the asynchronous or intermediate policy, the tcp predicate is

$$\alpha[x_0](t) = \beta[x_0](t) = x_0 + t \leq 5 \ \vee \ x_0 > 5.$$

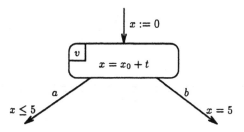

Fig. 6. A weak timeout

Now, starting with $x = 0$, the instant 5 is reachable for both policies. Moreover, it is not possible to go beyond time 5, and a and b may be performed *between times 0 and 5* and *at time 5* respectively. Hence, the resulting behavior is indeed that of a weak timeout.

However, such a behavior is still not specifiable using a synchronous policy. Indeed, action a being possible between times 0 and 5, it will be performed immediately, and the timeout will never expire.

Whether any hybrid system with a synchronous or intermediate scheduling policy may be described using only asynchronous policy remains an open question. A positive answer would simplify some theoretical developments. However, we believe that, at the description level, one should not impose a priori the asynchronous policy, which could sometimes yield more complicated descriptions, as it will be shown in the following example. In contrast, the examples presented in section 2.2 are such that their tcp predicates are the same for both synchronous and asynchronous policies and they coincide with those given.

Billiards Consider a billiard table of dimensions l and h, with a grey ball and a white ball (see Fig. 7).

Initially, the balls are placed at positions $b_g = (x_g, y_g)$ and $b_w = (x_w, y_w)$. The grey ball is knocked and starts moving with velocity v. If the ball reaches a vertical side then it rebounds, i.e., the sign of the horizontal velocity component v_x changes. The same occurs with the vertical velocity component v_y when the ball reaches a horizontal side. The combination of signs of velocity components gives four different movement directions.

The hybrid systems shown in Fig. 8 and Fig. 9 describe the movement of the grey ball for the billiards game for synchronous and asynchronous scheduling policies respectively. Each possible combination of directions is represented by a vertex. The rebounds correspond to the execution of transitions between vertices. Notice that the synchronous hybrid system is simpler due to the nature of the phenomenon specified. The guards of the asynchronous hybrid system can be obtained by strengthening appropriately the guards of the synchronous hybrid system. For instance, in the edge from vertex 1 to vertex 2, the triggering condition $x = l$ is restricted by $y \leq h$. For some state (x_0, y_0) such that $x_0 = l$ and $y_0 < h$, this transition must be taken

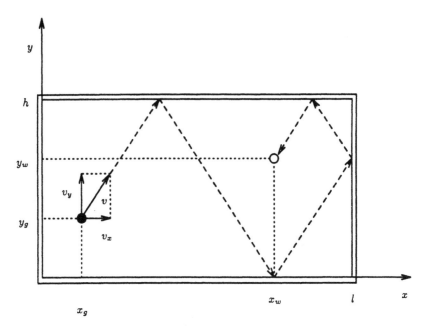

Fig. 7. Billiards game

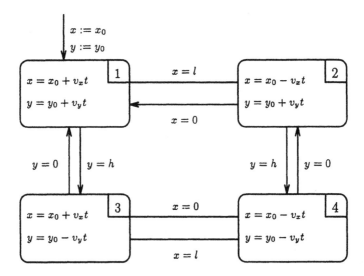

Fig. 8. The synchronous hybrid system for the movement of the grey ball

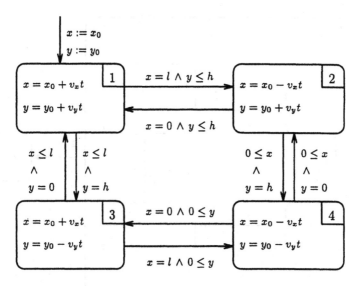

Fig. 9. The asynchronous hybrid system for the movement of the grey ball

because if time progresses by a quantity different from zero, both this transition and the transition from 1 to 3 will be disabled forever. If, for some state, $x_0 = l$ and $y_0 = h$, then the transitions from 1 to 2 and from 1 to 3 are enabled and they will be both disabled forever if time progresses by a non zero quantity. Thus one of them must be chosen non deterministically and executed.

We want now to modify the descriptions in Fig. 8 and Fig. 9 in order to represent the transition corresponding to the collision of the two balls. In Fig. 10 we show how the synchronous hybrid system of Fig. 8 can be modified to represent this transition.

An outgoing transition triggered by $x = x_w \wedge y = y_w$ is added to every vertex (this is represented following standard notations inspired from Statecharts by putting the description in a superstate 0 from which a transition with guard $x = x_w \wedge y = y_w$ is issued).

It is important to notice that a similar modification cannot be applied to the asynchronous hybrid system of Fig. 9. In asynchronous mode the execution of the transition labeled by $x = x_w \wedge y = y_w$ cannot be forced if there are other transitions issued from the same vertex that can be enabled later.

The asynchronous solution proposed in Fig. 11 consists in splitting each location into two sublocations (following again the notation of Statecharts). In location i, the sublocation $i.1$ corresponds to all states from which the collision is possible.

2.5 Remark

In previous papers about real-time systems [NSY91, NSY92, HNSY92], the *time can progress* predicate is not present. An *invariant* predicate $\text{inv}_v(X)$ is defined instead,

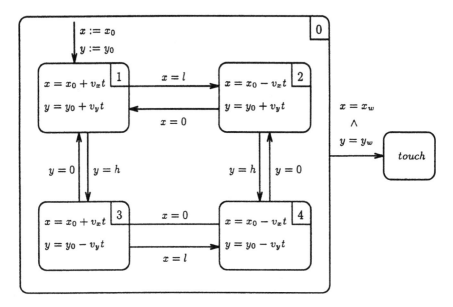

Fig. 10. The synchronous hybrid system with the collision transition

which expresses the possible values of variables X in vertex v. The operational semantics says that if $\text{inv}_v(X_0)$ is true, then it is possible to stay at vertex v for time t if and only if all the values taken by X during its evolution from X_0 for time t satisfy inv_v. In the context of hybrid systems, this is expressed by the rule

$$\frac{\forall t' \leq t \ \text{inv}_v(\varphi_v[X_0](t'))}{(v, X_0) \xrightarrow{t} (v, \varphi_v[X_0](t))}$$

To any invariant predicate inv_v corresponds then a tcp_v predicate defined by:

$$\text{tcp}_v[X_0](t) \equiv t = 0 \vee \forall t' \leq t \ \text{inv}_v(\varphi_v[X_0](t'))$$

Thus, any system defined using an invariant predicate may be easily redefined by means of tcp predicates. The converse is not necessarily true, as shown on the following example.

Consider the vertex v of Fig. 12, with two variables (actually timers) x and y. We choose the synchronous policy for v, that is,

$$\text{tcp}_v[x_0, y_0](t) \equiv (x_0 + t \leq 4 \vee x_0 > 4) \wedge (y_0 + t \leq 6) \vee y_0 > 6)$$

Depending on the values of x and y when entering the vertex, the possible behaviors are:

- If v is entered with $x \leq 4$ and $y \leq 6$, then v must be left when $x = 4$ or $y = 6$, whichever becomes true first;

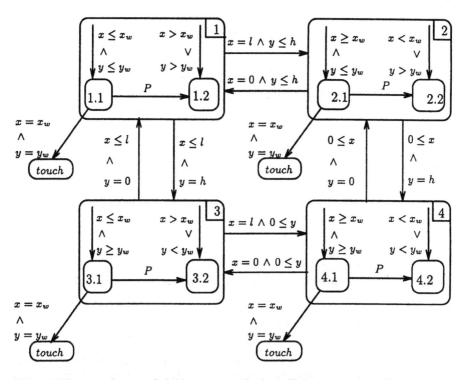

Fig. 11. The asynchronous hybrid system with the collision transition where P is the constraint $x = x_w \wedge y \neq y_w \vee x \neq x_w \wedge y = y_w$

- If v is entered with $x > 4$ and $y \leq 6$, then it must be left by crossing edges b as soon as the value of y becomes 6;
- If v is entered with $x \leq 4$ and $y > 6$, then it must be left by crossing edge a as soon as x becomes equal to 4;
- Finally, if $x > 4$ and $y > 6$ when entering v, then the vertex is never left, and time can progress indefinitely.

Such a behavior is hard to describe using an invariant predicate only, if it is possible to enter v with any value of the pair (x, y). The tcp_v predicate expresses the fact that time cannot progress in v from a state where $x \leq 4$ (resp. $y \leq 6$) to a state where $x > 4$ (resp. $y > 6$). This condition involves both the values of variables and the duration of time progress. In order to express such a condition with an invariant predicate, one has at least to add some variables to the system. This is possible for this simple example, but it may not necessarily be the case for any hybrid system.

In some sense, one can say that the inv_v predicate defines a *global* invariant on the values of the variables in v, whereas the tcp_v predicate defines a *union* of local invariants on the values of these variables in v: on the example, the values of x and

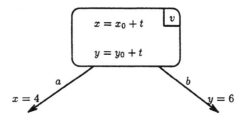

Fig. 12.

y must remain inside the region specified by one of the four invariants:

$$(x \leq 4 \wedge y \leq 6) \qquad (x > 4 \wedge y \leq 6) \qquad (x \leq 4 \wedge y > 6) \qquad (x > 4 \wedge y > 6)$$

We have adopted here the tcp approach because it seems more natural (though it is not proven to be more general), at least for defining scheduling policies.

3 Verification of hybrid systems

In this section we propose a general framework for analyzing hybrid systems which can be applied, in particular, to verify specifications expressed by formulas of a real-time temporal logic like TCTL [ACD90, HNSY92].

We point out that the method presented in [ACHH92] is related to ours in the sense that both rely on fixpoint characterizations of properties. Actually, both methods are related to the results presented in [HNSY92] about symbolic verification of real-time systems. The latter paper presents mainly two results.

- The first result concerns the fixpoint characterization of the temporal modalities of TCTL in terms of a dense next-time operator. The semantics of TCTL is defined for the same class of labeled transition systems used for the definition of the semantics of hybrid systems in Section 2. Thus, hybrid systems can be considered as models of TCTL and, conversely, TCTL can be used to specify the desired properties of hybrid systems.
- The second result states that for timed graphs [ACD90, NSY91, NSY92] the iterative computation of fixpoints always terminates and that the characteristic set of a formula is the disjunction of linear constraints on state variables. Unfortunately, the verification problem for hybrid systems is undecidable, therefore, in general, the iterative computation of fixpoints does not terminate.

In the rest of this section we present a general method for analyzing hybrid systems based on predicate transformers that compute the predecessors and successors of a given set of states. In Section 4 we show how these results can be applied in particular to hybrid systems with linear constraints.

3.1 Forward analysis

Given a predicate P on X and a vertex $v \in V$, we define the *forward time closure* of P at v as follows:

$$\overrightarrow{P}^v(X) \equiv \exists X_0, \exists t\, P(X_0) \wedge \mathtt{tcp}_v[X_0](t) \wedge X = \varphi_v[X_0](t)$$

That is, $\overrightarrow{P}^v(X)$ characterizes the set of states reachable from P by letting time progress at v.

It is easy to show that for any predicate P, vertex $v \in V$ and valuation X_0 of state variables, if $P(X_0)$ and there is an evolution $(v, X_0) \xrightarrow{t} (v, X_1)$ then $\overrightarrow{P}^v(X_1)$. Conversely, if $\overrightarrow{P}^v(X_1)$, there is X_0 such that $P(X_0)$ and $(v, X_0) \xrightarrow{t} (v, X_1)$ is an evolution.

Example 1. Consider for instance the Cat and Mouse example shown in Fig. 2. The mouse starts running from its initial position towards the hole and when it has run a distance x_δ, the cat is released and it chases the mouse. The first part of the run, when the mouse is running and the cat has not been released yet, corresponds to a time evolution of the system at location 0 as follows:

$$\overrightarrow{x = 0 \wedge y = 0}^0 \equiv$$
$$\equiv \exists x_0, y_0, t\, (x_0 = 0 \wedge y_0 = 0 \wedge x_0 + v_m t \leq x_\delta \wedge x = x_0 + v_m t \wedge y = y_0)$$
$$\equiv x \leq x_\delta \wedge y = 0$$

Given an edge $e = v_0 \xrightarrow{a, b, f} v_1$ and a predicate P, we define:

$$\mathtt{post}_e[P](X) \equiv \exists X_0\, P(X_0) \wedge b(X_0) \wedge X = f(X_0)$$

That is, $\mathtt{post}_e[P]$ characterizes the set of states reachable from P by executing a discrete transition corresponding to the edge e.

Clearly, for any $e = v_0 \xrightarrow{a, b, f} v_1$, predicate P and valuation X_0 of state variables, if $P(X_0)$ and there is a discrete transition $(v_0, X_0) \xrightarrow{a} (v_1, X_1)$ then $\mathtt{post}_e[P](X_1)$. Conversely, if $\mathtt{post}_e[P](X_1)$, there is X_0 such that $P(X_0)$ and $(v_0, X_0) \xrightarrow{a} (v_1, X_1)$ is a discrete transition.

Example 2. In the Cat and Mouse example, the cat is released when the mouse has run a distance x_δ, which corresponds to a transition from vertex 0 to 1. Let ℓ be a variable representing the locations. The resulting valuation of variables is the following:

$$\mathtt{post}[\ell = 0 \wedge x \leq x_\delta \wedge y = 0] \equiv$$
$$\equiv \ell = 1 \wedge \exists x_0, y_0\, (x_0 \leq x_\delta \wedge y_0 = 0 \wedge x_0 = x_\delta \wedge x = x_0 \wedge y = 0)$$
$$\equiv \ell = 1 \wedge x = x_\delta \wedge y = 0$$

Symbolic simulation A run of a hybrid system is a sequence of steps where each of them has two phases: the first one is a time progress phase and the second one is a discrete transition taking zero time. That is, a *run* is a sequence of states

$$(v_0, X_0)\,(v_1, X_1)\ldots(v_i, X_i)\ldots$$

satisfying that for all $i \geq 0$, either

- $(v_i, X_i) \xrightarrow{t_i} (v_i, X_i') \xrightarrow{a_i} (v_{i+1}, X_{i+1})$, or
- $(v_i, X_i) \xrightarrow{a_i} (v_{i+1}, X_{i+1})$.

Example 3. Consider the Cat and Mouse example where $v_m = 2$, $v_c = 3$, $x_\delta = 2$ and $D = 5$. The following sequence is an example of a run

$$(0, \langle 0, 0\rangle)\,(1, \langle 2, 0\rangle)\,(3, \langle 5, 4.5\rangle)$$

where $t_0 = 1$ and $t_1 = 1.5$.

For any predicate P_0, a *symbolic* run starting at P_0 is a sequence

$$(v_0, P_0)\,(v_1, P_1)\ldots(v_i, P_i)\ldots$$

such that for all $i \geq 0$, there is an edge e_i from v_i to v_{i+1} and $P_{i+1} = \text{post}_{e_i}[\overrightarrow{P_i}^{v_i}]$.
    That is, (v_{i+1}, P_{i+1}) characterizes the set of states reachable from a state (v_0, X_0) such that $P_0(X_0)$ after executing the sequence $e_1, \ldots, e_i$.

Example 4. The following is an example of symbolic run for the Cat and Mouse where the initial state predicate P_0 is $x = 0 \wedge y = 0$

$$(0, x = 0 \wedge y = 0)\,(1, x = 2 \wedge y = 0)\,(3, x = 5 \wedge y = 4.5)$$

and it characterizes exactly the run shown above.

Notice that P_{i+1} characterizes the set of states reachable right after the edge e_i is crossed without letting time pass. If we want to characterize all the intermediate states between two consecutive discrete state changes we should take

$$(v_0, \overrightarrow{P_0}^{v_0})\,(v_1, \overrightarrow{P_1}^{v_1})\ldots(v_i, \overrightarrow{P_i}^{v_i})\ldots$$

Example 5. For the symbolic run shown in Example 4 we have that the symbolic run characterizing all the intermediate states is

$$(0, x \leq 2 \wedge y = 0)\,(1, x \leq 5 \wedge y \leq 4.5)\,(3, x = 5 \wedge y = 4.5)$$

Reachability analysis Any state predicate can be represented by an m-tuple of predicates $(P_1, \ldots, P_m)$ corresponding to all the vertices $v_1, \ldots, v_m$ of V. The following proposition gives a characterization of the set of states reachable from a set I of possible initial states.

Proposition 1. *Let $I = (I_1, \ldots, I_m)$ be the set of initial states. Then, the set $P = (P_1, \ldots, P_m)$ of global states reachable from I is the least fixpoint of the equations:*

$$P_k = I_k \vee \bigvee_j \overrightarrow{\text{post}_e[P_j]}^{v_k}$$

for all $1 \leq k \leq m$, where $e = v_j \overset{a, b, f}{\Longrightarrow} v_k$.

This method can be applied to solve the reachability problem or to prove invariance properties of hybrid systems. A state predicate P is an invariant iff $\neg P$ is not reachable from the set of initial states. That is, if S is the least solution of the equations above, P is an invariant iff $S \wedge \neg P = \texttt{false}$.

The Gas Burner Consider the hybrid system of Fig. 13 given in [KPSY92] which is a simplified version of the Gas Burner example [HRR91]. There are three variables, namely x, y and z. Variable x increases at both vertices and it is reset by both edges. Variable y increases only when gas is leaking (at vertex 1), while variable z increases only when gas is not leaking (at vertex 2). That is, the values of y and z represent the times spent at vertices 1 and 2, respectively.

The set of initial states is defined by $I_1 \equiv (x = y = z = 0)$ and $I_2 \equiv \texttt{false}$. The set of reachable states is characterized by the least fixpoint of the following equations:

$$P_{1,i} = P_{1,i-1} \vee \overrightarrow{\text{post}_{21}[P_{2,i-1}]}^1$$
$$P_{2,i} = P_{2,i-1} \vee \overrightarrow{\text{post}_{12}[P_{1,i-1}]}^2$$

where $P_{1,0} = \overrightarrow{I_1}^1 = (x \leq 1 \wedge y - x = 0 \wedge z = 0)$ and $P_{2,0} = \texttt{false}$. For $i = 1$ we have that:

$$P_{1,1} = P_{1,0} \vee \overrightarrow{\text{post}_{21}[P_{2,0}]}^1$$
$$= P_{1,0}$$
$$P_{2,1} = P_{2,0} \vee \overrightarrow{\text{post}_{12}[P_{1,0}]}^2$$
$$= \overrightarrow{\text{post}_{12}[x \leq 1 \wedge y - x = 0 \wedge z = 0]}^2$$
$$= \overrightarrow{(x = 0 \wedge y \leq 1 \wedge z = 0)}^2$$
$$= (y \leq 1 \wedge z - x = 0)$$

Now, it is easy to show by induction that for all $i \geq 2$,

$$P_{1,i} = \begin{cases} P_{1,i-1} \vee x \leq 1 \wedge y - x \leq n \wedge 30n \leq z & \text{if } i = 2n, \\ P_{1,i-1} & \text{otherwise} \end{cases}$$

and

$$P_{2,i} = \begin{cases} P_{2,i-1} \lor y \le n+1 \land 30n \le z - x & \text{if } i = 2n+1, \\ P_{2,i-1} & \text{otherwise} \end{cases}$$

Hence, the least solution of the equations above is the pair (P_1, P_2), such that

$$P_1 = x \le 1 \land y = x \land z = 0 \lor \exists n \ge 1\,(x \le 1 \land y - x \le n \land 30n \le z)$$
$$P_2 = y \le 1 \land z = x \lor \exists n \ge 1\,(y \le n+1 \land 30n \le z - x)$$

This characterization of the reachability set can be used to verify some invariants about the gas burner system. For instance, the specification given in [HRR91] requires that for all intervals not shorter than 60, the percentage of leaking duration should not exceed 5%, that is, the system should satisfy the assertion $y + z > 60 \Rightarrow 20y \le y + z$. We can easily see that the predicate $(P_1 \lor P_2) \land \neg(y + z > 60 \Rightarrow 20y \le y + z)$ is unsatisfiable which implies that the gas burner satisfies the requirement.

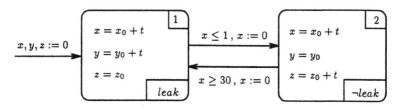

Fig. 13. The Gas Burner

3.2 Backward analysis

The predicate transformers already defined allow to compute the *successors* of the set of states satisfying a given predicate P. Dually, we can compute the *predecessors* of P. Given a predicate P and a vertex $v \in V$, we define the *backward time closure* of P at v as follows:

$$\overleftarrow{P}^{\,v}(X) \equiv \exists t\, \mathrm{tcp}_v[X](t) \land P(\varphi_v[X](t))$$

That is, $\overleftarrow{P}^{\,v}(X)$ characterizes the set of time predecessors of P at vertex v.

It is easy to show that for any predicate P, vertex $v \in V$ and valuation X_1 of state variables, if $P(X_1)$ and there is an evolution $(v, X_0) \xrightarrow{t} (v, X_1)$ then $\overleftarrow{P}^{\,v}(X_0)$. Conversely, if $\overleftarrow{P}^{\,v}(X_0)$, there is X_1 such that $P(X_1)$ and $(v, X_0) \xrightarrow{t} (v, X_1)$ is an evolution.

Example 6. Consider again the Cat and Mouse example and assume that the mouse has already run a distance x_δ while the cat has not been released yet. That is, the system is at location 0 with $x = x_\delta$ and $y = 0$. The set of time predecessors is computed as follows:

$$\overleftarrow{x = x_\delta \wedge y = 0}^{\,0} \equiv$$
$$\equiv \exists t\, (x + v_m t \leq x_\delta \wedge y = 0 \wedge x + v_m t = x_\delta \wedge y = 0)$$
$$\equiv x \leq x_\delta \wedge y = 0$$

Given an edge $e = v_0 \xrightarrow{a,\,b,\,f} v_1$ and a predicate P, we define

$$\mathbf{pre}_e[P](X) \equiv b(X) \wedge P(f(X))$$

That is, $\mathbf{pre}_e[P]$ characterizes the set of predecessors of P by executing a discrete transition corresponding to the edge e.

Example 7. Assume that the mouse has run a distance equal to x_δ and the cat has already been released. That is, the system is at location 1 with $x = x_\delta$ and $y = 0$. The set of predecessor states with respect to the releasing of the cat is the following:

$$\mathbf{pre}[\ell = 1 \wedge x = x_\delta \wedge y = 0] \equiv \ell = 0 \wedge x = x_\delta \wedge y = 0$$

Clearly, for any $e = v_0 \xrightarrow{a,\,b,\,f} v_1$, predicate P and valuation X_1 of state variables, if $P(X_1)$ and there is a discrete transition $(v_0, X_0) \xrightarrow{a} (v_1, X_1)$ then $\mathbf{pre}_e[P](X_0)$. Conversely, if $\mathbf{pre}_e[P](X_0)$, there is X_1 such that $P(X_1)$ and $(v_0, X_0) \xrightarrow{a} (v_1, X_1)$ is a discrete transition.

Reachability analysis Any state predicate can be represented by an m-tuple of predicates $(P_1, \ldots, P_m)$ corresponding to all the vertices $v_1, \ldots, v_m$ of V. The following proposition gives a characterization of the set of states from which it is possible to reach a set R of states.

Proposition 2. *Let $R = (R_1, \ldots, R_m)$ be the set of states. Then, the set of all predecessors of R is the least fixpoint of the equations:*

$$P_k = R_k \vee \bigvee_j \overleftarrow{\mathbf{pre}_e[P_j]}^{\,v_k}$$

for all $1 \leq k \leq m$, where $e = v_k \xrightarrow{a,\,b,\,f} v_j$.

This method can also be applied to solve the reachability problem or to prove invariance properties. Let I be the set of possible initial states and S be the least solution of the equations above for $R = \neg P$. The state predicate P is an invariant iff $I \wedge S = \mathbf{false}$.

Model checking For timed graphs, backward computation is the backbone of the model-checking algorithm for the real-time temporal logic TCTL given in [HNSY92]. Hereafter, we consider a version of TCTL where temporal operators have subscripts that limit their scope in time [ACD90].

Let $\mathcal{P}$ be a language of state predicates. The formulas of TCTL are defined by the following grammar:

$$\phi ::= P \mid \neg\phi \mid \phi_1 \vee \phi_2 \mid \phi_1 \exists \mathcal{U}_{\#n} \phi_2 \mid \phi_1 \forall \mathcal{U}_{\#n} \phi_2$$

where $P \in \mathcal{P}$, $n \in \mathbb{N}$ and $\# \in \{<, \leq, >, \geq, =\}$.

Intuitively, a state (v_0, X_0) satisfies the formula $\phi_1 \exists \mathcal{U}_{\#n} \phi_2$ if there exists a run starting at (v_0, X_0) where a state (v_1, X_1) satisfying ϕ_2 is reached at a time t such that $t \# n$ and for any state between (v_0, X_0) and (v_1, X_1), $\phi_1 \vee \phi_2$ continuously holds. Analogously, (v_0, X_0) satisfies $\phi_1 \forall \mathcal{U}_{\#n} \phi_2$ if for every run the above property holds.

The subscripts on the temporal operators limit their scope in time. For example, we can write $\mathrm{true} \exists \mathcal{U}_{<5} P$ to say that along some run, P becomes true before 5 time units.

We will use the typical abbreviations such as $\forall \Diamond_{\#n} \phi$ for $\mathrm{true} \forall \mathcal{U}_{\#n} \phi$, $\exists \Diamond_{\#n} \phi$ for $\mathrm{true} \exists \mathcal{U}_{\#n} \phi$, $\exists \Box_{\#n} \phi$ for $\neg \forall \Diamond_{\#n} \neg \phi$ and $\forall \Box_{\#n} \phi$ for $\neg \exists \Diamond_{\#n} \neg \phi$. The unrestricted temporal operators correspond to modalities subscripted by ≥ 0.

The symbolic verification method for TCTL developed in [HNSY92] is based on fixpoint characterizations of temporal modalities in terms of a binary "next-time" operator $\triangleright$. This operator can also be defined in the case of hybrid systems by means of the predicate transformers defined above.

Given two predicates P and P' on X, $P' \triangleright P$ characterizes the set of states from where it is possible to evolve during some time and then to (possibly) perform a discrete transition to reach a state satisfying P. We require that $P' \vee P$ holds during the evolution. That is, the $\triangleright$ operator can be seen as a "single-step until". Formally,

$$(P' \triangleright P)(X) \equiv \exists e, \exists t \, (\mathrm{pre}_e[P] \vee P)(\varphi_v[X](t)) \wedge$$
$$\mathrm{tcp}_v[X](t) \wedge \forall t' \leq t \, (P' \vee P)(\varphi_v[X](t'))$$

It is shown in [HNSY92] that the meaning of a modal operator of TCTL, i.e., the set of states satisfying it, can be computed iteratively from the meaning of its components by using $\triangleright$. Therefore, the characteristic sets of formulas are represented as state predicates.

- Let P_{ϕ_1} and P_{ϕ_2} be two state predicates representing the characteristic sets of ϕ_1 and ϕ_2, respectively. The characteristic set of $\phi_1 \exists \mathcal{U}_{\#n} \phi_2$ can be iteratively computed as $\bigvee_i P_i(0)$ with
  - $P_0(z) = z \# n \wedge P_{\phi_2}$, and
  - for all $i \geq 0$, $P_{i+1}(z) = P_i(z) \vee P_{\phi_1} \triangleright P_i(z)$,

  where z is a new state variable added to every vertex of the model. This variable represents a timer (i.e, its evolution law is $z = z_0 + t$) which is never reset nor tested.

- To check whether ϕ is an invariant the set of initial states of the system should be contained in the characteristic set of the the formula $\forall\Box\phi$. This characteristic set can be computed as $\bigwedge_i P_i$ with
  - $P_0 = P_\phi$, and
  - for all $i \geq 0$, $P_{i+1} = P_i \wedge \neg(\text{true} \rhd \neg P_i)$,

  where P_ϕ is a state predicate representing the characteristic set of ϕ.
- Another important real-time property stating that a given event eventually occurs within a certain time bound can be expressed in TCTL by a formula of the form $\forall\Diamond_{\leq c}\,\phi$, whose characteristic set can be computed as $\neg\bigvee_i P_i(0)$ with
  - $P_0(z) = z > c$, and
  - for all $i \geq 0$, $P_{i+1}(z) = P_i(z) \vee \neg P_\phi \rhd P_i(z)$,

  where P_ϕ is a state predicate representing the characteristic set of ϕ.

For timed graphs, iterative fixpoint computation always terminates. However, for hybrid systems in general, termination is no longer guaranteed.

In the following section we show how these symbolic model-checking techniques can be used to verify properties of constant-slope hybrid systems, in particular, those given in Section 2.

4 Hybrid systems with linear constraints

Given a set of variables $X = \{x_1, \ldots, x_n\}$, the language $\mathcal{P}$ of *linear constraints* on X is defined by the following syntax:

$$P ::= \sum_i \alpha_i x_i \# c \mid P_1 \wedge P_2 \mid \neg P \mid \exists x\, P$$

where $\# \in \{<, \leq, >, \geq, =\}$ and $\alpha_i, c \in \mathbb{Q}$.

It is easy to show that for any formula P there exists an equivalent formula of the form:

$$\bigvee_k \bigwedge_j \sum_i \alpha_{ij} x_i \# c_j$$

Let $\mathcal{P}_0$ be the set of formulas of this form.

The class of hybrid systems with linear constraints is such that:

- guards are linear constraints;
- assignments are of the form $x := \sum_i \alpha_i x_i$ or $x := c$ where $\alpha_i, c \in \mathbb{Q}$;
- for every vertex $v \in V$ and $x \in X$, the evolution function at v for the state variable x is $x = x_0 + \lambda_x t$, where $\lambda_i \in \mathbb{Q}$, that is variables evolve with *constant slopes*.

Clearly, for any $P \in \mathcal{P}$, since for all $v \in V$, φ_v is a linear function on X, $P \circ \varphi$ is a linear constraint. Hence, the following proposition holds.

Proposition 3. *For any $P, P' \in \mathcal{P}$, if $P, P' \in \mathcal{P}_0$ then $P' \rhd P \in \mathcal{P}_0$.*

Proposition 3 combined with the results presented in [HNSY92] about the symbolic model checking of TCTL shows that upper or lower approximations of the meanings of formulas can be computed as predicates of $\mathcal{P}_0$. This may lead to partial verification methods depending on the type of the formulas. In some cases, the iterative evaluation of the meaning of a formula can converge allowing full verification as it happens for the examples considered in the following section.

4.1 Examples

We present here the results obtained using the tool KRONOS [NSY92] for verifying some properties of the hybrid systems of Section 2. KRONOS is based on the symbolic model-checking algorithm developed in [HNSY92]. That is, in order to evaluate formulas with temporal modalities it computes fixpoints as it has been shown in Section 3.

The method applied requires the hybrid system to be a timed graph, therefore, for the examples considered below, formulas have been verified on equivalent timed graphs obtained by scaling constants.

The cat and the mouse It is easy to see that whenever

$$\frac{D - x_\delta}{v_m} \leq \frac{D}{v_c}$$

the mouse wins, that is, the time the mouse takes to reach the hole is smaller than or equal to the time taken by the cat. Otherwise, the cat wins. Clearly, in both cases, a terminal state is reached within a time less than or equal to $\frac{D}{v_m}$, either because the mouse reaches the hole first or because the cat catches the mouse before it reaches the hole.

Hereafter, to characterize locations we use a variable ℓ ranging over $\{0, 1, 2, 3\}$. Now, one of the following two formulas holds:

- *Eventually the mouse wins within $\frac{D}{v_m}$ time units*

$$[mw] \qquad \ell = 0 \wedge x = 0 \wedge y = 0 \Rightarrow \forall \Diamond_{\leq \frac{D}{v_m}} \text{ mousewins}$$

- *Eventually the cat wins within $\frac{D}{v_m}$ time units*

$$[cw] \qquad \ell = 0 \wedge x = 0 \wedge y = 0 \Rightarrow \forall \Diamond_{\leq \frac{D}{v_m}} \text{ catwins}$$

Let $v_m = 2$, $v_c = 3$, $x_\delta = 2$ and $D = 12$.

- Consider the formula $[mw]$. Computing the characteristic set of $\forall \Diamond_{\leq 6} (\ell = 2)$ with KRONOS terminates in 4 iterations and gives the following results, where z is the added timer and each $P_i(z)$ is computed according to the method described in Section 3.

$$P_0(z) = (\ell = 0 \wedge x \leq 2 \wedge 6 < z) \vee (\ell = 1 \wedge x \leq 12 \wedge y \leq 12 \wedge 6 < z) \vee$$
$$(\ell = 2 \wedge 6 < z) \vee (\ell = 3 \wedge 6 < z)$$

$$P_1(z) = (\ell = 0 \land x \le 2 \land x + 10 < 2z) \lor$$
$$(\ell = 1 \land x \le 12 \land y \le 12 \land x < 2z \land y + 6 < 3z) \lor$$
$$(\ell = 2 \land 6 < z) \lor \ell = 3$$
$$P_2(z) = (\ell = 0 \land x \le 2 \land x + 2 < 2z) \lor$$
$$(\ell = 1 \land y \le 12 \land (3x < 2y + 12 \lor x \le 12 \land x < 2z)) \lor$$
$$(\ell = 2 \land 6 < z) \lor \ell = 3$$
$$P_3(z) = (\ell = 0 \land x \le 2) \lor$$
$$(\ell = 1 \land y \le 12 \land (3x < 2y + 12 \lor x \le 12 \land x < z)) \lor$$
$$(\ell = 2 \land 6 < z) \lor \ell = 3$$
$$P_4(z) = P_3(z)$$

and the state predicate $\neg \bigvee_{i=0}^{3} P_i(0)$ representing the meaning of $\forall \Diamond_{\le 6} (\ell = 2)$ is

$$\ell = 1 \land x \le 12 \land 2y + 12 \le 3x \lor \ell = 2$$

The state predicate $\ell = 0 \land x = 0 \land y = 0$ characterizing the set of initial states does not imply the above predicate, then the system does not satisfy the property [mw] and the mouse is not the winner.

- Consider the formula [cw]. Computing the characteristic set of $\forall \Diamond_{\le 6} (\ell = 3)$ with KRONOS terminates in 3 iterations and gives the following results, where z is the added timer and each $P_i(z)$ is computed according to the method described in Section 3.

$$P_0(z) = (\ell = 0 \land x \le 2 \land 6 < z) \lor (\ell = 1 \land x \le 12 \land y \le 12 \land 6 < z) \lor$$
$$(\ell = 2 \land 6 < z) \lor (\ell = 3 \land 6 < z)$$
$$P_1(z) = (\ell = 0 \land x \le 2 \land x + 10 < 2z) \lor$$
$$(\ell = 1 \land x \le 12 \land y \le 12 \land x < 2z \land y + 6 < 3z) \lor$$
$$\ell = 2 \lor (\ell = 3 \land 6 < z)$$
$$P_2(z) = (\ell = 0 \land x \le 2 \land x + 2 < 2z) \lor$$
$$(\ell = 1 \land (x \le 12 \land (2y + 12 \le 3x \lor y \le 12 \land y + 6 < 3z))) \lor$$
$$\ell = 2 \lor (\ell = 3 \land 6 < z)$$
$$P_3(z) = P_2(z)$$

and the state predicate $\neg \bigvee_{i=0}^{2} P_i(0)$ representing the meaning of $\forall \Diamond_{\le 6} (\ell = 3)$ is

$$(\ell = 0 \land x \le 2) \lor (\ell = 1 \land y \le 12 \land 3x < 2y + 12) \lor \ell = 3$$

The state predicate $\ell = 0 \land x = 0 \land y = 0$ characterizing the set of initial states implies the above predicate, then the system satisfies the property [cw] and the cat is the winner.

Table 1 shows the number of iterations and the running times (measured in seconds) obtained with KRONOS on a SUN 4 Sparc Station for verifying the formulas [mw] and [cw] on the Cat and Mouse example with different values of the parameters.

parameters				formula	number of iterations	running times
v_m	v_c	x_δ	D			
2	3	2	12	$[mw]$	4	0.083
				$[cw]$	3	0.067
2	3	6	12	$[mw]$	3	0.067
				$[cw]$	4	0.083
30	5	630	3000	$[mw]$	4	0.083
				$[cw]$	3	0.067
80	100	2400	10000	$[mw]$	3	0.067
				$[cw]$	4	0.083
91	119	247	1557	$[mw]$	4	0.067
				$[cw]$	3	0.067

Table 1. Performances for the Cat and Mouse

A temperature control system Consider the temperature control system described in Section 2. The goal is to maintain the temperature of the coolant between lower and upper bounds θ_m and θ_M. If the temperature rises to its maximum θ_M and it cannot decrease because no rod is available, a complete shutdown is required.

Now, let $\Delta\theta = \theta_M - \theta_m$. Clearly, the time the coolant needs to increase its temperature from θ_m to θ_M is $\tau_r = \frac{\Delta\theta}{v_r}$, and the refrigeration times for rod 1 and rod 2 are $\tau_1 = \frac{\Delta\theta}{v_1}$ and $\tau_2 = \frac{\Delta\theta}{v_2}$, respectively.

The question is whether the system will ever reach the shutdown state. Clearly, if temperature rises at a rate slower than the time of recovery for the rods, i.e., $\tau_r \geq T$, shutdown is unreachable. Moreover, it can be seen that $2\tau_r + \tau_1 \geq T \wedge 2\tau_r + \tau_2 \geq T$ is a necessary and sufficient condition for never reaching the shutdown state (see Fig. 14).

The property stating that state shutdown is always unreachable corresponds to the following TCTL formula:

$$\ell = 0 \wedge \theta \leq \theta_M \wedge x_1 \geq T \wedge x_2 \geq T \Rightarrow \forall\Box\neg(\ell = 3)$$

or equivalently,

$$\ell = 0 \wedge \theta \leq \theta_M \wedge x_1 \geq T \wedge x_2 \geq T \Rightarrow \neg\exists\Diamond(\ell = 3)$$

- Let $v_r = 6$, $v_1 = 4$, $v_2 = 3$, $\theta_m = 3$, $\theta_M = 15$ and $T = 6$. In this case the condition $2\tau_r + \tau_1 \geq T \wedge 2\tau_r + \tau_2 \geq T$ holds. We compute using KRONOS the characteristic set of $\exists\Diamond\ell = 3$. The results obtained at each iteration are shown below, where each P_i has been computed according to the method described in Section 3.

$$P_0 \equiv \ell = 3$$
$$P_1 \equiv (\ell = 0 \wedge \theta \leq 15 \wedge 6x_1 < \theta + 21 \wedge 6x_2 < \theta + 21) \vee \ell = 3$$
$$P_2 \equiv (\ell = 0 \wedge \theta \leq 15 \wedge 6x_1 < \theta + 21 \wedge 6x_2 < \theta + 21) \vee$$
$$(\ell = 1 \wedge 3 \leq \theta \leq 15 \wedge 4x_2 + \theta < 19) \vee$$

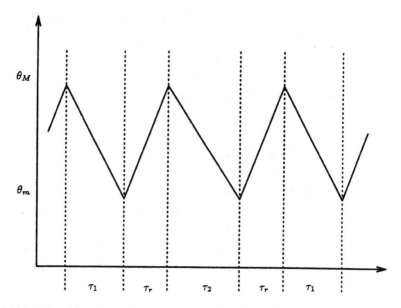

Fig. 14. Refrigeration times

$$(\ell = 2 \wedge 3 \leq \theta \leq 15 \wedge 3x_1 + \theta < 15) \vee \ell = 3$$
$$P_3 \equiv (\ell = 0 \wedge \theta \leq 15 \wedge (6x_1 < \theta + 21 \wedge 6x_2 < \theta + 21 \vee 6x_2 + 3 < \theta)) \vee$$
$$(\ell = 1 \wedge 3 \leq \theta \leq 15 \wedge 4x_2 + \theta < 19) \vee$$
$$(\ell = 2 \wedge 3 \leq \theta \leq 15 \wedge 3x_1 + \theta < 15) \vee \ell = 3$$
$$P_4 \equiv P_3$$

The state predicate $\neg \bigvee_{i=0}^{3} P_i(0)$ representing the meaning of $\neg \exists \Diamond (\ell = 3)$ is

$$\ell = 0 \wedge \theta \leq 15 \wedge (\theta + 21 \leq 6x_1 \wedge \theta \leq 6x_2 + 3 \vee \theta + 21 \leq 6x_2) \vee$$
$$\ell = 1 \wedge 3 \leq \theta \leq 15 \wedge 19 \leq 4x_2 + \theta \vee$$
$$\ell = 2 \wedge 3 \leq \theta \leq 15 \wedge 15 \leq 3x_1 + \theta$$

Since the state predicate $\ell = 0 \wedge \theta \leq 15 \wedge x_1 \geq 6 \wedge x_2 \geq 6$ characterizing the set of initial states implies the above predicate, the system satisfies the invariant as required.
- Suppose that we change the time of recovery to $T = 8$. Now, the condition $2\tau_r + \tau_1 \geq T \wedge 2\tau_r + \tau_2 \geq T$ is no longer satisfied. Again, we compute using KRONOS the characteristic set of $\exists \Diamond \ell = 3$. The results obtained at each iteration are shown below, where each P_i has been computed according to the method described in Section 3.

$$P_0 \equiv l = 3$$

$$P_1 \equiv (l = 0 \land \theta \le 15 \land 6x_1 < \theta + 33 \land 6x_2 < \theta + 33) \lor l = 3$$

$$P_2 \equiv (l = 0 \land \theta \le 15 \land 6x_1 < \theta + 33 \land 6x_2 < \theta + 33) \lor$$

$$(l = 1 \land 3 \le \theta \le 15 \land 4x_2 + \theta < 27) \lor$$

$$(l = 2 \land 3 \le \theta \le 15 \land 3x_1 + \theta < 21) \lor l = 3$$

$$P_3 \equiv (l = 0 \land \theta \le 15 \land (6x_1 + 3 < \theta \lor 6x_2 < \theta + 3 \lor$$

$$(6x_1 < \theta + 33 \land 6x_2 < \theta + 33))) \lor$$

$$(l = 1 \land 3 \le \theta \le 15 \land 4x_2 + \theta < 27) \lor$$

$$(l = 2 \land 3 \le \theta \le 15 \land 3x_1 + \theta < 21) \lor l = 3$$

$$P_4 \equiv (l = 0 \land \theta \le 15 \land (6x_1 + 3 < \theta \lor 6x_2 < \theta + 3 \lor$$

$$(6x_1 < \theta + 33 \land 6x_2 < \theta + 33))) \lor$$

$$(l = 1 \land 3 \le \theta \le 15 \land 4x_2 + \theta < 27) \lor$$

$$(l = 2 \land 3 \le \theta \le 15) \lor l = 3$$

$$P_5 \equiv (l = 0 \land \theta \le 15 \land (\theta + 33 \le 6x_2 \lor 6x_1 < \theta + 33 \lor 6x_2 < \theta + 3)) \lor$$

$$(l = 1 \land 3 \le \theta \le 15 \land 4x_2 + \theta < 27) \lor (l = 2 \land 3 \le \theta \le 15) \lor$$

$$l = 3$$

$$P_6 \equiv (l = 0 \land \theta \le 15 \land (\theta + 33 \le 6x_2 \lor 6x_1 < \theta + 33 \lor 6x_2 < \theta + 3)) \lor$$

$$(l = 1 \land 3 \le \theta \le 15) \lor (l = 2 \land 3 \le \theta \le 15) \lor l = 3$$

$$P_7 \equiv (l = 0 \land \theta \le 15) \lor (l = 1 \land 3 \le \theta \le 15) \lor (l = 2 \land 3 \le \theta \le 15) \lor$$

$$l = 3$$

$$P_8 \equiv P_7$$

The state predicate $\neg \bigvee_{i=0}^{7} P_i(0)$ representing the meaning of $\neg \exists \Diamond (\ell = 3)$ is

$$l = 0 \land \theta > 15 \lor$$

$$l = 1 \land (\theta < 3 \lor \theta > 15) \lor$$

$$l = 2 \land (\theta < 3 \lor \theta > 15)$$

and since the state predicate $\ell = 0 \land \theta \le 15 \land x_1 \ge 6 \land x_2 \ge 6$ characterizing the set of initial states does not imply the above predicate we have that shutdown is reachable.

Table 2 shows the number of iterations and the running times (measured in seconds) obtained with KRONOS on a SUN 4 Sparc Station for verifying the above formula on the Temperature Control System with different values of the parameters.

Billiards Consider the movement of the grey ball on the billard table. It is possible that the grey ball returns to the initial position with the initial direction. In this case the movement is periodic. A sufficient condition for the periodicity is that l, h, v_x and v_y are integers. The period T is calculated as follows:

$$T = 1\text{cm} \left(\frac{2l}{\gcd(v_x, 2l)}, \frac{2h}{\gcd(v_y, 2h)} \right)$$

parameters						number of	running
θ_m	θ_M	v_r	v_1	v_2	T	iterations	times
3	15	6	4	3	6	4	0.033
3	15	6	4	3	8	4	0.033
10	190	45	30	18	20	6	0.083
250	1100	34	25	10	80	4	0.033

Table 2. Performances for the Temperature Control System

Now, since the movement of the grey ball has a period T, in every interval longer than T it touches the white ball at least twice. That is, the grey ball cannot move for more than T time units without touching the white ball. We can express this property in TCTL as follows:

$$[periodT] \qquad \neg(\neg(x = x_w \wedge y = y_w)\exists \mathcal{U}_{>T}(x = x_w \wedge y = y_w))$$

We would like to characterize also all the positions where the grey ball may be placed in order to be able to touch the white ball once knocked with the stick. This set of points are characterized by the formula:

$$[touch] \qquad \exists \Diamond(x = x_w \wedge y = y_w)$$

and, since the movement of the grey ball has period T, the fact that the system satisfies the above formula is equivalent to the following one:

$$[touchT] \qquad \exists \Diamond_{\leq T}(x = x_w \wedge y = y_w)$$

Table 3 shows the number of iterations and the running times (measured in seconds) obtained with KRONOS on a SUN 4 Sparc Station for verifying the formulas [periodT], [touch] and [touchT] on the Billiards example with different values of the parameters.

5 Conclusion

The paper shows how ideas and results concerning the description and verification of timed systems developed in [HNSY92] can be extended to the case of hybrid systems.

The model proposed can be considered as an extension of timed automata. It generalizes the model for hybrid systems presented in [NSY91]. A novel idea is the use of the tcp predicate to determine how the discrete and continuous evolutions interact. The examples show that it is desirable for practical purposes to use standard tcp functions corresponding to predefined scheduling policies. Three such policies have been proposed and compared. Several questions remain open concerning the expressivity and comparison of classes of hybrid systems where a predefined scheduling policy is uniformly applied.

Concerning verification, the methods developed for timed automata are in principle applicable to hybrid systems modulo the decidability problem. These methods

parameters	formula	number of iterations	running times
l h x_0 y_0 x_w y_w			
13 20 0 0 10 16	[periodT]	55	7.77
	[touch]	55	6.69
	[touchT]	55	8.17
4 10 0 0 1 5	[periodT]	24	1.97
	[touch]	24	1.58
	[touchT]	24	1.90
3 4 0 0 1 3	[periodT]	10	0.56
	[touch]	10	0.40
	[touchT]	10	0.48

Table 3. Performances for the Billards

rely on an idea of "discretization" of the transition relation by defining "next-time" and "previous-time" operators. The paper shows how such operators can be defined and computed. Future work will concentrate on the specialization of these results to classes of hybrid systems for which effective full or partial verification methods can be developed.

References

[ACD90]   R. Alur, C. Courcoubetis, and D. Dill. Model-checking for real time systems. In *Proc. 5th Symp. on Logics in Computer Science*, pages 414–425, IEEE Computer Society Press, 1990.

[ACHH92] R. Alur, C. Courcoubetis, T. A. Henzinger, and Pei-Hsin Ho. Hybrid automata: an algorithmic approach to the specification and analysis of hybrid systems. In *Workshop on Theory of Hybrid Systems*, Lyngby, Denmark, October 1992.

[HMP91]   T.A. Henzinger, Z. Manna, and A. Pnueli. Timed transition systems. In *Proc. REX Workshop "Real-Time: Theory in Practice"*, Lecture Notes in Computer Science 600, Springer-Verlag, the Netherlands, June 1991.

[HNSY92] T.A. Henzinger, X. Nicollin, J. Sifakis, and S. Yovine. Symbolic model-checking for real-time systems. In *Proc. 7th Symp. on Logics in Computer Science*, IEEE Computer Society Press, 1992.

[HRR91]   K.M. Hansen, A.P. Ravn, and H. Rischel. Specifying and verifying requirements of real-time systems. *Proc. ACM SIGSOFT'91 Conf. on Software for Critical Systems*, 15(5):44–54, 1991.

[JLHM91] M. Jaffe, N. Leveson, M. Heimdahl, and B. Melhart. Software requierements analysis for real-time process-control systems. *IEEE Transactions on Software Engineering*, 17(3):241–258, 1991.

[KPSY92] Y. Kesten, A. Pnueli, J. Sifakis, and S. Yovine. Integration graphs: a class of decidable hybrid systems. In *Workshop on Theory of Hybrid Systems*, Lyngby, Denmark, October 1992.

[MMP91]   O. Maler, Z. Manna, and A. Pnueli. From timed to hybrid systems. In *Proc. REX Workshop "Real-Time: Theory in Practice"*, Lecture Notes in Computer Science 600, Springer-Verlag, the Netherlands, June 1991.

[NSY91]   X. Nicollin, J. Sifakis, and S. Yovine. From ATP to timed graphs and hybrid systems. In *Proc. REX Workshop "Real-Time: Theory in Practice"*, Lecture Notes in Computer Science 600, Springer-Verlag, the Netherlands, June 1991.

[NSY92]   X. Nicollin, J. Sifakis, and S. Yovine. Compiling real-time specifications into extended automata. *IEEE TSE Special Issue on Real-Time Systems*, 18(9):794–804, September 1992.

Integration Graphs:
A Class of Decidable Hybrid Systems [*]

Y. Kesten[1], A. Pnueli[1], J. Sifakis[2], S. Yovine[2]

[1] Department of Applied Mathematics and Computer Science
The Weizmann Institute of Science, Rehovot 76100, Israel
e-mail: yonit@wisdom.weizmann.ac.il
[2] Laboratoire de Génie Informatique
Institut IMAG B.P. 53X, 38041 Grenoble, France
e-mail:sifakis@imag.imag.fr

Abstract. *Integration Graphs* are a computational model developed in the attempt to identify simple Hybrid Systems with decidable analysis problems. We start with the class of *constant slope hybrid systems* (CSHS), in which the right hand side of all differential equations is an integer constant. We refer to continuous variables whose right hand side constants are always 1 as *timers*. All other continuous variables are called *integrators*. The first result shown in the paper is that simple questions such as reachability of a given state are undecidable for even this simple class of systems.

To restrict the model even further, we impose the requirement that no test that refers to integrators may appear within a loop in the graph. This restricted class of CSHS is called *integration graphs*. The main results of the paper are that the reachability problem of integration graphs is decidable for two special cases: The case of a single timer and the case of a single test involving integrators.

The expressive power of the integration graphs formalism is demonstrated by showing that some typical problems studied within the context of the Calculus of Durations and Timed Statecharts can be formulated as reachability problems for restricted integration graphs, and a high fraction of these fall into the subclasses of a single timer or a single dangerous test.

1 Introduction

Hybrid systems are systems that consist of a mixture of discrete and continuous components. Typically, the continuous components may represent a physical environment which obeys continuous rules of change, while the discrete components may represent a digital controller that senses and manipulates the environment. Characteristic examples are a computer system controlling a robot,

---

[*] This research was supported in part by the France-Israel project for cooperation in Computer Science and by the European Community ESPRIT Basic Research Action Project 6021 (REACT).

a manufacturing plant, or a transport system. Approaches to the specification, description, and analysis of hybrid systems were proposed in [MMP92], [NSY91], and [NOS+93].

An important question for the analysis and design of hybrid systems is identification of subclasses of such systems and corresponding restricted classes of analysis problems that can be settled algorithmically. In view of the success of model checking of finite-state systems and similar algorithmic approaches to the algorithmic analysis of reactive systems ([CES86], [BCM+90]) and timed systems ([Dil89], [ACD90]), it is only natural to search for similar decidable analysis problems for hybrid systems. This is the general aim of this paper.

The main results of this search is the identification of *integration graphs*, a class of hybrid systems that seem to avoid the main obstacles to decidability. Within this class, we give algorithmic solutions to the reachability problem of three important cases:

- Integer computations of an arbitrary integration graph.
- Integration graph with a single timer, and
- Integration graph with a single dangerous test along each computation.

Section 2 introduces the notion of *constant slope hybrid systems* (CSHS), which are hybrid systems all of whose differential equations have the form

$$\dot{x} = c,$$

for some integer constant c. Another restriction is that all guard (enabling) conditions of transitions are boolean combinations of linear inequalities with integer coefficients.

We give an example of a CSHS representing the Gas Burner Problem [CHR92], and explain the need for restrictions on tests applied to integrators. In subsection 2.1 we prove that, without these restrictions, the reachability problem becomes undecidable.

Section 3 introduces *integration graphs* which are CSHS's in which integrators (variables that have different slopes in different states) are not tested within loops. The section also introduces *finitary timed automata* (FTA), which is a slightly restricted class of timed automata [Dil89], [ACD90], and the notion of *duration formulas*. It shows that the reachability problem for integration graphs can be reduced to checking whether a duration formula is satisfied by a computation of an FTA.

Section 4 shows how to solve the duration satisfiability problem for integer computations of an FTA. The solution is based on constructing a set of equations that characterizes the length of time a computation spends in each automaton state.

Section 5 considers satisfiability of a duration formula by real computations of an FTA with a single timer. It provides an algorithm for solving this problem based on a similar set of characterizing equations.

Section 6 considers satisfiability of a disjunctive duration formula by real computations of an unrestricted FTA.

2 Constant Slope Hybrid Systems

In this section we introduce the class of constant slope hybrid systems. Many hybrid systems analyzed in the literature fall into this class. One of the advantages of this class is that the differential equations appearing in states can be trivially solved in closed form and yield solutions that are linear functions of time.

Let $\mathcal{P}$ be a finite set of propositions. Let N denote the natural numbers, R^+ – the non negative reals, and Z – the integers.

a *constant slope hybrid system* (CSHS) consists of the following components:

- S – A finite set of *locations*. In a graphical representation of the system, these are drawn as nodes of the graph.
- Λ – A proposition labeling function $\Lambda : S \mapsto 2^{\mathcal{P}}$, mapping each location $s \in S$ to the set of propositions that are true in s. For a state s and a boolean formula p over Λ, we write $s \models p$ to denote that p evaluates to *true* over $\Lambda(s)$.
- V – A finite set of *(data) variables*. These are the variables that change continuously within states and discretely via transitions.
- $\mathcal{R}$ – A rate labeling function $\mathcal{R} : S \times V \mapsto Z$, identifying for each location $s \in S$ and each variable $x \in V$ an integer $c = \mathcal{R}[s][x] \in Z$ which specifies the (constant) rate at which x changes continuously while being in location s. Thus, the differential equation for x within s is

$$\dot{x} = c,$$

  where $c = \mathcal{R}[s][x]$.
- $s_I \in S$ – An *initial location*.
- $S_f \subseteq S$ – A set of *final locations*.
- V_0 – Initial values. This is a tuple of values, representing the values of the variables V at the beginning of a computation. By default, a variable that is not assigned an explicit initial value has 0 as an initial value.
- E – A set of *edges*. Each edge $e \in E$ is associated with the following components:
    - A *source* location. This is the location from which the edge departs.
    - A *target* location. This is the location to which the edge connects.
    - An edge *guard* Γ. This is the condition under which the edge may be traversed. An edge guard is a linear equality/inequality of the form

$$\sum_{i=1}^{n} a_i \cdot x_i \sim c,$$

      where $a_i, c \in Z$, $x_i \in V$ and $\sim$ is one of the comparison relations $\{<, > , =, \neq, \leq, \geq\}$.
    - A multiple *assignment* of the form
      $$(y_1, \ldots, y_m) := (c_1, \ldots, c_m),$$
      where $y_i \in V$ and $c_i \in Z$. When edge e is taken, the variables $y_1, \ldots, y_m$ are assigned the values $c_1, \ldots, c_m$. We often write $Y := C$ as a schematic representation of the multiple assignment associated with the edge.

In the graphical representation of CSHS, we represent edge e by drawing an edge from the source location to the target location and label it by the *edge label*

$$\Gamma \ / \ (y_1, \ldots, y_m) := (c_1, \ldots, c_m).$$

In the case that Y, the set of assigned variables is empty, we use the simpler labeling Γ?

It is required that no edge departs from a final location.

In Fig. 1 we present a CSHS for the Cat and Mouse system [MMP92], representing the situation of a cat chasing a mouse, where the cat and the mouse run at constant velocities, v_c and v_m respectively, and the cat starts running Δ time units later than the mouse. Variables x_c and x_m measure the respective distances of the cat and the mouse from the wall. Variable y is a timer, used to measure the delay Δ in the start time of the cat.

Initially $x_c = x_m = X_0$, $y = 0$

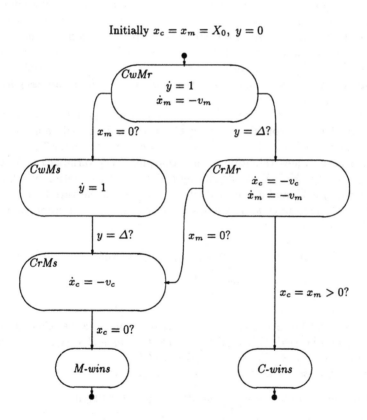

Fig. 1. Cat and Mouse System.

There are several differences between this presentation of the system and the one given in [MMP92]. A first difference is that [MMP92] uses Statechart notations to present control in a structured way. The Statechart presentation allocates concurrent superstates to the cat and the mouse. The CSHS presentation given here allows only flat unstructured control graph. Consequently, we have an individual node for every pair of concurrent states in the Statechart representation.

Another difference is that transitions in the hybrid systems model of [MMP92] are associated with lower and upper time-bounds, restricting the length of time the transition can be continuously enabled before it is taken. There is no such association in the CSHS model. Consequently, to represent the delay Δ of the cat before it starts running, we use an explicit timer y, initially preset to 0, and causing a transition as soon as it reaches the value Δ. In that, we follow the model of timed automata [AD90].

Behaviors and Computations

A behavior of a CSHS starts at the initial location s_I with all variables initialized to their initial values. As time progresses, the values of all variables increase according to the rates associated with the current location. At any point in time, the system can change location through an edge connecting location s to s' and labeled by $\Gamma/Y := C$, provided the current values of the variables satisfy the guard Γ. With the change of location, all variables in Y are assigned their respective right-hand sides C.

A *valuation* ν for V assigns a real value to each variable in V. Let $\mathcal{R}$ be a V-vector of rates (slopes), assigning to each $x \in V$ a real value $\mathcal{R}[x] \in \mathbb{Z}$, denoting the rate of growth of x. For a valuation ν, a rate vector $\mathcal{R}$, and $t \in \mathbb{R}^+$, $\nu + \mathcal{R} \cdot t$ denotes a new valuation ν' such that, for every variable $x \in X$, $\nu'[x] = \nu[x] + \mathcal{R}[x] \cdot t$. For $Y \subseteq X$, we denote by $\nu[Y \leftarrow C]$ the valuation which assigns $C[y]$ to every $y \in Y$, and agrees with ν over the rest of the variables.

A triple of the form $\langle s, \nu, t \rangle$, where s is a location, ν is a valuation, and $t \in \mathbb{R}^+$ is called a *situation*.

A *computation segment* of a CSHS is a sequence of situations

$$\langle s_0, \nu_0, t_0 \rangle, \langle s_1, \nu_1, t_1 \rangle, \ldots, \langle s_n, \nu_n, t_n \rangle$$

satisfying:

- [*Consecution*] For every i, $0 \leq i < n$, there is an edge $e \in E$ connecting s_i to s_{i+1} and labeled with $\Gamma/Y := C$ such that $\nu_i + \mathcal{R}[s_i](t_{i+1} - t_i)$ satisfies Γ and $\nu_{i+1} = (\nu_i + \mathcal{R}[s_i](t_{i+1} - t_i))[Y \leftarrow C]$.
- [*Time progress*] For all i, $0 \leq i < n$, $0 \leq t_i \leq t_{i+1}$.

A *computation* of a CSHS is a computation segment satisfying:

- [*Initiation*] $s_0 = s_I$ and $t_0 = 0$.
- [*Termination*] $s_n \in S_f$

A computation $\langle s_0, \nu_0, t_0 \rangle, \ldots, \langle s_n, \nu_n, t_n \rangle$ is called an *integer computation* if, for every $i \geq 0$, $t_i \in \mathbb{Z}$ (in fact $t_i \in \mathbb{N}$).

Reachability Problems

In this paper, we are mainly interested in reachability problems for constant slope hybrid systems. A typical reachability problem is: Problem

1 Reachability. Given a final location $s \in S_f$, is there a computation terminating at location s.

A concrete example of a reachability problem is one that can be asked about the Cat and Mouse system of Fig. 1.
    Problem

2 Security for Mouse. Under the assumption

$$\frac{X_0}{v_m} < \Delta + \frac{X_0}{v_c},$$

show that there is no computation that reaches location *C-wins*

As another reachability problem, we consider the Gas Burner system [CHR92]. Consider the CSHS presented in Fig. 2.

Initially $x = y = z = 0$

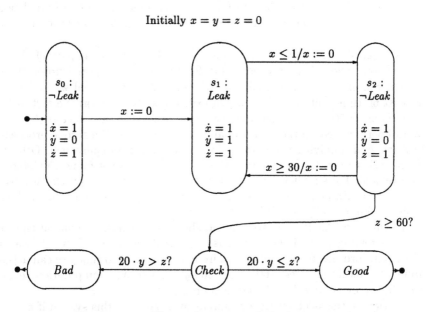

Fig. 2. H-GAS: The gas burner as a hybrid system.

Locations s_0, s_1, and s_2 represent a Gas Burner system that has these three control states. There is a proposition *Leak* which is true only at location s_1, representing a situation at which the system is leaking.
    The verification problem posed in [CHR92] can be formulated as follows.

Assuming

1. A continuous leaking period cannot extend beyond 1 time unit.
2. Two disjoint leaking periods are separated by a non-leaking period extending for at least 30 time units.

Prove:

- *Safety-Critical Requirement:* In any interval longer than 60, the *accumulated* leaking time is at most 5% of the interval length.

The CSHS of Fig 2 employs three variables as follows:

- Variable x measures the duration of time in each of the locations s_0, s_1, and s_2. It is reset to 0 on entry to each of these locations.
- Variable y measures the accumulated leaking time. It grows linearly in location s_1, and stays constant in any of the other locations.
- Variable z measures the total elapsed time.

Obviously, system H-GAS ensures assumptions 1 and 2. The only leaking location is s_1 and it is clear that no computation of the system can stay continuously in s_1 for more than 1 time unit and that, between two consecutive (but disjoint) visits to s_1, the system stays at the non-leaking location s_2 for at least 30 time units.

We can view locations s_0, s_1, and s_2, as the *operational* part of the representation. The other locations serve for testing the required property. The system can exit the operational part any time after an interval whose length is at least 60 time units has elapsed. On exit, the system performs a test at location *check*. If $20 \cdot y > z$, then the accumulated leaking time exceeds 5% of the overall period spent in the operational part. In that case, the system proceeds to location *Bad*, implying a violation of the safety-critical requirement. Otherwise it proceeds to *Good*, implying that the current run was not found to violate the requirement. For simplicity, we consider the safety-critical requirement only for *initial intervals*, i.e., intervals starting at $t = 0$. The extension of the method to arbitrary intervals is straightforward.

Having the edge from s_2 to *check* as the only exit from the operational part into the testing part of the system is not a real restriction. Since it is always possible to proceed from s_0 to s_1 and from s_1 to s_2 in zero time, we can actually apply the acceptance test $(60 \leq z \wedge 20 \cdot y \leq z)$ to any computation segment reaching s_0 or s_1 as well as to segments reaching s_2.

Obviously the safety-critical requirement is valid for this system if and only if location *Bad* is unreachable. This provides another example of an interesting reachability problem: namely, show that no computation of the system of Fig. 2 ever reaches location *Bad*.

2.1 Reachability is Undecidable for CSHS's

In this subsection we show that the reachability problem for CSHS's is undecidable.

The result is based on a reduction of an n-counter machine to a CSHS. The system emulating the n-counter machine only uses guards of the form:

$$u = c \quad \text{or} \quad u \neq c$$

where c is an integer constant .

The Construction

An n-counter machine can be described as a linear labeled program allowing the following basic commands:

- **go to ℓ,**
- **if $x_i = 0$ then go to ℓ_i else go to ℓ_j.**
- $x_i := x_i + 1$,
- $x_i := x_i - 1$, (this operation is undefined if $x_i = 0$)
- **stop**

Let P be a program for an n-counter machine with counters $x_1, \ldots, x_n$. Without loss of generality, assume that the first label of P is ℓ_0 and the last command (with label ℓ_t) is a *stop* command . We construct a CSHS S_P which emulates P, i.e., terminates precisely when P does. System S_P uses variables $x_1, \ldots, x_n$ and an additional variable y.

We represent S_P as a graph which has a location (node) for each label of program P. It may have additional locations.

It is not difficult to see how the *go-to* and conditional *go-to* commands can be implemented by edges connecting the corresponding nodes that may be labeled by $x_i = 0$ and $x_i \neq 0$ for the conditional transfer. The commands for incrementing and decrementing a counter x_i can be implemented by the following two subgraphs:

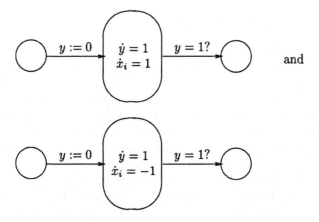

We claim that program P reaches the *stop* command at ℓ_t iff location ℓ_t is reachable in the CSHS S_P.

Conclusion

Since the halting problem for an n-counter machine is undecidable for $n \geq 2$, we conclude that the reachability problem is undecidable for CSHS's of the form considered here. In fact, since our construction uses an additional variable y, it is undecidable for systems having at least 3 variables.

Note that reachability is undecidable even if we restrict ourselves to integer computations.

3 Integration Graphs

Having realized that reachability is undecidable for CSHS's, we attempt to narrow the class of considered systems in the hope that reachability will be decidable for a more restricted class.

3.1 Definition of Integration Graphs

A variable of a CSHS that has the slope $+1$ at all locations, except perhaps at the final locations, is called a *timer*. All other variables are called *integrators*.

A lower bound of how much we have to restrict the class before reachability becomes decidable is provided by *timed automata* [AD90]. The differences between a timed automaton and a CSHS can be summarized as:

1. Timed automata do not allow integrators but only timers.
2. The guards allowed by timed automata are conditions of the forms

$$x \simeq c \quad \text{and} \quad x - y \simeq c$$

where $\simeq \in \{=, \leq, \geq\}$, x, y are variables (called *clocks* in [AD90]), and c is an integer constant.

Motivated by this comparison, consider a test

$$a_1 \cdot x_1 + \cdots + a_n \cdot x_n \sim b, \tag{1}$$

where $x_1, \ldots, x_n$ are variables, $a_1, \ldots, a_n$ and b are integer constants, and $\sim$ is one of $\{=, \neq, <, >, \leq, \geq\}$. Such a test is called *dangerous* if it refers to an integrator or does not have one of the forms listed in item 2 above.

The implication is that unbridled use of dangerous tests may lead to undecidability. The fact that tests that refer to more than two variables, or contain multiplicative factors with absolute value different from 1, lead to undecidability is proven in [AH90].

The fact that tests that refer to integrators are dangerous has been established in the undecidability result proven in Section 2. The construction used for the undecidability proof employs integrators to represent the registers x_i and tests them for being zero on edges representing conditional *go-to* commands.

Eliminating dangerous tests altogether is too harsh, since this will exclude systems such as the Cat and Mouse or the Gas Burner from the class we intend to study. For example, the test $20 \cdot y > z$ is dangerous for two reasons. It is not of one of the allowed forms, and it refers to the integrator y.

Instead, we strongly restrict the places where dangerous tests can appear in the graph of a CSHS. An edge in the graph representing a CSHS is called *cyclic* if it is part of a cycle in the graph. A CSHS is called an *integration graph* if

- Dangerous tests do not appear on cyclic edges.

This restriction ensures that there exists a bound K such that the number of times any computation encounters a dangerous test is bounded by K. In all proofs of undecidability, the constructed counter-examples rely on checking dangerous tests an unbounded number of times. Consequently, there is hope that reachability will be decidable for integration graphs.

It is not difficult to ascertain that both the Cat and Mouse system (that has no cycles at all) and the Gas Burner system are integration graphs.

It is straightforward to show that any integration graph is equivalent to a system, whose graph can be decomposed into a cyclic (the *looping*) part L with exits into an acyclic (the *testing*) graph T. This decomposition is such that L contains no dangerous tests, while T may contain some dangerous tests and all the final locations.

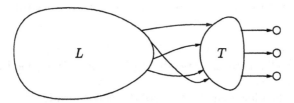

We can relax the definition of a timer, only requiring that it has a uniform slope at all states of L, the looping part of the integration graph. This allows us to assume that all slopes of all variables within T are 0.

Variables that appear in dangerous test are called *terminal variables* since, according to the integration graph restrictions, we are only interested in their values at the end of the computation.

For simplicity, we will consider only integration graphs in which terminal variables are not assigned values by any edge. It is not difficult to extend the results obtained to the more general case.

Single Integrator Tests

In the study of CSHS's, it is possible to restrict our attention to dangerous tests of the form

$$u \sim b$$

where u is an integrator and b is an integer constant. Observe that, for any dangerous test of the more general form

$$a_1 \cdot x_1 + \cdots + a_m \cdot x_m \sim b,$$

we can define a new integrator u whose slope at each location $s \in S$ is given by

$$\mathcal{R}[s][u] = a_1 \cdot \mathcal{R}[s][x_1] + \cdots + a_n \cdot \mathcal{R}[s][x_m]$$

For example, for the gas burner system, we can define a new integrator n whose value is intended to be $20 \cdot y - z$. The slopes of u at locations s_0, s_1 and s_2 are given by -1, 19 and -1, respectively. Then, instead of testing whether $20 \cdot y - z$ is positive, it is sufficient to check for $u > 0$.

3.2 Duration Properties of Finitary Timed Automata

In this subsection we consider the simpler model of timed automata, but ask more complicated questions than just reachability of some final location.

Finitary Timed Automata

We use a simplified version of timed automata ([Dil89], [AD90]), to which we refer as *finitary timed automata* (FTA). The simplification is that we are interested only in finite computations that reach some final location.

In our framework, an FTA can be presented as a CSHS with the following restrictions.

- All variables have the slope 1 in all states. Consequently, they are all timers, and we can eliminate the rate labeling function $\mathcal{R}$ from the description of a timed automaton.
- The set of final locations S_f consists of a single location s_f.
- The initial values of all variables are 0. Consequently, we need not specify the component V_0 in the description of an FTA.
- All guards are of the forms $l \leq x \leq u$ and $l \leq x - y \leq u$, where x, y are variables (timers) and $l, u \in \mathbb{Z}$.
- All assignments have the form $Y := 0$, i.e., 0 is the only assigned value. We refer to Y as the variables *reset* when the edge is taken.

Following these simplifications, an FTA can be described by the tuple

$$M : \langle S, \Lambda, V, s_I, s_f, E \rangle.$$

We often refer to edges as a tuple (s, s', Γ, Y), where s and s' are the source and destination locations, Γ is the guard, and Y is the set of variables reset by the edge.

Let $e = \langle s, s', \Gamma, Y \rangle \in E$. We say that e is a *resetting edge* if $Y = V$. Without loss of generality, we require that all edges arriving at s_f be resetting edges. For convenience we assume a virtual resetting edge e^- that enters s_I.

An FTA may be nondeterministic. It may contain two edges $e_1 = \langle s, s_1, \Gamma_1, Y_1 \rangle$ and $e_2 = \langle s, s_2, \Gamma_2, Y_2 \rangle$, $s_1 \neq s_2$, such that Γ_1 and Γ_2 are not mutually exclusive ($\Gamma_1 \wedge \Gamma_2$ is satisfiable). On the other hand, we require that every two locations, s_1 and s_2, have at most one edge connecting them.

A *trail* is a finite sequence

$$(s_0, t_0), \ldots, (s_n, t_n),$$

such that, for every i, $0 \le i < n$, $0 \le t_i \le t_{i+1}$.
If

$$\gamma = \langle s_0, \nu_0, t_0 \rangle, \langle s_1, \nu_1, t_1 \rangle, \ldots, \langle s_n, \nu_n, t_n \rangle$$

is a computation (segment), then

$$\tau = (s_0, t_0), (s_1, t_1), \ldots, (s_n, t_n)$$

is called the trail *corresponding to* γ. A trail that corresponds to some computation segment is called *realizable*. A trail that corresponds to a computation is called *complete*. Obviously, a complete trail is realizable. A trail is called an *integer trail*, if it corresponds to an integer computation. We denote by $\mathcal{T}$ the set of all complete trails of M, and by $Z(\mathcal{T}) \subseteq \mathcal{T}$, the set of integer trails of M. We use the shorthand notation $\tau = (\sigma, T)$, where $\sigma = s_0, s_1, \ldots, s_n$ and $T = t_0, t_1, \ldots, t_n$ are the location and time sequence, respectively, associated with the trail τ.

Duration Properties

The questions we intend to pose for finitary timed automata are expressed in a language that includes the propositional calculus augmented with the *duration function* $\int$ and linear inequalities. The version of duration function considered here was inspired by the use of a similar operator in duration calculus [CHR92]. However, the semantics given here to this operator differs from its semantic in [CHR92].

State formulas are defined in the usual way over the propositions in $\mathcal{P}$ and the boolean operators, and can be evaluated over single locations, using the interpretation assigned to them by the proposition labeling function Λ.

The duration function $\int$ is a temporal function interpreted over trails. Let φ be a state formula and

$$\tau = (s_0, t_0), \ldots, (s_n, t_n)$$

be a trail. The value of the *duration expression* $\int \varphi$ at position j, $0 \le j \le n$ of τ is defined as

$$val(\tau, j, \int \varphi) = \sum_{\substack{0 \le i < j \\ s_i \models \varphi}} (t_{i+1} - t_i)$$

Duration constraints are inequalities of the form:

$$\sum_{i=1}^{m} a_i \cdot \int \varphi_i \sim c$$

where $\sim \in \{<, >, =, \ne, \le, \ge\}$, $a_i, c \in \mathbb{Z}$ and φ_i are state formulas.

Duration formulas are boolean combinations of duration constraints.

Let $\tau = (s_0, t_0), \ldots, (s_n, t_n)$ be a trail, and ψ be a duration formula. We say that τ satisfies ψ, denoted $\tau \models \psi$, if ψ evaluates to true when all the duration expressions are evaluated at position n of τ. Let Θ be a set of trails. We say that ψ is *valid* over Θ, if for all $\tau \in \Theta$, $\tau \models \psi$. We say that ψ is *satisfiable* over Θ, if there exists a trail $\tau \in \Theta$ satisfying ψ. Let M be an FTA. We say that ψ is satisfiable (valid) over M if ψ is satisfiable (valid) over $\mathcal{T}$, the set of complete trails of M. Obviously, ψ is valid over M iff $\neg\psi$ is not satisfiable over M.

A duration property is called *conjunctive* if it is a conjunction of duration constraints. Similarly, a duration property is *disjunctive* if it is a disjunction of duration constraints. We use the notations

$$\bigvee\bigwedge\left(\sum_{i=1}^{m} a_i \cdot \int \varphi_i \sim c\right), \quad \bigwedge\left(\sum_{i=1}^{m} a_i \cdot \int \varphi_i \sim c\right), \quad \text{and} \quad \bigvee\left(\sum_{i=1}^{m} a_i \cdot \int \varphi_i \sim c\right)$$

to denote a (general) duration property, a conjunctive duration property, and a disjunctive duration property, respectively.

Decision Problems

Given an FTA M and a duration property ψ, we may ask the following questions:
Problem

3 Validity. Is ψ valid over M?

and
Problem

4 Satisfiability. Is ψ satisfied by some computation of M?

As indicated above, an algorithm for solving one of these problems can be used to solve the other. We will therefore concentrate on finding solutions to a satisfiability problem.

3.3 Reduction of Reachability to Satisfiability

The reachability problem for integration graphs can be reduced to the satisfiability problem for FTA's.

Let S be a given integration graph and $\hat{s} \in S_f$ one of its final locations. We are interested in the question whether $\hat{s}$ is reachable by some computation of S. According to the definition of integration graphs, S can be decomposed into the looping part L and the acyclic testing part T. Without loss of generality, we can assume that $\hat{s} \in T$.

As a first step, we construct an FTA M that represents the behavior of S ignoring all integrators and dangerous tests. Automaton M is obtained from S by the following transformation:

1. Delete from the integration graph S all locations and edges that cannot participate in a path from s_I to $\hat{s}$.

2. Replace all dangerous guards on the remaining edges by the trivial guard T (true).
3. Retain $\hat{s}$ as the only final location.

It is not difficult to see that M is an FTA with final location $\hat{s}$.

Next, we construct a duration formula ψ that expresses the condition for system S to be able to reach location $\hat{s}$. Our first task is to express the values of terminal variables at the end of a computation in terms of duration expressions.

Let $x \in V$ be a terminal variable of S. Let x_0 be the initial value specified for x in V_0 and $s_1, \ldots, s_m$ be all the locations of S in which the rate of growth of x, $\mathcal{R}[s][x] \neq 0$. Let $\mathcal{R}_1, \ldots, \mathcal{R}_m$ be the rates of growth of x in $s_1, \ldots, s_m$, respectively. We assume that each s_j has a proposition (as part of $\mathcal{P}$) that uniquely characterize it, i.e., is true at s_j and at no other location. We denote this proposition by at_s_j.

The value of x at the end of a computation can be expressed by the duration expression

$$x_f = x_0 + \mathcal{R}_1 \cdot \int at\_s_1 + \cdots + \mathcal{R}_m \cdot \int at\_s_m$$

This is based on the observation that any unit of time spent at location s_i contributes $\mathcal{R}_i$ to the final value of x.

Since T is acyclic, there are only finitely many paths $\pi_1, \ldots, \pi_k$ that a computation can follow within T until it reaches location $\hat{s}$. For each $i = 1, \ldots, k$, let the formula ψ_i be the conjunction of all the dangerous guards that appear in S on edges of π, replacing any occurrence of a terminal variable x by the expression x_f as defined above.

Finally, we let ψ be the disjunction $\psi_1 \vee \cdots \vee \psi_k$.

Claim 1 (Reduction) *Location $\hat{s}$ is reachable by a computation of S iff there exists a computation of M satisfying ψ.*

Examples of Reduction: Cat and Mouse

Consider applying the described reduction to the Cat and Mouse system. The decomposition of this system into L and T identifies the entire system (being acyclic) as T.

First, consider reachability of location *C-wins*. There is only one path leading to this location. In Fig. 3, we present the FTA obtained by the reduction. The duration formula whose satisfiability should be checked is

$$\psi_c : \left(X_0 - v_c \cdot \int CrMr = X_0 - v_m \cdot (\int CwMr + \int CrMr) \right) \wedge \left(X_0 - v_c \cdot \int CrMr > 0 \right),$$

where we use *CwMr* and *CrMr* as the propositions characterizing these states.

Next, we consider reachability of location *M-wins*. In Fig. 4, we present the automaton obtained by reduction of the Cat and Mouse system according to this location.

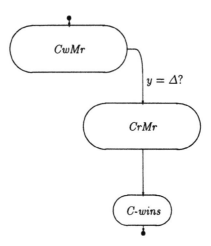

Fig. 3. Reduction of Cat and Mouse: Reachability of *C-wins*.

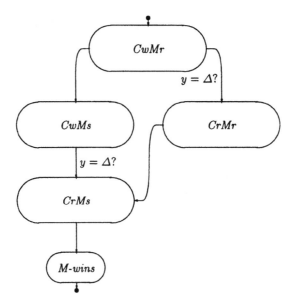

Fig. 4. Reduction of Cat and Mouse: Reachability of *M-wins*.

There are two paths leading to location *M-wins*. However, the conjunction of dangerous tests along these paths yields the same formula $x_c = 0 \land x_m = 0$. Consequently, we take

$$\psi_M : \left(X_0 - v_c \cdot (\int CrMr + \int CrMs) = 0 \right) \land \left(X_0 - v_m \cdot (\int CwMr + \int CrMr) = 0 \right)$$

Examples of Reduction: Gas Burner

There are two ways to reduce the Gas Burner system to an FTA. The difference between these two reductions is whether we consider variable z to be an integrator or a timer. In both cases, we are interested in reachability of location *Bad*. Note that any timer that is not tested or reset within the looping part L, can be promoted to be an integrator. Promoting a timer to an integrator may give rise to more duration constraints and less timer tests.

The first reduction presented in Fig. 5 considers z to be a timer. The corre-

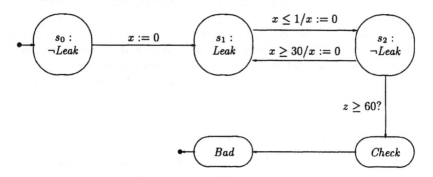

Fig. 5. Reduction of the Gas Burner: z Considered as a Timer.

sponding accessibility formula is given by:

$$\psi_1 : 20 \cdot \int at\_s_1 > \int at\_s_0 + \int at\_s_1 + \int at\_s_2.$$

In Fig. 6 we present a second reduction in which z is considered an integrator. The corresponding accessibility formula is given by:

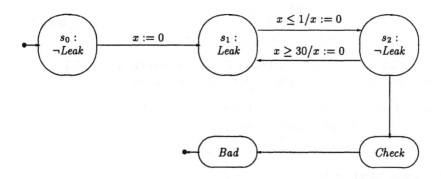

Fig. 6. Reduction of the Gas Burner: z Considered as an Integrator.

$$\psi_2 : 20 \cdot \int at\_s_1 > \int at\_s_0 + \int at\_s_1 + \int at\_s_2$$
$$\wedge \int at\_s_0 + \int at\_s_1 + \int at\_s_2 \geq 60$$

Conclusions

According to Claim 1, in order to find whether location $\hat{s}$ is reachable in the integration graph S, it is sufficient to check that the accessibility formula ψ is satisfiable over the FTA M_s. Consequently, we will concentrate on methods for solving the satisfiability problem of duration formulas over FTA's.

4 Duration Properties over Integer Computations

In the following, we present an algorithm for checking the satisfiability of duration properties over the integer computations of an FTA. Given an FTA $\hat{M}$, we first discuss the construction of a *0/1-unwinding* FTA M, whose set of computations is exactly the set of integer computations of $\hat{M}$. Next, we present an algorithm for the satisfiability of a (general) duration property over M.

Note that this solves the problem of reachability of integration graphs by integer computations.

4.1 The Integer Computations of an FTA

Let $M = \langle S, \Lambda, V, s_I, s_f, E \rangle$ and $\hat{M} = \langle \hat{S}, \hat{\Lambda}, \hat{V}, \hat{s}_I, \hat{s}_f, \hat{E} \rangle$ be FTA's and $\Phi : S \to \hat{S}$ be a function mapping locations of M to locations of $\hat{M}$. We say that Φ is a *homomorphism*, if

- $\forall s \in S, \Lambda(s) = \hat{\Lambda}(\Phi(s))$
- $\Phi(s_I) = \hat{s}_I \quad \Phi(s_f) = \hat{s}_f$
- $V \supseteq \hat{V}$
- For every $e = \langle s_1, s_2, \Gamma, Y \rangle \in E$,
    - Either, there exists an edge $\hat{e} = \langle \Phi(s_1), \Phi(s_2), \hat{\Gamma}, \hat{Y} \rangle \in \hat{E}$ such that
      $$\hat{Y} = Y \cap \hat{V}$$
      $\Gamma \to \hat{\Gamma}$, i.e., Γ implies $\hat{\Gamma}$
      In that case we write $\hat{e} = \Phi(e)$.
    - Or, $\Phi(s_1) = \Phi(s_2)$ and $Y \cap \hat{V} = \emptyset$. We can view this case as though e is mapped by Φ to a self-edge connecting $\Phi(s_1)$ to itself.

For a valuation $\nu : V \to \mathsf{R}$, we denote by $\nu|_{\hat{V}}$ its restriction to $\hat{V} \subseteq V$. The following is an immediate result of the above definition:

Claim 2 *Let $M = \langle S, \Lambda, V, s_I, s_f, E \rangle$ and $\hat{M} = \langle \hat{S}, \hat{\Lambda}, \hat{V}, \hat{s}_I, \hat{s}_f, \hat{E} \rangle$ be FTA's, and $\Phi : S \to \hat{S}$ be a homomorphism. Then if*

$$\gamma = \langle s_0, \nu_0, t_0 \rangle, \langle s_1, \nu_1, t_1 \rangle, \ldots, \langle s_n, \nu_n, t_n \rangle$$

is a computation of M, then

$$\Phi(\gamma) = reduce(\langle \Phi(s_0), \nu_0|_{\hat{V}}, t_0 \rangle, \langle \Phi(s_1), \nu_1|_{\hat{V}}, t_1 \rangle, \ldots, \langle \Phi(s_n), \nu_n|_{\hat{V}}, t_n \rangle)$$

is a computation of $\hat{M}$, where, for $\hat{\sigma} : \langle \hat{s}_0, \hat{v}_0, \hat{t}_0 \rangle, \ldots, \langle \hat{s}_n, \hat{v}_n, \hat{t}_n \rangle$, $reduce(\hat{\sigma})$ is obtained by removing from $\hat{\sigma}$ any situation $\langle \hat{s}_{i+1}, \hat{v}_{i+1}, \hat{t}_{i+1} \rangle$ such that $\hat{s}_{i+1} = \hat{s}_i$.

Let $\pi = e_0, s_0, \ldots, e_n, s_n$ be a path in M. We say that π is a *complete* path if $e_0 = e^-$ and $s_n = s_f$.

Given an FTA $\hat{M} = \langle \hat{S}, \hat{\Lambda}, \hat{V}, \hat{s}_I, \hat{s}_f, \hat{E} \rangle$, we say that an FTA $M = \langle S, \Lambda, V, s_I, s_f, E \rangle$ is a *0/1-unwinding* of $\hat{M}$ if

- There exists a surjective homomorphism $\Phi : S \to \hat{S}$.
- Every complete path π in M has a realizable trail tracing π.
- Every state $s \in S$ is associated with a fixed visit length $v_s \in \{0; 1\}$, such that each visit to s in a realizable trail lasts precisely v_s time units. That is, if

$$\ldots, (s, t_s), (s', t_{s'}), \ldots \in \mathcal{T}$$

  then $t_{s'} = t_s + v_s$.
- $reduce(\Phi(\mathcal{T})) = Z(\hat{\mathcal{T}})$

Claim 3 *For every* FTA, *there exists a 0/1-unwinding.*

The construction of the 0/1-unwinding of an FTA $\hat{M}$ is based on the region graph construction ([Dil89], [AD90]), where only a subset of the regions associated with $\hat{M}$ are used.

4.2 The Verification Algorithm

Let $M = \langle S, \Lambda, V, s_I, s_f, E \rangle$ be an FTA. For $e : \langle s, s', \Gamma, Y \rangle \in E$, we say that e *departs* from s and *arrives* at s'. For $s \in S$ and $e \in E$ departing from s, we define

- $succ(s)$ — the set of edges departing from s.
- $pred(s)$ — the set of edges arriving at s.
- $pred(e) = pred(s)$

Let $\tau = (s_0, t_0), \ldots, (s_n, t_n)$ be a trail. For every $s \in S$ and $e \in E$, we define

- n_s – the number of occurrences of location s in τ.
- m_e – the number of times e was taken in τ.
- r_e – the sequence number of e in the list of edges visited by τ, sorted according to the order of their first visit. That is, $r_e = k$ if e is the $k'th$ edge visited by τ.

Let $\hat{M} = \langle \hat{S}, \hat{\Lambda}, \hat{V}, \hat{s}_I, \hat{s}_f, \hat{E} \rangle$ be an FTA and $M = \langle S, \Lambda, V, s_I, s_f, E \rangle$ be a 0/1-unwinding of $\hat{M}$. Let

$$\psi = \bigvee \bigwedge \left(\sum_{i=1}^{m} a_i \cdot \int \varphi_i \sim c \right)$$

be a duration formula. The following set of constraints $C(M, \psi)$ is used to check the satisfiability of ψ over M.

$C(M, \psi) :$

$\bullet$ **Flow:** For every $s \in S - \{s_f\}$

$$n_s = \sum_{e_j \in pred(s)} m_j = \sum_{e_i \in succ(s)} m_i.$$

$\bullet$ **Initiation and Termination:**

$$n_{s_f} = \sum_{e_j \in pred(s_f)} m_j = m_{e^-} = 1$$

$\bullet$ **Accessibility:**

$$m_{e^-} = r_{e^-} = 1$$

For every edge $e \in E - \{e^-\}$

$$\left(m_e = r_e = 0 \right) \vee \left(\bigvee_{e' \in pred(e)} (0 < r_{e'} < r_e) \right)$$

$\bullet$ **Visit Durations:** For every $s \in S$

$$\Delta(s) = n_s \cdot v_s$$

$\bullet$ **The Duration Property:**

$$\bigvee \bigwedge_{i=1}^{m} \left(\sum a_i \cdot \sum_{\substack{s \in S \\ s \models \varphi_i}} \Delta(s) \sim c \right)$$

Proposition 4. *The set of constraints $C(M, \psi)$ has a solution iff ψ is satisfiable over M iff ψ is satisfiable by an integer computation of $\hat{M}$.*

The problem of finding an integer solution to $C(M, \psi)$ for the unknowns

$$n_s \geq 0, \quad m_e \geq 0, \quad r_e \geq 0$$

is a classical *integer linear programming* problem. It is shown in [GJ79] to be NP complete, but algorithms that are efficient for frequently occuring cases are known [Sal75].

5 Real Computations: Single Timer

In the previous section, we presented an algorithm for satisfiability of duration properties over the integer computations of an FTA. In this and the following section, we deal with satisfiability of duration properties over the entire set of an FTA's computations, including real computations.

First, we restrict the FTA to a single timer, checking satisfiability of general duration formulas. Note that this solves the problem of reachability of integration graphs with a single timer (and any number of integrators).

Next, in section 6 we give an algorithm for satisfiability over an unrestricted FTA, restricting the verified property to disjunctive durations.

5.1 Characterization of Complete Trails

Let $M = \langle S, \Lambda, V, s_I, s_f, E \rangle$ be an FTA satisfying:

- $V = \{x\}$.
- Every loop in M contains at least one resetting edge.

We refer to M as a *single timer* FTA.

Let $\pi = e_0, s_0, \ldots, e_n, s_n$ be a path in M. We call π an *rr-path* if e_0 and e_n are resetting edges. An rr-path is called *simple* if for every $i = 1, 2, \ldots n-1$, e_i is not a resetting edge. We denote by Π the (finite) set of all simple rr-paths in M. Since there is only one timer, the guard of each edge e_i has the form $l_i \leq x \leq u_i$.

Let $\pi = e_0, s_0, \ldots, e_n, s_n$ be a path in M. We say that

$$\tau = (s_0, t_0), \ldots, (s_n, t_n)$$

is a *trail tracing* π.

Let s be a location appearing one or more times in τ. We define the *visit duration* of the trail τ at location s to be

$$\Delta(s, \tau) \quad = \quad \sum_{0 \leq i < n, \; s_i = s} (t_{i+1} - t_i)$$

For a location s not visited by τ, we take $\Delta(s, \tau) = 0$. Thus, $\Delta(s, \tau)$ is the time spent by τ at location s accumulated over all the visits of τ at s.

When the trail τ is obvious from the context, we write $\Delta(s, \tau)$ simply as $\Delta(s)$. Note that if π is a simple path, τ can visit each $s \in S$ at most once.

Claim 5 *Let* $\pi = e_0, s_0, \ldots, e_n, s_n \in \Pi$ *and* $\tau = (s_0, t_0), \ldots, (s_n, t_n)$ *be a trail tracing* π. *Then* τ *is realizable iff*

$$\bigwedge_{i=1}^{n} \left(l_i \leq \sum_{j=0}^{i-1} \Delta(s_j) \leq u_i \right)$$

where, for every $i = 1, 2, \ldots, n$, $l_i \leq x \leq u_i$ *is the guard associated with* e_i.

Proof:

$\Rightarrow$ Assume that τ is realizable, then it corresponds to some computation segment

$$\gamma = \langle s_0, \nu_0, t_0 \rangle, \langle s_1, \nu_1, t_1 \rangle, \ldots, \langle s_n, \nu_n, t_n \rangle$$

such that:

- $\nu_0[x] = \nu_n[x] = 0$, and
- for all i, $0 \le i < n - 1$, $\nu_{i+1}[x] = \nu_i[x] + t_{i+1} - t_i = t_{i+1}$.

Clearly, for all i, $1 \le i \le n$, $l_i \le \nu_{i-1}[x] + t_i - t_{i-1} \le u_i$.

Since π is simple, $\Delta(s_j) = t_{j+1} - t_j$ for all j, $0 \le j \le n - 1$ and it is easy to show by induction that $\nu_{i-1}[x] + t_i - t_{i-1} = \sum_{j=0}^{i-1} \Delta(s_j)$ for all i, $1 \le i \le n$. Hence,

$$\bigwedge_{i=1}^{n} \left(l_i \le \sum_{j=0}^{i-1} \Delta(s_j) \le u_i \right)$$

$\Leftarrow$ Let

$$\gamma = \langle s_0, \nu_0, t_0 \rangle, \langle s_1, \nu_1, t_1 \rangle, \ldots, \langle s_n, \nu_n, t_n \rangle$$

such that

- $\nu_0[x] = \nu_n[x] = 0$, and
- for all i, $0 \le i < n - 1$, $\nu_{i+1}[x] = \nu_i[x] + \Delta(s_i) = \nu_i[x] + t_{i+1} - t_i$.

Clearly, for all i, $1 \le i \le n$, $\nu_{i-1}[x] + t_i - t_{i-1} = \sum_{j=0}^{i-1} \Delta(s_j)$ and $l_i \le \nu_{i-1}[x] + t_i - t_{i-1} \le u_i$.

Hence, γ is a computation segment and τ is realizable. ∎

Let $\pi = e_0, s_0, \ldots, e_n, s_n$ and $\pi' = e'_0, s'_0, \ldots, e'_k, s'_k$ be two rr-paths such that $e_n = e'_0$. We define the *fusion* of π and π', to be the path

$$\pi \circ \pi' = e_0, s_0, \ldots, e_n, s'_0, \ldots, e'_k, s'_k$$

Let $\tau = (s_0, t_0), \ldots, (s_n, t_n)$ be a realizable trail tracing the rr-path $\lambda = e_0, s_0, \ldots, e_n, s_n$. Obviously, λ can be uniquely presented as a concatenation $\pi_1 \circ \cdots \circ \pi_k$ of $k \ge 1$ simple rr-paths.

- For each $\pi \in \Pi$, let $n(\pi, \lambda)$ denote the number of times the simple rr-path π appears in the concatenation λ. If π does not appear in λ then $n(\pi, \lambda) = 0$.
- For each location s, let $\Delta^\pi(s, \tau)$ denote the accumulated visit duration of τ in location s when we restrict our attention to visits at s while tracing the simple rr-path π.

If both π_1 and π_2 with $\pi_1 \ne \pi_2$ visit location s and $\lambda = \pi_1 \circ \pi_2$, then $\Delta^{\pi_1}(s, \tau)$ refers to the first visit to s, while $\Delta^{\pi_2}(s, \tau)$ refers to the second visit. If π does not appear in λ then $\Delta^\pi(s, \tau) = 0$ for every $s \in S$.

Thus, the accumulated visit duration of τ in location s is

$$\Delta(s, \tau) = \sum_{\pi \in \Pi} \Delta^\pi(s, \tau)$$

When τ and λ are understood from the context, we write $n(\pi, \lambda)$ and $\Delta^\pi(s, \tau)$ simply as $n(\pi)$ and $\Delta^\pi(s)$. We refer to the sets $\{n(\pi) \mid \pi \in \Pi\}$ and $\{\Delta^\pi(s) \mid \pi \in \Pi, s \in S\}$ as the *attributes* of τ and π.

Claim 6 *Let λ be an rr-path and τ be a trail tracing λ. Then τ is realizable iff*

$$\bigwedge_{\pi \in \Pi} \bigwedge_{i=1}^{|\pi|} \left(n(\pi) \cdot l_i \leq \sum_{j=0}^{i-1} \Delta^{\pi}(s_j) \leq n(\pi) \cdot u_i \right)$$

where, for every $i = 1, 2, \ldots, |\pi|$, $l_i \leq x \leq u_i$ is the guard of edge e_i in π.

Proof: By induction on the length $|\lambda|$ of λ, namely, the number of simple rr-path in λ. For $|\lambda| = 1$ – follows from Claim 5. Let $\lambda = \lambda' \circ \pi$ and $\tau = \tau' \circ \tau_\pi$ where τ_π traces π. By the induction hypothesis, we have that τ' is realizable iff

$$\bigwedge_{\pi' \in \Pi} \bigwedge_{i=1}^{|\pi'|} \left(n(\pi', \lambda') \cdot l_i \leq \sum_{j=0}^{i-1} \Delta^{\pi'}(s_j, \tau') \leq n(\pi', \lambda') \cdot u_i \right) \tag{2}$$

$\Rightarrow$ Since τ is realizable it corresponds to some computation segment $\gamma = \gamma' \circ \gamma_\pi$ where

$$\gamma_\pi = \langle s_0, \nu_0, t_0 \rangle, \langle s_1, \nu_1, t_1 \rangle, \ldots, \langle s_{|\pi|}, \nu_{|\pi|}, t_{|\pi|} \rangle$$

and τ', τ_π correspond to γ', γ_π respectively. That is, τ_π is realizable and by Claim 5

$$\bigwedge_{i=1}^{|\pi|} \left(l_i \leq \sum_{j=0}^{i-1} \Delta(s_j, \tau_\pi) \leq u_i \right) \tag{3}$$

For $\pi' \neq \pi$
- $n(\pi') = n(\pi', \lambda) = n(\pi', \lambda')$, and
- $\Delta^{\pi'}(s) = \Delta^{\pi'}(s, \tau) = \Delta^{\pi'}(s, \tau')$;

and for $\pi' = \pi$
- $n(\pi) = n(\pi, \lambda) = n(\pi, \lambda') + 1$, and
- $\Delta^{\pi}(s) = \Delta^{\pi}(s, \tau) = \Delta(s, \tau_\pi) + \Delta^{\pi}(s, \tau')$.

Hence, (2) and (3) imply

$$\bigwedge_{\pi' \in \Pi} \bigwedge_{i=1}^{|\pi'|} \left(n(\pi') \cdot l_i \leq \sum_{j=0}^{i-1} \Delta^{\pi'}(s_j) \leq n(\pi') \cdot u_i \right)$$

$\Leftarrow$ Assume that

$$\bigwedge_{i=1}^{|\pi|} \left(n(\pi, \lambda) \cdot l_i \leq \sum_{j=0}^{i-1} \Delta^{\pi}(s_j, \tau) \leq n(\pi, \lambda) \cdot u_i \right) \tag{4}$$

Clearly, $\Delta(s, \tau_\pi) = \Delta^{\pi}(s, \tau) - \Delta^{\pi}(s, \tau')$ and from (4) and (2) we deduce that (3) holds.

From Claim 5, τ_π is realizable, corresponding to some computation segment γ_π. By the induction hypothesis τ' corresponds to some computation segment γ'. Hence, $\gamma = \gamma' \circ \gamma_\pi$ is a computation segment and τ is realizable. ∎

Let $e \in E$ be a resetting edge. We define the following sets of simple rr-paths associated with e

$$pred(e) = \{\pi \mid \pi = e_0, s_0, \ldots, e, s_n \in \Pi\}$$

$$succ(e) = \{\pi \mid \pi = e, s_0, \ldots, e_n, s_n \in \Pi\}$$

Let $\pi = e_0, s_0, \ldots, e_n, s_n \in \Pi$. Then

$$pred(\pi) = pred(e_0)$$

$$last(\pi) = s_n$$

$$first(\pi) = e_0$$

Let $\tau = \pi_0 \circ \cdots \circ \pi_n$ be a complete trail. For every $\pi \in \Pi$, we define

- $r(\pi)$ – the sequence number of π in the list of simple rr-paths visited by τ, sorted according to the order of their first visit.

The set of constraints (C) intended to characterize the attributes of a complete trail, is summarized in figure 7.

$(C):$

- **Flow:**
  For every reset edge $e \in E$ such that $e \neq e^-$ and e does not arrive at s_f,
  $$\sum_{\pi \in pred(e)} n(\pi) = \sum_{\pi \in succ(e)} n(\pi)$$

- **Initiation and Termination:**
  $$\sum_{first(\pi)=e^-} n(\pi) = \sum_{last(\pi)=s_f} n(\pi) = 1$$

- **Accessibility:**
  For every path $\pi = e_0, \ldots, s_n \in \Pi$ such that $e_0 = e^-$
  $$(r(\pi) = n(\pi) = 0) \vee (r(\pi) = n(\pi) = 1)$$
  For every other path $\pi \in \Pi$
  $$\left(r(\pi) = n(\pi) = 0\right) \vee \left(\bigvee_{\pi' \in pred(\pi)} (0 < r(\pi') < r(\pi)) \right)$$

- **Time Constraints:** For every $\pi \in \Pi$,
  $$\bigwedge_{i=1}^{|\pi|} \left(n(\pi) \cdot l_i \leq \sum_{j=0}^{i-1} \Delta^\pi(s_j) \leq n(\pi) \cdot u_i \right)$$
  For every $s \notin \pi, \quad \Delta^\pi(s) = 0$

Fig. 7. The Attributes of Complete Trails

The following proposition states that the constraint set (C) is a precise characterization of a complete trail.

Proposition 7. *A set of values for the variables $n(\pi)$, $\Delta^\pi(s)$ is an attribute set for a complete trail iff it is a solution for (C).*

Proof: The set $\{n(\pi) \mid \pi \in \Pi\}$ satisfies the first three constraints iff there exists an rr-path $\lambda : e^-, s_I, \ldots, s_f$, such that for every $\pi \in \Pi$, $n(\pi)$ is the number of occurrences of π in λ. Namely, the first three constraints characterize exactly the set of paths from s_I to s_f. Claim 6 guarantees that the set $\Omega = \{n(\pi), \Delta^\pi(s) \mid$ for every $\pi \in \Pi$ and $s \in S\}$ is an attribute set of a realizable trail iff Ω satisfies the last constraint. ∎

Corollary 8. *The duration formula*

$$\psi = \bigvee \bigwedge \left(\sum_{i=1}^{n} a_i \cdot \int \varphi_i \sim c \right)$$

is satisfiable over a single timer FTA M iff the set of equalities/inequalities consisting of (C) plus the constraint

$$\bigvee \bigwedge \left(\sum_{i=1}^{n} a_i \cdot \left(\sum_{\substack{\pi \in \Pi \\ s \models \varphi_i}} \Delta^\pi(s) \right) \sim c \right)$$

has a solution.

6 Real Computations: Disjunctive Durations

In section 4 we presented an algorithm for the satisfiability of general duration properties over the integer subset of an FTA's computation. In the following, we show that the same algorithm can be used to check satisfiability over the entire set of an FTA's computations, providing we restrict the property to disjunctive durations.

Note that this solves the problem of reachability of integration graphs with at most one dangerous test along each path.

6.1 Digitization of FTA Computations

A time sequence $T = t_0, \ldots, t_n$, $t_i \in \mathsf{R}^+$ is called an *integer* time sequence if $t_i \in \mathsf{Z}$, for every $i = 0, \ldots, n$. We denote by $[0, 1)$ the set of real numbers ϵ, satisfying $0 \leq \epsilon < 1$. Let $T = t_1, \ldots, t_n$ be a time sequence. For every $\epsilon \in [0, 1)$, we define the integer time sequence $[T]_\epsilon = [t_1]_\epsilon, \ldots, [t_n]_\epsilon$ as follows:

$$[t_i]_\epsilon = \lfloor t_i \rfloor \quad \text{if } t_i \leq (\lfloor t_i \rfloor + \epsilon), \qquad \text{otherwise } [t_i]_\epsilon = \lceil t_i \rceil$$

Claim 9 *Let M be an FTA, and $\tau = (\sigma, T) \in \mathcal{T}_M$ be a complete trail of M. Then, for every $\epsilon \in [0, 1)$, $[\tau]_\epsilon = (\sigma, [T]_\epsilon) \in \mathcal{T}_M$.*

In other words, the set of complete trails of an FTA, is *closed under digitization* (see [HMP92] for definitions).

Proof: Similar to the proof given for timed transition systems in [HMP92].

The proof relies on the assumption that all enabling conditions use the non-strict inequality $\leq$. Allowing strict inequality invalidates the claim.

6.2 Disjunctive Durations over Digitizable Computations

Lemma 10. *Let*

$$0 < a_1 \leq a_2 \leq \ldots \leq a_m \leq 1 \quad and \quad 0 < b_1 \leq b_2 \leq \ldots \leq b_n \leq 1$$

be two increasing sequences. If

$$\sum_{i=1}^{m} a_i \succ \sum_{j=1}^{n} b_j,$$

where $\succ \in \{\geq, >\}$, then there exists an $\epsilon \in [0, 1)$, such that

$$\sum_{i=1}^{m} [a_i]_\epsilon \succ \sum_{j=1}^{n} [b_j]_\epsilon.$$

The proof of this lemma is presented in the appendix.

Proposition 11. *Let M be an FTA and ψ be a disjunctive duration property. The formula ψ is satisfiable over M iff ψ is satisfiable over $Z(\mathcal{T}_M)$.*

Proof outline: Let $\tau = (s_0, t_0), \ldots, (s_n, t_n) \in \mathcal{T}_M$, and ψ be a disjunctive duration property. Let

$$\psi' = \sum_{i=1}^{m} a_i \cdot \int \varphi_i \sim c$$

be a single disjunct of ψ. Interpreting ψ' over τ, we get

$$\psi' = \left(\sum_{i=1}^{m} a_i \cdot \sum_{\substack{j=0 \\ s_j \models \varphi_i}}^{n-1} \left(t_{j+1} - t_j \right) \right) \sim c$$

which can be rewritten as

$$\sum_{j=0}^{n} c_j \cdot t_j \sim \sum_{j=0}^{n} d_j \cdot t_j + c \tag{5}$$

where $c_j, d_j \in \mathbb{N}$. For every expression $(x_j \cdot t_j)$ in equation 5, where $x \in \{c, d\}$ and

$$t_j = \lfloor t_j \rfloor + \delta_j, \qquad 0 \le \delta_j < 1$$

we rewrite the expression as follows

$$\overbrace{1 + \ldots + 1}^{x_j \cdot \lfloor t_j \rfloor - times} + \overbrace{\delta_j + \ldots + \delta_j}^{x_j - times}$$

turning equation 5 into the form used in Lemma 10. Thus, if ψ' is satisfiable over τ, there exists an $\epsilon \in [0, 1)$ such that ψ' is satisfiable over $[\tau]_\epsilon$, for every disjunct ψ' of ψ. Claim 9 completes the proof.

∎

The terms *closure under digitization* and *closure under inverse digitization* are introduced in [HMP92]. Note that a direct result of Lemma 10 is that conjunctive duration properties are closed under inverse digitization. Proposition 11 is another example of a more general observation proven in [HMP92].

Conclusion

Let M be an FTA and

$$\psi = \bigvee \left(\sum_{i=1}^{m} a_i \cdot \int \varphi_i \sim c \right)$$

be a disjunctive duration formula. The satisfiability of ψ over M can be checked using the algorithm described in section 4.

7 Discussion

This paper explores a subset of constant slope hybrid systems, searching for a decidable subset. The subset studied is that of Integration Graphs, which allow integrators but queries them only in a restricted way at the end of a computation.

For this class of systems, we have established decidability for integer computations, and two restricted cases of real computations: the case of a single timer, and that of a single dangerous query.

A question that remains open for further research is what can be said about the other cases; those involving several timers and a conjunction of dangerous queries.

References

[ACD90] R. Alur, C. Courcoubetis, and D.L. Dill. Model-checking for real-time systems. In *Proc. 5th IEEE Symp. Logic in Comp. Sci.*, 1990.

[AD90]   R. Alur and D.Dill. Automata for modelling real time systems. In *Proc. 17th Int. Colloq. Aut. Lang. Prog.*, 1990.

[AH90]   R. Alur and T.A. Henzinger. Real-time logics: Complexity and expressiveness. In *Proc. 5th IEEE Symp. Logic in Comp. Sci.*, 1990.

[BCM+90]   J.R. Burch, E.M. Clarke, K.L. McMillan, D.L. Dill, and J. Hwang. Symbolic model checking: 10^{20} states and beyond. Technical report, Carnegie Mellon University, 1990.

[CES86]   E.M. Clarke, E.A. Emerson, and A.P. Sistla. Automatic verification of finite state concurrent systems using temporal logic specifications. *ACM Trans. Prog. Lang. Sys.*, 8:244–263, 1986.

[CHR92]   Z. Chaochen, C.A.R Hoare, and A.P. Ravn. A calculus of durations. *Information Processing Letters*, 40(5):269–276, 1992.

[Dil89]   D. L. Dill. Timing assumptions and verification of finite-state concurrent systems. In *Automatic Verification Methods for Finite State Systems*, LNCS. Springer Verlag, New York, 1989.

[GJ79]   M. R. Garey and D. S. Johnson. *Computers and Intractability, a Guide to the theory of NP-Completeness*. W. H. Freeman and Company, 1979.

[HMP92]   T. Henzinger, Z. Manna, and A. Pnueli. What good are digital clocks? In W. Kuich, editor, *Proc. 19th Int. Colloq. Aut. Lang. Prog.*, volume 623 of *Lect. Notes in Comp. Sci.*, pages 545–558. Springer-Verlag, 1992.

[MMP92]   O. Maler, Z. Manna, and A. Pnueli. A formal approach to hybrid systems. In *Proceedings of the REX workshop "Real-Time: Theory in Practice"*, LNCS. Springer Verlag, New York, 1992.

[NOS+93]   X. Nicollin, A. Olivero, J. Sifakis, , and S. Yovine. An approach to the description and analysis of hybrid systems. In A. Ravn and H. Rischel, editors, *Workshop on Hybrid Systems*, Lect. Notes in Comp. Sci. Springer-Verlag, 1993.

[NSY91]   X. Nicollin, J. Sifakis, and S. Yovine. From ATP to timed graphs and hybrid systems. In *Real-Time: Theory in Practice*. Lec. Notes in Comp. Sci., Springer-Verlag, 1991.

[Sal75]   H.M. Salkin. *Integer Programming*. Addison-Wesley, 1975.

Appendix

In the following we present the proof of Lemma 10 (subsection 6.2).

Lemma 12. *Let*

$$0 < a_1 \leq a_2 \leq \ldots \leq a_m \leq 1 \quad and \quad 0 < b_1 \leq b_2 \leq \ldots \leq b_n \leq 1$$

be two increasing sequences. If

$$\sum_{i=1}^{m} a_i \; \succ \; \sum_{j=1}^{n} b_j,$$

where $\succ \; \in \{\geq, >\}$, *then there exists an* $\epsilon \in [0, 1)$, *such that*

$$\sum_{i=1}^{m} [a_i]_\epsilon \; \succ \; \sum_{j=1}^{n} [b_j]_\epsilon.$$

Proof: The proof is by induction on $n \geq 0$ for all m.

For $n = 0$, it is enough to take $\epsilon = 0$. Observing that, for every positive real c, $[c]_0 \geq c$, this is based on

$$\sum_{i=1}^{m} [a_i]_0 \geq \sum_{i=1}^{m} a_i \succ 0 = \sum_{j=1}^{0} [b_j]_0$$

Assume the lemma is true for n and show for $n + 1$. We assume that

$$\sum_{i=1}^{m} a_i \succ \sum_{j=1}^{n} b_j + b_{n+1} \tag{6}$$

Since $b_{n+1} > 0$, m must be positive. We consider several cases:

Case: $a_m > b_{n+1}$
In this case, we take any ϵ satisfying $b_{n+1} < \epsilon < a_m$. This yields

$$\sum_{i=1}^{m} [a_i]_\epsilon \geq 1 \succ 0 = \sum_{j=1}^{n+1} [b_j]_\epsilon.$$

Consequently, we assume from now on that $a_m \leq b_{n+1}$.

Case: $a_m = b_{n+1}$
Subtracting $a_m = b_{n+1}$ from both sides of inequality (6), we obtain

$$\sum_{i=1}^{m-1} a_i \succ \sum_{j=1}^{n} b_j.$$

Applying the induction hypothesis to this inequality, we obtain an ϵ such that

$$\sum_{i=1}^{m-1} [a_i]_\epsilon \succ \sum_{j=1}^{n} [b_j]_\epsilon.$$

Adding $[a_m]_\epsilon = [b_{n+1}]_\epsilon = 1$ to both sides yields

$$\sum_{i=1}^{m} [a_i]_\epsilon \succ \sum_{j=1}^{n+1} [b_j]_\epsilon.$$

Case: $a_m < b_{n+1}$, $\succ$ is $>$
Subtracting a_m from the left-hand side and the bigger b_{n+1} from the right-hand side of inequality (6), we obtain

$$\sum_{i=1}^{m-1} a_i > \sum_{j=1}^{n} b_j.$$

Applying the induction hypothesis to this equation, we obtain an ϵ such that

$$\sum_{i=1}^{m-1} [a_i]_\epsilon > \sum_{j=1}^{n} [b_j]_\epsilon. \tag{7}$$

This ϵ must be smaller than a_{m-1} because, otherwise, the left hand side of (7) evaluates to 0, and we get the contradictory inequality

$$0 > \sum_{j=1}^{n} [b_j]_\epsilon \geq 0.$$

Consequently, $\epsilon < a_{m-1} \leq a_m < b_{n+1}$ and we may add $[a_m]_\epsilon = [b_{n+1}]_\epsilon = 1$ to both sides of (7), obtaining

$$\sum_{i=1}^{m} [a_i]_\epsilon \succ \sum_{j=1}^{n+1} [b_j]_\epsilon.$$

The remaining cases will deal with $\geq$.

Case: $a_m < b_{n+1} < 1$
Taking ϵ that satisfies $b_{n+1} < \epsilon < 1$, we obtain

$$\sum_{i=1}^{m} [a_i]_\epsilon = 0 \geq 0 = \sum_{j=1}^{n+1} [b_j]_\epsilon.$$

From now on, we assume that $b_{n+1} = 1$.

Case: $b_n < a_m < b_{n+1} = 1$
Taking ϵ that satisfies $b_n < \epsilon < a_m$, we obtain

$$\sum_{i=1}^{m} [a_i]_\epsilon \geq 1 = \sum_{j=1}^{n+1} [b_j]_\epsilon.$$

Case: $a_m \leq b_n \leq b_{n+1} = 1$
Subtracting a_m from the left-hand side of (6) and the not smaller b_n from its right-hand side, we obtain (substituting 1 for b_{n+1})

$$\sum_{i=1}^{m-1} a_i \geq \sum_{j=1}^{n-1} b_j + 1.$$

Applying the induction hypothesis to this case (that has n elements on its right-hand side), we obtain an ϵ such that

$$\sum_{i=1}^{m-1} [a_i]_\epsilon \geq \sum_{j=1}^{n-1} [b_j]_\epsilon + 1. \tag{8}$$

This ϵ must be smaller than a_{m-1} because, otherwise, the left hand side of (8) evaluates to 0, and we get the contradictory inequality

$$0 \geq \sum_{j=1}^{n-1}[b_j]_\epsilon + 1 \geq 1.$$

Consequently, $\epsilon < a_{m-1} \leq a_m \leq b_n$ and we may add $[a_m]_\epsilon = [b_n]_\epsilon$ to both sides of (8), obtaining

$$\sum_{i=1}^{m}[a_i]_\epsilon \succ \sum_{j=1}^{n+1}[b_j]_\epsilon.$$

This concludes the proof.

∎

Hybrid Automata:
An Algorithmic Approach to the Specification and Verification of Hybrid Systems

Rajeev Alur[1], Costas Courcoubetis[2*], Thomas A. Henzinger[3**], Pei-Hsin Ho[3**]

[1] AT&T Bell Laboratories, USA.
[2] Department of Computer Science, University of Crete, Greece.
[3] Department of Computer Science, Cornell University, USA.

Abstract. We introduce the framework of *hybrid automata* as a model and specification language for hybrid systems. Hybrid automata can be viewed as a generalization of timed automata, in which the behavior of variables is governed in each state by a set of differential equations. We show that many of the examples considered in the workshop can be defined by hybrid automata. While the reachability problem is undecidable even for very restricted classes of hybrid automata, we present two semidecision procedures for verifying safety properties of *piecewise-linear* hybrid automata, in which all variables change at constant rates. The two procedures are based, respectively, on minimizing and computing fixpoints on generally infinite state spaces. We show that if the procedures terminate, then they give correct answers. We then demonstrate that for many of the typical workshop examples, the procedures do terminate and thus provide an automatic way for verifying their properties.

1 Introduction

More and more real-life processes, from elevators to aircrafts, are controlled by programs. These *reactive* programs are embedded in continuously changing environments and must react to environment changes in real time. Obviously, correctness is of vital importance for reactive programs. Yet traditional program verification methods allow us, at best, to approximate continuously changing environments by discrete sampling. A generalized formal model for computing systems is needed to faithfully represent both discrete and continuous processes within a unified framework. Hybrid automata present such a framework.

A *hybrid* system consists of a discrete program within an analog environment. Hybrid automata are generalized finite-state machines for modeling hybrid systems. As usual, the discrete transitions of a program are modeled by a change of the program counter, which ranges over a finite set of control locations. In addition, we allow for the possibility that the global state of a system changes

---

* Supported in part by the BRA ESPRIT project REACT.
** Supported in part by the National Science Foundation under grant CCR-9200794 and by the United States Air Force Office of Scientific Research under contract F49620-93-1-0056.

continuously with time according to the laws of physics. For each control location, the continuous activities of the environment are governed by a set of differential equations. We label each location also with an invariant condition that must hold while the control resides at the location, and each transition is labeled with a guarded set of assignments. This model for hybrid systems is inspired by the phase transition systems of [MMP92] and [NSY92], and can be viewed as a generalization of timed automata [AD90].

The current paper pursues three objectives. First, hybrid automata are defined and their suitability for specification is demonstrated through some paradigmatic examples. Second, the verification problem for hybrid automata is studied and shown to be intrinsically difficult even under severe restrictions. Third, and most importantly, we successfully verify interesting properties of truly hybrid system behaviors. We note that Nicollin et al. have independently developed an approach similar to ours [NOSY].

For verification purposes, we restrict ourselves to *linear* hybrid automata. In a linear hybrid automaton, for each variable the rate of change with time is constant — though this constant may vary from location to location — and the terms involved in the invariants, guards, and assignments are required to be linear. An interesting special case of a linear hybrid automaton is a *timed automaton* [AD90]. In a timed automaton each continuously changing variable is an accurate clock whose rate of change with time is always 1. Furthermore, in a timed automaton all terms involved in assignments are constants, and all invariants and guards only involve comparisons of clock values with constants. Even though the reachability problem for linear hybrid automata is undecidable, it is PSPACE-complete for timed automata. In this paper, we show that some of the algorithms for the analysis of timed automata can be extended to obtain semidecision procedures for solving the verification problem for linear hybrid automata. In particular, we consider the fixpoint computation method presented in [HNSY92] and the minimization procedure for timed automata presented in [ACH+92]. Both methods perform a reachability analysis over the infinite state space of a timed automaton by computing with sets of states. We show that the primitive steps of the two algorithms can be performed relatively easily even in case of linear hybrid automata and, thus, both methods can be generalized. The crucial observation is that each set of states computed by the algorithms is definable by a linear formula; that is, it is a union of convex polyhedra. However, as we move from timed automata to linear hybrid automata, the termination of the two procedures is no longer guaranteed.

Both methods we consider can be used to prove invariant properties of linear hybrid systems. We illustrate these methods on three examples, and in each case the procedures terminate. The first example involves a water level monitor. It is a truly hybrid system, since the water level increases and decreases continuously in phases. We show how to prove that the water level always remains within the specified bounds. The second example proves the mutual exclusion property of a real-time mutual exclusion protocol. Earlier algorithmic methods based on timed automata fail when the bounds on the various delays are not known. We show how to perform a symbolic analysis so as to deduce constraints between the various bounds. Our third example involves leakage in a gas burner. This is an example of a so-called *integrator system* in which we are required to prove a bound on the ratio of two durations.

2 Modeling Hybrid Systems

We define a formal model and a specification language for hybrid systems.

2.1 Hybrid traces

An *interval* is a nonempty convex subset of the nonnegative real line R^+. Intervals may be open, halfopen, or closed; bounded or unbounded. The left end-point of an interval I is denoted by l_I and the right end-point, for bounded I, is denoted by r_I. Two intervals I_1 and I_2 are *adjacent* if (1) $r_{I_1} = l_{I_2}$, and (2) either I_1 is right-open and I_2 is left-closed, or I_1 is right-closed and I_2 is left-open. An *interval sequence* $I_0 I_1 I_2 \ldots$ is a finite or infinite sequence of intervals that partitions R^+:

1. Any two neighboring intervals I_i and I_{i+1} are adjacent.
2. For all $t \in R^+$, there is some interval I_i with $t \in I_i$.

In particular, I_0 is left-closed and $l_{I_0} = 0$. The last interval of any finite interval sequence is unbounded. The interval sequence $\bar{I}_1$ *refines* the interval sequence $\bar{I}_2$ if $\bar{I}_1$ is obtained from $\bar{I}_2$ by splitting some intervals. We henceforth identify an interval sequence $\bar{I}$ with its refinement closure $\{\bar{J} \mid \bar{J} \text{ refines } \bar{I}\}$. Clearly, for any finite set $\mathcal{I}$ of interval sequences there is an interval sequence $\bigcap \mathcal{I}$ that refines all sequences in $\mathcal{I}$.

   Let V be a finite set of real-valued variables. A *state* is an interpretation of all variables in V. We write Σ for the set of states. A *trace* is a function from R^+ to Σ. Equivalently, a trace τ is a collection of functions $\tau(x)$ from R^+ to R, one for each variable $x \in V$. We say that the trace τ has property P if all of its constituent functions $\tau(x)$, for $x \in V$, have property P. We will use the following properties of functions:

- A function $f : R^+ \to R$ is *piecewise smooth* if there exists an interval sequence $\bar{I}^f = I_0 I_1 I_2 \ldots$ and a sequence $f_0 f_1 f_2 \ldots$ of C^∞-functions such that the restriction of f to each interval I_i coincides with the restriction of f_i to I_i. Each restriction of f to an interval I_i is called a *phase* of f. The phases of a piecewise-smooth trace τ are the restrictions of τ to the intervals of the sequence $\bar{I}^\tau = \bigcap \{\bar{I}^{\tau(x)} \mid x \in V\}$.
- A piecewise-smooth function $f : R^+ \to R$ is *piecewise linear* if each phase of f is linear.
- A piecewise-linear function $f : R^+ \to R$ is a *step function* if the slope of all phases of f is 1.

- A piecewise-linear function $f \colon \mathsf{R}^+ \to \mathsf{R}$ is a *clock function* if the slope of all phases of f is 1.
- A piecewise-linear function $f \colon \mathsf{R}^+ \to \mathsf{R}$ is a *skewed-clock function* if there is some constant $k \in \mathsf{R}$ such that the slope of all phases of f is k.
- A piecewise-linear function $f \colon \mathsf{R}^+ \to \mathsf{R}$ is an *integrator function* if the slope of each phase of f is either 0 or 1.

A *timed trace* τ is a trace each of whose constituent functions $\tau(x)$, for $x \in V$, is either a step function or a clock function. A set S of traces is *fusion-closed* if for all traces $\tau_1, \tau_2 \in S$ and all $t_1, t_2 \in \mathsf{R}^+$, if $\tau_1(t_1) = \tau_2(t_2)$, then $\tau \in S$ for the trace τ with $\tau(t) = \tau_1(t)$ for all $t \leq t_1$ and $\tau(t) = \tau_2(t + t_2 - t_1)$ for all $t > t_1$.

2.2 Hybrid automata

We model a hybrid system as fusion-closed set S of piecewise-smooth traces. Each trace $\tau \in S$ represents a possible behavior of the system over real time. The piecewise smoothness of τ ensures that in any bounded interval of time, there are only finitely many discontinuous state changes. The fusion closure of S ensures that each state contains all information necessary to determine the future evolution of the system.

We define sets of traces by graphs whose edges represent discrete transitions and whose vertices represent continuous activities. A *hybrid system* $A = (V_D, Q, \mu_1, \mu_2, \mu_3)$ is given by six components:

- A finite set V_D of real-valued *data variables*. A *data state* is an interpretation of all variables in V_D. We write Σ_D for the set of data states.
- A finite set Q of vertices called *locations*. We use the variable $pc \notin V_D$ as a *control variable* that ranges over the set Q of locations (properly encoded in R), and let $V = \{pc\} \cup V_D$. Thus, $\Sigma = Q \times \Sigma_D$; that is, a (system) state is a pair (ℓ, σ) consisting of a location $\ell \in Q$ and a data state $\sigma \in \Sigma_D$.
- A labeling function μ_1 that assigns to each location in Q a set of possible *activities*. Each activity is a C^∞-function from R^+ to Σ_D.
- A labeling function μ_2 that assigns to each location $\ell \in Q$ an *exception* set $\mu_2(\ell) \subseteq \Sigma_D$. The system control must leave location ℓ before an exception $\mu_2(\ell)$ occurs. The complement $\Sigma_D - \mu_2(\ell)$ is called the *invariant* of the location ℓ.
- A labeling function μ_3 that assigns to each pair $e \in Q^2$ of locations a *transition relation* $\mu_3(e) \subseteq \Sigma_D^2$. We require that for all locations $\ell \in Q$ and all data states $\sigma \in \Sigma_D$, $(\sigma, \sigma) \in \mu_3(\ell, \ell)$. The state (σ', ℓ') is called a *successor* of the state (ℓ, σ) iff $(\sigma, \sigma') \in \mu_3(\ell, \ell')$.

At any time instant, the state of a hybrid system specifies a control location and values for all data variables. The state can change in two ways: (1) by an instantaneous transition that changes the entire state according to the successor relation, and (2) by elapse of time that changes only the values of data variables in a continuous manner according to the activities of the current location. The exceptions of a hybrid system enforce the progress of the underlying discrete transition system: some transition must be taken before an exception occurs. Typical exceptions are timeouts and sensor readings that trigger a discrete state change.

Formally, a *run* of the hybrid system A is a finite or infinite sequence

$$\rho: \ \mapsto_{\sigma_0} \ (\ell_0, I_0, f_0) \ _{\sigma_0'} \rightarrow_{\sigma_1} \ (\ell_1, I_1, f_1) \ _{\sigma_1'} \rightarrow_{\sigma_2} \ (\ell_2, I_2, f_2) \ _{\sigma_2'} \rightarrow \ \cdots \quad (\dagger)$$

of data states $\sigma_i, \sigma_i' \in \Sigma_D$, locations $\ell_i \in Q$, intervals I_i, and activities f_i such that

1. for all $i \geq 0$, the state $(\ell_{i+1}, \sigma_{i+1})$ is a successor of the state (ℓ_i, σ_i');
2. $I_0 I_1 I_2 \ldots$ is an interval sequence;
3. for all $i \geq 0$, the activity f_i is in $\mu_1(\ell_i)$, and (1) $f_i(0) = \sigma_i$ and $f_i(r_{I_i} - l_{I_i}) = \sigma_i'$, and (2) for all $t \in I_i$, $f_i(t - l_{I_i}) \notin \mu_2(\ell_i)$.

Each run ρ of A uniquely determines a trace τ_ρ: for all $i \geq 0$ and $t \in I_i$, let $\tau(t) = (\ell_i, f_i(t - l_{I_i}))$. Observe that, for all $i > 0$, if I_i is left-closed, then the state at time l_{I_i}, that is, at the time of transition from state σ_{i-1}' to state σ_i, is defined to be σ_i. On the other hand, if I_i is left-open, then the state at time l_{I_i} is defined to be σ_{i-1}'.

By $\mathcal{S}_A$ we denote the set of all traces τ_ρ that correspond to runs ρ of the system A. The set $\mathcal{S}_A$ is fusion-closed, because at any time instant during a run, the configuration of the system is completely determined by the location in which the control resides and the values of all data variables.

Linear hybrid systems A *linear term* α over a set of variables V is a linear combination of the variables in V with rational coefficients. A *linear formula* ϕ over V is a boolean combination of inequalities between linear terms over V.

The hybrid system $A = (V_D, Q, \mu_1, \mu_2, \mu_3)$ is *linear* if its activities, exceptions, and transition relations can be defined by linear expressions over the set V_D of data variables:

1. For all locations $\ell \in Q$, the possible activities are linear functions defined by a set of differential equations of the form $x' = k_x$, one for each data variable $x \in V_D$, where k_x is a rational constant:
   $f \in \mu_1(\ell)$ iff all $t \in \mathsf{R}^+$ and $x \in V_D$, $f(t)(x) = f(0)(x) + k_x \cdot t$.
   We write $\mu_1(\ell, x) = k_x$ to define the activities of the linear hybrid system A.
2. For all locations $\ell \in Q$, the exception is defined by a linear formula ϕ over V_D: $\sigma \in \mu_2(\ell)$ iff $\sigma(\phi)$.
3. For all pairs $e \in Q^2$ of locations, the transition relation $\mu_3(e)$ is defined by a guarded set of assignments

$$\phi \ \rightarrow \ \{x := \alpha_x \mid x \in V_D\},$$

where ϕ is a linear formula over V_D and each α_x is either a linear term over V_D or "?":
   $(\sigma, \sigma') \in \mu_3(e)$ iff $\sigma(\phi)$ and for all $x \in V_D$, either $\alpha_x =?$ or $\sigma'(x) = \sigma(\alpha_x)$.

An assignment of the form $x := ?$ indicates that the value of the variable x is changed nondeterministically to an arbitrary value. We write $\mu_3(e, x)$ for the term α_x.

Various special cases of linear hybrid systems are of particular interest:

- If $\mu_1(\ell, x) = 0$ for each location $\ell \in Q$, then x is a *discrete variable*. Thus a discrete variable changes only when the location of control changes. A *discrete* system is a linear hybrid system all of whose data variables are discrete variables.
- A discrete variable x is a *proposition* if $\mu_3(e, x) \in \{0, 1\}$ for all pairs $e \in Q^2$. If all the data variables are propositions, then a linear hybrid automaton is same as a finite-state system whose states are labeled with propositions.
- If $\mu_1(\ell, x) = 1$ for each location ℓ and $\mu_3(e, x) \in \{0, x\}$ for each pair $e \in Q^2$, then x is a *clock*. Thus the value of a clock variable increases with time uniformly; a transition of the automaton either resets it to 0, or leaves it unchanged. A (finite-state) *timed* system is a linear hybrid system all of whose data variables are propositions and clocks.
- If there is a constant $k \in R$ such that $\mu_1(\ell, x) = k$ for each location ℓ and $\mu_3(e, x) \in \{0, x\}$ for each pair $e \in Q^2$, then x is a *skewed clock*. Thus a skewed clock is similar to a clock variable except that it changes with time at some (fixed) rate different from 1. A *multirate timed* system is a linear hybrid system all of whose data variables are propositions and skewed clocks. An n-*rate* timed system is a multirate timed system whose skewed clocks proceed at n different rates.
- If $\mu_1(\ell, x) \in \{0, 1\}$ for each location ℓ and $\mu_3(e, x) \in \{0, x\}$ for each pair $e \in Q^2$, then x is an *integrator*. Thus an integrator is like a clock that can be stopped and restarted, and can measure accumulated durations. An *integrator* system is a linear hybrid system all of whose data variables are propositions and integrators.
- A discrete variable is a *parameter* if $\mu_3(e, x) = x$ for all pairs $e \in Q^2$. Thus a parameter is a symbolic constant which can be used, for instance, in the guards of the transitions. For different special types of linear hybrid automata defined above, we can define its parameterized version also. For instance, a *parameterized timed* system is a linear hybrid system all of whose data variables are propositions, parameters, and clocks.

Clearly, if A is linear (discrete; timed; multirate timed; integrator) system, then all traces in S_A are piecewise linear (step traces; timed traces; skewed-clock traces; integrator traces, respectively).

Graphical representation Instead of using exceptions, we label locations with their invariants. We suppress location labels of the form $x' = 0$ for activities and *true* for invariants. For transition labels, we suppress the guard *true* and assignments of the form $x := x$. Reflexive transitions with the label *true* are suppressed altogether.

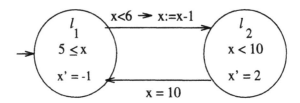

Fig. 1. Graphical representation

As an example, consider the linear hybrid system of Figure 1 with the single data variable x. This system has two locations, ℓ_1 and ℓ_2. In location ℓ_1, the value of x decreases at a constant rate of 1. The transition from ℓ_1 to ℓ_2 may be taken at any time after the value of x has fallen below 6, and it must be taken before the value of x falls below 5. When the transition is taken, the value of x is instantaneously decreased by 1. Once in location ℓ_2, the rate of x starts to increase at the constant rate of 2. The transition back to location ℓ_1 is taken exactly when the value of x hits 10. Indeed, at the very time instant when $x = 10$, the control of the system is already in location ℓ_1, because location ℓ_2 has the invariant $x < 10$.

Initial and acceptance conditions We can turn a hybrid system A into an automaton by adding initial and acceptance conditions. The initiality criterion is given by a labeling function μ_4 that assigns to each location $\ell \in Q$ an *initial condition* $\mu_4(\ell) \subseteq \Sigma_D$. The Muller acceptance criterion is given by a collection $F \subseteq 2^Q$ of *acceptance sets* of locations. The run ρ (†) of the *hybrid Muller automaton* (A, μ_4, F) is *accepting* if

4. $\sigma_0 \in \mu_4(\ell_0)$;
5. either ρ is finite with final location ℓ_n and $\{\ell_n\} \in F$, or $\rho_\infty \in F$ for the set ρ_∞ of locations that are visited infinitely often during ρ (i.e, ρ_∞ is the set $\{\ell \mid \ell = \ell_i$ for infinitely many $i \geq 0\}$).

The hybrid automaton (A, μ_4, F) is linear if A is a linear hybrid system and for all locations $\ell \in Q$, the initial condition is defined by a linear formula ϕ over V_D (i.e., $\sigma \in \mu_4(\ell)$ iff $\sigma(\phi)$).

Parallel composition A hybrid system typically consists of many components operating concurrently and coordinating with each other. Such a system can be constructed from the descriptions of its components using a product operation. Let $A_1 = (V_D^1, Q^1, \mu_1^1, \mu_2^1, \mu_3^1)$ and $A_2 = (V_D^2, Q^2, \mu_1^2, \mu_2^2, \mu_3^2)$ be two hybrid systems. The *product* $A_1 \times A_2$ of A_1 and A_2 is the hybrid system $(V_D^1 \cup V_D^2, Q^1 \times Q^2, \mu_1, \mu_2, \mu_3)$ such that

- An activity f belongs to $\mu_1(\ell_1, \ell_2)$ iff the restriction of f to the data variables V_D^1, denoted by $f|_{V_D^1}$, is in $\mu_1^1(\ell_1)$, and $f|_{V_D^2}$ is in $\mu_1^2(\ell_2)$;

- A data state σ over $V_D^1 \cup V_D^2$ is in $\mu_2(\ell_1, \ell_2)$ iff $\sigma|_{V_D^1}$, the projection of σ onto the variables V_D^1, is in $\mu_2^1(\ell_1)$, or $\sigma|_{V_D^2}$ is in $\mu_2^2(\ell_2)$;
- $(\sigma, \sigma') \in \mu_3((\ell_1, \ell_2), (\ell_1', \ell_2'))$ iff $(\sigma|_{V_D^1}, \sigma'|_{V_D^1}) \in \mu_3(\ell_1, \ell_1')$ and $(\sigma|_{V_D^2}, \sigma'|_{V_D^2}) \in \mu_3(\ell_2, \ell_2')$.

It is not hard to see that traces of the product system are precisely those hybrid traces whose projections are traces of the component systems. It follows that the product of two linear hybrid systems is again linear, etc. An accepting run of a product automaton must meet the initial and acceptance conditions of both component automata.

2.3 Examples of hybrid systems

We model a thermostat, a water level monitor, a clock-based mutual-exclusion protocol, and a leaking gas burner as hybrid systems.

Temperature controller Our first example describes a nonlinear hybrid system. The temperature of a plant is controlled through a thermostat, which continuously senses the temperature and turns a heater on and off [NSY92]. The temperature is governed by differential equations. When the heater is off, the temperature, denoted by the variable x, decreases according to the exponential function $x(t) = \theta e^{-Kt}$, where t is the time, θ is the initial temperature, and K is a constant determined by the plant; when the heater is on, the temperature follows the function $x(t) = \theta e^{-Kt} + h(1 - e^{-Kt})$, where h is a constant that depends on the power of the heater. Suppose that initially the temperature is M degrees and the heater is turned off. We wish to keep the temperature between m and M degrees. The resulting system can be described by the hybrid automaton of Figure 2 (note the representation of the initial condition $x = M$). The automaton has two locations: in location ℓ_0, the heater is turned off; in location ℓ_1, the heater is on.

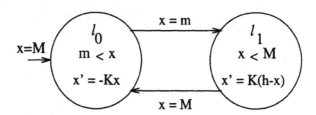

Fig. 2. Temperature controller

Water level monitor Our next example describes a linear system. The water level in a tank is controlled through a monitor, which continuously senses the water level and turns a pump on and off. Unlike the temperature, the water level changes as a piecewise-linear function over time. When the pump is off, the water level, denoted by the variable y, falls by 2 inches per second; when the pump is on, the water level rises by 1 inch per second. Suppose that initially the water level is 1 inch and the pump is turned on. We wish to keep the water between 1 and 12 inches. But from the time that the monitor signals to change the status of the pump to the time that the change becomes effective, there is a delay of 2 seconds. Thus the monitor must signal to turn the pump on before the water level falls to 1 inch, and it must signal to turn the pump off before the water level reaches 12 inches.

The linear hybrid automaton of Figure 3 describes a water level monitor that signals whenever the water level passes 5 and 10 inches, respectively. The automaton has four locations: in locations ℓ_0 and ℓ_1, the pump is turned on; in locations ℓ_2 and ℓ_3, the pump is off. The clock x is used to specify the delays: whenever the automaton control is in location ℓ_1 or ℓ_3, the signal to switch the pump off or on, respectively, was sent x seconds ago. In the next section, we will prove that the monitor indeed keeps the water level between 1 and 12 inches.

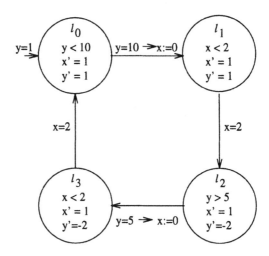

Fig. 3. Water level monitor

Mutual-exclusion protocol This example describes a parameterized multi-rate timed system. We present a timing-based algorithm that implements mutual exclusion for a distributed system with skewed clocks. Consider an asynchronous shared-memory system that consists of two processes P_1 and P_2 with atomic read and write operations. Each process has a critical section and at each time instant, at most one of the two processes is allowed to be in its critical section.

Mutual exclusion is ensured by a version of Fischer's protocol [Lam87], which we describe first in pseudocode. For each process P_i, where $i = 1, 2$:

```
repeat
    repeat
        await k = 0
        k := i
        delay b
    until k = i
    Critical section
    k := 0
forever
```

The two processes P_1 and P_2 share a variable k and process P_i is allowed to be in its critical section iff $k = i$. Each process has a private clock. The instruction **delay** b delays a process for at least b time units as measured by the process's local clock. Furthermore, each process takes at most a time units, as measured by the process's clock, for a single write access to the shared memory (i.e., for the assignment $k := i$). The values of a and b are the only information we have about the timing behavior of instructions. Clearly, the protocol ensures mutual exclusion only for certain values of a and b. If both private processor clocks proceed at precisely the same rate, then mutual exclusion is guaranteed iff $a < b$.

To make the example more interesting, we assume that the two private clocks of the processes P_1 and P_2 proceed at different rates, namely, the local clock of P_2 is 1.1 times faster than the clock of P_1. The resulting system can be modeled by the product of the two hybrid automata presented in Figure 4.

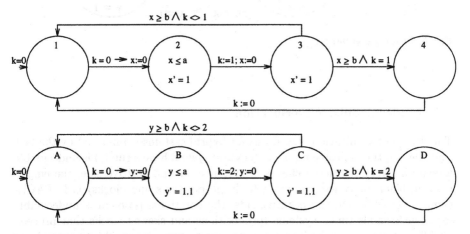

Fig. 4. Mutual-exclusion protocol

Each of the two automata models one process, with the two critical sections being represented by the locations 4 and D. The private clocks of the processes P_1 and P_2 determine the rate of change of the two skewed-clock variables x and y, respectively. In the next section, we will prove that mutual exclusion is guaranteed if $a = 2$ and $b = 3$: in this case, it will never happen that the control of P_1 is in location 4 while the control of P_2 is in location D.

Leaking gas burner Now we consider an integrator system. In [CHR91], the duration calculus is used to prove that a gas burner does not leak excessively. It is known that (1) any leakage can be detected and stopped within 1 second and (2) the gas burner will not leak for 30 seconds after a leakage has been stopped. We wish to prove that the accumulated time of leakage is at most one twentieth of the time in any interval of at least 60 seconds. The system is modeled by the hybrid automaton in Figure 5. The automaton has two locations: in location ℓ_1, the gas burner leaks; ℓ_2 is the nonleaking location. The integrator t records the cumulative leakage time; that is, the accumulated amount of time that the system has spent in location ℓ_1. The clock x records the time the system has spent in the current location; it is used to specify the properties (1) and (2). The clock y records the total elapsed time. In the next section, we will prove that $y \geq 60 \rightarrow 20t \leq y$ is an invariant of the system.

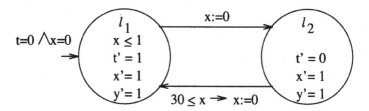

Fig. 5. Leaking gas burner

2.4 Undecidability of verification

The design of verification algorithms for hybrid systems is impaired by the fact that the emptiness problem ("Does a hybrid system have a run?") is undecidable already for very restricted classes of systems. On the positive side, the emptiness problem for timed automata (only propositions and clocks) is PSPACE-complete [AD90]. On the negative side, the emptiness problem is undecidable for asynchronous timed systems (propositions and skewed clocks that proceed at different rates) and for integrator systems (propositions and integrators).

To obtain strong undecidability results, we restrict the classes of multirate timed systems and integrator systems further. A linear hybrid system is *simple* if all linear atoms in exceptions and transition guards are of the form $x \leq k$, and all assignments are of the form $x := x$ or $x := k$, for $x \in V_D$ and $k \in Z$.

In particular, for n-rate timed systems the simplicity condition prohibits the comparison of clocks with different rates.

Theorem 1. *The emptiness problem is undecidable for 2-rate timed systems and for simple integrator systems.*

Proof. The first part of the theorem follows from the undecidability of the halting problem for nondeterministic 2-counter machines (NCMs). Given any two distinct clock rates, a 2-rate timed system can encode the computations of an NCM. Suppose we have three "accurate" clocks of rate 1 and two skewed clocks x_1 and x_2 of rate 2. Then we can encode the values of two counters in the i-th machine configuration by the values of x_1 and x_2 at accurate time i: the counter value n is encoded by the clock value $1/2^n$.

The accurate clock y is reset whenever it reaches 1 and thus marks intervals of length 1. It is obvious how a counter can be initialized to 0 and tested for being 0. Hence it remains to be shown how a counter can be incremented and decremented. To increment the counter represented by the skewed clock x from time i to time $i+1$, start an accurate clock z with x in the interval $[i-1, i)$ and reset z when it reaches 1; then nondeterministically reset x in the interval $[i, i+1)$ and test $x = z$ at time $i + 1$. To decrement the counter represented by the skewed clock x from time i to time $i + 1$, nondeterministically start an accurate clock z in the interval $[i - 1, i)$ and test $x = z$ at time i; when z reaches 1 in the interval $[i, i + 1)$, reset x. Given an NCM M, we can so construct a 2-rate timed system that has a run iff M halts. (Indeed, using acceptance conditions, we can construct a 2-rate timed automaton that has a run iff a counter is 0 infinitely often along some run of M; this shows that the emptiness problem is Σ_1^1-complete for 2-rate timed automata [HPS83].)

The second part of the theorem follows from an undecidability result for timed systems with memory cells [Čerāns]. ∎

We point out that the emptiness problem is decidable for simple n-rate timed automata. This is because any simple asynchronous timed automaton can be transformed into a timed automaton by (1) factoring into 1-process timed automata, (2) scaling all 1-process timed automata to the same clock rate, and (3) constructing the product. An analogous result holds for real-time temporal logics [WME92].

3 Verification Procedures

Consider a linear hybrid system $A = (V_D, Q, \mu_1, \mu_2, \mu_3)$. Given a linear formula ϕ over V_D, we wish to determine whether ϕ is an *invariant* of A:

"Is ϕ true in all states that occur on some trace of A?"

Recall that $\Sigma = Q \times \Sigma_D$ is the state space of A. We define the following reachability relations between states:

– *Time step.* For all locations $\ell \in Q$ and data variables $x \in V_D$, let $\mu_1(\ell, x) = k_x$ and let $\mu_2(\ell) = \phi$. For all data states $\sigma \in \Sigma_D$ and nonnegative reals $\delta \in \mathbb{R}^+$, define $(\sigma + \delta) \in \Sigma_D$ to be the data state that assigns to each data variable $x \in V_D$ the value $\sigma(x) + \delta \cdot k_x$. Then $\Rightarrow_l \subseteq \Sigma^2$ is the smallest relation such that

$\quad$ if $\neg(\sigma + \delta')(\phi)$ for all $0 \leq \delta' < \delta$, then $(\ell, \sigma) \Rightarrow_l (\ell, \sigma + \delta)$;

$\Rightarrow_r \subseteq \Sigma^2$ is the smallest relation such that

$\quad$ if $\neg(\sigma + \delta')(\phi)$ for all $0 < \delta' \leq \delta$, then $(\ell, \sigma) \Rightarrow_r (\ell, \sigma + \delta)$;

and $\Rightarrow_{lr} \subseteq \Sigma^2$ is the smallest relation such that

$\quad$ if $\neg(\sigma + \delta')(\phi)$ for all $0 \leq \delta' \leq \delta$, then $(\ell, \sigma) \Rightarrow_{lr} (\ell, \sigma + \delta)$.

In other words, $\Rightarrow_{lr} = \Rightarrow_l \cap \Rightarrow_r$.

– *Transition step.* The relation $\triangleright \subseteq \Sigma^2$ is the smallest relation such that for all pairs $(\ell, \ell') \in Q^2$ and all data states $\sigma_1, \sigma_2 \in \Sigma_D$,

$\quad$ if $(\sigma_1, \sigma_2) \in \mu_3(\ell, \ell')$, then $(\ell, \sigma_1) \triangleright (\ell', \sigma_2)$.

Note that the relation $\triangleright$ is reflexive.

– *Single step.* For all states $\sigma_1, \sigma_2 \in \Sigma$,

$\quad \sigma_1 \Rightarrow \sigma_2$ iff there exist $\sigma_1', \sigma_2' \in \Sigma$ with either $\sigma_1 \Rightarrow_{lr} \sigma_1' \triangleright \sigma_2' \Rightarrow_r \sigma_2$
$\quad$ or $\sigma_1 \Rightarrow_l \sigma_1' \triangleright \sigma_2' \Rightarrow_{lr} \sigma_2$.

– The *reachability relation* $\Rightarrow^* \subseteq \Sigma^2$ is the transitive closure of the single-step relation $\Rightarrow$.

The single-step reachability relation can be extended to sets of states. For a state $\sigma \in \Sigma$ and a set $R \subseteq \Sigma$ of states, let $\sigma \Rightarrow R$ iff $\sigma \Rightarrow \sigma'$ for some $\sigma' \in R$; for two sets $R_1, R_2 \subseteq \Sigma$ of states, define $R_1 \Rightarrow R_2$ iff $\sigma \Rightarrow R_2$ for some $\sigma \in R_1$. Again, $\Rightarrow^*$ denotes the transitive closure of $\Rightarrow$.

Given two sets $R_i, R_f \subseteq \Sigma$ of states, we wish to find out if the reachability relation $R_i \Rightarrow^* R_f$ holds. A solution to this reachability problem allows the verification of safety properties of the hybrid system A. Suppose the initial condition is given by the labeling function μ_4, then take R_i to be the set defined by $(\ell, \sigma) \in R_i$ iff $\sigma \in \mu_4(\ell)$. To check whether a linear formula ϕ is an invariant of A, we consider the set R_f of "bad" states: $\sigma \in R_f$ iff $\neg\phi(\sigma)$. Now ϕ is an invariant iff the reachability relation $R_i \Rightarrow^* R_f$ does not hold.

From the undecidability of the emptiness problem, it follows that the reachability problem is undecidable for linear hybrid automata. As the state space Σ of A is generally infinite, we will attempt to work on a quotient of the state-transition graph $(\Sigma, \Rightarrow)$. Our method will succeed only if there is a finite quotient of the state space in which states are identified whenever they are "equivalent" with respect to the given reachability problem (R_i, R_f). This problem can be attacked in many different ways:

– We can choose from two definitions of state equivalence. We can move "forward" from the initial set R_i and identify two states whenever they can be reached from R_i by the same sequence of single steps. Alternatively, we can move "backward" from the final set R_f and identify two states whenever they can reach R_f by the same sequence of single steps.

– Working forward from R_i (backward from R_f), we can choose to add one
  equivalence class of states at a time until either the current set intersects
  with R_f (or R_i, respectively) or no new states can be added. We refer to
  this category of verification methods as *fixpoint* methods, because the computation can be viewed as the iterative approximation of a fixpoint that
  defines the class of reachable states.

  Alternatively, we can start with an initial partition of the state space and
  refine it until it respects the equivalence relation, and thus can be used for
  checking reachability. The verification methods in this category are called
  minimization methods, because the computation can be viewed as constructing a bisimulation relation, namely, the minimal (coarsest) state partition
  that respects single-step reachability.

In this paper, we present one instance of the fixpoint computation approach and
one instance of the minimization approach. Both procedures rely on the same
set of primitive operations.

3.1 Fixpoint computation

We define a backward fixpoint computation procedure that solves the reachability problem (R_i, R_f) provided it terminates. The procedure starts with the
set $R_{cur} = R_f$ and repeatedly adds states from which any state in R_{cur} can be
reached. The procedure terminates with the answer YES (indicating that R_f is
reachable from R_i) if at some stage an initial state in R_i is added, and it terminates with the answer NO if no new states can be added. The procedure may,
of course, not terminate at all; it is a semidecision procedure for the reachability
problem of linear hybrid systems.

Backward fixpoint computation:

$R_{old} := \emptyset; \ R_{cur} := R_f$
while $R_{cur} \cap R_i = \emptyset$ and $R_{cur} \not\subseteq R_{old}$ **do**
    $R := pre(R_{cur})$
    $R_{old} := R_{old} \cup R_{cur}; \ R_{cur} := R$
od
return $R_{cur} \cap R_i \neq \emptyset$.

The crucial step is the computation of the state set

$$pre(R) = \{\sigma \in \Sigma \mid \sigma \Rightarrow R\}.$$

This computation is possible due to the fact that all state sets encountered by
the procedure are definable by linear formulas, and hence, for two state sets R
and R', the problems of deciding whether $R \subseteq R'$ and whether $R \cap R' = \emptyset$
have algorithmic solutions. A *data region* $R_D \subseteq \Sigma$ is a set of data states of the
form $\{\sigma \in \Sigma_D \mid \sigma(\phi)\}$, for some linear formula ϕ over V_D. A *region* $(\ell, R_D) \subseteq \Sigma$

consists of a location $\ell \in Q$ and a data region R_D. The union $R \subseteq \Sigma$ of regions (ℓ, R_D^ℓ), one for each location $\ell \in Q$, is called a *region family*. The following central lemma ensures that for linear hybrid systems, all required state sets are computable:

Lemma 2. *If $R \subseteq \Sigma$ is a region family, then so is $pre(R)$.*

Proof. It suffices to show that if R_D is a data region, then so is the set

$$pre_e(R_D) \;=\; \{\sigma \in \Sigma_D \mid (\ell_1, \sigma) \Rightarrow (\ell_2, R_D)\},$$

for each pair $e = (\ell_1, \ell_2)$ of locations. Let ψ be the linear formula that defines R_D. We construct a linear formula $pre_e(\psi)$ that defines the set $pre_e(R_D)$. If ψ contains n variables, we can think of ψ as defining a set of points in n-dimensional space. This set is an n-dimensional polyhedron whose bounding hyperplanes are linear functions with rational coefficients.

First let us extend the time-step and transition-step relations to linear formulas. For any linear formula ψ and location ℓ of A:

$$\triangleright(e, \psi) \;=\; \{\sigma \in \Sigma_D \mid \exists \sigma' \in \Sigma_D.\,(\psi(\sigma') \wedge (\ell_1, \sigma) \triangleright (\ell_2, \sigma'))\},$$

$$\rightarrow (\ell, \psi) \;=\; \{\sigma \in \Sigma_D \mid \exists \sigma' \in \Sigma_D.\,(\psi(\sigma') \wedge (\ell, \sigma) \rightarrow (\ell, \sigma'))\}$$

for $\rightarrow \in \{\Rightarrow_l, \Rightarrow_r, \Rightarrow_{lr}\}$. Then the linear formula $pre_e(\psi)$ is the following disjunction:

$$pre_e(\psi) \;=\; \Rightarrow_{lr} (\ell_1, \triangleright(e, \Rightarrow_r (\ell_2, \psi))) \vee \Rightarrow_l (\ell_1, \triangleright(e, \Rightarrow_{lr} (\ell_2, \psi))).$$

The transition-step relation $\triangleright$ can be computed by substitution. Let ϕ be the guard of $\mu_3(e)$ and for all $x \in V_D$, let $\mu_3(e, x) = \alpha_x$. Then:

$$\triangleright(e, \psi) \;=\; \phi \wedge (\psi[x := \alpha_x]),$$

where the linear formula $\psi[x := \alpha_x]$ is obtained by replacing all occurrences of x in ψ with α_x.

The time-step relations can be computed by quantifier elimination. For all locations ℓ of A, let $\mu_2(\ell) = \phi_\ell$. If $\mu_1(\ell, x) = k_x$ for all $x \in V_D$, then the linear formulas $\psi +_\ell \delta$ and $\psi -_\ell \delta$ result from ψ by replacing all occurrences of x with $x + k_x \cdot \delta$ or $x - k_x \cdot \delta$, respectively. Then:

$$\Rightarrow_r (\ell, \psi) \;=\; (\exists \delta \geq 0.\,(\psi \wedge \neg\phi_\ell) +_\ell \delta \wedge \forall 0 < \epsilon < \delta.\,\neg\phi_\ell +_\ell \epsilon),$$

$$\Rightarrow_l (\ell, \psi) \;=\; (\psi \wedge \neg\phi_\ell) \vee (\exists \delta > 0.\,\psi +_\ell \delta \wedge \forall 0 \leq \epsilon < \delta.\,\neg\phi_\ell +_\ell \epsilon),$$

and $\Rightarrow_{lr} (\ell, \psi) = \neg\phi_\ell \wedge \Rightarrow_r (\ell, \psi)$. It remains to be shown how the quantifiers can be eliminated from these formulas. We first convert all quantifiers into existential form and translate all quantifier-free subformulas into disjunctive normal form. Since existential quantifiers distribute over disjunction, it suffices construct a linear formula over V that is equivalent to the formula $\exists \delta \in \mathsf{R}.\,\varphi$, where φ is a conjunction of linear inequalities over $V \cup \{\delta\}$. Note that the formula $\exists \delta.\,\varphi$ defines a *convex* rational polyhedron. To eliminate the existential quantifier, (1) solve all inequalities for δ and (2) construct the conjunction of all δ-free inequalities that are implied by transitivity. ∎

As for timed systems [HNSY92], the fixpoint method can be extended to check properties of linear hybrid systems that are specified in real-time extensions of branching-time logics such as CTL.

3.2 Minimization

Let π be a partition of the state space Σ into regions. A region $R \in \pi$ is *stable* if

$$\forall R' \in \pi. (R \Rightarrow R' \text{ implies } \forall \sigma \in R. \sigma \Rightarrow R').$$

The partition π is a *bisimulation* if every region of π is stable. The partition π *respects* R_f if for every region $R \in \pi$, either $R \subseteq R_f$ or $R \cap R_f = \emptyset$. Observe that if a partition π that respects R_f is a bisimulation, then it can be used for reachability analysis: to see if R_f is reachable from R_i, check if there exists a path from some π-region R_1 such that $R_1 \cap R_i \neq \emptyset$ to some π-region R_2 such that $R_2 \subseteq R_f$. Our objective is to construct the coarsest bisimulation provided it is finite. For this purpose, we can adopt algorithms for performing a simultaneous reachability and minimization analysis of transition systems [BFH90, LY92].

The minimization procedure of [BFH90] is given below. Let $\pi_0 = \{(\ell, true) \mid \ell \in Q\}$ be the initial partition of Σ into regions — one region per location. The initial partition is refined into $\pi_1 = \pi_0 \cap \{R_f, \Sigma - R_f\}$ so that it respects R_f.

Minimization:

$\pi := \pi_1; \ \alpha := \{R \mid R \cap R_i \neq \emptyset\}; \ \beta := \emptyset$
while $\alpha \neq \beta$ **do**
    choose $R \in (\alpha - \beta)$
    let $\alpha' := split(R, \pi)$
    if $\alpha' = \{R\}$ **then**
        $\beta := \beta \cup \{R\}$
        $\alpha := \alpha \cup \{R' \in \pi \mid R \Rightarrow R'\}$
    else
        $\alpha := \alpha - \{R\}$
        if $\exists R' \in \alpha'$ such that $R' \cap R_i \neq \emptyset$ **then**
            $\alpha := \alpha \cup \{R'\}$
        fi
        $\beta := \beta - \{R' \in \pi \mid R' \Rightarrow R\}$
        $\pi := (\pi - \{R\}) \cup \alpha'$
    fi
od
return there is $R \in \alpha$ such that $R \subseteq R_f$.

Starting from π_1, the procedure selects a region R and checks if R is stable with respect to the current partition; if not, then R is split into smaller regions. Additional book-keeping is needed to record which regions are reachable from the region containing the initial state. In the following procedure, π is the current partition, α is the set of π-regions that have been found reachable from (the region of) the initial state, and β is the set of π-regions that have been found stable with respect to π. The function $split(R, \pi)$ splits the π-region R into subregions that are "more" stable with respect to π: $split(R, \pi) := \{R', R - R'\}$ if there is some region $R'' \in \pi$ such that the region $R' = pre(R'') \cap R$ is a proper subset of R, and $split(R, \pi) := \{R\}$ otherwise. Since the operation $split$ is computed using pre, all state sets encountered by the minimization procedure are again definable by linear formulas. The procedure terminates if the coarsest bisimulation has only a finite number of equivalence classes.

If the minimization procedure terminates, we obtain a finite bisimulation of Σ with respect to $\Rightarrow$. As with timed automata [ACH+92], the resulting reachability graph can be used to solve also the emptiness problem for A, even in presence of acceptance conditions, and for model checking branching-time properties. The minimization procedure may be replaced by the more efficient procedure presented in [LY92], which can also be implemented using the primitive pre.

3.3 Verification examples

In the following, we demonstrate that both the fixpoint computation procedure and the minimization procedure terminate in many cases.

Minimization: water level monitor Let A be the hybrid automaton defined in Figure 3. We use the minimization procedure to prove that the formula $1 \leq y \leq 12$ is an invariant of A. It follows that the water level monitor keeps the water level between 1 and 12 inches.

By $\langle \psi \rangle$, for a linear formula ψ over V, we denote the set of all states (ℓ, σ) such that σ satisfies $\psi[pc := \ell]$. Let the set R_i of initial states be so defined by

$$R_i = \langle pc = 0 \wedge x = 0 \wedge y = 1 \rangle$$

and let the set R_f of "bad" states be defined by

$$R_f = \langle y < 1 \vee y > 12 \rangle.$$

The initial partition is $\pi_0 = \{(\ell, true) \mid \ell \in \{\ell_0, \ell_1, \ell_2, \ell_3\}\}$. We next partition each region of the initial partition into "good" and "bad" states: $\pi_1 = \{$

$$
\begin{aligned}
C_{00} &= \langle pc = 0 \wedge 1 \leq y \leq 12 \rangle, & C_{01} &= \langle pc = 0 \wedge (y < 1 \vee y > 12) \rangle, \\
C_{10} &= \langle pc = 1 \wedge 1 \leq y \leq 12 \rangle, & C_{11} &= \langle pc = 1 \wedge (y < 1 \vee y > 12) \rangle, \\
C_{20} &= \langle pc = 2 \wedge 1 \leq y \leq 12 \rangle, & C_{21} &= \langle pc = 2 \wedge (y < 1 \vee y > 12) \rangle, \\
C_{30} &= \langle pc = 3 \wedge 1 \leq y \leq 12 \rangle, & C_{31} &= \langle pc = 3 \wedge (y < 1 \vee y > 12) \rangle\}.
\end{aligned}
$$

The bad states are in the regions C_{i1}, for $i \in \{0, 1, 2, 3\}$. Since the initial region R_i is contained in C_{00}, let $\alpha = \{C_{00}\}$. Considering $R = C_{00} \in \alpha$, we find that $split(C_{00}, \pi_1) = \{$

$$
\begin{aligned}
C_{000} &= \langle pc = 0 \wedge 1 \leq y < 10 \rangle, \\
C_{001} &= \langle pc = 0 \wedge 10 \leq y \leq 12 \rangle\}.
\end{aligned}
$$

Therefore, $\pi_2 = \{C_{000}, C_{001}, C_{01}, C_{10}, C_{11}, C_{20}, C_{21}, C_{30}, C_{31}\}$. Now $R_i \subseteq C_{000}$, so take $\alpha = \{C_{000}\}$ and $\beta = \emptyset$. Considering $R = C_{000}$, we find that it is stable with respect to π_2. Thus $\alpha = \alpha \cup \{R' \in \pi \mid R \Rightarrow R'\} = \{C_{000}, C_{001}, C_{10}\}$ and $\beta = \{C_{000}\}$. Since $R = C_{001}$ is also stable in π_2 and is not reaching any new states not in α, α remains the same and $\beta = \{C_{000}, C_{001}\}$. However, considering $R = C_{10}$, we obtain $split(C_{10}, \pi_2) = \{$

$$C_{100} = \langle pc = 1 \wedge 0 \leq x < 2 \wedge 1 \leq y \leq 12 \rangle,$$
$$C_{101} = \langle pc = 1 \wedge x \geq 2 \wedge 1 \leq y \leq 12 \rangle\}.$$

These two regions together with the regions in π_2, except for C_{10}, constitute π_3. The new β is obtained by removing $\{R' \in \pi \mid R' \Rightarrow R\} = C_{000}$ from the old β. The new α becomes $\{C_{000}, C_{001}\}$. Now $R = C_{000}$ is stable in π_3. Hence $\alpha = \{C_{000}, C_{001}, C_{100}\}$ and $\beta = \{C_{000}, C_{001}\}$. Since $R = C_{100}$ is stable in π_3, we have $\alpha = \{C_{000}, C_{001}, C_{100}, C_{101}, C_{20}\}$ and $\beta = \{C_{000}, C_{001}, C_{100}\}$. $R = C_{101}$ is also stable in π_3, so $\beta = \{C_{000}, C_{001}, C_{100}, C_{101}\}$ and α remains unchanged. Considering $R = C_{20}$, we obtain $split(C_{20}, \pi_3) = \{$

$$C_{200} = \langle pc = 2 \wedge 5 < y \leq 12 \rangle,$$
$$C_{201} = \langle pc = 2 \wedge 1 \leq y \leq 5 \rangle\}.$$

Now π_4 contains C_{200} and C_{201}, and thus C_{100} must be reconsidered. It is split into $split(C_{100}, \pi_4) = \{$

$$C_{1000} = \langle pc = 1 \wedge 0 \leq x < 2 \wedge 3 < y \leq 12 \wedge 3 < y - x \leq 12 \rangle,$$
$$C_{1001} = \langle pc = 1 \wedge 0 \leq x < 2 \wedge 1 \leq y \leq 5 \wedge 1 \leq y - x \leq 3 \rangle\}.$$

Thus π_5 contains C_{1000} and C_{1001}. After finding that C_{000}, C_{1000} and C_{200} all are stable, we finally have $\alpha = \{C_{000}, C_{001}, C_{1000}, C_{200}, C_{201}, C_{30}\}$ and $\beta = \{C_{000}, C_{001}, C_{1000}, C_{200}\}$. So let $R = C_{201}$. It is stable, so $\beta = \beta \cup \{C_{200}\}$ and α does not change. Then $R = C_{30}$ is partitioned into $\{$

$$C_{300} = \langle pc = 3 \wedge 0 \leq x < 2 \wedge 1 \leq y \leq 12 \rangle,$$
$$C_{301} = \langle pc = 3 \wedge x \geq 2 \wedge 1 \leq y \leq 12 \rangle\}.$$

C_{200} has to be considered again. It is stable with respect to the current partition. Then $R = C_{300}$ is considered and $split(C_{300}, \pi_6) = \{$

$$C_{3000} = \langle pc = 3 \wedge 0 \leq x < 2 \wedge 1 \leq y \leq 12 \wedge 5 \leq y + 2x < 14 \rangle,$$
$$C_{3001} = \langle pc = 3 \wedge 0 \leq x < 2 \wedge 1 \leq y < 5 \wedge 1 \leq y + 2x < 5 \rangle\}.$$

We must consider C_{200} again. It turns out that it is still stable. After considering $R = C_{3000}$, we have $\beta = \{C_{000}, C_{001}, C_{1000}, C_{200}, C_{201}, C_{3000}\}$ and $\alpha = \alpha \cup \{C_{000}\}$. Now the partition is

$$\pi_7 = \{C_{000}, C_{001}, C_{01}, C_{1000}, C_{1001}, C_{101}, C_{11},$$
$$C_{200}, C_{201}, C_{21}, C_{3000}, C_{3001}, C_{301}, C_{31}\}.$$

Since C_{000} is stable in π_7, we have $\alpha = \beta = \{C_{000}, C_{001}, C_{1000}, C_{200}, C_{201}, C_{3000}\}$. Notice that no region in α contains any bad states from R_f. Therefore, the invariant property has been verified.

Fixpoint computation: mutual-exclusion protocol Let A be the product of the two hybrid automata defined in Figure 4, for $a = 2$ and $b = 3$. We use the fixpoint computation procedure to prove that the formula $pc \neq (4, D)$ is an invariant of A. It follows that the protocol ensures mutual exclusion.

Let $R_i = \langle pc = (1, A) \rangle$ be the region of initial states and let $R_f = \langle pc = (4, D) \rangle$ be the region of "bad" states. Let R^i denote the value of $R = pre(R_{cur})$ after the i-th iteration of the algorithm. Then $R^0 = R_f$ and

$$
\begin{aligned}
R^1 &= pre(R_{cur}) \\
&= \langle (pc_1 = 3 \wedge pc_2 = D \wedge 1 = k) \vee (pc_1 = 4 \wedge pc_2 = C \wedge 2 = k) \rangle,
\end{aligned}
$$

where pc_i, for $i = 1, 2$, denotes the i-th component of the control variable pc. We keep computing $R^{i+1} = pre(R^i)$ as long as R^i and R_i are disjoint:

$$
\begin{aligned}
R^2 = \langle &(pc_1 = 2 \wedge pc_2 = D \wedge x \leq 2) \vee (pc_1 = 4 \wedge pc_2 = B \wedge y \leq 2) \\
&\vee (pc_1 = 3 \wedge pc_2 = B \wedge y \leq 2 \wedge 13 \leq 11x - 10y \wedge 1 = k) \\
&\vee (pc_1 = 2 \wedge pc_2 = C \wedge x \leq 2 \wedge 8 \leq -11x + 10y \wedge 2 = k) \\
&\vee (pc_1 = 2 \wedge pc_2 = C \wedge x \leq 2 \wedge 30 \leq 10y \wedge 2 = k) \\
&\vee (pc_1 = 3 \wedge pc_2 = B \wedge y \leq 2 \wedge 3 \leq x \wedge 1 = k) \rangle
\end{aligned}
$$

$$
\begin{aligned}
R^3 = \langle &(pc_1 = 1 \wedge pc_2 = D \wedge 0 = k) \vee (pc_1 = 4 \wedge pc_2 = A \wedge 0 = k) \\
&\vee (pc_1 = 2 \wedge pc_2 = C \wedge 2 = k \wedge x \leq 2 \wedge 8 \leq -11x + 10y) \\
&\vee (pc_1 = 3 \wedge pc_2 = B \wedge 1 = k \wedge y \leq 2 \wedge -20 \leq -10y \wedge 13 \leq 11x - 10y) \\
&\vee (pc_1 = 2 \wedge pc_2 = B \wedge x \leq 2 \wedge y \leq 2 \wedge 13 \leq -10y) \rangle
\end{aligned}
$$

$$
\begin{aligned}
R^4 = \langle &(pc_1 = 4 \wedge pc_2 = D) \vee (pc_1 = 3 \wedge pc_2 = D \wedge 0 = k) \\
&\vee (pc_1 = 4 \wedge pc_2 = C \wedge 0 = k) \vee (pc_1 = 3 \wedge pc_2 = D \wedge 1 = k) \\
&\vee (pc_1 = 4 \wedge pc_2 = C \wedge 2 = k) \\
&\vee (pc_1 = 2 \wedge pc_2 = A \wedge 8 \leq -11x \wedge 0 = k) \\
&\vee (pc_1 = 2 \wedge pc_2 = B \wedge y \leq 2 \wedge 8 \leq -11x) \\
&\vee (pc_1 = 2 \wedge pc_2 = B \wedge x \leq 2 \wedge y \leq 2 \wedge 13 \leq -10y) \\
&\vee (pc_1 = 1 \wedge pc_2 = B \wedge 0 = k \wedge y \leq 2 \wedge 13 \leq -10y) \rangle
\end{aligned}
$$

$$
\begin{aligned}
R^5 = \langle &(pc_1 = 3 \wedge pc_2 = D \wedge 1 = k) \vee (pc_1 = 4 \wedge pc_2 = C \wedge 2 = k) \\
&\vee (pc_1 = 2 \wedge pc_2 = D \wedge x \leq 2) \vee (pc_1 = 4 \wedge pc_2 = B \wedge y \leq 2) \\
&\vee (pc_1 = 2 \wedge pc_2 = A \wedge 8 \leq -11x \wedge 0 = k) \\
&\vee (pc_1 = 1 \wedge pc_2 = B \wedge y \leq 2 \wedge 13 \leq -10y \wedge 0 = k) \\
&\vee (pc_1 = 2 \wedge pc_2 = C \wedge 8 \leq -11x \wedge 38 \leq -11x + 10y \wedge 0 = k) \\
&\vee (pc_1 = 3 \wedge pc_2 = B \wedge y \leq 2 \wedge 13 \leq 11x - 10y \wedge 1 = k) \\
&\vee (pc_1 = 3 \wedge pc_2 = B \wedge y \leq 2 \wedge 3 \leq x \wedge 1 = k) \\
&\vee (pc_1 = 2 \wedge pc_2 = C \wedge x \leq 2 \wedge 8 \leq -11x + 10y \wedge 2 = k) \\
&\vee (pc_1 = 2 \wedge pc_2 = C \wedge x \leq 2 \wedge 3 \leq y \wedge 2 = k) \\
&\vee (pc_1 = 3 \wedge pc_2 = B \wedge y \leq 2 \wedge 13 \leq -10y \wedge 46 \leq 11x - 10y \wedge 0 = k) \rangle
\end{aligned}
$$

$$R^6 = \langle (pc_1 = 1 \wedge pc_2 = D \wedge 0 = k) \vee (\wedge pc_1 = 4 \wedge pc_2 = A \wedge 0 = k)$$
$$\vee (pc_1 = 2 \wedge pc_2 = D \wedge x \le 2) \vee (pc_1 = 4 \wedge pc_2 = B \wedge y \le 2)$$
$$\vee (pc_1 = 2 \wedge pc_2 = C \wedge 8 \le -11x \wedge 38 \le -11x + 10y \wedge 0 = k)$$
$$\vee (pc_1 = 3 \wedge pc_2 = B \wedge y \le 2 \wedge 13 \le 11x - 10y \wedge 1 = k)$$
$$\vee (pc_1 = 3 \wedge pc_2 = B \wedge y \le 2 \wedge 3 \le x \wedge 1 = k)$$
$$\vee (pc_1 = 2 \wedge pc_2 = C \wedge x \le 2 \wedge 8 \le -11x + 10y \wedge 2 = k)$$
$$\vee (pc_1 = 2 \wedge pc_2 = C \wedge x \le 2 \wedge 3 \le y \wedge 2 = k)$$
$$\vee (pc_1 = 2 \wedge pc_2 = B \wedge x \le 2 \wedge y \le 2 \wedge 13 \le -10y)$$
$$\vee (pc_1 = 3 \wedge pc_2 = B \wedge y \le 2 \wedge 13 \le -10y \wedge 46 \le 11x - 10y \wedge 0 = k)\rangle$$

Since $R^6 \subseteq R^3 \cup R^5$, a fixpoint is found in 6 iterations. Notice that the fixpoint $\bigcup_{0 \le i \le 5} R^i$ contains no initial states from R_i. Therefore, the invariant property has been verified.

Fixpoint computation: leaking gas burner Let A be the integrator system defined in Figure 5. We use the fixpoint computation procedure to prove that the formula $y \ge 60 \rightarrow 20t \le y$ is an invariant of A. It follows that the gas burner leaks at most one twentieth of the time in any interval of at least 60 seconds. Let $R_i = \langle pc = 1 \wedge t = 0 \wedge x = 0 \rangle$ be the region of initial states and let $R_f = \langle y \ge 60 \wedge 20t > y \rangle$ be the region of "bad" states. Let R^i again denote the value of $R = pre(R_{cur})$ after the i-th iteration of the algorithm. Then:

$$R^0 = R_f$$

$$R^1 = \langle (pc = 2 \wedge -19 < 20t - y \wedge 11 < 20t + x - y \wedge 2 < t \wedge 0 \le t \wedge 0 \le x)$$
$$\vee (pc = 1 \wedge -19 < 20t - 19x - y \wedge 2 < t - x \wedge -1 \le -x \wedge 0 \le x)\rangle$$

$$R^2 = \langle (pc = 1 \wedge -8 < 20t - 19x - y \wedge 1 < t - x \wedge -1 \le -x \wedge 0 \le x)$$
$$\vee (pc = 2 \wedge -19 < 20t - y \wedge 2 < t \wedge 11 < 20t + x - y \wedge 0 \le x)\rangle$$

$$R^3 = \langle (pc = 2 \wedge -8 < 20t - y \wedge 1 < t \wedge 22 < 20t + x - y \wedge 0 \le x)$$
$$\vee (pc = 1 \wedge -8 < 20t - 19x - y \wedge 1 < t - x \wedge -1 \le -x \wedge 0 \le x)\rangle$$

$$R^4 = \langle (pc = 1 \wedge 0 < t - x \wedge 3 < 20t - 19x - y \wedge -1 \le -x \wedge 0 \le x)$$
$$\vee (pc = 2 \wedge -8 < 20t - y \wedge 1 < t \wedge 22 < 20t + x - y \wedge 0 \le x)\rangle$$

$$R^5 = \langle (pc = 2 \wedge 0 < t \wedge 3 < 20t - y \wedge 33 < 20t + x - y \wedge 0 \le x)$$
$$\vee (pc = 1 \wedge 0 < t - x \wedge 3 < 20t - 19x - y \wedge -1 \le -x \wedge 0 \le x)\rangle$$

$$R^6 = \langle (pc = 1 \wedge -1 < t - x \wedge 14 < 20t - 19x - y \wedge -1 \le -x \wedge 0 \le x)$$
$$\vee (pc = 2 \wedge 0 < t \wedge 3 < 20t - y \wedge 33 < 20t + x - y \wedge 0 \le x)\rangle$$

$$R^7 = \langle (pc = 2 \wedge 14 < 20t - y \wedge 44 < 20t + x - y \wedge 0 \le t \wedge 0 \le x)$$
$$\vee (pc = 1 \wedge -1 < t - x \wedge 14 < 20t - 19x - y \wedge -1 \le -x \wedge 0 \le x)\rangle$$

$$R^8 = \langle (pc = 1 \wedge 25 < 20t - 19x - y \wedge -1 \le t - x \wedge -1 \le -x \wedge 0 \le x)$$
$$\vee (pc = 2 \wedge 14 < 20t - y \wedge 44 < 20t + x - y \wedge 0 \le t \wedge 0 \le x)\rangle$$

$$R^9 = \langle (pc = 2 \wedge 25 < 20t - y \wedge 55 < 20t + x - y \wedge 0 \le t \wedge 0 \le x)$$
$$\vee (pc = 1 \wedge 25 < 20t - 19x - y \wedge -1 \le t - x \wedge -1 \le -x \wedge 0 \le x)\rangle.$$

Since $R^9 \subseteq R^8$, a fixpoint is found in 9 iterations. As the fixpoint $\bigcup_{0 \leq i \leq 8} R^i$ contains no initial states from R_i, the invariant property has been verified.

Acknowledgements. Amir Pnueli and Joseph Sifakis have influenced the ideas contained in this paper through numerous discussions.

References

[ACH⁺92] R. Alur, C. Courcoubetis, N. Halbwachs, D.L. Dill, and H. Wong-Toi. Minimization of timed transition systems. In *CONCUR 92: Theories of Concurrency*, Lecture Notes in Computer Science 630, pages 340–354. Springer-Verlag, 1992.

[AD90] R. Alur and D.L. Dill. Automata for modeling real-time systems. In M.S. Paterson, editor, *ICALP 90: Automata, Languages, and Programming*, Lecture Notes in Computer Science 443, pages 322–335. Springer-Verlag, 1990.

[BFH90] A. Bouajjani, J.C. Fernandez, and N. Halbwachs. Minimal model generation. In R.P. Kurshan and E.M. Clarke, editors, *CAV 90: Automatic Verification Methods for Finite-state Systems*, Lecture Notes in Computer Science 531, pages 197–203. Springer-Verlag, 1990.

[Čerāns] K. Čerāns. Decidability of bisimulation equivalence for parallel timer processes. In *CAV 92: Automatic Verification Methods for Finite-state Systems*, Lecture Notes in Computer Science. Springer-Verlag. To appear.

[CHR91] Z. Chaochen, C.A.R. Hoare, and A.P. Ravn. A calculus of durations. *Information Processing Letters*, 40(5):269–276, 1991.

[HNSY92] T.A. Henzinger, X. Nicollin, J. Sifakis, and S. Yovine. Symbolic model checking for real-time systems. In *Proceedings of the Seventh Annual Symposium on Logic in Computer Science*, pages 394–406. IEEE Computer Society Press, 1992.

[HPS83] D. Harel, A. Pnueli, and J. Stavi. Propositional dynamic logic of regular programs. *Journal of Computer and System Sciences*, 26(2):222–243, 1983.

[Lam87] L. Lamport. A fast mutual exclusion algorithm. *ACM Transactions on Computer Systems*, 5(1):1–11, 1987.

[LY92] D. Lee and M. Yannakakis. Online minimization of transition systems. In *Proceedings of the 24th Annual Symposium on Theory of Computing*. ACM Press, 1992.

[MMP92] O. Maler, Z. Manna, and A. Pnueli. From timed to hybrid systems. In J.W. de Bakker, K. Huizing, W.-P. de Roever, and G. Rozenberg, editors, *Real Time: Theory in Practice*, Lecture Notes in Computer Science 600, pages 447–484. Springer-Verlag, 1992.

[NOSY] X. Nicollin, A. Olivero, J. Sifakis, and S. Yovine. An approach to the description and analysis of hybrid systems. This volume.

[NSY92] X. Nicollin, J. Sifakis, and S. Yovine. From ATP to timed graphs and hybrid systems. In J.W. de Bakker, K. Huizing, W.-P. de Roever, and G. Rozenberg, editors, *Real Time: Theory in Practice*, Lecture Notes in Computer Science 600, pages 549–572. Springer-Verlag, 1992.

[WME92] F. Wang, A.K. Mok, and E.A. Emerson. Real-time distributed system specification and verification in asynchronous propositional temporal logic. In *Proceedings of the 12th International Conference on Software Engineering*, 1992.

Hybrid Systems: the SIGNAL approach

Albert BENVENISTE, Michel LE BORGNE, Paul LE GUERNIC*

Abstract. Hybrid Systems are models of systems operating in real-time and handling events as well as "continuous" computations. The SIGNAL formalism for Hybrid Systems is presented in this paper. Its expressive power is discussed, and a general method, to associate various formal systems with it, is presented with applications to SIGNAL compilation and proof system.

1 Introduction: Real-Time and Hybrid Systems

Hybrid Systems are models of systems operating in real-time and handling events as well as "continuous" computations. As an example, let us discuss the case of an aircraft control system. Measurements are received from sensors and processed by the control loops to produce commands as outputs for the actuators: this involves various kinds of numerical computations. Switching from one operating mode to another one can be performed automatically or by the pilot: in both cases, events are received that control the various computations in some discrete event mode. For safety purposes, on-line failure detection and reconfiguration is performed by taking advantage of the redundancies in the aircraft system: actuators and sensor failure detection procedures are numerical computations that produce alarms and various detections which in turn result in reconfiguring the operating mode. From this follows that discrete events and computations are tightly coupled in a fairly symmetric way. Also, in aircraft systems, response times are often critical. They depend on the particular tasks being performed at the considered instant. On the other hand, timeouts due to timing constraints can influence the task being performed. To summarize, in real-time hybrid systems, both aspects of "what" should be implemented and "how fast" it should run must be considered. Similarly, aspects of "discrete control" and "analog computations" deeply interact, and both interact also with response times via the mechanism of timeouts. The term of *Hybrid System* recently emerged [7, 21] to refer to formalisms providing an attempt to cover the above mentioned issues in some unified way.

For instance, various kinds of transition systems are proposed in [15, 16] to combine qualitative and quantitative aspects of real-time. Similarly, [21] proposes to generalize a similar approach to Hybrid Systems. In both cases, the expressive power of the models is explicitly described in the axioms and inference rules of the semantics, so that expanding this expressive power is achieved by making

---
* A.B and P. L.G. are with INRIA-IRISA, M. L.B. is with IRISA-University, Campus Beaulieu, 35042 RENNES Cedex, FRANCE, name@irisa.fr. **Keywords:** hybrid systems, real-time.

the model more complicated. In this paper, we analyze an alternative formalism for Hybrid Systems that has been firstly proposed in [7, 6]. The main features of this formalism are:

1. it is both simple and "universal" as far as expressive power for specification is concerned: discrete events and "analog computations" are encompassed, and timing constraints can be specified; it does not rely on the notion of transition system, but handles "traces" or "behaviours", see [6];
2. it has various formal systems associated with it that can be used to verify or even synthesize some properties related to the above mentioned issues.

While "universal" expressive power is achieved relatively easily as we shall see later, only limited reasoning capability can be expected in turn (otherwise, undecidability problems would result). A major objective of this paper is to present a systematic approach to derive formal systems associated with our universal formalism, which concentrate on some particular property, e.g., discrete event features, timing constraints, etc...

The paper is organized as follows. Section 2 is devoted to the presentation of the SIGNAL language with emphazis on hybrid systems: general principles of synchrony are discussed, then the kernel of SIGNAL is presented, and finally a mathematical model of hybrid systems (with discrete but dense time) is presented and used to establish the semantics of SIGNAL. In Section 3 hybrid systems description using SIGNAL is discussed, with a careful analysis of SIGNAL's expressive power: we first concentrate on pure synchronization systems and show how a single-token buffer can be specified, then we investigate timed systems, and finally we discuss an example of a "gas burner" control — a typical hybrid systems control problem. Section 4 is devoted to the design of formal calculi for SIGNAL : we first discuss the case of pure synchronisation systems and show how implementation from specifications results for the buffer example, then we provide a general approach to derive formal calculi for hybrid discrete/analog systems, and finally we present guidelines for the development of formal calculi for discrete systems that are themselves hybrid (e.g., synchronisation + timers). A conclusion summarizes the major features of our approach as well as the very nature of the questions raised by hybrid systems.

2 Hybrid Systems and the SIGNAL language

2.1 Some features of Hybrid Systems

A first major issue is that of the very nature of time for Hybrid Systems.

A first question arises, namely *should continuous time be used in Hybrid Systems ?* Computer systems are always discrete time systems, and analog to digital conversion is performed at the output of sensors or before the input of actuators *according to the response time — or frequency bandwidth — of the various components of the system,* i.e., sensors, actuators, and the plant in consideration. Signal processing or control engineers refer to this as the "Shannon sampling rate principle". Consequently, continuous time is mainly useful for simulation purposes, where ordinary or partial differential equations are typical analog model. However, simulating ODE's or PDE's is again performed via discretization. Hence, while using continuous time may be of interest in the study of hybrid systems (the mathematics of continuous time are sometimes simpler), this is not mandatory per se. Hence we choosed to concentrate on discrete time Hybrid Systems (see also the "gas burner example" and subsequent discussion in Subsection 3.4). However, the nature of this (discrete) time still remains to be discussed.

Complex applications such as the one mentioned above are inherently distributed in nature. Hence every subsystem possesses its own time reference, namely the ordered collection of all the communications or actions this subsystem performs: in sensory based control systems, each sensor posses its own digital processing with proper sampling rate, actuators generally have a slower sampling rate than sensors, and moreover the software devoted to monitoring only reacts to various kinds of alarms that are triggered internally or externally. Hence the nature of time in Hybrid Systems is by no means universal, but rather local to each subsystem, and consequently multiform. This very fundamental remark justifies the kind of model for Hybrid System we use in this paper.

Our model handles infinite sequences of data of arbitrary type with a certain kind of restricted asynchronism. Assume that each sequence, in addition to the normal values it takes in its range, can also take a special value representing the *absence* of data at that instant. The symbol used for absence is $\perp$. Therefore, an infinite time sequence of data (we shall refer to informally as a *signal* in this discussion, this notion will be properly formalized later) may look like

$$1, -4, \perp, \perp, 4, 2, \perp, \ldots \tag{1}$$

which is interpreted as the signal being absent at the instants $n = 3, 4, 7, \ldots$ etc. *Systems specified via constraints on signals of the form (1) will be termed Hybrid Systems (Hybrid System).* A typical way of specifying such constraints will be to write equations relating different signals. The following questions are immediate from this definition:

(1) If a single signal is observed, should we distinguish the following samples from each other?

$$\{1, -4, \perp, \perp, 4, 2, \perp, \ldots\}, \ \{\perp, 1, \perp, -4, \perp, 4, \perp, 2, \perp, \ldots\}, \ \{1, -4, 4, 2, \ldots\}$$

Consider an "observer"[2] who monitors this single signal and does nothing else. Since he is assumed to observe only *present* values, there is no reason to distinguish the samples above. In fact, the symbol $\perp$ is simply a tool to specify the *relative* presence or absence of a signal, given an *environment*, i.e. other signals that are also observed. Jointly observed signals taking the value $\perp$ simultaneously for any environment will be said to *possess the same clock,* and they will be said to possess different clocks otherwise. Hence clocks may be considered

---

[2] in the common sense, no mathematical definition is referred to here

as equivalence classes of signals that are present simultaneously. This notion of time makes no reference to any "physical" universal clock: time is rather local to each particular subset of signals in consideration.

(2) How to interconnect two Hybrid Systems? Consider the following two Hybrid Systems specified via equations:

$$y_n = \text{ if } x_n > 0 \text{ then } x_n \text{ else } \perp \tag{2}$$

and the usual addition on sequences, namely

$$z_n = y_n + u_n \tag{3}$$

In combining these Hybrid Systems, it is certainly preferable to match the successive occurrences $y_1, y_2, \ldots$ in (3) with the corresponding *present* occurrences in (2) so that the usual meaning of addition be met. But this is in contradiction with the bruteforce conjunction of equations (2,3)

$$y_n = \text{ if } x_n > 0 \text{ then } x_n \text{ else } \perp$$
$$z_n = y_n + u_n$$

which yields $z_n = \perp + u_n$ whenever $x_n \leq 0$. In subsection 2.3 a denotational model for Hybrid Systems firstly introduced in [7] and improved in [6] is presented, it provides an adequate answer to the question of how to properly interconnect equations (2,3). This model is then used to establish the semantics of the SIGNAL language we introduce informally in the following section. To summarize, our formalism will provide a *multiform* but *coherent* notion of time. Other formalisms using the same approach to handle time are the so-called *synchronous languages* [5, 12, 14].

2.2 SIGNAL-kernel

We shall introduce only the primitives of the SIGNAL language, and drop any reference to typing, modular structure, and various declarations; the interested reader is referred to [20]. SIGNAL handles (possibly infinite) sequences of data with time implicit: such sequences will be referred to as *signals*. At a given instant, signals may have the status *absent* (denoted by $\perp$) and *present*. If x is a signal, we denote by $\{x_n\}_{n\geq 1}$ the sequence of its values when it is present. Signals that are always present simultaneously are said to have the same *clock*, so that clocks are equivalence classes of simultaneously present signals. Instructions of SIGNAL are intended to relate clocks as well as values of the various signals involved in a given system. We term a system of such relations *program*; programs may be used as modules and further combined as indicated later.

A basic principle in SIGNAL is that a single name is assigned to every signal, so that in the sequel, identical names refer to identical signals. The kernel-language SIGNAL possesses 6 instructions, the first of them being a generic one.

```
 (i)   R(x1,...,xp)
 (ii)  y := x $1 init x0
(iii)  y := x when b
 (iv)  y := u default v
  (v)  P | Q
 (vi)   P !! x1,...,xp
```

(i) direct extension of instantaneous relations into relations acting on signals:

$$R(x1, ..., xp) \iff \forall n : R(x1_n, ..., xp_n) \text{ holds}$$

where $R(...)$ denotes a relation and the index n enumerates the instants at which the signals xi are present. Examples are functions such as $z := x+y$ ($\forall n : z_n = x_n + y_n$). A byproduct of this instruction is that *all referred signals must be present simultaneously, i.e. they must have the same clock*. This is a generic instruction, i.e. we assume a family of relations is available. If $R(...)$ is the universal relation, i.e., it contains all the p-tuples of the relevant domains, the resulting SIGNAL instruction only constrains the involved signals to have the same clock: the so obtained instruction will be written $x \mathbin{\hat{}}= y$ and only forces the listed signals to have the same clock.

(ii) shift register.

$$y := x \$1 \text{ init } x0 \iff \forall n > 1 : y_n = x_{n-1}, y_1 = x0$$

Here the index n refers to the values of the signals when they are *present*. Again this instruction forces the input and output signals to have the same clock.

(iii) condition (b is boolean): y equals x when the signal x and the boolean b are available and b is true; otherwise, y is absent; the result is an event-based undersampling of signals. Here follows a diagram summarizing this instruction:

$$
\begin{array}{l}
x : 1\ 2 \perp \perp 3\ 4 \perp \perp 5\ 6\ 9 \ldots \\
b : t\ f\ t \perp f\ t\ f \perp \perp f\ t \ldots \\
y : 1 \perp \perp \perp \perp 4 \perp \perp \perp \perp 9 \ldots
\end{array}
$$

(iv) y merges u and v, with priority to u when both signals are simultaneously present; this instruction is the key to oversampling as we shall see later. Here follows a table summarizing this instruction:

$$
\begin{array}{l}
u : 1\ 2 \perp \perp 3\ 4\ \perp \perp 5 \perp 9 \ldots \\
v : \perp \perp \perp 3\ 4\ 10 \perp 8\ 9\ 2 \perp \ldots \\
y : 1\ 2 \perp 3\ 3\ 4 \perp 8\ 5\ 2\ 9 \ldots
\end{array}
$$

Instructions (i-iv) specify the elementary programs.

(v) combination of already defined programs: signals with common names in P and Q are considered as identical. For example

```
(| y := zy + a
 | zy := y $1 x0 |)
```

denotes the system of recurrent equations:

$$y_n = zy_n + a_n \qquad (4)$$
$$zy_n = y_{n-1}, \ zy_1 = \text{x0}$$

On the other hand, the program

```
(| y := x when x>0
 | z := y+u |)
```

yields

$$\text{if } x_n > 0 \text{ then } \begin{cases} y_n = x_n \\ z_n = y_n + u_n \end{cases} \qquad (5)$$
$$\text{else } y_n = u_n = z_n = \perp$$

where (x_n) denotes the sequence of present values of x. Hence the communication | causes $\perp$ to be inserted whenever needed in the second system z:=y+u. This is what we wanted for the example (2,3). Let us explain this mechanism more precisely. Denote by $u_1, u_2, u_3, u_4, \cdots$ the sequence of the present values of u (recall that y, z are present simultaneously with u). Then, according to point (1) of the discussion at the beginning of this section, $u_1, u_2, u_3, u_4, \cdots$ is equivalent to its following expanded version:

$$u : \quad \perp, u_1, \perp, \perp, u_2, u_3, \perp, \perp, \perp, u_4, \perp, \cdots,$$

for any finite amount of "$\perp$"s inserted between successive occurrences of u. Assuming all signals of integer type, suppose the following sequence of values is observed for x:

$$x : \quad -2, +1, -6, -4, +3, +8, -21, -7, -2, +5, -9, \cdots$$

Then the amount of inserted "$\perp$"s for the above expanded version of u turns out to fit exactly the negative occurrences of x: this flexibility in defining u allows us to match the present occurrences of u with the present occurrences of y, i.e., the positive occurrences of x. This mechanism is formalized in the model we present in the next subsection.

(vi) restriction of program P to the mentioned list of signals: other signals involved in P are local and are not visible when communication is considered.

2.3 An elementary model for Hybrid Systems and the semantics of SIGNAL

In this subsection, we present a simple mathematical model for Hybrid Systems, and use it to give the semantics of SIGNAL.

Behaviours We are given a finite set **X** of *signals,* signals are generically denoted by **x**, **y**, ... With each signal **x** is associated a domain of values $V(\mathbf{x})$, we call it the *type* of **x**, and each domain $V(\mathbf{x})$ contains a distinguished symbol $\perp$ to be interpreted as the status "absent". Let $\mathcal{X}$ denote the set

$$\mathcal{X} = \mathbf{X} \times \mathbf{N}$$

where **N** denotes the set of integers, and let

$$V = \coprod_{\mathbf{x} \in \mathbf{X}} V(\mathbf{x})$$

where $\coprod$ denotes the disjoint union.

Definition 1 behaviours. A behaviour $\mathcal{B}$ on **X** is a map :

$$\mathcal{B} : \mathcal{X} \mapsto V$$

such that $\mathcal{B}(\mathbf{x}, n) \in V(\mathbf{x})$ for each $\mathbf{x} \in \mathbf{X}$ and $n \in \mathbf{N}$. We will write $\mathbf{x}_n$ to denote the value $\mathcal{B}(\mathbf{x}, n)$.

In other words, a behaviour is an infinite sequence of events, where an event is the assignment to each signal of a value or the status "absent".

Hybrid Systems and their combinators Let $\mathcal{B}$ and $\mathcal{B}'$ be two behaviours on **X**. $\mathcal{B}'$ is said to be a *compression* of $\mathcal{B}$ (or, vice-versa, $\mathcal{B}$ is a *dilation* of $\mathcal{B}'$) if there exists some increasing subsequence (n_k) of positive integers such that

(i) in $\mathcal{B}$: if $n \notin (n_k)$ then $\mathbf{x}_n = \perp$ for all **x** ;
(ii) the map $\{(\mathbf{x}, k) \mapsto \mathcal{B}(\mathbf{x}, n_k)\}$ is equal to $\mathcal{B}'$.

Hence $\mathcal{B}$ is obtained from $\mathcal{B}'$ by inserting finitely many "silent instants", i.e. instants of the form (i) above.

Definition 2 Hybrid Systems. A Hybrid System on **X** is a set of behaviours on **X**, which is invariant with respect to dilations and compressions. Hybrid Systems are generically denoted by Σ.

Compression– and dilation–invariance means the following :

$$\forall \mathcal{B} \in \Sigma , \quad \text{dilat}(\mathcal{B}) \text{ and } \text{compress}(\mathcal{B}) \in \Sigma$$

where "dilat$(\mathcal{B})$" and "compress$(\mathcal{B})$" denote any dilation or compression of $(\mathcal{B})$.

COMMENT : a Hybrid System is thus a set of "legal" behaviours. The key conditions that Hybrid Systems should be compression– and dilation–invariant is extremely important : it expresses that *discrete time has no physical interpretation, only the global interleaving of behaviours is relevant.* In fact, being equal up to a finite combination of dilations and compressions defines an equivalence relation on behaviours, and we are interested in equivalence classes of this relation, as pointed out in the informal discussion of subsection 2.1. Referring to this discussion, an equivalence class of behaviours modulo compression/dilation has several components, one for each signal : these are the *signals* we informally discussed in subsection 2.1. Also, the compression/dilation–invariance of Hybrid Systems, together with the way we define below their composition, allows our model to provide (multiform) *dense but discrete* time, since the insertion of additional instants between instants is the basic mechanism for systems composition defined below, see the discussion and the end of the preceding subsection, and [6] for a deeper discussion of this aspect.

The following operators can be defined on Hybrid Systems. Let Σ be a Hybrid System on $\mathbf{X}$, and consider $\mathbf{Y} \subset \mathbf{X}$. We can interpret behaviours $\mathcal{B}$ in a curried form $\mathbf{X} \mapsto (\mathbf{N} \mapsto V)$. In doing so, the set of the restrictions to $\mathbf{Y}$ of the legal behaviours of Σ spans a unique Hybrid System, we call it the *restriction* of Σ to $\mathbf{Y}$ and denote it by

$$\Sigma_{!!}\mathbf{Y}$$

Consider now, for $i = 1, 2$, two Hybrid Systems Σ_i on $\mathbf{X}_i$. We call *composition* of Σ_1 and Σ_2, denoted by

$$\Sigma_1 | \Sigma_2$$

the largest[3] Hybrid System on $\mathbf{X_1} \cup \mathbf{X_2}$ having its restriction to $\mathbf{X}_1$ (resp. $\mathbf{X}_2$) contained in Σ_1 (resp. Σ_2).

The definition of SIGNAL According to the preceding subsection, in order to specify a Hybrid System on a given set of signals, it is enough to describe some subset of all behaviours that can be built upon this set of signals, and the desired specification will be obtained by including all dilations and compressions of the listed behaviours. Describing such subsets can be done by *listing a family of constraints on the set of all behaviours*. This is what we do next.

Instruction (i): R(x1, ... xp)

$$\forall n \in \mathbf{N}_+, \ \forall i : \mathrm{xi}_n \neq \perp$$
$$\wedge \ \forall n \in \mathbf{N}_+ \qquad : R(\mathrm{x1}_n, ..., \mathrm{xp}_n)$$

Recall that the notation xi_n denotes the value carried by the signal with name xi at the n-th instant of the considered behaviour.

Instruction (ii): y := x \$1 x0

$$\forall n \in \mathbf{N}_+ : \mathrm{x}_n \neq \perp$$
$$\wedge \ \forall n > 1 \quad : \mathrm{y}_n = \mathrm{x}_{n-1}$$
$$\wedge \qquad\qquad \mathrm{y}_1 = \mathrm{x}0$$

---

[3] with respect to the order by inclusion defined on Hybrid Systems

Instruction (iii): y := x **when** b

$$\forall n \in \mathbf{N}_+, \ y_n = \begin{cases} \text{if } x_n \neq \perp \text{ and } b_n = true \text{ then } x_n \\ \text{else } \perp \end{cases}$$

Instruction (iv): y := u **default** v

$$\forall n \in \mathbf{N}_+, \ y_n = \begin{cases} \text{if } u_n \neq \perp \text{ then } u_n \\ \text{else if } u_n = \perp \text{ and } v_n \neq \perp \text{ then } v_n \\ \text{else } \perp \end{cases}$$

Instruction (v): P | Q

We already defined the operator | on Hybrid Systems.

Instruction (vi): P !! x1,...,xp

We already defined the restriction on Hybrid Systems. The reader is referred to [6] for deeper studies related to this kind of model; in particular, it is proved there that SIGNAL provides the most general synchronisation mechanisms that are required for dense but discrete time hybrid systems.

3 Expressing Hybrid Systems in SIGNAL

To make our presentation easier, we shall first present some of the most useful macros that are provided in full SIGNAL. An **event** type signal T (or "pure" signal) is an always *true* boolean signal. Hence "**not** T" denotes the boolean signal with clock T which always carries the value *false*. Given any signal X,

    T := event X

defines the **event** type signal T whose occurrences are simultaneous with those of X: it represents the clock of X. The variation

    T := when B

of the **when** operator defines the **event** type signal T which is present whenever the boolean signal B is present and has the value *true* and delivers nothing otherwise; it is equivalent to "T := B **when** B". Constraints may be defined on the clocks of signals; in this paper, the following notations are used:

X ^= Y   X and Y have the same clock;

X ^< Y   X is no more frequent than Y, which is equivalent to X ^= (X **when event** Y)

3.1 Systems involving synchronization and logic : the buffer example

We discuss here how a single token buffer can be specified. Here are the corresponding requirements :

(r-i) we have a memory that can be written or read, and otherwise remains unchanged;

(r-ii) writing is allowed only when the memory is empty;

(r-iii) reading is allowed only when the memory is full.

Let us proceed on expressing these requirements using SIGNAL (references to each requirement will be made explicit). Consider a memory with content M, which can be written (signal WRITE) and read (signal READ):

```
(r-i)    (1)    (| M := WRITE default (M $1 init any)
         (2)     | READ := M when (event READ) |)
```

The first instruction expresses that the memory M is refreshed when WRITE is received, otherwise the previous value (M$1) is kept. Note that the clock of M is not entirely determined, it only has to be more frequent than that of WRITE. The second equation expresses that, when reading is wanted (**event** READ), it actually occurs and provides us with a READ signal carrying the value of M. Consequently, the clock of M has to be more frequent that that of READ. To proceed further on, let us encode the status (being written or being read) of the memory as follows:

```
(3)    FULL := (event WRITE) default (not (event READ))
```

i.e., FULL takes the value true when WRITE occurs, otherwise it takes the value false when READ occurs, otherwise it is absent. Now suppose that writing in the memory is allowed only when the previous value of the memory has been read. This constraint is expressed by the following equation:

```
(r-ii)   (4)   WRITE ^= when (not (FULL $1))
```

which prevents two successive occurrences of FULL to be true. Conversely, if we want any written value to be read at most once, we have to write:

```
(r-iii) (5)   READ ^= when (FULL $1)
```

which prevents two successive occurrences of FULL to be false. Finally, putting these three additional equations together specifies the single token buffer according to the above requirements.

3.2 Timed systems

Timed systems are a particular case of hybrid systems. We shall informally discuss how such systems can be expressed in SIGNAL.

Some macros related to timing We shall first introduce additional macros, some of them are listed below:

```
Y :=  X in ]S,T]            (i)
N := #X in ]S,T]            (ii)
```

In expression (i), S and T are both pure signals. This expression delivers those present X's which occur strictly after some occurrence of S and before the next occurrence of T (with T included). Here follows a picture explaining this: the horizontal line depicts the (discrete) time and its direction of growing, and the additional line figures the subset of instants in consideration. Also a simultaneous occurrence of S and T is depicted, and its effect is shown (no time window is open).

$$\begin{array}{ccccccccc} & & & & & & & \text{S} & \\ \text{T} & \text{S} & \text{T} & \text{S} & \text{S} & \text{T} & \text{T} & \text{T} & \text{S} \cdots \end{array}$$

Expression (ii) is a *timer* which counts the occurrences of X within the mentioned interval and is reset to zero every S; this signal is delivered exactly when equation (i) delivers its output.

Here follows a diagram showing the behaviour of the first macro; we added the boolean signal "in_]S,T]" which is delivered when X,S, or T is present, and is *true* within intervals]S,T] and *false* otherwise:

$$
\begin{array}{l}
\text{X} : 1\ 2\ \bot\ \bot\ 3\ 4\ 1\ \bot\ 5\ 6\ 9 \ldots \\
\text{S} : \bot\ \bot\ t\ t\ \bot\ \bot\ \bot\ \bot\ \bot\ t\ \bot \ldots \\
\text{T} : t\ \bot\ \bot\ \bot\ \bot\ t\ \bot\ \bot\ \bot\ \bot\ t \ldots \\
\text{in\_]S,T]} : f\ f\ f\ t\ t\ t\ f\ \bot\ f\ f\ t \ldots \\
\text{Y} : \bot\ \bot\ \bot\ \bot\ 3\ 4\ \bot\ \bot\ \bot\ \bot\ 9 \ldots
\end{array}
$$

Expanding these macros and their variations on the shape of the considered intervals ([S,T[, [S,T], etc...) into the primitive SIGNAL statements is easily done. For instance, (i) is defined via the following SIGNAL module:

```
(|(| IN_S_T ^= (S default T default (event X))
   |(| HITTING_S_T := (not T) default S default IN_S_T
   | IN_S_T := HITTING_S_T $1 init false
   |)
   |)
 | Y := X when IN_S_T
 |)
```

The hierarchy of submodules is depicted by the imbrication of of "(|...|)". This program is composed of two blocks. The meaning of the second one (last equation) is immediate, thus we concentrate on the first one which purpose is to produce the boolean IN_S_T (corresponding to "in_]S,T]" discussed above). The first equation indicates when this signals has to be delivered. The block composed of the equations 2 and 3 delivers the value of IN_S_T: the boolean signal HITTING_S_T corresponds to "in_[S,T[". Note that

```
T ^= T in ]S,T]
```

expresses that no event T is lost by keeping only its occurrences occurring within]S,T], i.e., T can only follow S. Such SIGNAL equations can be used to specify particular interleavings of events.

Specifying timing constraints Consider again the previously discussed single token buffer. Assume now that this single token buffer is being used as a mailbox by some other module and it is desired that, when stored, *a message must be read within some specified delay*. A corresponding SIGNAL specification is as follows:

```
(| event READ ^< TICK
 | N := #TICK in ]WRITE,READ]
 | N < MAX_TIME
 |)
```

where `MAX_TIME` is some maximum allowed delay. The first statement expresses that the mailbox can be checked at any `TICK` instant. The second equation counts the delay between writing and reading the message in terms of `TICK`s. Finally, the last statement expresses that the constraint (`N < MAX_TIME`) must be satisfied. This mechanism is typically used in the description of (discrete time) timed automata [17], where transitions are guarded by expressions involving duration intervals.

3.3 A sketchy discussion of the "gas burner" example

This example is borrowed from [22] but we shall take into account some of the interesting remarks pointed out in [18]. We consider a much simplified version of a computer controlled (on-off) gas burner. This is a safety-critical system as an accident may occur if an excessive amount of unburned gas leaks to the environment. Small gas leaks cannot be avoided during ignition. A burning flame may also be blown out causing some gas to leak before the failure is detected[4]. The gas burner is controlled by a thermostat and the gas is ignited by an ignition transformer. The informal requirements are :

1. For safety, gas concentration must stay within some specified limit.
2. Heat request off shall result in the flame being off after 60 seconds.
3. Heat request shall after 60 seconds result in gas burning unless an ignite or flame failure has occurred.

In what follows, time is discretised according to the response time of the various components — discrete controller, sensors, dynamics of the combustion — of the system. The discrete part of controller consists of the description of different phases (see for example Figure 2 of [22]). To shorten the discussion, we shall omit the detailed discussion of these, since synchronisation and logical aspects have been already discussed. We shall rather concentrate on the hybrid part of the controller where analog quantities and discrete events are mixed.

---

[4] since how such a detection is performed is *not* indicated in [22], we shall consider two different cases, depending on whether gas concentration is or is not used for this purpose.

Model of the plant We proceed first on modelling the plant. To simplify our description and since we want to concentrate on the hybrid part of the controller, we shall only model the plant when the gas is on. We model its behaviour by the following mathematical equation :

$$q_n = \max\{ (q_{n-1} + \mu.0_{F_n} - \nu + \xi_n) , 0 \} \tag{6}$$

where

q_n denotes the gas concentration at time n,
ν denotes the dissipation rate and is a constant,
μ denotes the injection rate,
F_n denotes the value of the predicate *"flame is on"* at time n,
ξ_n denotes some (small, possibly random) unobserved disturbance to take into account faults in injection, burning, etc.

As for the additional notations, 0_A where A is a predicate is the function which takes the value 0 when the predicate is true, and 1 otherwise. Equation (6) expresses that, at each instant, the gas concentration decreases by a quantity of ν due to dissipation, increases by a quantity of μ when gas is on but flame is off (typically $\mu \gg \nu$), and ξ_n accounts for model errors. The model is directly encoded in SIGNAL as follows :

```
(| QQ = (Q $ 1 init q0) + ((MU when not FLAME) default 0) + NU + XI
 |  Q = max(0){QQ}
 |)
```

with the following correspondence between the objects of equation (6) and those of this program[5] :

$$Q \leftrightarrow q_n$$
$$\text{(Q \$ 1 init q0)} \leftrightarrow q_{n-1}, \; q_0 = q0$$
$$\text{((MU when not FLAME) default 0)} \leftrightarrow \mu.0_{F_n}$$
$$\text{NU} \leftrightarrow \nu$$
$$\text{XI} \leftrightarrow \xi_n$$
$$\text{max(0)\{...\}} \leftrightarrow \max\{ \dots, 0\}$$

(we use here the externally defined function `max(0){...}`). This model can be used for simulation purposes.

The hybrid part of the controller Request 2 ("heat request off shall result in the flame being off after 60 seconds") is typical of what we already discussed in subsection 3.2, and we shall discuss in the next section how corresponding verification or synthesis could be performed. Let us concentrate on Request 1 ("for safety, gas concentration must stay within some specified limit"). Expressing this in SIGNAL is immediate

---

[5] we omit here to mention the distinction between signals and parameters, this is obviously handled in the language, see [20].

```
Q < MAX_Q
```

where `MAX_Q` is some parameter [6], but this does not help very much indeed! So, what interesting can be done?

Assume first that gas concentration is sensed according to the equation

$$y_n = q_n + v_n$$

where v_n is some zero mean measurement noise. Then an obvious rule to stop delivering the gas would be

$$y_n > \text{MAX\_Q} \tag{7}$$

and the corresponding SIGNAL expression is

```
GAS_OFF := when (Y > MAX_Q)
```

where Y denotes the measurement y_n and `GAS_OFF` is the event ("pure signal") stopping the injection of gas. This is the best one can do in this case, but this does not guarantee that requirement 1 would be met: noise v_n causes trouble to this requirement, and nothing can be done against it.

If gas concentration is not directly sensed, but only flame failure is detected (again with some unavoidable imperfection), we shall proceed differently, by reconstructing an estimate of the unmeasured gas concentration; this is called an *observer* in the litterature of automatic control. This estimate $\hat{q}_n$ will then be used within a stopping rule of the form (7). The desired observer is simple in our case, namely:

$$\hat{q}_n = \max\left\{ \left(\hat{q}_{n-1} + \mu.0_{\widehat{F}_n} - \nu\right), 0 \right\} \tag{8}$$

Let us comment this equation:

$\hat{q}_n$ is the estimated gas concentration
$\widehat{F}_n$ is the *estimated* value of the predicate "*flame is on*"; we do not know the exact value of this predicate since flame failure is detected with some unavoidable error;
if we compare this equation (8) with equation (6) which was used in the model of the gas burner, ξ_n is not used here any more: this is fully consistent since ξ_n is unknown and random and there is no reason here to add noise in the controller.

Finally, collecting these equations together, we obtain the "hybrid" part of the controller, informally described as follows:

$$\hat{q}_n = \max\left\{ \left(\hat{q}_{n-1} + \mu 0_{\widehat{F}_n} - \nu\right), 0 \right\} \tag{9}$$

$$\widehat{F}_n = G_n \text{ before flame failure is detected} \tag{10}$$

$$\hat{q}_n > \text{MAX\_Q} \Rightarrow \text{stop gas injection} \tag{11}$$

---

[6] Recall that instruction (i) of SIGNAL is a generic one, i.e., we skip defining formally usual functions and operators on reals or integers such as the $+, \times, <$, etc.

where G_n is the value of the predicate *"gas is on"*. In this description of the controller, equation (9) is the *"observer"*, i.e., a "noiseless" copy of model (6), equation (10) uses the flame failure detection to estimate the current state of the ignition, and (11) is the rule to stop gas injection. Equations (9,10,11) provide an informal description since the sequencing of the events is not easily described using mathematical equations. A precise SIGNAL description of the hybrid part of the controller is given in Table 1. In this program, SAMPLING denotes the

```
(| EST_FLAME = SAMPLING in ] GAS_ON , EST_FLAME_FAILURE ]
 |)
% EST_FLAME delivers the instants where the estimated value
% of the predicate "flame is on" is true

(| EST_QQ = (EST_Q $ 1 init q0) +
            ((MU when not EST_FLAME) default 0) + NU
 |   EST_Q = max(0){EST_QQ}
 |)
% EST_Q is the estimated value of Q

(| GAS_OFF = when (EST_Q > MAX_Q)
 |)
% this is the rule for stopping gas injection

(| SAMPLING ^= EST_Q
 | SAMPLING ^> GAS_ON
 | SAMPLING ^> GAS_OFF
 | SAMPLING ^> EST_FLAME_FAILURE
 |)
% these are synchronisation equations expressing that
% everything occurs at the sampling instants
```

Table 1. SIGNAL *program of the hybrid controller*

sampling event, recall that events are considered as boolean signals that are always true when present. This SIGNAL program describes the hybrid part of the controller, and the SIGNAL compiler generates code for its execution.

Discussion The reader may have noticed the significant difference with [22] [18] in discussing this example. We did not follow the computer science point of view of formal verification, but we rather considered the gas burner example as a *control problem*, where model uncertainties and measurement noise should be considered. This is especially important since one always wants to use a simplified model of the plant (exact combustion models are partial differential equations of huge complexity, even for this simple gas burner!). Since model uncertainties and measurement noise must be taken into account, formal verification of hard constraints should be replaced by (typically statistical) estimates of how much overshooting of the desired bounds may actually occur. The gas stopping rule we have proposed is typical of what a control engineer would do in such a situation. A mathematical analysis of the controller we have thus designed is beyond the scope of this paper, we refer the reader to some classics of control (the introductory chapter of [2]), and the basic references for the failure detection problem and its uncertainty aspects [3, 4].

3.4 Summary

Challenges raised by Hybrid Systems As we have discussed in this section, Hybrid Systems raise questions of very diverse nature, namely :

(i) significant complexity due to a deep interplay of discrete events handling and analog processing[7]; this complexity makes *programming* of Hybrid Systems a difficult and demanding task, see [1] for a discussion of this point;

(ii) designing algorithms for information processing associated with Hybrid Systems; as we have discussed, uncertainties and measurement noises should be considered at this point;

(iii) designing control algorithms associated with Hybrid Systems; again, issues of robustness versus performance are central in this activity;

(iv) expressing overall requirements for hybrid systems and verifying the previous two items against these requirements; this is much demanding a task since usual frameworks are much different when different aspects of discrete events, analog information processing, and analog control of Hybrid Systems are concerned.

In designing SIGNAL as a tool for Hybrid Systems, our objective was rather modest : mainly question (i) was considered and we view other questions (ii)–(iv) as being tackled using different techniques. Obviously, providing powerful formal techniques with having point (i) as a main objective will indirectly help the designer for his tasks (ii)–(iv) as well as the implementation of the final computer controlled system. This analysis motivates the discussion to follow.

Basic problems As the above examples show, SIGNAL programs generally express *constraints* on the behaviours of their involved signals. This makes the composition of SIGNAL programs fairly obvious. On the other hand, SIGNAL programs will generally attempt to specify real-time systems that are *transducers*, i.e., possess inputs that drive them and produce outputs. Hence implementing a SIGNAL specification consists in constructing a transducer producing all

---

[7] Obviously, this was not the case in the toy gas burner example, but such a complexity is typical ot real life application, see for instance our experience of developing in SIGNAL a whole speech recognition system.

solutions to the considered system of SIGNAL equations. Getting a transducer form out of a SIGNAL specification written as constraints on signals requires a powerful compiler. This compiler must be able to "solve" the SIGNAL systems of equations in some way to transform them into some input/output map. So it has to be a sort of a "formal calculus system". One of the objectives of this paper is to explain informally how such formal calculi can be derived. By the way, other services are immediately provided such as proofs, since there is no distinction between properties to be checked (these are constraints) and programs on which properties must be checked (these are also constraints).

Discussion: expressive power and formal reasoning capability As illustrated by the examples above, SIGNAL can be used to specify all key features we mentioned as being relevant to Hybrid Systems: computations, events and logic, timing constraints, and their mutual interaction. As far as the current SIGNAL compiler is concerned, the following should be noticed:

- the single token buffer specification involves only synchronization and logic and is thus fully handled by the SIGNAL compiler in the very same way as temporal logic does : implementation is synthesized directly from specifications;
- in contrast, expressions such as "X := Y+Z" are handled via rewriting (each occurrence of the left hand-side can be replaced by the expression in the right hand-side), no formal property is handled about real signals, nor about their associated operations; obviously, this technique does not prevent us from compiling programs involving such type of numerical expression, thus preserving efficiency at runtime;
- finally, no formal calculus about quantitative time is performed by the current SIGNAL formal system, in particular it cannot be proved that performing two successive responses in less than $10\mu s$ results in an overall response time of less than $20\mu s$; we shall however draw in the following section a general method to develop this and other useful formal calculi.

Hence the distinction between specifying and verifying must be emphasized: while the SIGNAL formalism has general expressive power [6], the SIGNAL formal system has limited (although quite powerful) capabilities. In [6, 7] we present a mathematical model for Hybrid Systems and use it to establish the semantics of SIGNAL. In particular it is shown in [6] that SIGNAL has maximum expressive power for Hybrid Systems description. In the next section we show how to derive formal systems for reasoning about Hybrid Systems defined by SIGNAL.

4 Deriving formal calculi for Hybrid Systems

We first discuss the simple case of "pure" SIGNAL, i.e., of programs involving only synchronization and logic, i.e., **event** and **logical** data types. Then we discuss how this simple case can be generalized.

4.1 "Pure" SIGNAL: programs involving only synchronization and logic

Three labels are required to encode the status *absent, true, false*. The finite field $\mathcal{F}_3$ of integers modulo 3 is used for this purpose[8] via the coding:

$$\text{absent} \leftrightarrow 0, \ \text{true} \leftrightarrow +1, \ \text{false} \leftrightarrow -1$$

Using this coding, we define a mapping from syntactic SIGNAL expressions to equations in $\mathcal{F}_3$ (recall that all signals are of type `event`, `logical`). This mapping is shown in table 2. In this table, the first instruction is a sample of a SIGNAL instruction of type (i), other ones are encoded similarly. In the coding of the second instruction, α denotes the current value of the internal state of the delay. Here and in the sequel, the generic notation x' denotes the next value of x. In particular, the coding of the instruction B := A\$1 involves two successive values of the state α; this equation expresses that, when a is received ($a^2 = 1$) it is fed into the next value α' of the state otherwise it is unchanged; then b receives the current state when a is received. Finally, clock(P) is the clock calculus of P and $\cup$ denotes conjunction.

SIGNAL equation	$\mathcal{F}_3$ coding (or "clock calculus")
A or B = event A	$a^2 = b^2$, $ab(a-1)(b-1) = 0$
B := A\$1	$\alpha' = (1 - a^2)\alpha + a$, $b = a^2\alpha$
y := x when b	$y = x(-b - b^2)$
y := u default v	$y = u + (1 - u^2)v$
P|Q	clock(P) $\cup$ clock(Q)

Table 2. Encoding "pure" SIGNAL programs

Let us apply this technique to the "single token bugffer" example. To simplify the notations, we denote the various signals by their first letter, e.g., w for `WRITE`, etc... Applying the rules of table 2 to each successive instruction of the program of subsection 3.1, as referred to with their respective numbers, yields

$$
\begin{array}{rl}
(1) & \mu' = (1 - m^2)\mu + m \ , \ zm = m^2\mu \\
(1) & m = w + (1 - w^2)zm \\
(2) & r = mr^2 \\
(3) & f = w^2 - (1 - w^2)r^2 \\
(\text{FULL } \$1) & \xi' = (1 - f^2)\xi + f \ , \ zf = f^2\xi \\
(4) & w^2 = zf - f^2 \\
(5) & r^2 = -zf - f^2
\end{array}
\tag{12}
$$

---

[8] elements of $\mathcal{F}_3$ are written $\{0, +1, -1\}$

A little algebra allows us to rewrite (12) as follows (comments are written for each equation):

$$m^2 = f^2 = 1 \qquad \text{fastest clock, always present by convention}$$
$$f' = -f \qquad \text{FULL is a flip-flop}$$
$$zm' = m$$
$$w^2 = -f - 1 \qquad \text{WRITE} \Leftrightarrow \text{box full} \qquad\qquad (13)$$
$$r^2 = f - 1 \qquad \text{READ} \Leftrightarrow \text{box empty}$$
$$r = mr^2$$
$$m = w + (1 - w^2)zm \quad \text{M stores the written value}$$

While (12) was given as a fixpoint equation, i.e., in an implicit form, (13) is in explicit form: reading the equations from top to down yields an execution mode. This explicit form reveals that the single-token is *not* a transducer: we cannot consider that WRITE acts as an input and READ is the output. Instead, FULL acting as a flip-flop drives the synchronization, and the additional input is the *value* carried by WRITE, not its clock. We say that (13) is a *solved form* of (12). Deriving (13) from (12) amounts to applying elimination techniques to polynomial functions over $\mathcal{F}_3$ and can be done automatically. When the specified system actually was equivalent to a transducer, calculating the solved form provides us with this equivalent transducer. A subset of this formal calculus serves as a basis for the SIGNAL compiler, and a full version of it is available in the SIGALI proof system [13] [19].

4.2 Developing formal calculi for general Hybrid Systems

Consider the following SIGNAL program:

```
(| N := R default (ZN-1)
 | ZN := N$1
 | R ^= when (ZN=1)
 |)
```

In this program, R is assumed to be a strictly positive integer signal, and ZN has initial value 1. The behaviour of this program is depicted in table 3. This

input R: 3 $\perp$ $\perp$ 4 $\perp$ $\perp$ $\perp$ 1 5 $\perp$ $\perp$ $\perp$ $\perp$ etc...
output N: 3 2 1 4 3 2 1 1 5 4 3 2 1 etc...

Table 3. UPSAMPLING in SIGNAL

program serves as a basic mechanism for data dependent *upsampling* of the input signal R. It is a particular and powerful feature of the SIGNAL formalism that programs with upsampling can be specified, see [6] for an extensive discussion about this aspect. Two domains are encountered, namely the positive integers R, N, ZN, and the boolean (ZN=1).

Hence for general Hybrid Systems or SIGNAL programs the situation is drastically more difficult: infinite domains are involved such as integers, reals, etc... So the systems of equations corresponding to general SIGNAL programs will be *approximately* solved using a technique we shall describe now informally. We describe now how the above UPSAMPLING program is actually handled by the current SIGNAL compiler. While most of the formalisms for real-time or Hybrid Systems hide computations inside actions that are viewed as black-boxes [21, 15, 16], we refuse to do so, since such actions turn out to influence the real-time behaviour (comparing a real signal to a threshold can be a mechanism to produce interruptions). The following general procedure is proposed to overcome this difficulty:

1. Select the domains for which you want to provide a formal system; we already discussed this issue in the two preceding sections.
2. Equations involving other domains can be handled in the weakest way, namely via syntax-based rewriting[9]; syntax-based rewriting algorithms amount to handling directed graphs; but these graphs vary dynamically according to the clock of the considered instant, we call them *dynamical graphs*.

The interested reader is referred to [7, 8] for details and formal definitions of dynamical graphs, we concentrate here on an informal discussion of the UPSAMPLING example. The coding of the three instructions of this program is given in table 4 (again we denote by B the boolean expression (ZN=1)). In this table, the notation $x \to y$ means that y depends on x when both are present, i.e., $x^2 y^2 = 1$: this statement is used if y must be evaluated using x. Similarly, the notation $x \xrightarrow{h^2} y$ means that y depends on x when $x^2 y^2 h^2 = 1$ holds : this statement is used if y must be evaluated using x when h is present; this latter notation is for instance used in the coding of the **default** instruction. Some comments

clock calculus	conditional dependency graph
$n^2 = r^2 + (1 - r^2)zn^2$	$n^2 \to N, r^2 \to R, zn^2 \to ZN,$ $R \to N \xleftarrow{1-r^2} ZN, n^2 r^2 \to N$
$zn^2 = n^2$	
$b^2 = zn^2$	$b^2 \to b, ZN \to b$
$r^2 = -b - b^2$	

Table 4. Encoding the UPSAMPLING program

follow. In Table 4, capitals are used (e.g., N) to refer to values of present signals,

---

[9] Again, let us emphasize that this technique is fully compatible with compilation.

while variables of the clock calculus are written in lower cases. Values of present signals depend on their clock (e.g., $n^2 \to$ N etc.). In this table, the second line of the conditional dependency graph is the coding of the **default**: the double dependency R $\to$ N $\xleftarrow{1-r^2}$ ZN is that between values (N takes the value of R when the latter is present, and otherwise it takes the value of ZN), and the second one $n^2 r^2 \to$ N expresses that, to compute the output N it is needed to know the clock deciding which dependency will be in force at the considered instant.

This coding is hybrid in nature: two different algebras are involved, namely the algebra of polynomial expressions in $\mathcal{F}_3$ variables, and labelled directed graphs. This is an algebra which is hybrid in nature, and to make it tractable we proceed as follows : based on the actual *syntactic* form of the equations in $\mathcal{F}_3$, we associate with each of these equations a graph according to the same principles as before. Performing this for the UPSAMPLING program yields

$$r^2 \to n^2 \xleftarrow{1-r^2} zn^2 : n^2 \text{ uses } r^2 \text{ for its evaluation and otherwise it uses } zn^2$$
$$b \to r^2 : r^2 \text{ uses } b \text{ for its evaluation}$$

Combining this graph with the preceding one yields for instance the circuit

$$b \to r^2 \to n^2 = b^2 \to b$$

which is effective at a clock equal to $b^2 r^2 = r^2 \neq 0$, i.e., a circuit in the graph occurs at some non empty clock. Thus we were not provided so far with a partial order, hence no evaluation scheme is derived at this point.

To overcome this *we use the calculus of $\mathcal{F}_3$ to replace the present equations of the clock calculus by other ones that are equivalent, in such a way that the concatenation of the resulting graphs be globally circuitfree.* For instance, we can replace the clock calculus of the UPSAMPLING program by its solved form

$$zn^2 = n^2 = b^2 \ , \ r^2 = -b - b^2$$

which yields the graph with a single branch $b \to r^2$. The resulting global graph is finally (arrows without input nodes denote inputs of the evaluation scheme):

$$\text{(input clock)} \quad \to b^2 = n^2 = zn^2 \to b \ , \ b^2 \to \text{ZN}$$
$$\text{(input: memory content)} \quad \to \text{ZN} \to b \to r^2$$
$$\text{(input: value of R)} \quad r^2 \to \text{R} \ , \quad \to \text{R}$$
$$\text{ZN} \xrightarrow{1-r^2} \text{N} \ , \ b^2 \to \text{N} \ , \ r^2 \to \text{N} \ , \ \text{R} \to \text{N}$$

No circuit is exhibited, so that a partial order can be associated with this graph and evaluation scheme follows accordingly. It is interesting to note that the input clock is that of the output N, and that only the value carried by R is an input: UPSAMPLING runs according to the demand driven mode.

Again, the method we presented here informally can be extended to other algebras. The current SIGNAL compiler is powered with a very fast implementation of the above procedure, see [20, 11]. Finally, the reader is referred to [8] for a formal presentation of the SIGNAL formal system in its present form.

4.3 Extension to other domains: quantitative real-time

We consider again the UPSAMPLING program, but we assume now that its input R is a positive *bounded* integer signal. Since all signals have now finite domains nothing really new happens compared to the elementary case of "pure" SIGNAL, for instance we may translate bounded integers into vectors of booleans. However we find it more convenient to use codings that are tightly tailored to each of these domains, since the resulting coding will generally be more compact, thus memory saving and efficient calculations should result; in particular, the very efficient formal calculi developed for $\mathcal{F}_3$ [20, 11] generalize to any $\mathcal{F}_p$.

Thus the boolean signal (ZN=1) is encoded using $\mathcal{F}_3$ as before. Similarly, the finite field $\mathcal{F}_p$ of integers modulo p can be used with p large enough to encode the integer signals R, N, ZN with the following mapping:

$$\text{absent} \leftrightarrow 0, \ 1 \leftrightarrow 1, \cdots, \ p-1 \leftrightarrow p-1$$

In this coding, clocks of integer signals are recovered as follows

$$\text{R absent} \leftrightarrow r^{p-1} = 0 \ , \ \text{R present} \leftrightarrow r^{p-1} = 1$$

¿From these remarks, the coding of the UPSAMPLING program follows (we denote by B the boolean signal (ZN=1)):

$$n = r + (1 - r^{p-1})(zn - 1) \quad \text{(i) first instruction}$$

$$\begin{aligned} \nu' &= (1 - n^{p-1})\nu + n \\ zn &= n^{p-1}\nu \end{aligned} \quad \text{(ii) second instruction} \quad (14)$$

$$\begin{aligned} b &= zn^{p-1}\left(1 + (zn-1)^{p-1}\right) \quad \text{(iii) B as a function of ZN} \\ r^{p-1} &= -r - r^2 \quad \text{(iv) last instruction} \end{aligned}$$

However in doing this, a new difficulty appears. Equations (i,ii) are polynomial equations within $\mathcal{F}_p$ and thus can be handled in a way similar to that of pure SIGNAL programs. Unfortunately, equation (iii) is a function mapping some $\mathcal{F}_p$-expression into $\mathcal{F}_3$; similarly, equation (iv) is a function mapping some $\mathcal{F}_3$-expression into $\mathcal{F}_p$. This is again a hybrid coding similar to that of the preceding subsection. We can handle it in two different ways.

1. Since B,ZN,R only are involved in the "hybrid" equations (iii,iv), we may first eliminate ν and N from equations (i,ii). This is equivalent to projecting the dynamical system (i,ii) onto the components (zn, r) only, and is generally performed using elimination techniques in $\mathcal{F}_p$ via efficient algorithms as mentioned before. In our case this yields the unique equation

$$zn' = r + (1 - r^{p-1})(zn - 1)$$

which we handle in combination with (iii,iv) via exhaustive scanning. In this way of doing, efficient algorithms can be called for to project each homogeneous subsystem (here (i,ii)) onto its interfaces to other homogeneous ones (here B,ZN,R). Then the problem reduces to that of standard model checking techniques on the joint behaviour of the set of all such interfaces.

2. Perform first as before the reduction to the interfaces of the homogeneous subsystems. The remaining "hybrid" equations are then approximately solved using the graph method of the preceding subsection. This is a less powerful but likely faster method of compilation.

As sketched on this example, formal calculi can be developed for properties relevant to synchronization, logic, and quantitative timing. In particular, (qualitative and quantitative) *real-time specifications for discrete event systems can be synthesized.*

5 Conclusion

As we have discussed in this paper, Hybrid Systems raise questions of very diverse nature, namely :

(i) significant complexity due to a deep interplay of discrete events handling and analog processing; this complexity makes *programming* of Hybrid Systems a difficult and demanding task;

(ii) designing algorithms for information processing associated with Hybrid Systems; uncertainties and measurement noises should be considered at this point;

(iii) designing control algorithms associated with Hybrid Systems; again, issues of robustness versus performance are central in this activity;

(iv) expressing overall requirements for hybrid systems and verifying the previous two items against these requirements; this is much demanding a task since usual frameworks are much different when different aspects of discrete events, analog information processing, and analog control of Hybrid Systems are concerned.

In designing SIGNAL as a tool for Hybrid Systems, our objective was rather modest : mainly question (i) was considered and we view other questions (ii)–(iv) as being tackled using different techniques. Obviously, providing powerful formal techniques with having point (i) as a main objective will indirectly help the designer for his tasks (ii)–(iv) as well as the implementation of the final computer controlled system. This analysis motivated the way SIGNAL has been conceived and designed. According to these objectives, SIGNAL is a formalism to specify general Hybrid Systems (with discrete but possibly dense time). SIGNAL is currently used to specify and program real-time systems [20] according to the principles of synchrony [5]. We have discussed the expressive power of SIGNAL and have illustrated its expressive generality, mathematical studies are also available in [6] to support this claim.

Also, formal reasoning capabilities are highly useful to help handling issues of combinatorial complexity : they allow to verify consistency of the specification (no contradiction in synchronisation, logic, and data paths), they provide synthesis from specifications as far as the above properties are considered, and, last but not least, they allow modular development of large applications. Accordingly, we have presented an original and general method for deriving formal calculi for Hybrid Systems. The central notion of this method is that of *dynamical graph* and is used in the current SIGNAL compiler to handle synchronization, logic, and data dependencies.

Several directions for future research are currently pursued. Improving the efficiency and power of the formal system that handles $\mathcal{F}_3$–based dynamical graphs is a key issue to fast and efficient compilation [19]. The SIGALI system [13] is an important step toward this direction. Deriving efficient systems to handle bounded integers will open the route to quantitative real-time: a major issue is to handle the tradeoff efficiency/generality of such formal calculi. Finally, as lengthly discussed in [6], the SIGNAL formalism is already very close to models of stochastic processes: adding a *single* instruction to SIGNAL provided us with the SIGNalea extension [9]. SIGNalea is able to specify and handle various probabilistic real-time systems such as queuing networks or uncertain real-time information processing systems.

ACKNOWLEDGEMENT. *Hans Rischel is gratefully acknowledged for providing both a careful reading of and improvements to a former version of the manuscript.*

References

1. K.J. ASTRÖM, A. BENVENISTE (chairman), P.E. CAINES, G. COHEN, L. LJUNG, P. VARAIYA, "Facing the challenge of computer science in the industrial applications of control, a joint IEEE-IFAC project, advance report", IRISA, 1991, see also the two reports by A. B. and G. C. in *IEEE Control Systems,* vol 11, No 4, 87-94, June 1991.

2. K.J. ASTRÖM, B. WITTENMARK, *Adaptive Control,* Addison-Wesley series in Electrical Eng.: Control Eng., Addison-Wesley, 1989.

3. M. BASSEVILLE, A. BENVENISTE Eds, *Detecting changes in signals and systems,* LNCIS vol 77, Springer Verlag, 1986.

4. M. BASSEVILLE, I.V. NIKIFOROV, *Detection of Abrupt Changes - Theory and Applications,* Prentice Hall Information and Systems Sciences Series, Prentice Hall, to appear 1993.

5. A. BENVENISTE, G. BERRY, "Real-Time systems design and programming", *Another look at real-time programming,* special section of *Proc. of the IEEE,* vol. 9 n° 9, September 1991, 1270–1282.

6. A. BENVENISTE, P. LE GUERNIC, Y. SOREL, M. SORINE, "A denotational theory of synchronous communicating systems", *Information and Computation,* vol. 99 n°2, August 1992, 192–230.

7. A. BENVENISTE, P. LE GUERNIC, "Hybrid Dynamical Systems Theory and the SIGNAL Language", *IEEE transactions on Automatic Control,* 35(5), May 1990, pp. 535–546.

8. A. BENVENISTE, P. LE GUERNIC, C. JACQUEMOT, "Synchronous programming with events and relations: the SIGNAL language and its semantics", *Science of Computer Programming,* 16 (1991) 103–149.

9. A. BENVENISTE, "Constructive probability and the SIGNalea language", INRIA res. rep. n° 1532, 1991.

10. B. BUCHBERGER, "Gröbner Bases: An Algorithmic Method in Polynomial Ideal Theory" N.K. Bose (ed.), Multidimensional Systems Theory, 184-232, D. Reidel Publishing Company.

11. L. BESNARD, "Compilation de SIGNAL : horloges, dépendances, environnement", Thesis, IFSIC-IRISA, 1991.

12. F. BOUSSINOT, R. DE SIMONE, "The ESTEREL language", *Another look at real-time programming*, special section of *Proc. of the IEEE*, vol. 9 n° 9, September 1991, 1293-1304.

13. B. DUTERTRE, "Spécification et preuve de Systèmes Dynamiques", PhD thesis, University of Rennes I, IRISA, December 2, 1992.

14. N. HALBWACHS, P. CASPI, D. PILAUD, "The synchronous dataflow programming language LUSTRE", *Another look at real-time programming*, special section of *Proc. of the IEEE*, vol. 9 n° 9, September 1991, 1305-1320.

15. T.A. HENZINGER, Z. MANNA, A. PNUELI, "An Interleaving Model for Real-time", Jersalem Conf. on Information Technology 1990, IEEE Computer Society Press.

16. T.A. HENZINGER, Z. MANNA, A. PNUELI, "Temporal proof methodologies for Real-time systems", POPL'91.

17. T.A. HENZINGER, Z. MANNA, A. PNUELI, "Timed Transition Systems", in J.W. de Bakker, K. Huizing, W-P de Roever, and G. Rozenberg Eds., *Real-Time : theory in practice*, LNCS vol 600, 226-251, Springer Verlag, 1992.

18. L. LAMPORT, "Hybrid Systems in TLA", draft provided in the handouts of *Workshop on Theory of Hybrid Systems*, Tech. Univ. of Denmark, Lyngby, Denmark, 19-21 October, 1992.

19. M. LE BORGNE, A. BENVENISTE, P. LE GUERNIC, "Polynomial Ideal Theory Methods in Discrete Event, and Hybrid Dynamical Systems", in *Proceedings of the 28th IEEE Conference on Decision and Control*, IEEE Control Systems Society, Volume 3 of 3, 1989, pp. 2695-2700.

20. P. LE GUERNIC, T. GAUTIER, M. LE BORGNE, C. LE MAIRE, "Programming real-time applications with SIGNAL", *Another look at real-time programming*, special section of *Proc. of the IEEE*, vol. 9 n° 9, September 1991, 1321-1336.

21. X. NICOLLIN, J. SIFAKIS, S. YOVINE, "From ATP to Timed Graphs and Hybrid Systems", REX workshop "Real-Time, theory in practice", Mook, The Netherlands, June 3-7, 1991.

22. A.P. RAVN, H. RISCHEL, K.M. HANSEN "Specifying and verifying requirements of real-time systems", to appear in *IEEE Trans. on Software Eng.*, Jan. 1993, pp. 41-55.

A Dynamical Simulation Facility for Hybrid Systems

Allen Back, John Guckenheimer and Mark Myers

Center for Applied Mathematics
Cornell University, Ithaca, New York 14850

Abstract. This paper establishes a general framework for describing hybrid dynamical systems which is particularly suitable for numerical simulation. In this context, the data structures used to describe the sets and functions which comprise the dynamical system are crucial since they provide the link between a natural mathematical formulation of a problem and the correct application of standard numerical algorithms. We describe a partial implementation of the design methodology and use this simulation tool for a specific control problem in robotics as an illustration of the utility of the approach for practical applications.

A Definition of Hybrid Systems

The last several decades have witnessed an explosive development in the theory of dynamical systems, much of which is oriented toward equations of the form

$$\dot{x} = F(x, p) \tag{1}$$

where $F : V \subset \mathbf{R}^n \times \mathbf{R}^k \to \mathbf{R}^n$ is a smooth map defined on an open, connected subset of Euclidean space and possibly dependent upon a vector of parameters, $p \in \mathbf{R}^k$. However, many areas of application frequently involve hybrid systems: dynamical systems which require a mixture of discrete and continuously evolving events. Natural examples of such situations are mechanical systems which involve impacts, control systems that switch between a variety of feedback strategies, and vector fields defined on manifolds described by several charts. In this note we present a uniform approach for representing hybrid systems which applies to a wide variety of problems and, in addition, is particularly well suited for computer implementation. Simulation tools which exploit the approach that we describe are being developed in a computer program called **dstool** [1].

Our theoretical objective may be viewed as the natural extension of Equation (1) in two fundamental ways: First, we wish to generalize the underlying domain of F, denoted V above, to include open sets composed of many components where the vector field may vary discontinuously from component to component. Second, we want to accomodate the situation that discrete events depending upon the phase space coordinates or time may occur in the flow of the dynamical system. It is worth pointing out that these goals represent fundamental changes in the theoretical character of the dynamics. By satisfying the first requirement, we greatly increase the class of systems under study, but must generalize the underlying theory that supports the simulation and analysis of such systems. The second extension demands a formulation that treats the discrete event components, or some representation of them, as equal members of the underlying phase space with the continuous parts.

Assume the problem domain may be decomposed into the form:

$$V = \bigcup_{\alpha \in I} V_\alpha$$

where I is a finite *index set* and V_α is an open, connected subset of $\mathbf{R}^n$. We shall refer to each element in this union as a *chart*. Each chart has associated with it a (possibly time-dependent) vector field, $f_\alpha : V_\alpha \times \mathbf{R} \to \mathbf{R}^n$. Notice that the charts are not required to be disjoint. Moreover, for $\alpha, \beta \in I$ we do not require continuity, or even agreement, of the vector fields on the intersection set $V_\alpha \cap V_\beta$. We introduce further structure by requiring that for each $\alpha \in I$, the chart V_α must enclose a *patch*, an open subset U_α satisfying $\overline{U}_\alpha \subset V_\alpha$. The boundary of U_α is assumed piecewise smooth and is referred to as the *patch boundary*. Together, the collection of charts and patches is called an *atlas*.

To implement this model, a concrete representation of the patch boundaries must be selected. For each $\alpha \in I$ we define a finite set of *boundary functions*, $h_{\alpha,i} : V_\alpha \to R$, $i \in J_\alpha^{bf}$, and real numbers called *target values*, $C_{\alpha,i}$, for $i \in J_\alpha^{bf}$ that satisfy the condition: For $x \in V_\alpha$ where $\alpha \in I$, we require

$$x \in U_\alpha \text{ if and only if } h_{\alpha,i}(x) - C_{\alpha,i} > 0 \text{ for all } i \in J_\alpha^{bf} \ .$$

Thus, a patch is to be considered the domain on which a collection of smooth functions are positive. The boundary of a patch is assumed to lie within the set:

$$\bigcup_{i \in J_\alpha^{bf}} h_{\alpha,i}^{-1}(\{C_{\alpha,i}\}) \qquad \text{for } \alpha \in I \ .$$

We remark that the target values $C_{\alpha,i}$ may depend on the initial point of the trajectory segment as well as the vector of system parameters. This dependence enables a change of state to occur after a specified time has elapsed.

Since ultimately we work with floating point arithmetic, we shall specify an explicit tolerance for the values of $h_{\alpha,i}$ so that the operational definition of the boundary is given by:

$$x \in \partial U_\alpha \text{ if } |h_{\alpha,i}(x) - C_{\alpha,i}| < \epsilon_{\alpha,i} \text{ for some } i \in J_\alpha^{bf} \ .$$

Note, however, that points not in ∂U_α may satisfy this condition, and it may need to be supplemented with inequalities describing which portions of a boundary are described by a level set of $h_{\alpha,i}$. Conceptually, the evolution of the system is viewed as a sequence of trajectory segments where the endpoint of one segment is connected to the initial point of the next by a transformation. It follows that time may be divided into contiguous periods, called *epochs*, separated by instances where *transition functions* are applied at times referred to as *events*. The transition functions are maps which send a point on the boundary of one patch to a point in another (not necessarily different) patch in the atlas. The "best" definition of the transition functions is still not apparent. Events only occur at "exit" points from the boundary of a patch, so it is not essential that transition functions be defined at other boundary points of a patch. We define subsets S_α of the patch boundaries, called *transition sets* and transformations $T_\alpha : S_\alpha \to V \times I$ satisfying $\pi_1(T_\alpha(x)) \in \overline{U}_{\pi_2(T_\alpha(x))}$ where π_1, π_2 are the natural projection of $V \times I$ onto its factors. Thus $\pi_1(T_\alpha)$ is the "continuous" part and $\pi_2(T_\alpha)$ is the "discrete" part of a transition function (mapping indices). We require further that there is a finite partition $\bigcup W_{\alpha,j}$ of S_α with the properties:

i) $\quad \pi_2 \circ T_\alpha$ is constant on each $W_{\alpha,j}$

ii) $\quad \phi_{\alpha,j} = \pi_1 \circ T_\alpha\big|_{W_{\alpha,j}}$ is the restriction of a smooth
$\qquad$ mapping defined in a neighborhood of $\overline{W}_{\alpha,j}$.

Figure (1) illustrates the definitions presented so far. Within this framework, an orbit in the flow of a hybrid dynamical system which begins at a time t_0 and terminates at t_f may be completely described. A *trajectory* for Equation (1) is a curve $\gamma : [t_0, t_f] \to V \times I$ together with an increasing sequence of real numbers $t_0 < t_1 < \cdots < t_m = t_f$ that satisfies three properties:

- Each time interval (t_i, t_{i+1}) corresponds to an epoch and there exists a designated α so that $\gamma(t)$ lies entirely in $\overline{U}_\alpha \times \{\alpha\}$ for all $t \in (t_i, t_{i+1})$.
- For $t \in [t_i, t_{i+1})$ and the unique α specified above, $t \to \pi_1(\gamma(t))$ is an integral curve of the vector field f_α.
- $\lim_{t \to t_{i+1}^-} \pi_1(\gamma(t)) = y$ exists, $y \in S_\alpha$ and $T_\alpha(y) = \lim_{t \to t_{i+1}^+} \gamma(t)$.

A hybrid system will be called *complete* if every trajectory can be continued for $t \to \infty$. This implies that every point of ∂U_α reached by a trajectory lies in the transition set S_α of U_α.

One of our goals in the definition of hybrid systems has been to retain the applicability of as many "standard" numerical algorithms as possible. Thus, a design objective has been to ensure that algorithms for tasks such as numerical integration and root finding can be used without modification. This explains the seemingly pedantic distinction that we make between charts and patches. Consider the task of locating the intersection of a trajectory segment with a patch boundary. If an integration step carries a trajectory outside of a patch, then interpolation methods to find the point of intersection may require that we are able to evaluate functions defining the vector field and the patch boundary at the computed point outside the patch. If we check that an integration step remains inside the chart containing the patch, then we are able to apply standard algorithms. If a computed point lies outside the chart, then remedial action like diminishing a step size can be taken.

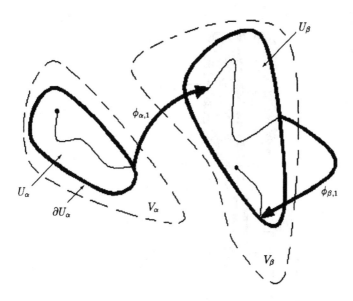

Fig. 1. Schematic diagram illustrating the basic components of a hybrid dynamical system.

An Example: Raibert's Hopper

A partial implementation of the approach for modeling hybrid dynamical systems presented above has been accomplished in the **dstool** program [1]. This Section describes results for a particular control problem involving the dynamical behavior and stability of a "hopping" robot, constrained to move in a single (vertical) dimension. The study, both theoretical and experimental, of motion and balance in legged robots has a long history [4]; the model we describe here was proposed recently by Koditschek [3] as a partial abstraction of a physical machine built and tested by Raibert [5].

Raibert's robot consists of two main components, a *body* which houses the control mechanism and a compressible *leg* upon the base of which all of the impact dynamics of the robot with its environment occurs. The leg is constructed as a pneumatic cylinder whose pressure is subject to closed feedback control by opening and closing valves connected to two reservoirs of compressed gas. A schematic cartoon showing the important features of the device relevant to Koditschek's model is shown in Figure (2). To activate the physical device, the robot is dropped a short distance above a flat surface and, for some values of the control parameters and after an initial transient phase, the vertical motion of the robot becomes periodic. Indeed, this behavior is remarkably stable under perturbation; following a gentle impact, the motion of the machine will autonomously return to its periodic "hopping" state.

The dynamics of the hopping robot may be described in terms of four distinct *phases*. Let y denote the height of the robot body relative to the surface

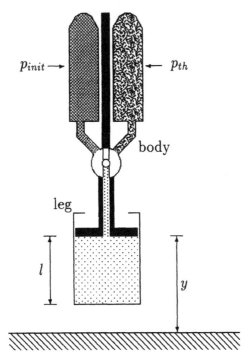

Fig. 2. Schematic of a simplified model for Raibert's hopping robot.

above which the mechanism is dropped. The device is in its *flight* phase whenever the leg is not in contact with the surface, and the governing dynamics are presumed to be Newton's equations for the one-dimensional motion of a particle in a gravitational field with constant acceleration, g. During this phase, the leg is kept at its fully extended length, l, by a gas pressure inside the pneumatic cylinder, p_{init}, maintained from one of the gas reservoirs. The robot enters the *compression* phase as soon as the leg makes contact with the surface, and the weight of the body begins to compress the gas in the pneumatic cylinder. This compression is modeled by a nonlinear spring, whose force is inversely proportional to the robot height with a spring constant, η, related to the initial gas pressure, p_{init}. Associated with the action of the nonlinear spring is a simple mechanical damping force with coefficient of friction denoted by γ. The *thrust* phase begins at the point of maximal leg compression or, equivalently, at the minimum value of the y coordinate achieved in the compression phase. Let t_b denote the relative time at which this minimum occurs. At the onset of the thrust phase, a valve opens to admit gas from the second reservoir into the pneumatic cylinder at a pressure p_{th}, which continues for a <u>fixed</u> period of time, δ, and

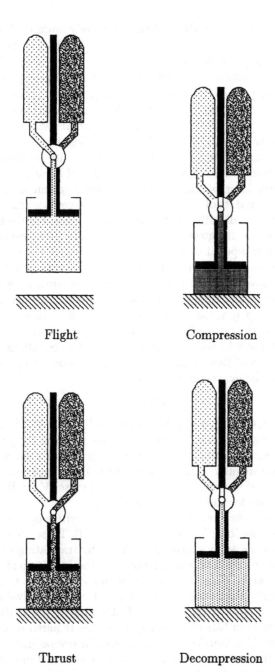

Flight   Compression

Thrust   Decompression

Fig. 3. Sequence of phases which comprise the robot's hopping dynamics.

exerts a constant force, τ, which accelerates the body upward. The thrust phase ends at the time $t_b + \delta$ corresponding to a body height y_{et}. At this point the valve closes, and initiates the final *decompression* phase where the robot body continues to move upward under the action of the nonlinear pneumatic spring. The new nonlinear spring constant, τy_{et}, in effect for this phase is a function of both the pressure of the gas injected into the cylinder during the thrust phase and the height at which the thrust phase was completed. Once the height of the body reaches the length of the fully extended leg, the decompression phase ends and the mechanism reenters the flight phase. Figure (3) illustrates the four phases which comprise the behavior of the dynamical model.

The crucial step in preparing a hybrid dynamical system for numerical simulation and analysis in **dstool** involves identifying the components of the model in terms of the formalism described in the previous Sections. For the hopping robot, it is natural to express the governing dynamics as a two-dimensional vector field on a phase space composed of four patches, each patch corresponding to one of the phases shown in Figure (3). Thus, the index set, $\{1, 2, 3, 4\}$, corresponds to the flight, compression, thrust and decompression patches, respectively. The vector field is non-autonomous and dependent upon a six-dimensional vector of parameters, $p = (l, g, \eta, \gamma, \tau, \delta)$, whose components were described earlier. The definition of the patch regions, U_α, are derived by considering the simple required conditions which must be imposed upon $(y, \dot{y}, t)$. The equations of motion follow from application of Newtonian force balance laws on each patch. The model which results may be expressed as,

$$\ddot{y} = f(y, \dot{y}, t; \alpha, p) \tag{2}$$

where, $\alpha \in \{1, 2, 3, 4\}$, $(y, \dot{y}, t) \in \mathbf{R}^+ - \{0\} \times \mathbf{R}^2$ and $p \in \mathbf{R}^6$. Table (1) summarizes the important information required to characterize the patches together with the equations of motion which are in effect within each patch. We note that f_2 and f_4 are singular whenever $y = 0$ and all the equations are physically unreasonable for $y < 0$; thus, we may take the open set $V_\alpha = \mathbf{R}^+ - \{0\} \times \mathbf{R}$ as an appropriate chart for each $\alpha \in \{1, 2, 3, 4\}$.

Recall that the definition of a patch was the positive open set for a collection of boundary functions: $h_{\alpha,i}(y, \dot{y}, t; p) - C_{\alpha,i} > 0$. These boundary functions are obtained from the set notation of Table (1) by translating the definition of intervals into statements involving functions of the independent variables y, $\dot{y}$ and t and target values, $C_{\alpha,i}$. It is important to note that this translation is not unique. For example, the relations $y > l$ and $y - l > 0$ are equivalent expressions regarding the same mathematical object. **Dstool** allows the analyst to take advantage of repetition in lists of boundary functions required to define a hybrid dynamical system. If a function, h, can be used in several instances to express boundary functions, with or without different target constants $(C_{\alpha,i})$, a *single block of code* can be used to define the boundary condition in each case. Thus, the user may minimize the work required to install the model by reuse of the definitions of boundary functions. As a concrete illustration, Table (2) provides a list of boundary functions and target constants for the legged robot, corresponding to the open sets defined in Table (1). While there are nine boundary

Phase	f	Patch
Flight	$f(y,\dot{y},t;1,p) = -g$	$U_1 = \{(y,\dot{y})|y > l\}$
Compression	$f(y,\dot{y},t;2,p) = \eta/y - \gamma\dot{y} - g$	$U_2 = \{(y,\dot{y})|y < l, \dot{y} < 0\}$
Thrust	$f(y,\dot{y},t;3,p) = \tau - \gamma\dot{y} - g$	$U_3 = \{(y,\dot{y})|y < l, \dot{y} > 0,$ $t_b < t < t_b + \delta\}$
Decompression	$f(y,\dot{y},t;4,p) = \tau y_{et}/y - \gamma\dot{y} - g$	$U_4 = \{(y,\dot{y})|y < l, \dot{y} > 0,$ $t > t_b + \delta\}$

Table 1. Definition of the vector field components for Raibert's "hopping" robot.

conditions required, only six C-language software functions must be written to accomodate this list. Admittedly, for the purposes of this example the savings is small but in larger dynamical systems, especially those composed of many similar subcomponents, the advantages gained in code size and error reduction may be significant.

Phase	Index Key (α)	Index (i)	$h_{\alpha,i}$	$C_{\alpha,i}$
Flight	1	1	$h_{1,1}(y,\dot{y},t;p) = y$	l
Compression	2	1	$h_{2,1}(y,\dot{y},t;p) = -y$	$-l$
		2	$h_{2,2}(y,\dot{y},t;p) = -\dot{y}$	0
Thrust	3	1	$h_{3,1}(y,\dot{y},t;p) = -y$	$-l$
		2	$h_{3,2}(y,\dot{y},t;p) = \dot{y}$	0
		3	$h_{3,3}(y,\dot{y},t;p) = -t$	$-(t_b + \delta)$
Decompression	4	1	$h_{4,1}(y,\dot{y},t;p) = -y$	$-l$
		2	$h_{4,2}(y,\dot{y},t;p) = \dot{y}$	0
		3	$h_{4,3}(y,\dot{y},t;p) = t$	$(t_b + \delta)$

Table 2. Boundary functions for the patches used in the model for Raibert's "hopping" robot.

Together Tables (1) and (2) contain the information required to install the hybrid model for Raibert's legged robot into the **dstool** program. For details regarding the installation of dynamical systems into **dstool** see [1, 2]. For this example, eight C-language procedures that provide information required by the program must be written by the **dstool** user. The routines may be described as follows:

- **model_init()**: This procedure is called when the dynamical system is first initialized to provide basic information such as the dimension and names of the variables, Jacobian information, and plotting ranges.
- **info()**: A routine is required to inform the data structure object manager of the name of each procedure used to define the components of the atlas and transition functions.
- **equations()**: The vector field equations, $f(y, \dot{y}, t; \alpha, p)$ are specified in this procedure for each chart in the atlas.
- **patch_def()**: Information which defines each patch, U_α, is provided by this routine including target constants and tolerances to be used.
- **bndry_fncts()**: Each boundary function which participates in the definition of the patches contained in the atlas is specified by this procedure.
- **trans_fncts()**: The set of transition functions to be applied at the various boundaries of the patches is collected together in this routine.
- **traj_init()**: At the beginning of a trajectory segment, the user may require certain conditions or parameters to be checked and used to set characteristics of the model before propagation begins. This routine provides this facility.
- **size()**: One important responsibility expected of the data structure object manager is the allocation and deallocation of system memory. To fulfill this task, the manager must be provided with information concerning the size of each element in the atlas. This routine may be called on demand to provide these dimensions.

Once these routines are installed in **dstool**, the program may be used to propagate trajectories using standard versions of integration algorithms. Collections of trajectories generated using different initial conditions may be displayed together to form phase portraits, which may be used to elucidate global structures and bifurcations which vary with system parameters. Figure (4) displays a periodic orbit for the hopping robot. The upper plot shows the phase portrait corresponding to thrust force, τ, of 41.86 units. The trajectory sense indicated is clockwise, with the robot height, y, displayed on the abscissa and the velocity along the ordinate. The four mechanical phases which contribute to the cycle are clearly evident: The flight phase occurs above the point $y = l$, followed by the compression of the pneumatic spring, a short period of thrust and finally the full expansion of the robot leg. The lower plot in this Figure displays the height of the robot body as a function of time. Each revolution of the limit cycle corresponds to a single robot "hop", and the maximum height attained is constant. This is an example of an *attracting* limit cycle; if initial conditions are chosen which do not lie on the periodic orbit and integrated forward in time the resulting orbit will approach the cycle asymptotically in the limit. Clearly, if one presumes this model to represent the dynamics of the physical device, the remarkable stabil-

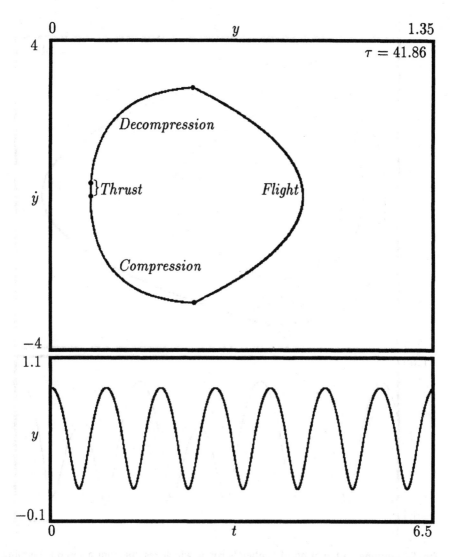

Fig. 4. Periodic "hopping" of Raibert's robot. Figure (4a) shows the two-dimensional phase portrait, the coordinate y along the ordinate and $\dot{y}$ on the abcissa. The height of the body, y, plotted as a function of time is shown in Figure (4b).

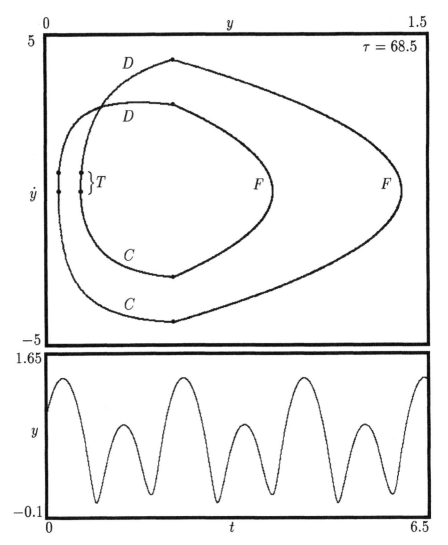

Fig. 5. Limping gait motion of Raibert's robot which results from a period doubling bifurcation. Figure (5a) shows the two-dimensional phase portrait, the coordinate y along the ordinate and $\dot{y}$ on the abcissa. The labels F, C, T and D denote the flight, compression, thrust and decompression phases of the mechanical device. The height of the body, y, plotted as a function of time is shown in Figure (5b).

ity of Raibert's hopper may be interpreted as a direct consequence of the large domain of attraction for this limit cycle.

As the thrust parameter, τ, is increased a *period-doubling* bifurcation of this periodic orbit occurs. Figure (5) shows an example of a period-2 orbit which results at a value of $\tau = 68.5$. Again, the upper plot shows a phase portrait with y and $\dot{y}$ displayed along the axes. Notice that a single limit cycle consists of two components, each component involving one compete set of the four mechanical phases. Moreover, one subcomponent of the limit cycle may be characterized by the observation that the maximum height attained during its flight phase is only about half that of the second subcomponent. A physical device displaying this behavior would take a short hop, followed by a higher hop and repeat this sequence indefinitely. Once again, this periodic orbit can be shown to be stable with a large domain of attraction, so that if such a physical device were perturbed - bumped - after an initial transient period it would return to this two-stage hopping behavior. The presence of this period-doubling bifurcation has been previously reported by Vakakis and Burdick [6] in numerical studies and Koditschek [3] in simplified analytical approximations. These period-two trajectories are confirmed by experimental observations by Raibert, who coined the term "limping gait" to characterize the corresponding motions.

Conclusions

We have presented a methodology for studying dynamical systems whose phase spaces involve both continuous and discrete components, representing a large and important class of models. A software program for studying dynamical systems has been enhanced to accomodate hybrid models of the genre described, and this tool has been shown to reproduce established behavior for a selected problem in robotics. Further development of software tools for investigating hybrid systems is underway and this initial implementation supports these efforts in two ways. First, this software can serve as a platform for evaluating the methodology presented here on problems of current research interest. The representation of hybrid systems may then be refined and extended. Second, the study of such systems will undoubtedly require new numerical algorithms which can be developed and carefully tested using programs based on the initial efforts described in this paper. We briefly describe two directions of current activity.

In sophisticated hybrid dynamical systems, the geometry of the various components which comprise the phase space and boundary sets becomes important. We are interested in extending the model described in the first Section to accomodate the complex relationships which result near the boundaries of the patches. We foresee the need to further decompose the patch boundaries into sets where the action of the transition functions may vary, according to requirements dictated by their relative geometry or significance as an entity in the model. Within this decomposition, conditions for discrete state changes might result from combinations of Boolean functions based on both discrete state information and phase space coordinates. Thus, the appropriate action of the transition functions near the intersection of patch boundaries or for a trajectory nearly parallel to a boundary, for example, can be discriminated.

The body of literature describing numerical algorithms for vector fields represent by Equation (1) is large and growing rapidly. In some cases, adapting these procedures for use with hybrid dynamical systems should require little more than safeguarding their behavior near patch boundaries; locating fixed points using Newton-type methods is such an example. In other cases, extension is more difficult. Consider, for example the problem of approximating a periodic orbit in the phase space of a hybrid vector field that spans more than a single patch. Simulation may be used, as it was in the Example problem above, to search for an initial condition in the phase space that returns to the starting condition after some fixed period. Another method widely used for Equation (1) is to solve an appropriate two-point boundary value problem using parallel shooting or collocation methods. However, in the case of a hybrid system on multiple patches, this naturally yields a multipoint boundary-value problem without the smoothness properties often assumed at the interior points. Thus, the extension and development of numerical procedures for hybrid systems poses new and interesting problems for analysis.

Acknowledgements

The authors acknowledge Dr. Anil Nerode of Cornell University and the Mathematical Sciences Institute whose suggestions provided the original impetus for this work. We also thank Dr. Daniel Koditschek at Yale University for a number of valuable discussions concerning the dynamics of Raibert's hopper. This research was partially supported by the National Science Foundation and the Defense Advanced Research Projects Agency.

References

1. Back, A., Guckenheimer, G., Myers, M., Wicklin, F. and P. Worfolk, 'dstool: Computer Assisted Exploration of Dynamical Systems,' *AMS Notices*, April, 1992.
2. Guckenheimer, J., Myers, M.R., Wicklin, F.J. and P.A. Worfolk, 'dstool: A Dynamical System Toolkit with an Interactive Graphical Interface', Center for Applied Mathematics, Cornell University, Ithaca, NY, 1991.
3. Koditschek, D.E. and M. Buhler, 'Analysis of a Simplified Hopping Robot,' to appear in the *International Journal of Robotics Research*, 1991.
4. Raibert, M.H., *Legged Robots That Balance*, MIT Press, Cambridge, Mass., 1986.
5. Raibert, M.H. and H.B. Brown, Jr., 'Experiments in Balance with a 2D One-Legged Hopping Machine,' *ASME Journal for Dynamical Systems, Measurement and Control*, Vol. 106, 1984, pps. 75-81.
6. Vakakis, A.F. and J.W. Burdick, 'Chaotic Motions in the Dynamics of a Hopping Robot,' *Procedings of the IEEE International Conferance on Robotics and Automation*, Cincinnati, OH, 1990, pps. 1464-1469.

Event Identification and Intelligent Hybrid Control

Michael Lemmon *, James A. Stiver, and Panos J. Antsaklis

Department of Electrical Engineering
University of Notre Dame
Notre Dame, IN 46556, USA

Abstract. Hybrid dynamical systems consist of two types of systems, a continuous state system called the *plant* and a discrete event system called the *supervisor*. Since the plant and supervisor are different types of systems, an interface is required to facilitate communication. An important issue in the design of hybrid control systems is the determination of this interface. Essentially, the interface associates logical *symbols* used by the supervisor with nonsymbolic *events* representative of the plant's behaviour. This chapter discusses a method for learning a hybrid system interface where symbols and events are bound in a way which is compatible with the goal of plant stabilization. The method is called *event identification* and provides an on-line method for adapting hybrid dynamical systems in the face of unforseen plant variations.

1 Introduction

Hybrid dynamical systems provide a convenient tool for the analysis and design of supervisory control systems. A supervisory control system arises when a discrete event system is used to supervise the behaviour of a plant by the issuance of logical control directives. The hybrid system framework shown below in figure (1) clearly illustrates this architecture. The specific architecture illustrated below is based on the model used in [Antsaklis 1993] which appears in this volume. The notational conventions adopted in this chapter will be found in [Antsaklis 1993].

In figure (1), note that an interface is included to facilitate communication between the two different types of systems. This interface consists of two subsystems known as the *generator* and *actuator*. The generator transforms the plant's observation vector $z \in \Re^p$ into a symbol $\tilde{z}$ which is drawn from an alphabet $\tilde{Z}$. The actuator transforms control symbols $\tilde{r}$ output by the supervisor into control vectors $r \in \Re^m$ which are used by the plant. The control symbols are drawn from an alphabet $\tilde{R}$.

In view of the preceding discussion, it is apparent that the supervisor is used to control the plant. It is also apparent that the supervisor is a symbol manipulation system whose logical symbols have "meanings" grounded in nonsymbolic

---

* The partial financial support of the National Science Foundation (IRI91-09298 and MSS92-16559) is gratefully acknowledged

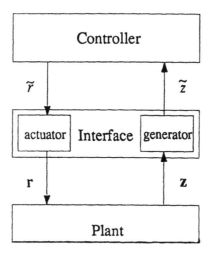

Fig. 1. Hybrid Dynamical System Architecture

external "events". In other words the supervisor's symbols can be assigned interpretations which are generally representative of important categories of plant behaviour. A plant's behaviour refers to its state trajectory, $x(t)$. Behavioural categories are therefore collections of trajectories with some "common" property. State trajectories of autonomous systems, however, are determined by the initial state. This observation suggests that a class or category of behavioural trajectories can be adequately represented by a subset of the plant's state space. The term *event* is therefore used to denote a distinguished subset of the plant's state space. The term *symbol* is used to denote the logical label associated with an event. The association of a symbol with an event will be referred to as a symbol/event *binding*.

For example, a certain set of temperatures and pressure measurements may be indicative of a potential system failure. In this case, we would like to associate the "nonsymbolic" measurements with a "symbolic" label called "FAILURE". Therefore the supervisor's computations represent the manipulation of abstractions about the plant's current state. The use of such high-level abstractions (representations) of system state to control the system is sometimes called "intelligent" control.

This notion of intelligence, however, is singularly unsatisfying. Note that the action of the controller relies on the prior interpretation assigned to the plant and control symbols. Therefore the "intelligence" of the system lies in the interpretation of these symbols. The "intelligent" choices, however, were made by the human designer, not by the machine. Therefore it is the designer, rather than the machine which is intelligent. This same fundamental argument has been previously leveled against production based inference as a model for human cognition [Searle 1984]. Essentially, it asserts that the "blind" manipulation of symbols is not sufficient to render a system intelligent.

The reduction of the plant to an effective DES plant model, represents one way of designing so-called "intelligent" controllers. This approach to design was discussed briefly in a companion chapter [Antsaklis 1993]. However, this design approach represents the precise disembodiment of controller symbol and event, which was immediately discussed above. In this regard, an approach to supervisory control which assumes a priori symbol/event bindings cannot be considered an "intelligent" control system. Intelligence will only arise when the system is capable of determining its own event/symbol bindings. This requires that any intelligent system solve what may be called the *event identification* problem. The relationship between this "event identification" problem and more traditional issues in artificial intelligence such as the symbol grounding problem [Harnad 1990] is discussed in one of the closing sections of this chapter.

Whether or not the symbolic manipulations of a computational system constitute intelligence can, no doubt, be argued endlessly. There is, however, a much more pragmatic reason for considering such a system undesirable. If we consider those applications for which supervisory control systems are intended, it is immediately apparent that supervision is meant for complex and and unpredictable systems. For such systems, prior plant knowledge or complete plant knowledge may be impossible. This means that "events" which are defined with respect to an assumed plant structure, may change unexpectedly. If this is the case, then it is well within the realm of possiblity for our not-so intelligent supervisor to happily chunk away and produce of stream of nonsensical control symbols. The reason this occurs, of course, is because the supervisor really doesn't understand the significance of the symbols it is manipulating. If we wish to call this un-intelligent processing, that is fine. The end result is the same, however, a system whose autonomy is limited by the designer's initial assignment of symbol bindings. Therefore, a more pragmatic reason for requiring event identification of "intelligent" control systems is that it will undoubtably lead to increased system autonomy. The issue of autonomy in intelligent control was discussed thoroughly in the introduction. It is the need for such autonomy that really motivates the requirement for event identification in hybrid systems. As will be pointed out in one of the closing sections, this ability is also consistent with notions of "intelligence" stemming the symbolic and subsymbolic AI communities [Chalmers 1992].

The preceding discussion therefore indicates that an important problem in hybrid system control is the identification of events. How does one choose events which are consistent with the desired control objectives? Is it possible for the system to identify its own set of "optimal" events. This chapter presents one example of how such event identification can be accomplished. The problem of event identification can be viewed in a variety of contexts. For example, consider a system which has the general architecture shown in figure (1). Assume that the plant uses a collection of control policies, so that the plant's differential equation has the form

$$\dot{\mathbf{x}} = \sum_{i=1}^{m} r_i f_i(\mathbf{x}) \tag{1}$$

where $\mathbf{x}$ is the state space and $\mathbf{r}$ is an m-vector of "coordination" coefficients, r_i. The individual vector fields can be seen as "control policies" which are co-ordinated through the specification of the vector $\mathbf{r}$. In figure (1), it can now be seen that the binding of plant/control symbols with subsets of the state space determines the behaviour of this system. One side of the problem, involves determining plant symbol and event bindings which allow a deterministic or quaside-terministic plant DES (see [Antsaklis 1993]). The solution of this problem yields a design for the interface's event generator. A system which can learn a set of bindings consistent with deterministic behaviour will have gone a long way in learning to control itself. Another side of the problem focuses on learning the symbol bindings between the control symbols, $\tilde{r}$, and vectors $\mathbf{r}$. The solution to this problem yields a design for the interface's actuator. System which are capable of forming the event/symbol bindings consistent with with control objectives (i.e. determinism or controllability) will go a long way towards making truly "intelligent" control systems exhibiting a high degree of autonomy.

The following sections provide a specific example of a hybrid system which can automatically learn event/symbol bindings. The example system is a variable structure system and the symbol bindings are learned with regard to invariant sets generated by the plant's dynamics. Early work on this was done in [Lemmon 1992] and refined in [Lemmon 1993a] with regard to the binding of plant symbols. Considerations on the binding of control symbols were discussed in [Lemmon 1993b]. In all of this work it was shown that bindings could be learned in finite time with a sample complexity that scales in a polynomial manner with plant complexity. The remainder of this section is organized as follows. Section (2) discusses the example problem which is referred to in this section as the invariant subspace identification (ISID) problem. Section (3) introduces the learning algorithm. This algorithm consists of two procedures called the oracle and the update procedure. These procedures are derived in sections (4) and (5). The convergence and complexity properties of this learning procedure are discussed in section (6). An example of this algorithm's use is illustrated in section (7). The importance of the following example is that it provides a concrete example of a hybrid system which learns to "identify" its own events in a computationally efficient manner. Some issues and concerns associated with this example are discussed in section (8). The presented algorithm also provides a novel perspective on the relationship between intelligence and control. The central issue in this perspective is the so-called "symbol grounding" problem [Harnad 1990]. This novel perspective on "intelligence" will be discussed in section (9).

2 Invariant Subspace Identification (ISID) Problem

The hybrid system under consideration is assumed to have a very special form. Specifically, it will be assumed that the plant's dynamics are represented by the following differential equations.

$$\dot{\mathbf{x}} = \sum_{i=1}^{2} r_i f_i(\mathbf{x}) \tag{2}$$

where $\mathbf{x} \in \Re^n$ is the state vector, f_1 and f_2 are smooth mappings from $\Re^n$ onto $\Re^n$. It is also assumed that the vector $\mathbf{r} = (r_1, r_2)^t$ takes on the values of $(0,0)^t$, $(1,0)^t$ or $(0,1)^t$. The resulting plant is therefore a variable structure system [Utkin 1977].

The interface generator for this hybrid system will be formed with respect to two events, c^+ and c^-.

$$c^+ = \left\{ \mathbf{x} \in \Re^n : \mathbf{s}^t \mathbf{x} > -|\alpha| \right\} \tag{3}$$

$$c^- = \left\{ \mathbf{x} \in \Re^n : \mathbf{s}^t \mathbf{x} < |\alpha| \right\} \tag{4}$$

where α is a real number and $\mathbf{s}$ is an n-dimensional real vector. These two events form overlapping linear halfspaces and will be called *covering events*. The symbols $\tilde{c}^+$ and $\tilde{c}^-$ are bound with the events c^+ and c^-, respectively. The covering generates three distinct plant events which are represented by the symbols, $\tilde{z}_1$, $\tilde{z}_2$, and $\tilde{z}_3$. The plant state either lies in the deadzone formed by the intersection of c^+ and c^- or else it lies only in one of the halfspaces. Therefore the plant symbols issued by the generator will be either $\tilde{z}_1 = \{\tilde{c}^+, \tilde{c}^-\}$, $\tilde{z}_2 = \{\tilde{c}^+\}$, or $\tilde{z}_3 = \{\tilde{c}^-\}$.

It will be assumed that the supervisor is an identity mapping which simply passes on the plant symbol to the interface actuator. The actuator will then associate each symbol with a control vector $\mathbf{r}$ as follows

$$\mathbf{r} = \begin{cases} (1\,0)^t & \text{if } \{\tilde{c}^+\} \\ (0\,0)^t & \text{if } \{\tilde{c}^+, \tilde{c}^-\} \\ (0\,1)^t & \text{if } \{\tilde{c}^-\} \end{cases} \tag{5}$$

These assumptions for the plant, actuator, generator, and supervisor yield a hybrid system which is, essentially, a variable structure control system. The dynamics of the plant under the supervisor's control are represented by the following set of switching differential equations

$$\dot{\mathbf{x}} = \begin{cases} f_1(\mathbf{x}) & \text{if } \mathbf{s}^t \mathbf{x} < -|\alpha| \\ 0 & \text{if } |\mathbf{s}^t \mathbf{x}| < |\alpha| \\ f_2(\mathbf{x}) & \text{if } \mathbf{s}^t \mathbf{x} > |\alpha| \end{cases} \tag{6}$$

The nature of the system shown in equation (6) is such that the system's structure changes discontinuously as the state crosses over surfaces defined by the equation $\mathbf{s}^t \mathbf{x} = \pm \alpha$. Such a surface is commonly called a *switching surface*.

One objective in variable structure control is to drive the plant state onto the hyperplane, $H_\mathbf{s}$, and keep it in the neighborhood of that surface. Define the surface $H_\mathbf{s}$ as

$$H_\mathbf{s} = \left\{ \mathbf{x} \in \Re^n : \mathbf{s}^t \mathbf{x} = 0 \right\} \tag{7}$$

This is a hyperplane passing through the origin, with normal vector **s**. The neighborhood of this surface is represented by the set formed by intersecting the two events c^+ and c^-. Since the control objective is to drive the system state into $c^+ \cap c^-$ and keep it there, it is important that this set be an attracting invariant set with respect to the controlled plant's dynamics as shown in equation (6). The $H_\mathbf{s}$ which is invariant with respect to the plant's dynamics will be referred to as a *sliding mode*.

An invariant subset with respect to a transformation group $\{\Phi_t\}$ is defined as follows.

Definition 1. The set $H \subset \Re^n$ will be a *Φ-invariant* of the transformation group $\{\Phi_t : \Re^n \to \Re^n\}$ if and only if for any $\mathbf{x} \in H$, $\Phi_t(\mathbf{x}) \in H$ for all $t > 0$.

Of more interest are sets which are attracting invariants of the flow.

Definition 2. The set $H \subset \Re^n$ will be an *attracting Φ-invariant* of the transformation group $\{\Phi_t : \Re^n \to \Re^n\}$ if and only if for any $\mathbf{x} \in H$, there exists a finite $T > 0$ such that $\Phi_t(\mathbf{x}) \in H$ for all $t > T$.

In our example, the transformation groups are the family of transition operators generated by the differential equations (6). These transformations, Φ_t, represent a collection of automorphisms over the state space which are sometimes called the "flow" of the dynamical system.

Unfortunately, not all choices of **s** will leave the target event, $c^+ \cap c^-$, invariant. Those hyperplanes which yield invariant target events can be determined directly from the set of vector fields $\{f_1, f_2\}$ representing the system's control policies. Examples of this computation can be found in nonlinear systems theory [Olver 1986]. However, this computation requires explicit equations for these control policies and there are numerous applications where such prior knowledge is unavailable. Uncertainty in the precise form of the control policies can arise from unpredicted variations in the plant's structure. Uncertainty can also arise in highly complex systems where the state space's high dimensionality precludes complete prior knowledge of the distributions. In such situations, it is necessary that the invariants be determined directly from the system's observed behaviour. Since the hybrid system's event covering is defined with respect to these invariants, we see that the problem of finding such invariants is essentially the problem of event identification. In other words, we need to identify a collection of covering events which are invariant with respect to the available control policies f_1 and f_2. This problem is referred to in this chapter as the invariant subspace identification (ISID) problem. The algorithms discussed in the following sections provide one way of solving the ISID problem by direct (active) experimentation.

3 Invariant Subspace Identification Algorithm

Inductive inference is a machine learning protocol in which a system learns by example. It has found significant practical and theoretical uses in learning Boolean functions by example [Angluin 1983], proving poly-time complexity of linear programming [Khachiyan 1979] and combinatorial optimization [Groetschel 1988] algorithms, developing finite time procedures [Rosenblatt 1962] for training linear classifiers, and estimating sets bounding unknown parameters [Dasgupta 1987]. In this section, an inductive protocol for learning an $n-1$-dimensional invariant subspace of a variable structure system is formally stated.

The inductive protocol developed in this chapter can be seen as consisting of three fundamental components;

- an experiment for generating examples,
- a query to an algorithm called the membership oracle,
- and an update algorithm for modifying the system's current controller (i.e. switching surface).

These components are used to iteratively adjust the system's current estimate for the invariant subspace $H_\mathbf{s}$. Figure (2) illustrates the relationship between these three algorithm components.

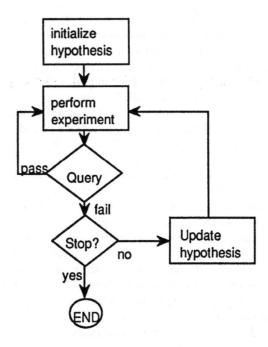

Fig. 2. Flow chart for an inductive inference protocol solving the ISID problem

The algorithm begins by forming an initial hypothesis about the system's sliding mode. This hypothesis takes the form of an n-dimensional vector $\mathbf{s}$ and an n by n symmetric matrix, $\mathbf{Q}$. The vector represents a unit normal to a switching surface, $H_{\mathbf{s}}$, which is hypothesized to be a sliding mode. The matrix represents a convex cone which is known to contain those vectors normal to a sliding modes of the system. Because the matrix, $\mathbf{Q}$, is associated with a convex cone in $\Re^n$, it will have one negative eigenvalue and $n - 1$ positive eigenvalues. For the purposes of this section it will therefore be convenient to make the following notational conventions. Let $\mathbf{e}_i$ be the ith eigenvector of $\mathbf{Q}$ and let λ_i be its associated eigenvalue. Assume that the eigenvalues and eigenvectors are ordered so that $\lambda_i > \lambda_{i+1}$ for all i. Define an n by $n - 1$ matrix, $\mathbf{E}$, whose columns are the eigenvectors of $\mathbf{Q}$ with positive eigenvalues. This matrix will be called $\mathbf{Q}$'s positive eigenvector matrix. Also form an $n - 1$ by $n - 1$ diagonal matrix, $\mathbf{L}$, from the positive eigenvalues of $\mathbf{Q}$. This matrix will be called the positive eigenvalue matrix. Both matrices are shown below.

$$
\mathbf{E} = \begin{pmatrix} \mathbf{e}_1 \ \mathbf{e}_2 \ \cdots \ \mathbf{e}_{n-1} \end{pmatrix}, \mathbf{L} = \begin{pmatrix} \lambda_1 & 0 & \cdots & 0 \\ 0 & \lambda_2 & \cdots & 0 \\ \cdot & \cdot & \cdots & \cdot \\ 0 & 0 & \cdots & \lambda_{n-1} \end{pmatrix}. \tag{8}
$$

The normalized eigenvalue matrix will be defined as $\mathbf{R} = \mathbf{L}/|\lambda_n|$.

After forming the initial hypothesis, the algorithm's first component, the experiment, is performed. This component involves the active measurement of the system's state and state velocity. The second algorithm component uses these experimental measurements to make a declaration on the validity of the hypothesis that the switching surface $H_{\mathbf{s}}$ is indeed a sliding mode. The declaration is made by a Boolean functional called the invariance oracle. The oracle's response is either 0 or 1 with a semantic interpretation which depends on the precise form of the Boolean functional. In the applications considered below, a response of 0 has a semantic meaning of TRUE, thereby indicating that the hypothesis *is* consistent with the measured data. A response of 1 has a semantic meaning of MAYBE, thereby indicating that the hypothesis *may not be* consistent with the measured data. If the answer is TRUE then nothing is done. If the answer is MAYBE, however, then the current hypothesis is modified using the algorithm's third component, the update algorithm.

The update algorithm uses a modification of the central-cut ellipsoid method [Shor 1977] to recompute the symmetric matrix $\mathbf{Q}$ and the vector $\mathbf{s}$. In modifying the hypothesis after the oracle's FALSE declaration, the update procedure attempts to generate a new hypothesis which is consistent with prior experimental data. This basic cycle of experiment, query, and update continues until an attracting invariant subspace is found.

The ISID algorithm can now be formally stated.

Invariant Subspace Identification (ISID) Algorithm

1. **Initialize:** Initialize an n by n symmetric matrix, $\mathbf{Q}$, which has $n - 1$ positive

eigenvalues and 1 negative eigenvalue such that if $H_\mathbf{z}$ is a sliding mode, then $\mathbf{z}^t \mathbf{Q} \mathbf{z} < 0$. Compute the eigendecomposition of $\mathbf{Q}$.

2. **Form Hypothesis:** Set the system's current switching surface, $\mathbf{s}$, equal to the negative eigenvector, $\mathbf{e}_n$, of $\mathbf{Q}$.
3. **Experiment:** Measure the system's state and state velocity, $\mathbf{x}$ and $\dot{\mathbf{x}}$.
4. **Query:** Compute the invariance oracle's response,

$$I_1(\mathbf{x}, \dot{\mathbf{x}}, \mathbf{s}) = \begin{cases} 0 & \text{if } (\mathbf{s}^t \mathbf{x})(\mathbf{s}^t \dot{\mathbf{x}}) < 0 \\ 1 & \text{otherwise} \end{cases} \tag{9}$$

5. **Update Hypothesis:** If the oracle returns 1 (MAYBE), then recompute $\mathbf{Q}$ using the following equations,

$$\mathbf{c} = \text{sgn}(\mathbf{e}_n^t \mathbf{x}) \mathbf{E}^t \dot{\mathbf{x}}, \tag{10}$$

$$\mathbf{b} = \frac{\mathbf{R}^{-1} \mathbf{c}}{\sqrt{\mathbf{c}^t \mathbf{R}^{-1} \mathbf{c}}}, \tag{11}$$

$$\mathbf{a} = -\frac{1}{n} \mathbf{b}, \tag{12}$$

$$\overline{\mathbf{R}}^{-1} = \frac{(n-1)^2}{(n-1)^2 - 1} \left(\mathbf{R}^{-1} - \frac{2}{n} \mathbf{b} \mathbf{b}^t \right), \tag{13}$$

$$\mathbf{x}_a = \mathbf{E} \mathbf{a} + \mathbf{e}_n, \tag{14}$$

$$\overline{\mathbf{Q}} = (\mathbf{I} - \mathbf{e}_n \mathbf{x}_a^t) \mathbf{E} \overline{\mathbf{R}} \mathbf{E}^t (\mathbf{I} - \mathbf{x}_a \mathbf{e}_n^t), \tag{15}$$

Set $\mathbf{Q}$ equal to $\overline{\mathbf{Q}}$ and recompute the eigendecomposition of $\mathbf{Q}$.
6. If the oracle returns 0 (TRUE), then do nothing.
7. **Loop:** go to step 2.

4 Invariance Oracles

This section derives the oracle used by the ISID algorithm and formally stated in equation (9). The oracle is a Boolean functional which evaluates a sufficient condition for the set $H_\mathbf{s}$ to be attracting and invariant.

Consider a set X called the sample set and let M be a measurable subset of X. The membership oracle is defined as follows.

Definition 3. Given a sample set X and a measurable set $M \subset X$, the *membership oracle* for M is a mapping, $O : X \to \{0, 1\}$, such that for any $\mathbf{x} \in X$,

$$O(\mathbf{x}) = \begin{cases} 0 & \text{if and only if } \mathbf{x} \in M \\ 1 & \text{if and only if } \mathbf{x} \notin M \end{cases} . \tag{16}$$

The membership or M-oracle can be thought of as a decision machine determining whether or not an example is a member of set M. An example which is an element of M will be called a positive M-example. If M^c is the complement of M, then a positive M^c-example will sometimes be called a negative example.

In this regard, the M-oracle's response 0 or 1 can be interpreted as a TRUE or FALSE declaration, respectively, concerning the membership of the example.

In certain cases, complete membership information may not be practical. It is therefore desirable to consider a weaker form of the M-oracle.

Definition 4. Given a sample set X and a measurable set $M \subset X$, the mapping, $O : X \to \{0,1\}$, is called an *incomplete M-oracle* if there exists another measurable set N such that $M \subseteq N$ and the mapping, O, is an N-oracle.

The incomplete M-oracle is a weaker version of the M-oracle since it only declares that the example is not an element of M. It does not make any declaration about an example's membership in M. In this regard, an incomplete oracle's response of 0 or 1 can be interpreted as a response of MAYBE or FALSE declaration, respectively, on the example's membership in M.

An *invariance oracle* will be a Boolean functional which declares whether or not a given subspace, $H_{\mathbf{S}}$, is attracting and Φ-invariant. Therefore the first step in defining an "invariance" oracle is to determine a test by which invariance can be determined. Sufficient conditions for attracting Φ-invariant sets form the basis of these tests. The following theorem provides a specific example of such a test.

Theorem 5. *Let $\mathbf{s}$ be a given n-dimensional real vector and let $\dot{\mathbf{x}}$ be given by equation (2). If the following condition*

$$\left(\mathbf{s}^t \mathbf{x}\right) \left(\mathbf{s}^t \dot{\mathbf{x}}\right) < 0, \tag{17}$$

is satisfied for all $\mathbf{x} \notin H_{\mathbf{S}}$, then the subspace, $H_{\mathbf{S}}$, is an attracting Φ-invariant set.

Proof: Define the functional $V(\mathbf{x}) = \frac{1}{2} \left(\mathbf{s}^t \mathbf{x}\right)^2$. Clearly, $V > 0$ for all $\mathbf{x}$. By the theorem's assumption, $\dot{V} < 0$, for all $\mathbf{x} \notin H_{\mathbf{S}}$. Therefore by theorem 8 in [Utkin 1977], $H_{\mathbf{S}}$ must be a sliding mode and is therefore an attracting Φ-invariant set of the flow. $\bullet$

It should be apparent that equation (17) can be recast as a logical function making a declaration about the consistency of the measured state and state velocity with the hypothesis that $H_{\mathbf{S}}$ is a sliding mode. This, then motivates the following definition for an "invariance" oracle.

Definition 6. The Boolean functional, $I_1 : \Re^{3n} \to \{0,1\}$, defined by equation (9) will be called an *invariance oracle*.

Let A denote a subset of $\Re^n$ consisting of those n-dimensional vectors $\mathbf{s}$ for which $H_{\mathbf{S}}$ is attracting and Φ-invariant. This set, A, will be referred to as the set of attracting and invariant subspaces. The following theorem states that the invariance oracle, I_1, is an incomplete A^c-oracle where A^c is the complement of set A.

Theorem 7. *Let A be the set of attracting invariant subspaces. If the function $I_1 : \Re^{3n} \to \{0,1\}$ is an invariance oracle, then it is an incomplete A^c-oracle.*

Proof: Let A_1 be a set of n-dimensional vectors **s** such that $I_1 = 0$ for any **x** and $\dot{\mathbf{x}}$ given by equation (2). By definition (3), I_1 must be an A_1^c-oracle. By theorem 1, any element of A_1 must also be an attracting Φ-invariant set. Therefore $A_1 \subset A$, which implies $A^c \subset A_1^c$. This therefore establishes I_1 as an incomplete A^c-oracle according to definition (4). •

The set A_1 defined in the above proof will be referred to as the set of attracting invariant subspaces which are *declarable* by the invariance oracle, I_1. Note that this set is smaller than A, the set of all attracting invariant subspaces. For this reason, the oracle is incomplete and its response of 0 or 1 is a declaration of TRUE or MAYBE, respectively, concerning the membership of **s** in A.

In the remainder of this chapter, the data collection gathered by an experiment will be denoted as $\mathcal{X} = \{\mathbf{x}, \dot{\mathbf{x}}\}$. It is assumed, that these measurements have no measurement noise, so that the oracle's declarations are always correct. An invariance oracle which always makes the correct declaration for a given data collection, $\mathcal{X}$, will be called a *perfect oracle*. In practical oracle realizations, the assumption that the invariance oracle is perfect may be too optimistic due to measurement uncertainty. This realization prompts the definition of an *imperfect oracle* as an oracle whose declarations are incorrect with a given probability. The distinction between perfect and imperfect oracles is critical, because inductive protocols based on oracle queries can fail disasterously with imperfect oracles. The convergence results of section (6) only apply to perfect invariance oracles. Precisely how to manage failures due to imperfect oracles is an important issue for future study. A preliminary indication of how to handle this problem will be discussed in section (8).

5 Ellipsoidal Update Method

The ISID algorithm uses an update procedure which recursively adjusts an estimate for the set A_1 of attracting invariant subspaces declarable by I_1. The proposed updating procedure is therefore a set-estimation algorithm which is closely related to set-membership identification algorithms [Dasgupta 1987] [Deller 1989]. It is also related to analytical techniques used in proving polynomial oracle-time complexity for certain optimization algorithms [Groetschel 1988] [Khachiyan 1979]. The common thread between both of these related areas is the use of the ellipsoid method [Shor 1977] [Bland 1981], which the following discussion also uses to great advantage.

An important property of A_1 (the set of declarable subspaces) is provided in the following lemma.

Lemma 8. A_1 *is a convex cone centered at the origin.*

Proof: Consider a specific collection of measurements as $\mathcal{X} = \{\mathbf{x}, \dot{\mathbf{x}}\}$. Define $C_{\mathcal{X}}$ as

$$C_{\mathcal{X}} = \{\mathbf{s} \in \Re^n : I_1(\mathcal{X}, \mathbf{s}) = 0\}. \tag{18}$$

A_1 will therefore be given by

$$A_1 = \bigcap_{\mathcal{X}} C_{\mathcal{X}}. \tag{19}$$

Since the oracle's response for a given $\mathbf{s}$ is independent of the vector's magnitude, $C_{\mathcal{X}}$ must be a cone centered at the origin. Since $C_{\mathcal{X}}$ is formed by the intersection of two halfspaces (see inequality (17)), it must also be convex. A_1 is therefore the intersection of a collection of convex cones centered at the origin and must therefore be one itself. •

The significance of the preceding lemma is that it suggests A_1 may be well approximated by sets which are themselves convex cones centered at the origin. A particularly convenient selection of approximating cones are the so-called ellipsoidal cones.

An "ellipsoidal cone" cone is defined as follows,

Definition 9. The *ellipsoidal cone*, $C_e(\mathbf{Q})$, is

$$C_e(\mathbf{Q}) = \{\mathbf{s} \in \Re^n : \mathbf{s}^t\mathbf{Q}\mathbf{s} < 0\}, \tag{20}$$

where $\mathbf{Q}$ is an n by n symmetric matrix with $n-1$ positive eigenvalues and one negative eigenvalue.

In the update procedure to be derived below, an ellipsoidal cone, $C_e(\mathbf{Q})$, will be used as an initial estimate for A_1. The current hypothesis is that the subspace normal to the negative eigenvector of $\mathbf{Q}$ is an attracting invariant set. If any data collection, $\mathcal{X}$, results in the oracle, I_1, declaring 1 (MAYBE), the query is said to have *failed*. The information from that failed query can be used to identify a set of subspaces which cannot possibly lie in A_1. This set will be referred to as the "inconsistent" set of subspaces generated by $\mathcal{X}$. The following lemma provides one characterization of these sets.

Lemma 10. *Let* $C_e(\mathbf{Q})$ *be an ellipsoidal cone with negative eigenvector,* $\mathbf{e}_n$. *Let* $\mathcal{X}$ *be a data collection for which a perfect invariance oracle,* I_1, *declares a failure,* $I_1(\mathcal{X}, \mathbf{e}_n) = 1$. *If* $A_1 \subset C_e(\mathbf{Q})$, *then* $A_1 \subset C_e(\mathbf{Q}) \cap H(\mathcal{X}, \mathbf{e}_n)$ *where*

$$H(\mathcal{X}, \mathbf{e}_n) = \left\{\mathbf{s} \in \Re^n : \mathbf{s}^t\dot{\mathbf{x}} < \text{sgn}(\mathbf{e}_n^t\mathbf{x})\mathbf{e}_n^t\mathbf{x}\right\}. \tag{21}$$

The set $H(\mathcal{X}, \mathbf{e}_n)$ *will be called the* inconsistent set *generated by* $\mathcal{X}$.

Proof: If a perfect invariance oracle I_1 returns 1 for $\mathcal{X}$ given the subspace represented by $\mathbf{e}_n$, then the following inequality holds.

$$\left(\mathbf{e}_n^t\dot{\mathbf{x}}\right)\left(\mathbf{e}_n^t\dot{\mathbf{x}}\right) > 0. \tag{22}$$

Note that for all $\mathbf{z}$ such that $\mathbf{z}^t\mathbf{x} \geq \mathbf{e}_n^t\mathbf{x}$, it can be inferred by the comparison principle that $\mathbf{z}^t\mathbf{x} \geq \mathbf{e}_n^t\mathbf{x}$. Similar arguments apply if the inequalities are reversed. Therefore any subspace, $H_{\mathbf{z}}$, such that

$$\mathbf{z}^t\dot{\mathbf{x}} \geq \text{sgn}(\mathbf{e}_n^t\mathbf{x})\mathbf{e}_n^t\dot{\mathbf{x}}, \tag{23}$$

cannot possibly be an attracting invariant set. The collection of such subspaces form the complement of the halfspace $H(\mathcal{X}, \mathbf{e}_n)$ defined in the theorem. Since A_1 is assumed to lie in $C_e(\mathbf{Q})$, it must therefore lie in the intersection of $C_e(\mathbf{Q})$ with $H(\mathcal{X}, \mathbf{e}_n)$. •

The significance of the preceding lemma is that the inconsistent set is an n-dimensional halfspace in $\Re^n$. To discuss this more fully, we first need to introduce the linear varieties of $n - 1$-dimensional subspaces.

Definition 11. Let S be an $n - 1$-dimensional subspace of $\Re^n$ and let $\mathbf{x}$ be an n-dimensional real vector. The *linear variety* of S generated by $\mathbf{x}$ is the set

$$V(S, \mathbf{x}) = \{\mathbf{s} + \mathbf{x} : \mathbf{s} \in S\}. \tag{24}$$

The following lemma shows that the inconsistent set forms a halfspace in the linear variety, $V(\mathrm{sp}(\mathbf{E}), \mathbf{e}_n)$, where $\mathrm{sp}(\mathbf{E})$ is the span of the $n - 1$ positive eigenvectors of $\mathbf{Q}$.

Lemma 12. *Let $C_e(\mathbf{Q})$ be an ellipsoidal cone with negative eigevector, $\mathbf{e}_n$. Let $V(\mathrm{sp}(\mathbf{E}), \mathbf{e}_n)$ be a linear variety of the subspace spanned by the positive eigenvectors of $\mathbf{Q}$. If the inconsistent set $H(\mathcal{X}, \mathbf{e}_n)$ is as defined in lemma (10), then the set $H(\mathcal{X}, \mathbf{e}_n) \cap V(\mathrm{sp}(\mathbf{E}), \mathbf{e}_n)$ is an $n - 1$ dimensional halfspace.*

Proof: Any vector $\mathbf{s}$ which lies in $V(\mathrm{sp}(\mathbf{E}), \mathbf{e}_n)$ can be written as

$$\mathbf{s} = \mathbf{E}\mathbf{w} + \mathbf{e}_n \tag{25}$$

where $\mathbf{w}$ is an $n - 1$ dimensional real vector. If $\mathrm{sgn}(\mathbf{e}_n^t \mathbf{x}) > 0$ then inserting $\mathbf{s}$ into equation (21) yields

$$\mathbf{w}^t \mathbf{E}^t \dot{\mathbf{x}} + \mathbf{e}_n^t \dot{\mathbf{x}} < \mathbf{e}_n^t \dot{\mathbf{x}} \tag{26}$$

which implies that $\mathbf{w}^t \mathbf{E}^t \dot{\mathbf{x}} < 0$. This, of course, determines a halfspace in the linear variety. A similar equation can be obtained if $\mathrm{sgn}(\mathbf{e}_n^t \mathbf{x}) < 0$. These considerations lead to the following,

$$\mathbf{w}^t \left[\mathrm{sgn}(\mathbf{e}_n^t \mathbf{x}) \mathbf{E}^t \dot{\mathbf{x}} \right] < 0 \tag{27}$$

which is an equation for the halfspace in the linear variety generated by the inconsistent set. ●

The geometry implied by the preceding lemmas is illustrated in figure (3). The characterization of the ellipsoidal cone and inconsistent sets provided by these lemmas forms the basis for the following theorem. This theorem states the equations used in obtaining a bounding ellipsoidal cone for A_1 from a prior bounding cone and the inconsistent set generated by $\mathcal{X}$. The proof of this theorem is a straightforward application of the central-cut ellipsoid method [Shor 1977].

Theorem 13. *Let $C_e(\mathbf{Q})$ be an ellipsoidal cone with negative eigenvector $\mathbf{e}_n$ such that $A_1 \subset C_e(\mathbf{Q})$. Let $\mathcal{X}$ be a data collection for which $I_1(\mathcal{X}, \mathbf{e}_n) = 1$. There exist ellipsoidal cones, $C_e(\underline{\mathbf{Q}})$ and $C_e(\overline{\mathbf{Q}})$, such that*

$$C_e(\underline{\mathbf{Q}}) \subset H(\mathcal{X}, \mathbf{e}_n) \cap C_e(\mathbf{Q}) \subset C_e(\overline{\mathbf{Q}}). \tag{28}$$

Furthermore if $\mathbf{R} = \mathbf{L}/|\lambda_n|$ where $\mathbf{L}$ is the positive eigenvalue matrix and if $\mathbf{E}$ is the positive eigenvector matrix of $\mathbf{Q}$, then $\overline{\mathbf{Q}}$ is given by equations (10) through (15) and $\underline{\mathbf{Q}}$ is given by equations (10) through (15) where $\underline{\mathbf{R}}^{-1} = \overline{\mathbf{R}}^{-1}/(n-1)^2$ is used in place of $\overline{\mathbf{R}}^{-1}$ in equation (10).

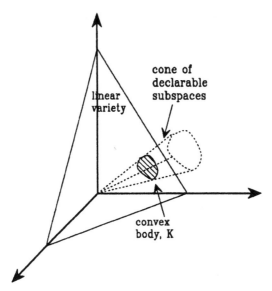

Fig. 3. The set of subspaces, A_M, declarable by a perfect invariance oracle, I_M, forms a convex cone centered at the origin. The intersection of A_M with a linear variety of an $n - 1$-dimensional subspace will be a bounded $n - 1$-dimensional convex body, K.

Proof: From lemma 2, the intersection of cone $C_e(\mathbf{Q})$ with $V(\mathrm{sp}(\mathbf{E}), \mathbf{e}_n)$ is an ellipsoid of the following form

$$E(\mathbf{R}^{-1}, 0) = \{\mathbf{w} \in \Re^{n-1} : \mathbf{w}^t \mathbf{R} \mathbf{w} < 1\}. \tag{29}$$

From lemma 4, the intersection of the inconsistent set, $H(\mathcal{X}, \mathbf{e}_n)$ and $V(\mathrm{sp}(\mathbf{E}), \mathbf{e}_n)$ will be an $n - 1$-dimensional halfspace, H, given by

$$H = \{\mathbf{w} \in \Re^{n-1} : \mathbf{w}^t \mathbf{c} < 0\}, \tag{30}$$

where $\mathbf{c} = \mathrm{sgn}(\mathbf{e}_n^t \mathbf{x}) \mathbf{E}^t \dot{\mathbf{x}}$. Therefore the intersection of $C_e(\mathbf{Q})$, $V(\mathrm{sp}(\mathbf{E}), \mathbf{e}_n)$, and $H(\mathcal{X}, \mathbf{e}_n)$ will be an $n - 1$-dimensional convex body, K.

It is well known that any bounded convex body can be contained within a unique ellipsoid of minimal volume called the Lowner-John ellipsoid [John 1984]. For convex bodies formed by single cuts of an ellipse, however, the Lowner-John ellipse can be computed in closed form [Groetschel 1988] [Bland 1981]. In particular, let K be the convex body formed by the intersection of an ellipse

$$E(\mathbf{A}, \mathbf{a}) = \left\{ \mathbf{x} \in \Re^n : (\mathbf{x} - \mathbf{a})^t \mathbf{A}^{-1} (\mathbf{x} - \mathbf{a}) \leq 1 \right\} \tag{31}$$

with a halfspace, $\{x : c^t x < c^t a\}$, then the Lowner-John ellipse, $E(\overline{A}, \overline{a})$ is given by

$$\overline{a} = a - \frac{n^2}{n+1} b, \tag{32}$$

$$\overline{A} = \frac{n^2}{n^2 - 1} \left(A - \frac{2}{n+1} bb^t \right), \tag{33}$$

$$b = \frac{Ac}{\sqrt{c^t Ac}}. \tag{34}$$

Computing the Lowner-John ellipsoid for $K = E(R^{-1}, 0) \cap H$ will yield the ellipsoid $E(\overline{R}^{-1}, a)$ where $\overline{R}$ and a are as given in the theorem. Figure (4) illustrates the geometry implied by the central-cut ellipsoid method.

The $n - 1$-dimensional Lowner-John ellipsoid generates an n-dimensional ellipsoidal cone. Let s be any point in the cone generated by the ellipsoid $E(\overline{R}^{-1}, a)$. There exists an $\alpha \in \Re$ such that αs is in the linear variety, $V(\mathrm{sp}(E), e_n)$. The α for which this is true must satisfy the orthogonality condition,

$$0 = e_n^t (\alpha s - e_n) \tag{35}$$

$$= \alpha e_n^t s - 1, \tag{36}$$

which implies that $\alpha = 1/e_n^t s$.

Since, $s = Ew + e_n$, the ellipsoid equation for $E(\overline{R}^{-1}, a)$ is

$$1 > (w - a)^t \overline{R}(w - a) \tag{37}$$

$$> (s - x_a)^t E^t \overline{R} E(s - x_a), \tag{38}$$

where $x_a = Ea + e_n$. The vector s in this equation must, of course, lie in the linear variety generated by e_n, $V(\mathrm{sp}(E), e_n)$. From our preceding discussion, any vector in the cone can be pulled back to the variety by appropriate renormalization with α. This then implies that if s is any vector in the cone, then

$$\left(\frac{s}{s^t e_n} - x_a \right)^t E^t \overline{R} E \left(\frac{s}{e_n^t s} - x_a \right) < 1. \tag{39}$$

Multiplying through by $|s^t e_n|^2$, we obtain

$$s^t \left[\left(I - e_n x_a^t \right) E^t \overline{R} E \left(I - x_a e_n^t \right) - e_n e_n^t \right] s < 0. \tag{40}$$

This inequality determines an ellipsoidal cone and the term within the square brackets is $\overline{Q}$.

$\underline{Q}$ is obtained by noting that if $E(\overline{R}^{-1}, a)$ is a Lowner-John ellipsoid for K, then $E(\overline{R}^{-1}/(n-1)^2, a)$ is an ellipsoid contained within K. By repeating the preceding construction with this smaller ellipsoi, the equation for $\underline{Q}$ is obtained.

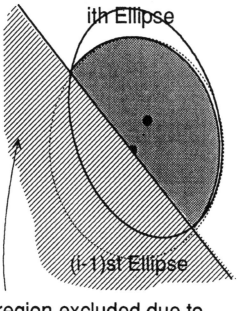

region excluded due to
failed invariance test

Fig. 4. Lowner-John ellipsoid for convex body formed by a central cut ellipsoid.

6   Convergence and Complexity

This section shows that the ISID algorithm generates a sequence of ellipsoidal cones whose negative eigenvectors must eventually lie in A_1. In particular, it is shown that if A_1 is non-empty then the ISID algorithm must converge after a finite number of MAYBE (1) declarations by the invariance oracle. It is further shown that under certain conditions the convergence time scales as $o(n^{2.5})$ where n is the plant's state space dimension. The section therefore proves that the ISID algorithm has finite oracle-time convergence and polynomial oracle-time complexity where oracle-time is measured by the number of MAYBE declarations made by the invariance oracle.

To prove the convergence of the ISID algorithm requires that there be some measure of the ellipsoidal cone's size or "volume". The set function used to define this volume is given below.

Definition 14. Let $C_e(\mathbf{Q})$ be an ellipsoidal cone and let the eigenvalues of $\mathbf{Q}$

be ordered as $\lambda_i > \lambda_{i+1}$ $(i = 1, \ldots, n)$. The *volume* of cone $C_e(\mathbf{Q})$ is defined to be

$$\text{vol} C_e(\mathbf{Q}) = \sqrt{\prod_{i=1}^{n-1} \frac{|\lambda_n|}{\lambda_i}}. \tag{41}$$

The volume of an ellipsoid, $E(\mathbf{A}, \mathbf{a})$, will be proportional to the square root of the determinant of $\mathbf{A}$. Since the determinant of $\mathbf{A}$ is simply the product of its eigenvalues, it should be clear that the preceding definition is using the volume of the $n-1$-dimensional ellipsoid contained in the linear variety $V(\text{sp}(\mathbf{E}), \mathbf{e}_n)$ as the "volume" of the cone.

The following theorem shows that the ISID algorithm must locate an attracting invariant subspace after a finite number of failed queries to a perfect invariance oracle.

Theorem 15. *Initialize the ISID algorithm with an ellipsoidal cone whose volume is unity and which is known to contain A_1. Let ϵ denote the volume of the smallest ellipsoidal cone containing A_1. If n is the state space dimension, then the ISID algorithm will determine an attracting invariant subspace after no more than $2(n-1) \ln \epsilon^{-1}$ failed queries to a perfect invariance oracle.*

Proof: Consider the ellipsoidal cone $C_e(\mathbf{Q}_i)$ after the ith failed invariance test. Let $\mathbf{E}$ and $\mathbf{L}$ be the positive eigenvector and eigenvalue matrices of $\mathbf{Q}_i$, respectively. The volume of this ellipsoid will be given by

$$\text{vol} C_e(Q_i) = \sqrt{\frac{1}{\prod_{j=1}^{n-1} \lambda_j(\mathbf{R})}}, \tag{42}$$

where $\lambda_j(\mathbf{R})$ is the jth positive eigenvalue of $\mathbf{R}$ and $\mathbf{R} = \mathbf{L}/|\lambda_n|$. Consider the ellipsoidal cone obtained using equations (10) through (15) of section (3). The symmetric matrix characterizing this cone is $\overline{\mathbf{Q}} = \mathbf{X}^t \mathbf{Y} \mathbf{X}$ where

$$\mathbf{X} = \begin{pmatrix} \mathbf{E}^t(\mathbf{I} - \beta \mathbf{e}_a \mathbf{e}_n^t) \\ \mathbf{e}_n^t \end{pmatrix}, \tag{43}$$

$$\mathbf{Y} = \begin{pmatrix} \mathbf{R} & 0 \\ 0 & -1 \end{pmatrix}. \tag{44}$$

where $\beta = \|\mathbf{x}_a\|$. Applying the orthogonal transformation,

$$\mathbf{P} = \begin{pmatrix} \mathbf{E} & \mathbf{e}_n \end{pmatrix}, \tag{45}$$

to $\mathbf{X}$, yields

$$\mathbf{P}^t \mathbf{X}^t = \begin{pmatrix} \mathbf{I} & 0 \\ -\beta \mathbf{e}_a^t \mathbf{E} & 1 \end{pmatrix} \tag{46}$$

where $\beta = \|\mathbf{x}_a\|$ and $\beta \mathbf{e}_a = \mathbf{x}_a$. Recall that $\mathbf{x}_a$ is the center of the updated ellipsoid in the linear variety $V(\text{sp}(\mathbf{E}), \mathbf{e}_n)$. For convenience, let $\mathbf{v}^t = -\beta \mathbf{e}_a^t \mathbf{E}$.

Since the eigenvalues of $\overline{\mathbf{Q}}$ are unchanged by an orthogonal transformation, the eigenvalues of $\mathbf{P}^t\mathbf{X}^t\mathbf{Y}\mathbf{X}\mathbf{P}$ can be used to compute the volume of $\overline{\mathbf{Q}}$. This transformed matrix has the form

$$\mathbf{P}^t\overline{\mathbf{Q}}\mathbf{P} = \mathbf{P}^t\mathbf{X}^t\mathbf{Y}\mathbf{X}\mathbf{P} \tag{47}$$

$$= \begin{pmatrix} \overline{\mathbf{R}} & \overline{\mathbf{R}}\mathbf{v} \\ \mathbf{v}^t\overline{\mathbf{R}} & \mathbf{v}^t\overline{\mathbf{R}}\mathbf{v} - 1 \end{pmatrix}. \tag{48}$$

Note that $\overline{\mathbf{R}}$ is an $n-1$ by $n-1$ leading principal submatrix of $\mathbf{P}^t\overline{\mathbf{Q}}\mathbf{P}$, so the eigenvalues of the two matrices satisfy the following interlacing property [Golub 1983].

$$\lambda_n(\overline{\mathbf{Q}}) \le \lambda_{n-1}(\overline{\mathbf{R}}) \le \lambda_{n-1}(\overline{\mathbf{Q}}) \le \cdots \le \lambda_2(\overline{\mathbf{Q}}) \le \lambda_1(\overline{\mathbf{R}}) \le \lambda(\overline{\mathbf{Q}}). \tag{49}$$

Since it is known that $\lambda_n(\overline{\mathbf{Q}})$ is negative (by the definition of an ellipsoidal cone), it can be shown that

$$\lambda_n(\mathbf{P}^t\mathbf{X}^t\mathbf{Y}\mathbf{X}\mathbf{P}) \le \sigma_n^2(\mathbf{P}^t\mathbf{X}^t)\lambda_n(\mathbf{Y}), \tag{50}$$

where $\sigma_n(\mathbf{P}^t\mathbf{X}^t)$ is the smallest singular value of $\mathbf{P}^t\mathbf{X}^t$ and $\lambda_n(\mathbf{Y})$ is the negative eigenvalue of $\mathbf{Y}$ [Golub 1983]. Note that this eigenvalue must be negative one (by construction of $\mathbf{Y}$). Also note that the singular value must satisfy the following inequality for any $\mathbf{x} \in \Re^n$,

$$\sigma_n^2(\mathbf{P}^t\mathbf{X}^t) \le \frac{\mathbf{x}^t\mathbf{P}^t\mathbf{X}^t\mathbf{X}\mathbf{P}\mathbf{x}}{\mathbf{x}^t\mathbf{x}}. \tag{51}$$

In particular, if we let $\mathbf{x} = (0 \cdots 01)^t$, then the smallest singular value must be less than unity. It can therefore be concluded that $|\lambda_n(\overline{\mathbf{Q}})| < 1$.

With the preceding results, it can be concluded that

$$\text{vol}C_e(\overline{\mathbf{Q}}) = \sqrt{\prod_{j=1}^{n-1} \frac{\lambda_n(\overline{\mathbf{Q}})}{\lambda_j(\overline{\mathbf{Q}})}} \tag{52}$$

$$\le \sqrt{\prod_{j=1}^{n-1} \frac{1}{\lambda_j(\overline{\mathbf{R}})}} \tag{53}$$

$$\le e^{-\frac{1}{2(n-1)}} \text{vol}C_e(\mathbf{Q}). \tag{54}$$

Inequality (54) is a consequence of the bound on the absolute value of the negative eigenvalue as well as the interlacing property (Eq. (49)). This inequality is simply the volume of an ellipsoid $E(\overline{\mathbf{R}}^{-1}, \mathbf{a})$. Recall, however, that $\overline{\mathbf{R}}$ is obtained from $\mathbf{R}$ using the central-cut ellipsoid method. The relationship between the volumes of these ellipsoids is given by the last line of the inequality. This last inequality [Groetschel 1988] is

$$\frac{\text{vol}E(\overline{\mathbf{A}}, \overline{\mathbf{a}})}{\text{vol}E(\mathbf{A}, \mathbf{a})} = e^{-1/2n}, \tag{55}$$

which bounds the rate at which ellipsoid volumes decrease when the central-cut ellipsoid method is used.

Since the initial ellipsoidal cone's volume is unity, then the ellipsoidal cone's volume after the Lth failed query must be bounded as follows,

$$\text{vol}C_e(\mathbf{Q}_L) \leq e^{-\frac{L}{2(n-1)}}. \tag{56}$$

However, $C_e(\mathbf{Q}_L)$ cannot be smaller than ϵ by assumption, therefore the number of failed queries, L, must satisfy

$$\epsilon \leq e^{-\frac{L}{2(n-1)}}. \tag{57}$$

Rearranging this inequality to extract L shows that the number of failed invariance queries can be no larger than the bound stated by the theorem. •

The following corollary for the preceding theorem establishes the polynomial oracle-time complexity of the ISID algorithm.

Corollary 16. *Assume that A_1 is a set which is contained within an ellipsoidal cone characterized by a matrix, $\mathbf{Q}$, whose normalized positive eigenvalues satisfy the inequality*

$$\frac{|\lambda_n|}{\lambda_i} > \gamma \tag{58}$$

for $1 > \gamma > 0$ and $i = 1, \ldots, n-1$. Under the assumptions of theorem (15), the ISID algorithm will determine an attracting invariant subspace after no more than $2(n-1)^2 \ln(n-1) + (n-1)^2 \ln \gamma^{-1}$ MAYBE declarations by the invariance oracle.

Proof: Because of the constraints on $\mathbf{Q}$, the volume of the smallest bounding ellipsoid will be no greater than $\gamma^{(n-1)/2}(n-1)^{-n+1}$. Inserting this into the bound of theorem (15) yields the asserted result. •

The significance of the preceding corollary is apparent when we consider how such restrictions on the eigenvalues of $\mathbf{Q}$ might arise. In particular, if the ISID algorithm is realized in finite precision arithmetic, then γ is proportional to the least significant bit of the realization. In this regard, the result shows that for finite precision implementations, there is an upper bound on the number of queries which the system can fail before exceeding the realization's precision. In particular, this result then shows that the bound scales as $o(n^{2.5})$ where n is the number of plant states. This result thereby establishes the polynomial oracle-time complexity of the algorithm.

7 Example: AUV Stabilization

This section discusses an application of the ISID algorithm to the stabilization of an autonomous underwater vehicle's (AUV) dive plane dynamics. This problem represents an example of the ISID algorithm's use as an adaptive variable structure control algorithm.

AUV dynamics are highly nonlinear and highly uncertain systems. Nonlinearities arise from hydrodynamic forces, uncompensated buoyancy effects, as well as cross-state dynamical coupling. Uncertainties arise due to environmental effects such as changing current and water conditions as well as poorly known mass and hydrodynamic properties. When an AUV retrieves a large object, for example, the drag and mass of this object may substantially modify the vehicle's buoyancy and drag coefficients. Such changes cannot be accurately modeled beforehand and can therefore have a disasterous effect on the success of the AUV's mission. In these situations, it would be highly desirable to develop an algorithm which can quickly and efficiently relearn the stabilizing controller for the system. The ISID algorithm represents one method for achieving this goal.

The following simulation results illustrate how the ISID algorithm can quickly stabilize an AUV's dive plane dynamics. The simplified equations of motion for vehicle (pitch) angle of attack, θ, in the dive plane as a function of velocity, v, may be written as

$$\ddot{\theta} = K_1\dot{\theta} + K_2\theta|\theta| + K_3\theta|v| + u_\theta, \tag{59}$$

$$\dot{v} = -v + K_4|\theta|v + u_v, \tag{60}$$

where K_1, K_2, K_3, and K_4 are hydrodynamic force coefficients. u_v and u_θ represent control forces applied in the velocity and angle of attack channels, respectively. These equations clearly show how nonlinearities enter the dynamics through the hydrodynamic cross coupling between θ and v. Uncertainty arises from the simple fact that the hydrodynamic coefficients may be poorly known. In general, these coefficients will be complex functions of vehicle geometry, speed, orientation, and water conditions. Consequently, they can never be completely characterized because there are too many degrees of freedom. Figure (5) illustrates the geometry implied by the equations of motion.

Figure (6) illustrates the behaviour of an AUV without active attitude control. The figure shows the 3-d state space trajectory for a vehicle with initial condition $\theta = 1$ and $v = 1$. The commanded state is $\theta = 0$ and $v = 2$. Without active attitude control, $u_\theta = 0$, and $u_v = -v + 2$. In this example, the system is hydrodynamically stable so that natural system damping can be used to eventually null the angle of attack. The figure shows that by using this control strategy, the vehicle exhibits large oscillations in θ and v before settling to the commanded state. For this particular system, the results therefore indicate that the angle of attack should be actively nulled to improve trajectory tracking.

Variable structure control (VSC) has emerged as a powerful technique for controlling AUV's with uncertain dynamics [Yoerger 1985]. In the following simulations, a hierarchical variable structure controller [DeCarlo 1988] with boundary layer was designed. The controls, u_θ and u_v, have the following form

$$u_\theta = \begin{cases} 1 & \text{if } \mathbf{s}_\theta^t\mathbf{x} \leq -\epsilon \\ \frac{\mathbf{s}_\theta^t\mathbf{x}}{\epsilon} & \text{if } -\epsilon < \mathbf{s}_\theta^t\mathbf{x} < \epsilon \\ -1 & \text{if } \mathbf{s}_\theta^t\mathbf{x} > \epsilon \end{cases}, \tag{61}$$

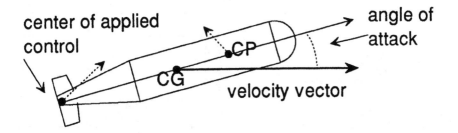

center of applied control

angle of attack

CP

CG

velocity vector

Fig. 5. Autonomous Underwater Vehicle Diveplane Dynamics

$$u_v = \begin{cases} h(\mathbf{s}_\theta^t \mathbf{x}) & \text{if } \mathbf{s}_v^t \mathbf{x} \le -\epsilon \\ \frac{\mathbf{s}_v^t \mathbf{x}}{\epsilon} & \text{if } -\epsilon < \mathbf{s}_v^t \mathbf{x} < \epsilon \; , \\ -h(\mathbf{s}_\theta^t \mathbf{x}) & \text{if } \mathbf{s}_v^t \mathbf{x} > \epsilon \end{cases} \tag{62}$$

where $\epsilon > 0$ denotes the width of the boundary layer and $\mathbf{x} = (\theta, \dot{\theta}, v)^t$ is the state vector. The function $h : \Re \to \{0, 1\}$ is assumed to have the following form

$$h(x) = \begin{cases} 0 \text{ if } |x| > \epsilon \\ 1 \text{ otherwise} \end{cases} , \tag{63}$$

and is used to implement a control hierarchy in which the system nulls angle of attack prior to nulling commanded velocity errors. The n-dimensional vectors $\mathbf{s}_\theta$ and $\mathbf{s}_v$ represent hyperplanes called switching surfaces just as was originally shown in equation (2) of the introduction.

The initial design of variable structure controllers can usually be partitioned into two phases. The first phase consists of determining the switching surfaces on which the system trajectories exhibit the desired transient response. The second phase determines the control strategies (gain levels) which insure that the switching surfaces are attracting invariant sets. Such switching surfaces are called sliding modes, since the system state is eventually captured by and slides along the switching surface. The need for adaptive identification of these surfaces arises when the system's structure changes in an unpredictable manner as when the vehicle retrieves a bulky package. In order to preserve system autonomy, the two phase design procedure cannot be followed, since the system's control strategies were fixed at the system's initial design. Consequently, the only part of the controller which can be modified is the switching surface and this modification must be done adaptively on the basis of the system's observed behaviour.

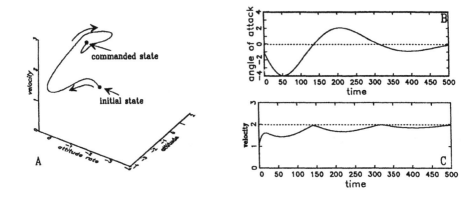

Fig. 6. Simulated AUV dive with no active nulling of angle of attack, θ. **A**: 3-d phase space trajectory, **B**: angle of attack, θ, time history, **C**: velocity, v, time history.

The simulation results shown in figures (7), (8), and (9) illustrate precisely how the ISID algorithm can be used to "relearn" the system's sliding modes. Figure (7) shows the AUV's performance (same initial conditions as shown in figure (6)) with the hierarchical sliding mode controller after a system failure causes the initially chosen switching surfaces to no longer be invariant sets. As can be seen, the sliding controller is actually unstable with the system exhibiting large oscillations in θ. Figures (8) and (9) show the system's behaviour during two "learning" sessions with the ISID algorithm. A learning session involves starting the vehicle at the initial condition and then commanding it over to the desired state. The first learning session is shown in figure (8). This particular example exhibited four adjustments to the sliding surface. On the last adjustment, the sliding condition is satisfied and the system slides easily to the commanded state. Figure (9) shows the system's response during the second training session. In this case, it is clear that learning is complete. There are no readjustments of the sliding surface and the system wastes little effort in bringing the system to its commanded state.

Perhaps the most remarkable thing about this example is the apparent speed with which the sliding surface is learned. In these simulations, only 4 failed invariance tests were required before finding a sliding mode. This low number of failed tests was observed in other simulation runs where the system's initial conditions were randomly varied. When compared with existing methods for learning nonlinear controllers [Narendra 1990] [Barto 1983] [Jacobs 1991], this approach appears to be exceptionally fast.

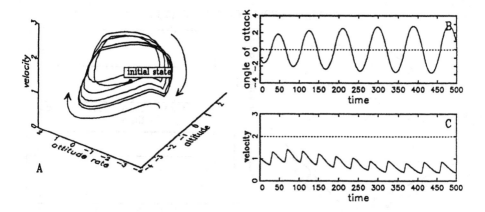

Fig. 7. Simulated AUV dive with hierarchical sliding control in which sliding mode constraints are violated. **A**: 3-d phase space trajectory, **B**: angle of attack, θ, time history, **C**: velocity, v, time history.

FIRST TRAINING SESSION RESULTS

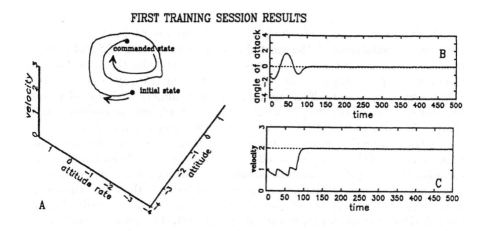

Fig. 8. Simulated AUV dive where ISID algorithm is used to relearn hierarchical sliding mode controller (First Learning Session). **A**: 3-d phase space trajectory, **B**: angle of attack, θ, time history, **C**: velocity, v, time history.

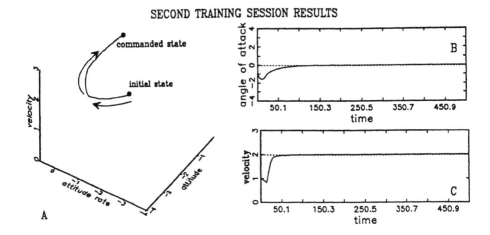

SECOND TRAINING SESSION RESULTS

Fig. 9. Simulated AUV dive where ISID algorithm is used to relearn hierarchical sliding mode controller (Second Learning Session). **A**: 3-d phase space trajectory, **B**: angle of attack, θ, time history, **C**: velocity, v, time history.

8 Significant Issues

The final theorem of section (6) is significant for two reasons. First it shows that the invariant subspaces can be located after a *finite* number of failed queries. In sliding mode control, such subspaces are used to stabilize the system as was shown in the preceding example. Therefore, theorem (15) says that a system only needs to perceive itself as "unstable" a finite number of times before system stability is re-established. This result stands in stark contrast to other results [Barto 1983] [Narendra 1990] [Jacobs 1991] where system stability can only be iteratively "learned" after a prohibitively long training period. The second important aspect of the preceding results is that the theorem's bound implies that the algorithm has polynomial time complexity. This means that as systems become more and more complex (i.e. larger state spaces), the time required to learn the system invariants will grow at a modest rate. In other words, the proposed ISID algorithm may represent a practical method for adaptive control and identification of highly complex nonlinear dynamical systems.

It should be noted, however, that these bounds are *not* with respect to system time, but rather with respect to failed oracle time. This is an important distinction for it is quite possible that there may be a long period of time between consecutive oracle declarations of failure. Consequently, convergence of the ISID algorithm can be extremely long in "system" time and may, in fact, never converge at all. At first glance, this observation may seem to cast doubt upon the value of theorem (15). Upon closer consideration, however, it provides further insight into the method. Recall that the oracle will always declare failures if the system trajectory is diverging from the current subspace, H_S. In other words, if the system is exhibiting "unstable" behaviour, the switching surface is modified. For the times between failures, the system appears to be stable and there is, therefore, no reason to change the switching surfaces. From this viewpoint, it can be seen that the bound of theorem (15) is very meaningful since it is measured with respect to the only quantity of physical interest to the system; the number of times the system "stumbles".

This point should be contrasted to parameter and set-membership identification [Dasgupta 1987] [Deller 1989] algorithms. In these cases, the important measure of parameter convergence is system time (i.e., the total number of experiments), since we are interested in obtaining accurate estimates as quickly as possible. Obviously for the parameter identification problem, the bounds computed by the preceding theorem would be useless unless the time between consecutive failures could be bounded also. That is not the situation, however, in the ISID problem which is primarily an adaptive "control" problem. The fact that these oracle-time bounds are meaningful for the ISID problem is an important point distinguishing this application from other more traditional applications of inductive inference protocols.

Finally, it must be observed that the preceding theorem assumes a perfect invariance oracle. In practice, oracles will not be perfect and the question is then what can be done to minimize the problems generated by an imperfect oracle. The answer is also provided by the preceding theorem. Theorem (15) provides a hard bound on the number of failed oracle queries. If the system generates more failures than implied by the bound, then a failure must either have occured in the oracle or else in the system itself. In either case, the theorem's finite time bound provides a natural criterion for failure detection and the subsequent reinitialization of the identification process. If the rate of oracle failure is known to be small (i.e. failure probability is small), then the natural course of action is to reinitialize the ISID algorithm and try again. The preceding discussion therefore implies the existence of effective and practical methods for dealing with the identification failures caused by imperfect oracles. In particular, if we model an oracle's imperfection as a probabilistic failure rate, then it should be possible to discuss the ISID algorithm's learning abilities within the so-called "probably almost correct" (PAC) framework used by a variety of researchers in the inductive inference community [Valiant 1984]. A full study of techniques for optimally managing the failures introduced by an imperfect oracle is well beyond the scope of the current chapter and represents an important topic for further inquiry.

9 Symbol Grounding and Event Identification

Formal computational systems are often interfaced to the external world. Such "hybrid" systems are used to control or interpret events with that external world.

The example discussed in this chapter is one example of such a hybrid system. Since the supervisor uses high-level abstractions to control the plant, such controllers are often referred to as "intelligent".

As noted in the section's introduction, this notion of intelligence is somewhat limited. If high-level decision making is to constitute intelligence, then this would imply that many symbol systems would be intelligent systems. This notion is, of course, at the heart of symbolic Artificial Intelligence research and it has its detractors. John Searle [Searle 1984] disputed the AI notion of intelligent machines with his now famous Chinese room argument. In this thought experiment it was noted that a prerequisite for "intelligence" is semantic content and that such content is unavailable to a purely symbolic system. For this reason, a computer can never be intelligent thereby debunking the traditional AI assumptions concerning the computational basis of human cognition.

At the heart of Searle's complaint is the notion of a symbol's meaning. This problem is also referred to as the symbol grounding problem [Harnad 1990]. Symbol grounding refers to methods by which symbols of a formal system acquire semantic content or "meaning". Such meaning generally has two interpretations. Meaning can be acquired through the association of symbols with nonsymbolic items in the external world. This acquired meaning is referred to as a symbol's extrinsic meaning. However, symbols also acquire content through internal associations with other symbols. This form of content might be referred to as intrinsic meaning.

An example of extrinsic meaning is seen in the association of the symbolic token "ELEPHANT" with those sensory inputs produced by seeing an elephant. The association of these sensory experiences with the symbolic token is determined by experience. The "meaning" of the symbol is determined by its nonsymbolic associations and is therefore external to the symbol system, itself. Consequently, we refer to such meaning as "extrinsic".

A symbol system, as Searle asserts, is not sufficient for an intelligent machine. Extrinsic meaning simply replaces the symbolic token with nonsymbolic tokens. The system has no real "understanding" of what those tokens signify so that the resulting computation is "meaningless". A good example of this type of "unintelligent" association is seen in intelligent control systems which make extensive use of DES models. In these cases, the "meaning" of the logical constructs is determined in an a priori way by the DES modeler. These intelligent choices for symbol/event bindings therefore imply that it is the modeler, rather than the system, which is intelligent.

In order for a system to be intelligent it must not only use high level abstractions, it must be able to generate them from internally generated construction principles. Symbols which arise in this manner may be grounded in nonsymbolic entities, but the meaning of these entities is determined internally, i.e. with respect to some intrinsic systems principle. In this regard, the symbols of such a system are intrinsically grounded. It is this form of "intrinsic" semantically meaning, which Searle asserted as a prerequisite for intelligence.

Clearly, conventional symbolic AI does not intrinsically ground its symbols.

It has been argued that the more recent connectionist or subsymbolic AI concepts embody some form of internal grounding[Chalmers 1992]. In view of these preceding remarks concerning symbol grounding and intelligence, it might now be appropriate to discuss the preceding ISID algorithm in light of the symbol grounding problem. Does the ISID algorithm produce event/symbol bindings which are "intrinsically" or "extrinsically" grounded. If the bindings are wholly external, then the resulting control system cannot be "intelligent" in the sense proposed by Searle.

In reviewing the modeling framework used in this paper, it is apparent that all plant symbols, $\tilde{z}$, are grounded with respect to a specific subset of the state space. At first glance, one might conclude then that this is an external grounding. However, the true test of external grounding is to see what happens if the event/symbol bindings change. In other words, if we shuffle the associations between symbols and nonsymbolic entities, does the operation of the supervisor change? If the ISID algorithm is not used, then clearly the bindings are unchanged. However, the ISID algorithm uses a computational algorithm (i.e. the invariance oracle) to decide whether or not the current event/symbol bindings satisfy or are consistent with the "internal" principle of control invariance. Therefore, if the initial event/symbol bindings change so that the resulting symbol groundings are inconsistent with the invariance oracle, then the supervisor changes the symbol bindings by redefining the "events". In other words, there is an internally coded principle guiding the symbol grounding process. Under this viewpoint, we can then assert that the ISID algorithm produces symbols with intrinsic semantic content.

The intrinsic content embodied by the invariance oracle is, of course, hard-wired into the system. The choice of what this oracle is, represents a choice by the system designer. There can be other oracle structures used, in which different internal event principles are used. Therefore, in some sense, it is still through the ingenuity of the designer that this system appears to have "intelligent" processing. However, the fact remains that this system is endowing its symbols with a meaning which is derived by the system internally. This fact is still true regardless of where that "internally" coded principle came from. In this regard, we could consider the use of the ISID algorithm as resulting in an "intelligent" control system in the sense proposed by J. Searle.

These notions also provide some additional perspective on what is "intelligent" control. Intelligence is often a vaguely defined concept referring to the use of high-level decision making processes in control. In the preceding section, it has been argued that this is not sufficient. Intelligence is not a behaviour, but a property of a system. For intelligence, a system must not only use symbolic abstractions, it must formulate its own symbol bindings with regard to specific internal principles. The ISID algorithm provides a way by which traditional hybrid systems might accomplish this intelligence through event identification.

10 Concluding Remarks

This chapter has shown how hybrid dynamical systems can be used to dichotomize the symbolic and nonsymbolic parts of a supervisory control system. The significance of this dichotomy is that it clearly identifies one of the key challenges facing hybrid supervisory control. This challenge concerns the way in which symbols used by the supervisor are associated with *meaningful* events occuring in the plant. The problem of relating symbols and events has been called *event identification*, the associations are referred to as *symbol/event bindings*. This chapter has presented a method for *learning* event bindings in a way which insures the stabilizability of the plant. This method represents a novel approach to adaptive control in which *inductive inference of controller structure is emphasized over statistical inference of controller parameters*. In this regard, the proposed method provides a radical departure from conventional model reference adaptive control. The advantage of this new approach is that is allows the formulation of learning algorithms which converge to a stabilizing set of bindings after a finite number of updates. Another significant aspect is that this convergence time is bounded in a polynomial manner by plant complexity, thereby recommending this approach as a practical method for the adaptive stabilization of large scale plants. Finally, this new learning algorithm sheds light on the meaning of *intelligent control*. Specifically, the notions developed in this chapter allow the development of a working characterization of intelligent control which is consistent with current viewpoints held by the cognitive psychology and subsymbolic AI communities concerning the computational basis of human cognition.

References

[Angluin 1983] D. Angluin, C.H. Smith, "Inductive Inference: Theory and Methods." *Computing Surveys*, 15(3):237-269, September 1983.

[Antsaklis 1993] P.J. Antsaklis, J.A. Stiver, M.D. Lemmon, "Hybrid System Modeling", this volume.

[Barto 1983] A.G. Barto, R. S. Sutton, and C. W. Anderson), "Neuronlike Elements that can Solve Difficult Learning Control Problems", *IEEE Trans. on Systems, Man, and Cybernetics*, Vol 13:835-846.

[Bland 1981] R.G. Bland, D. Goldfarb, M.J. Todd, "The Ellipsoid Method: a Survey", *Operations Research*, 29:1039-1091, 1981.

[Chalmers 1992] D. J. Chalmers, "Subsymbolic Computation and the Chinese Room", in *The Symbolic and Connectionist Paradigms: closing the gap*, ed: John Dinsmore, Lawrence Erlbaum Associates, pp. 25-48, 1992.

[Dasgupta 1987] Dasgupta and Huang, "Asymptotically Convergent Modified Recrusive Least-Squares with Data-Dependent Updating and Forgetting Factor for Systems with Bounded Noise", *IEEE Trans. on Information Theory*, IT-33(3):383-392.

[DeCarlo 1988] R. A. DeCarlo, S. H. Zak, and GP Matthews (1988), "Variable Structure Control of Nonlinear Multivariable Systems: a Tutorial", *Proceedings of the IEEE*, Vol. 76(3):212-232.

[Deller 1989] J. R. Deller, "Set Membership Identification in Digital Signal Processing", *IEEE ASSP Magazine*, Vol. 6:4-20, 1989.

[Golub 1983] G. Golub, C. Van Loan, *Matrix Computation*, Johns Hopkins University Press, Baltimore, Maryland, 1983

[Groetschel 1988] Groetshel, Lovasz, and Schrijver, *Geometric Algorithms and Combinatorial Optimization*, Springer-Verlag, 1988.

[Harnad 1990] S. Harnad, "The Symbol Grounding Problem", *Physica D*, vol. **42**, pp. 335-446, 1990.

[Jacobs 1991] R. A. Jacobs, M. I. Jordan, S. J. Nowlan, and G. E. Hinton, "Adaptive Mixtures of Local Experts", *Neural Computation*, Vol 3(1):79-87.

[John 1984] John, *Fritz John: Collected Papers (1948)*, Birkhauser, 1984.

[Khachiyan 1979] L. G. Khachiyan, "A Polynomial Algorithm in Learn Program", (english translation), *Soviet Mathematics Doklady*, 20:191-194, 1979.

[Lemmon 1992] M. D. Lemmon, "Ellipsoidal Methods for the Estimation of Sliding Mode Domains of Variable Structure Systems", *Proc. of 26th Annual Conference on Information Sciences and Systems*, Princeton N.J., March 18-20, 1992.

[Lemmon 1993a] M. D. Lemmon, "Inductive Inference of Invariant Subspaces", *Proceedings of the American Control Conference*, San Francisco, California, June 2-4, 1993.

[Lemmon 1993b] M. D. Lemmon, J. A. Stiver, P. J. Antsaklis, "Learning to Coordinate Control Policies of Hybrid Systems", *Proceedings of the American Control Conference*, San Francisco, California, June 2-4, 1993.

[Narendra 1990] K. S. Narendra and K. Parthsarathy, "Identification and Control of Dynamical Systems Using Neural Networks", *IEEE Trans. Neural Networks*, Vol 1(1):4-27.

[Olver 1986] P. J. Olver (1986), *Applications of Lie Groups to Differential Equations*, Springer-Verlag, New York, 1986

[Rosenblatt 1962] F. Rosenblatt), *Principles of Neurodynamics*, Spartan books, Washington D.C.

[Searle 1984] J. R. Searle, *Minds, brains, and science*, Cambridge, MA: Harvard University Press, 1984.

[Shor 1977] N. Z. Shor, "Cut-off Method with Space Extension in Convex Programming Problems", (english translation), *Cybernetics*, 13:94-96, 1977.

[Utkin 1977] V.I. Utkin, "Variable Structure Systems with Sliding Modes", *IEEE Transactions on Automatic Control*, Vol. AC-22:212-222.

[Valiant 1984] L. Valiant, "A Theory of the Learnable", *Comm. of ACM*, Vol 27(11):1134-1142

[Yoerger 1985] D. R. Yoerger and J. J. E. Slotine, "Robust Trajectory Control of Underwater Vehicles", *IEEE Journal of Oceanic Engineering*, Vol OE-10(4):462-470.

Multiple Agent Hybrid Control Architecture

Anil Nerode[1]* and Wolf Kohn[2]

[1] Mathematical Sciences Institute
Cornell University, Ithaca, New York 14850
e-mail: anil@math.cornell.edu
[2] Intermetrics Corporation
Bellevue, Washington
e-mail: wfk@minnie.bell.inmet.com

Introduction

Multiple Agent Declarative Control Architecture is a software system for real–time implementation of reactive, intelligent, and distributed controllers. This architecture is based on a theory developed by the authors over the last three years. This theory extends the concepts, principles, and algorithms of Single Agent Declarative Control, [14] and [20], and merges them with the principles of concurrent computing [2] and dynamical hybrid systems [28], [26], [10], [11]. The multiple agent architecture consists of a collection of these agents and an inter–agent connectivity network. A detailed description of this architecture and its functionality is given in [27].

We have developed a sample shop floor simulation. This is a simple factory with 2 workcells and an agent for each cell. The simulation controls Planning, Scheduling, and Control (PSC) with respect to the flow through these cells. Of course an actual factory would have many workcells and a network of paths between them.

The use of such agents is not limited to controlling PSC. Agents can be built to control design, analysis, prototyping, equipment programming, and even maintenance and accounting. The control of all such tasks is accomplished through the coordinated inferencing of actions from agent knowledge bases.

Architectural Highlights

The multiple agent declarative control architecture provides several key capabilities which are highlighted below:

Reactive: The theorem proving function of each agent of the architecture operates according to a "first principles" feedback paradigm. This functionality extends the feedback principles to the decision levels in the manufacturing process and allows for status- dependent tuning of actions.

Adaptive: The knowledge base of each agent is open and modified by received sensory data. Theorem failure triggers both tuning and corrective action. This

* Research supported by ORA Corp., DARPA-US ARMY AMCCOM (Picatinny Arsenal,NJ) contract DAAA21-92-C-0013, the U. S. Army Research Office contract DAAL03-91-C-0027.

functionality maintains the validity of the knowledge base of the agents and is the basic mechanism implementing the feedback paradigm discussed above.

Distributed with Coordination: The theorem proving is carried out distributed over the agents. The coordination scheme is implicit and without supervisor. This functionality is central for ensuring that, although each agent controls a local aspect of PSC, the global logic integrity of the system is maintained over time.

Dynamic Hierarchization: The architecture operate simultaneously at several different levels of abstraction. The knowledge elements that characterize PSC are naturally distributed over many levels of abstraction. The current state of each level is a function of process status. Therefore, in order for the agent controllers to maximize the use of available knowledge, the hierarchization must be tuned to the current status.

Figure of Merit: The behavior of the closed loop distributed system is determined by proving that there exist a command trajectory that minimizes a goal functional. The selection of that functional depends both on short and long term agent objectives, which themselves depend on current status.

Real–Time: Constraints for real–time performance are explicit and part of the knowledge base. This is important because in a manufacturing environment, the timing of normal or unexpected events is usually correlated with the status of the manufacturing process. Therefore the structure of the real time constraints cannot be fully instantiated at design time.

The central mechanism for providing these capabilities is an on–line restricted automated theorem prover associated with each agent. The architecture consists of a Knowledge Base which stores the goal for the agent, the system constraints, the inputs, and the inference operations. A Planner generates the theorem which represents goal. For some agents, this goal will govern the behavior local to that agent. For other agents, the goal will also include behaviors global to the system. The Inferencer proves the theorem. If the theorem is true, control actions, computed during inferencing, are issued to the plant. If the theorem is false, an Adapter processes the failed terms in the theorem for replacement or modificaiton. Data from other agents is provided to the Planner for incorporation as constraints into the theorem and passes through a Knowledge Decoder for entry into the Knowledge Base.

The next section presents an overview of the simulation and observations. It is followed by an extensive detailed description of the components in the architecture and the adaptation of these components to PSC. Conclusions are listed at the end.

Simulation and Observation

The simulation and control are both executed on a workstation using three Quintus Prolog processes: one for the producer controller agent, one for the consumer controller agent, and finally one for the the the simulation.

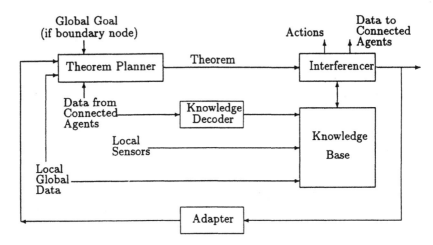

Fig. 1. Agent Controller

The simulation randomly generates a flow of orders which are absorbed into the system goal. For each order, the order number, delivery time requirements, quantification, and priority are tracked. The Planner, in constructing the behavior which satisfies the goal in the form of a true theorem, includes terms for identification of the processes and also tools and materials required by the Producer and Consumer for manufacturing the part. A sensor on the holding bin between the Producer and the Consumer provides monitoring of the inventory of work in progress. A similar sensor on the bin after the Consumer indicates the quantity of finished parts. Other sensors provid production rate feedback. The simulation is executed having as a goal minimizing consumer unsatisfied demand.

As the simulation ran, the system could be observed to react to changes in inventory, both finished and in–process, by tuning the processing rate of the Producer and the Consumer. As a result, the delivery of on–time orders was maximized.

Multiple Agent Declarative Control Architecture

Each agent of the declarative control architecture is provided with a real–time theorem prover whose domain is a theory, called relaxed variational theory. This theory is an encoding of the relaxed variational theory developed by Young [44]. We overview some of its main features in Section 3. In this section we discuss the main features of the architecture.

The architecture is composed of two building blocks. These are the agent Controllers and The Controller Network. These are illustrated in Figures 1 and 2 respectively.

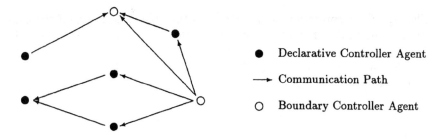

Fig. 2. Controller Network

— *Architectural Elements of the a control agent*

A control agent is composed of five functional elements — a *Knowledge Base*, a *Theorem Planner*, an *Inferencer*, a *Knowledge Decoder* and an *Adapter*. We discuss their functionalities next.

— *Knowledge Base*

The Knowledge Base consists of a set of equational first order logic clauses. The syntax of clauses is similar to Prolog. Each clause is of the form

$$\text{Head} < \text{Body} \tag{1}$$

where Head is a functional form, $p(x, \ldots, x_n)$, and where $x_1, x_2, \ldots, x_n$ are variables or parameters in the domain of the controller. The symbol $<$ stands for logical implication.

The Body of a clause is a conjunction of one or more terms,

$$e_1 \wedge e_2 \wedge \ldots \wedge e_m \tag{2}$$

where $\wedge$ is the logical "and". Each term in (2) is either an equational form, an inequational form, a covering form, or a clause head. The generic structure of these forms is illustrated in Table 1 below.

Form	Structure	Meaning
equational	$w(x_1, \ldots, x_n) = v(x_1, \ldots, x_n)$	equal
inequational	$w(x_1, \ldots, x_n) \neq v(x_1, \ldots, x_n)$	not equal
covering	$w(x_1, \ldots, x_n) \leq v(x_1, \ldots, x_n)$	recursion, chaining

Table 1

In Table 1, w and v are polynomial forms with respect to a finite set of operations whose definitional and property axioms are included in the knowledge base.

The logical interpretation of (1) and (2) is that the Head is *true* if the conjunction of the terms of Body are jointly *true* for instances of the variables in the clause head.

The domain D in which the variables in a clause head take values is a Cartesian product of the form

$$D = G \times S \times X \times A \tag{3}$$

where G is the space of goals, S is the space of sensory data, X is the space of controller states and A is the space of actions. The structure of these spaces is application–dependent. Their structure is characterized by clauses in the knowledge base.

The denotational semantics of each clause in the knowledge base is one of the following:

1 – a conservation principle,
2 – an invariance principle,
3 – a constraint principle.

Conservation principles are clauses about balance of a particular process in the dynamics of the system, the goals or the computational resources. For instance, in the manufacturing example described later one, a conservation clause describes the material balance in the manufacturing process as viewed by each of the two agents.

As another example, consider the following clause representing conservation of computational resources:

comp(Load, ProcessOp_count, Limit) < process(process_count)
$\qquad \wedge$ process_count $\cdot$ Load1 $-$ Op_count Load
$\qquad \wedge$ Load1 $\leq$ Limit
$\qquad \wedge$ comp(Load1, Process, Op_count, Limit).

where Load corresponds to the current computational burden, measured in VIPS (Variable Instantiations Per Second), Process is a clause considered for execution, and Op_count is the current number of terms in process.

In general, conservation principles always involve recursion. These are not necessarily expressed by a single clause as happens in the example above, but may involve chaining through several clauses.

Invariance principles are clauses establishing constants of motion in a generalized sense. These principles include stationarity, which plays a pivotal role in the formulation of the theorems proved by the architecture, and geodesics.

The importance of invariant principles lies in the opportunity their observation provides for the detection of unexpected events. For example, in an adiabatic chemical process, the enthalpy is constant under normal operating conditions. An equational clause that states this invariance has a ground form that is constant. Any deviation from this value, represents deviation from normality, and can signal the need for corrective actions.

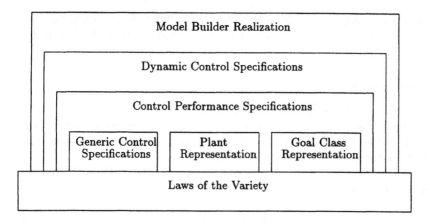

Fig. 3. Knowledge Base Organization

Constraint principles are clauses representing engineering limits to actuators or sensors and, most importantly, behavioral policies. For instance, in the chemical reactor example the characteristics of the speed of response of values controlling the input products or temperature are given by empirical graphs (e.g. pressure vs velocity) and a strategy for interpolation.

The clause database is organized in a nested hierarchical structure as illustrated in Figure 3. The bottom of this hierarchy contains equations that characterize the algebraic structure on which the terms equational forms are defined. This is an equational variety.

At the next level of the hierarchy, three types of clauses are stored: Generic Control Specifications, the Plant Representation, and the Goal Class Representation.

Generic control specifications are clauses expressing the general desired behavior of the system. They include statements about controllability, observability, complexity, and robustness generic to the class of declarative rational controllers. These specifications are written as clauses using the eqwuational Horn clause format described earlier.

The Plant Representation is given by clauses characterizing the dynamic behavior and structure of the plant, which includes the dynamic behavior of sensors and actuators. These clauses are written as conservation principles for the dynamic behavior and as invariance principles for the structure. As for the generic control specifications, they are constructed by combining variety laws in the equational Horn clause format.

The next level of the hierarchy involves the Control Performance Specifications. These are typically problem–dependent criteria and constraints. They are written in equational Horn clause format. They include generic constraints such as speed and time of response, and qualitative properties of state trajectories [44]. Dynamic control specifications are equational Horn clauses whose bodies are modified as a function of the sensor and goal commands (see Figure 3).

Finally, model builder realization clauses constitute a recipe for building a procedural model for variable instantiation and theorem proving.

— *Theorem Planner*

The function of the Theorem Planner, which is domain–specific, is to generate, for each update interval, a symbolic statement of the desired behavior of the system, as viewed by the agent (j, say) throughout the interval. The theorem statement that it generates has the following form:

"There exists an action A_j in the control interval $[t, t + D)$ such that A_j minimizes the functional

$$\int_t^{t+D} L_j(X, C) dA_j(X) \tag{4}$$

subject to the following constraints:

$$g_j(S, X[t + D]) \approx \text{ local goal for the interval,}$$
$$\sum_m C_m^j(X, t) \approx V_{\text{inter}_j}(X, t), \text{ and} \tag{5}$$
$$\int dA_j \approx 1."$$

In (4), L_j is called the local Lagrangian of the system as viewed by agent j for the current interval of control $[t, t + D)$. This function, which maps the Cartesian product of the state and control spaces into the real line with a rational topology of intervals (small topology), captures the dynamics, constraints and requirements of the system as viewed by agent j. The relaxed Lagrangian function is a *continuous* projection in the topology defined by the Knowledge Base (see [28]), in the coordinates of the agent of the global Lagrangian function L that characterizes the system as a whole.

In (4), X represents the state of the system as viewed by the agent, C is a variable taking values in the countable set of possible actions (the control space) agent j can generate. These actions are constructed by a mixing process from a basic set of primitive actions. This mixing process is described next.

The term A_j in (1) is a Radon probability measure [43] on the set of primitive command or control actions the agent can execute for the interval $[t, t + D)$. It measures, for the interval, the percentage of time C is to spend in each of the primitive actions. This is the nature of the mixing process. The central function of the control agent is to determine this mixture of actions for each control interval. This function is carried out by each agent by inferring from the current status of the knowledge base whether a solution of the optimization problem stated by the current theorem exists, and, if so, to generate corresponding actions and state updates.

The expressions in (5) constitute the constraints imposed in the relaxed optimization problem solved by the agent. The first one is the local goal constraint expressing the general value of the state at the end of the current interval. The second represents the constraints imposed on the agent by the other agents in the network. Finally, the third one indicates that A_j is a probability measure.

Under relaxation, and the appropriate selection of the domain (see [43], [14], [44]), the optimization problem stated in (4) and (5) is a **convex** optimization problem. This is important because it guarantees that if a solution exists, it is essentially unique. It also guarantees the computational effectiveness of the inference method that the agent uses for proving the theorem.

The construction of the theorem statement given by (4) and (5) is the central task carried out in the Planner. It characterizes the desired behavior of the system as viewed by the agent in the current interval so that its requirements are satisfied and the system "moves" towards its goal in an optimal manner.

— Adapter

The functional in (4) includes a term, referred to as the "catch–all" potential, which is not associated with any particular clauses in the Knowledge Base. Its function is to measure unmodelled dynamic events. This monitoring function is carried out by the Adapter which implements a generic "commutator principle". Under this principle, if the value of the catch–all potential is empty, then the current theorem statement is regarded as adequately modelling the status of the system.

On the other hand, if the theorem fails, this means that there is a mismatch between the current statement of the theorem and current system status. The catch–all potential carries the equational terms of the theorem that caused the failure. These terms are negated and conjuncted together by the Inferencer according to the commutation principle, itself defined by equational clauses in the Knowledge Base. It is stored in the Knowledge Base as an adaptation dynamic clause. The Adapter then generates a symbolic potential, which is characterized by the adaptation clause and corresponding tuning constraints. This potential is added to the Lagrangian criterion for the theorem characterizing the interval.

The new potential symbol and tuning constraints are sent to the Planner. The Planner generates a modified Local criterion and goal constraint. The new theorem thus constructed represents the adaptive behavior of the system. This is the essence of reactive structural adaptation in our model.

We pause to address the issue of **robustness**. The Adapter mechanism of Declarative Control provides the system with a generic, computationally effective, means to recover from *failures or unpredictable events*. Theorem failures are symptoms of mismatches between what the agent thinks the system looks like and what it really looks like. The adaptation clause incorporates knowledge into the agent's Knowledge Base and represents a recovery strategy. The Inferencer, discussed next, effects this strategy as part of its normal operation.

— Inferencer

The Inferencer is an on–line equational theorem prover. The class of theorems it can prove are represented by statements of the form of (4) and (5). We will first present a brief overview of its functionalities.

The class of theorems proved by the inferencer are each expressed by an existentially quantified conjunction of equational terms of the form:

$$\exists Z \mid W_1(Z,p)rel_1V_1(Z,p) \wedge \cdots \wedge W_n(Z,p)rel_nV_n(Z,p) \tag{6}$$

where Z is a tuple of variables each taking values in some domain D, p is a list of parameters in D, and $\{W_i, V_i\}$ are polynomial terms in the semiring polynomial algebra [6], [30], [8]:

$$\tilde{D}\langle\Omega\rangle \tag{7}$$

with

$$\tilde{D} = \{D, \langle+, \cdot, 1, 0\rangle\}$$

a semiring algebra with additive unit 0 and multiplicative unit 1.

In (6) rel_j, $i = 1, \ldots, n$ are binary relations on the polynomial algebra. Each rel_j can be either an equality relation ($\approx$), inequality relation ($\neq$), or a partial order relation. In a given theorem, more than one partial order relation may appear. In each theorem, *at least one* of the terms is a partial order relation defining a complete lattice on the algebra. It has a minimal element.

Given a theorem statement of the form of (6) and a Knowledge Base of equational clauses, the inferencer determines whether the statement logically follows from the clauses in the Knowledge Base. If so, as a side effect of the proof, it generates a non–empty subset of tuples with entries in D giving values to Z. These entries determine the agent's actions. Thus, a side effect is instantiation of the agent's decision variables.

In (7), Ω is a set of primitive unary operations $\{f_i\}$, called the infinitesimal operators. Each f_i maps the semiring algebra, whose members are power series involving the composition of operators, on Z to itself:

$$f_i : \tilde{D}\langle\langle Z \rangle\rangle \rightarrow \tilde{D}\langle\langle Z \rangle\rangle. \tag{8}$$

These operators are characterized by axioms in the Knowledge Base and are problem dependent. In formal logic, the implemented inference principle can be stated as follows: Let Σ be the set of clauses in the Knowledge Base. Let $\vdash$ represent implication. Then, proving the theorem means to show that it logically follows from Σ, i.e.,

$$\Sigma \vdash \text{Theorem}. \tag{9}$$

The proof is accomplished by sequences of applications of the following inference axioms:

- equality axioms
- inequality axioms
- partial order axioms
- compatibility axioms
- convergence axioms
- knowledge base axioms
- limit axioms

Theorem equations

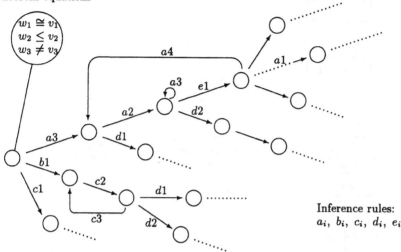

Fig. 4. Conceptual Structure of the Proof Automaton

The specifics of these inference axioms can be found in [19], [14], [20] where it is shown that each of the inference principles can be expressed as an *operator* on the Cartesian product:

$$\tilde{D}\langle W \rangle \times \tilde{D}\langle W \rangle. \tag{10}$$

Each inference operator transforms a relational term into another relational term. The inferencer applies sequences of inference oeprators on the equational terms of the theorem until these terms are reduced to either a set of ground equations of the form of (9), or determine that no such ground form exists.

$$Z_i \approx \alpha_i, \quad \alpha_i \in \tilde{D}. \tag{11}$$

We now provide an overview of the mechanism by which the inferencer carries out the procedure described above. The inferencer *builds* a procedure for variable goal instantiation: a *locally finite automaton*. We refer to this automaton as the Proof Automaton. This important feature is unique to our approach. The proof procedure is customized to the particular theorem statement and knowledge base instance it is currently handling.

The structure of the proof automaton generated by the inferencer is illustrated in Figure 4.

In Figure 4, the initial state represents the equations associated with the theorem. In general, each state corresponds to a derived equational form of the theorem through the application of a chain of inference operators to the initial state that is represented by the path

$$S_0 \xrightarrow{\text{Inf}_1} S_1 \xrightarrow{\text{Inf}_2} S_2 \longrightarrow \cdots \xrightarrow{\text{Inf}_k} S_k$$

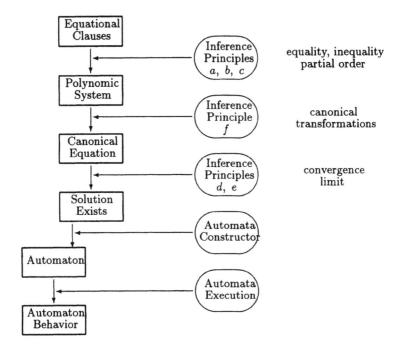

Fig. 5. Summary of Inferencer Procedure

Each edge in the automaton corresponds to one of the possible inferences. A state is *terminal* if its equational form is a tautology, or it corresponds to a canonical form whose solution form is stored in the Knowledge Base.

In transversing the automaton state graph, values or expressions are assigned to the variables. In a terminal state, the equational terms are all ground states (see (11)). If the automaton contains at least one path starting in the initial state and ending in a terminal state, then the theorem is true with respect to the given Knowledge Base, and the resulting variable instantiation is a valid one. If this is not the case, the theorem is false.

The function of the complete partial order term, present in the conjunction of each theorem provable by the inferencer, is to provide a guide for constructing the proof automaton. This is done by transforming the equational terms of the theorem into a *canonical fixed point equation*, called the Kleene–Schutzenberger Equation (KSE) [6], [8], which constitutes a *blueprint* for the construction of the proof automaton. The general form of KSE is:

$$Z = E(p) \cdot + T(p). \tag{12}$$

In (12), E is a square matrix, with each entry a rational form constructed from the basis of inference operators described above, and T is a vector of equational forms from the Knowledge Base. Each non–empty entry, E_{ij} in E corresponds to the edge in the proof automaton connecting states i and j. The binary operator between $E(p)$ and Z represents the "apply inference to" operator. Terminal states are determined by the non–empty terms of T. The p terms are custom parameter values in the inference operator terms in $E(\cdot)$.

A summary of the procedure executed by the inferencer is presented in Figure 5.

Note that the construction of the automaton is carried out from the canonical equation and not by non–deterministic application of the inference rules. This approach reduces the computational complexity of the canonical equation (low polynomic) and is far better than applying the inference rules directly (exponential).

The automaton is simulated to generate instances of the state, action and evaluation variables using an automaton decomposition procedure which requires $n \log_2 n$ time, where $n \approx \#$ of states of the automaton. This procedure implements the recursive decomposition of the automaton into a cascade of parallel unitary (one initial and one terminal state) automata. Each of the automata of this decomposition is executed independently of the rest. The behavior of the resulting network of automata is identical with the behavior obtained from the original automaton, but with *feasible time complexity*.

How is the inferencer used by an agent? It fulfills two functions. First, it generates a proof for the system behavior theorem of each agent generated by the Planner ((4), (5)). Second, it is the central element in the Knowledge Decoder. We now describe its function for proving the behavior theorem. Later, we will overview its function as part of the Knowledge Decoder.

To show how the inferencer is used to prove the Planner theorem, (4), (5), first, we show how this theorem is transformed into a pattern of the form of (6). Since (4), (5) is a convex optimization problem, a necessary and sufficient condition for optimality is provided by the following dynamic programming formulation:

$$V_j(Y, \tau) \approx \inf_{A_j} \int_\tau L_j(X, C) dA_j(X(\sigma))$$

$$\partial V_j / \partial t = \inf_{A_j} H(Y, \partial Y_j / \partial Y, A_j) \tag{13}$$

$$X(t) = Y$$

$$\tau \in [t, t + D).$$

In (13), the function V_j, called the optimal cost–to–go function, characterizes minimality starting from any arbitrary points inside the current interval to its end. The second equation is the corresponding Hamilton Jacobi Bellman equation for the problem stated in (4) and (5) with H the Hamiltonian of the relaxed problem. This formulation provides the formal coupling between deductive theorem proving and optimal control theory. The inferencer allows the real–time optimal solution of the formal control problem resulting in intelligent distributed real–time control of the multiple–agent system. The central idea for solving (10) is to expand the cost–to–go function, in a *rational power series*, $V(.\,,.)$, in the algebra:

$$\widetilde{\langle\langle (Y, \tau) \rangle\rangle}. \tag{14}$$

Replacing V for V_j in the second equation in (13) gives two results. One is a set of polynomial equations for the coefficients of V. The other is a partial order expression encoding the optimality requirement. Because of the convexity and rationality of V, the number of equations needed to characterize the coefficients of V is finite. The resulting string of conjunctions of coefficient equations and the optimality partial order expression is of form (6).

In summary, for each agent, the inferencer operates according to the following procedure.

Step 1: Load the current theorem (4), (5).

Step 2: Transform the theorem to an equational form (6) via (13).

Step 3: Execute the proof according to Figure 5.

If the theorem logically follows from the Knowledge Base (i.e., it is true), the inferencer procedure outlined will terminate at Step 3 with actions $A(t + D)$. If the theorem does not logically follow from the Knowledge Base, the Adapter is activated, and the theorem is modified by the Theorem Planner according to the strategy outlined above. This mechanism is the essence of reactivity in the agent. Because of relaxation and convexity, this mechanism ensures that the controllable set of the domain is strictly larger than it would be without this correction strategy.

— *Knowledge Decoder*

The function of the Knowledge Decoder is to translate knowledge data from the network into the agent's Knowledge Base by updating the inter–agent specification clauses. These clauses characterize the second constraint in (2). Specifically, they express the constraints imposed by the rest of the network on a single agent. They also characterize the global–to–local transformations (see [1]). Finally, they provide the rule for building a *generalized multiplier* for incorporating the inter–agent constraints into a complete unconstrained criterion, which is then used to build the cost–to–go function in the first expression in (13).

The Knowledge Decoder has a built–in inferencer used to infer the structure of the multiplier and transformations by a procedure similar to the one described in (13). Specifically, the multiplier and transformations are expanded in rational power series in the algebra defined in (14). Then, the necessary conditions for duality are used to determine conjunctions of equational forms and a partial order expression to construct a theorem of the form of (6) to generate the multiplier.

The conjunction of equational forms for each global–to–local transformation is constructed by apply the following invariant embedding principle:

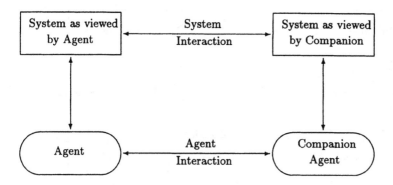

Fig. 6. Dipole Network Equivalent

"For each agent, the actions at given time t in the current interval, as computed according to (10), are the same actions computed at t when the formulation is expanded to include the previous, current, and next intervals."

Using the transitivity and convexity of the criterion the principle can be analytically extended to the entire horizon. The invariant embedding equation has the same structure as the dynamic programming equation given in (13), but with a global criterion and a global Hamiltonian instead of the corresponding local ones.

The local–to–global transformations are obtained by *inverting* the global–to–local transformations, obtained by expressing the invariant embedding equation, as an equational theorem of the form of (6). These inverses exist because of convexity of the relaxed Lagrangian and rationality of the power series.

It is important at this point to interpret the functionality of the Knowledge Decoder of each agent in terms of what it does. The multiplier described above has the effect of *aggregating* the rest of the system and the other agents into an equivalent companion system and companion agent, respectively. This is illustrated in Figure 6.

The aggregation model (Figure 6) describes how each agent perceives the rest of the network. This unique feature allows us to characterize the *scalability* of the architecture in a unique manner: In order to determine computational complexity of an application, we have only to consider the agent with the highest complexity (i.e., the agent with a local most complex criterion) and its companion.

— *Inter–agent Network*

The inter–agent communication network's main feature is to transfer inter–agent constraints among agents according to a protocol written in an equational Horn clause language. These constraints include application dependent data and, most importantly, *inter–agent synchronization*. The inter–agent synchronization strategy is very simple. An agent is synchronous with respect to the network

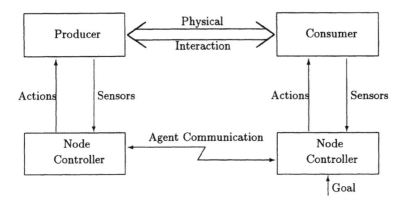

Fig. 7. Two–Agent Manufacturing System

if its inter–agent constraint multiplier is continuous with respect to the current instance of its active knowledge. Since the equational Horn clause format allows for the effective test of continuity (which is implicitly carried out by the Inferencer in the Knowledge Decoder), a failure in the Knowledge Decoder theorem results in a corrective action toward re–establishing continuity with respect to the topology defined by the current instance of the knowledge base (see [28]).

The specification of the geometry of the network, as a function of time, is dictated primarily by *global observability*. By global observability we mean the closure of the knowledge of the system as whole relative to the scope of system reactivity. One of the central research tasks of this study is to provide knowledge in the equational clause format to characterize global observability. Preliminary analysis suggests that controllability, observability and goal reachability are closely dependent on global observability among the agents. This concludes the description of our architecture.

Preliminary Results

In this section we overview some preliminary results about the structure and functionality of the architecture. A number of important theoretical extensions have been achieved. They provide some of the initial elements of a methodology for designing multiple–agent controllers.

• Formulation

We have established a precise formulation of the multiple–agent problem in terms of Declarative Control. We have identified the architectural elements and functions required for its solution. Our approach to PSC characterizes the problem via a Knowledge Base of equational rules incorporating the dynamics, constraints, and requirements of the factory system being controlled (manufacturing workstation, conveyor, processes, manpower, scheduling and planning systems, etc.). Previous prototype projects formulated the single–agent control problem for workstations in terms of Declarative Control and developed machinery to implement the inference mechanisms required. The central objective in our current research has been to define what multiple–agent control without umpire means in a shop floor control setting. We added communication actions to our model of inference and hypothesized that communications between controllers should be represented in terms of inferences each controller can make about its intent and its assumptions about other controllers. See Figure 7.

• Representation

We developed a canonical way to represent networks of controllers. We showed that, given a connectivity graph with N nodes (controllers) and the corresponding agent's knowledge bases, a network of $2N$ agents can be constructed with the same input–output characteristics, so that each agent interacts only with one companion agent, whose Knowledge Base is an abstraction of the knowledge in the network. Thus, in general, the multiple–agent controller for any topology is reduced to a set of agent pairs.

We call the agent of the pair that represents network information the Thevenin agent, after the author of a similar theorem in electrical network theory. The proof carried out by the Thevenin agent generates, as a side effect, coordination rules that define what to communicate and how often to communicate to other agents. These rules also define what the controller needs from the network to maintain intelligent control of its physical plant.

We developed a canonical way to represent the state of an agent by representing the agent's theorem–proving process as a data structure called a "proof automaton". The inference process within a controller is represented as an adjacency matrix of a proof automaton. Matrix elements represent rule applications to states in the automaton. States having sentences about the network form a submatrix that pulls communication from those specific other agents. We derived the required communication protocol among agents this way. The protocol is pulled in the sense that what an agent sends to other agents is defined by what the other agent uses in its proof automaton.

The inference process is represented as a recursive variational problem in which the criterion is an integral of a function called the logic Lagrangian over time. The logic Lagrangian maps the Cartesian product of equational rules and inference principles to the real line, thus effectively providing a hill–climbing heuristic for the inference strategy of the theorem prover. In Declarative Control the inference steps play a role analogous to action signals in conventional control.

We showed that for any measure of performance of the system, assuming unbounded bandwidth and computational resources, one obtains performance better than, or at least equal to that of the Multiple–agent version by constructing a single super–controller whose Knowledge Base contains all the knowledge of the system.

Finally, we showed that movement of material through a factory can always be characterized in terms of a production rate function and associated Volterra operator. The model we developed assumes one controlling agent for each workstation in the workcell. That controller needs an evaluation function on its own set of possible actions to measure the effect of each possible action on system performance. Other applications of Declarative Control had identified norms on average flow as a workable evaluation function, so we sought to define a flow function in terms of movement of material through a factory. The granularity of our plant model allowed flow definition down to workpiece setup and tool selection. Our controllers assumed a single operation at each workstation with no setup time. We used the model at this level to characterize unsatisfied demand in a time interval as arrival–departures. We used this result and standard figures of merit (e.g., tardiness, throughput) to define a hill–climbing function for search in the proof automaton.

We demonstrated the heterogeneous capabilities of the system by incorporating an external procedure into the Knowledge Base. The declarative semantics of the corresponding algorithm, in this case the scheduler algorithm, was described with equational terms and added to the knowledge base of each agent. The theorem treats the entire scheduler algorithm axiomatically. A description of the interface is also provided.

These are implemented in Quintus Prolog as three communicating processes (a producer controller agent, a consumer controller agent, and a simulation process). (See Figure 7.)

In order to study controller behavior, we created several files representing streams of orders. Each order included an order name, a part quantity, a process plan, a material code, an order arrival time (when the consumer controller is notified), a due time, and a priority. The controllers were sensitive to priority and due time. The order base was generated by randomly distributing order size, priority, and due time.

The consumer controller received order arrival messages from a file and communicated with the producer controller to enter the order workpieces into the producer queue. The producer performed one of four operations, using one of four cutters, and routed the workpiece to the consumer holding area. The holding area had an occupancy limit, which was one of the key model constraints. If the simulator detected a holding area size constraint, the control regime had failed. Otherwise the producer and consumer provided each other with the relevant portions of proof automata at the beginning of each control cycle. In our model each controller worried about downstream holding area availability, and so the protocol was reduced to one item of information each controller sent to its neighbors: how much room was available in the holding area. If a workstation broke, no network effect was observed until the holding area for that workstation was saturated.

A simulator was built to generate workstation states and provide controllers with evolving sensor updates and actions. The simulator had a finer model of the workstations than either controller. This allowed the controllers to demonstrate some adaptability with respect to knowledge uncertainty of their own physical plant. We simulated small segments of orders (10 to 80) with significant variation in order size, due dates, and priorities. We noted that the system managed to react to high–priority orders by splitting them (it was not built with any order splitting strategy — the figure–of–merit measurements forced this behavior).

In our implementation, the proof automata had several hundred states, and terminal states had 50 to 100 associated data items. This sizing resulted from a model with order priority, order size, due date, and holding area as the key operative constraints.

We will extend this model in essentially two directions. In the upstream direction, the model could contain means for anticipated order arrivals. The other direction would be to provide finer models of the physical plants involved, potentially down to the level of physical device control (e.g., numerically controlled machining, thermal and robot controllers, etc.). Our preliminary analysis suggests that the model size (number of states of the proof automaton) is cubic in the number of constraints; addition of a constraint such as cutter speed rate limits would increase automata size by about 50%.

Conclusions

In this section we present a few conclusions pertaining to the suitability of our proposed architecture for implementing, designing and analyzing PSC.

The two–agent architecture has been shown to be sound and complete. Further research is needed to prove and test a network with more than two agents.

The architecture is easily extensible. Additional rules can be easily added to reflected changes in goals or constraints. This feature allows easy and flexible customization of the agent behavior and goals.

Robustness was demonstrated by showing that small variations in goals produced small variations in behavior. The architecture also responded appropriately to introduced errors, such as process shutdown and interrupted material flow, making adaptations to the goals and constraints to reflect these changes.

The simulation also demonstrated the architecture's ability to incorporate externally coded algorithms, a scheduler in this case.

References

1. R. Aris, *The Mathematical Theory of Diffusion and Reaction in Permeable Catalysts*, vol. 2, Calendon Press, Oxford, 1975.

2. M. Ben-Ari, *Principles of Concurrent Programming*, Prentice-Hall, 1990.

3. K. M. Chandy and J. Misra, *An Introduction to Parallel Program Design*, Addison-Wesley, 1988.

4. N. Coleman, An Emulation-Simulation for Intelligent Controls, Proc. of the Workshop on Software Tools for Distributed Intelligent Control Systems, 37-52, July 17-19, 1990, Pacifica, California.

5. R. T. Dodhiawala, V. Jagoenthan, and L. S. Baum, Erasmus System Design: Performance Issues, Proc. of the Workshop on Blackboard Implementation Issues, AIII, Seattle, WA, July, 1987.

6. S. Eilenberg, *Automata, Languages, and Machines* (vol. A), Academic Press, New York, 1974.

7. H. Garcia and A. Ray, Nonlinear Reinforcement Schemes for Learning Automata, Proc. 29th IEEE CDC, vol. 4, 2204-2207, Honolulu, Hawaii, Dec 5-7, 1990.

8. J. Goldstine, A Rational Theory of AFL's, LCNS 71, 271-281, 1979.

9. R. L. Grossman and R. G. Larson, Viewing Hybrid Systems as Products of Control Systems and Automata, Proc. IEEE 31st CDC,vol. 3, 2953-2955, Tucson, 1992.

10. J. Guckenheimer, A. Back, M. Myers, A Dynamical Simulation Facility for Hybrid Systems, MSI Tech. Report 92-6, Cornell University, 1992.

11. J. Guckenheimer and A. Nerode, Simulation for Hybrid and Nonlinear Control, Proc. IEEE 31st CDC, vol. 3, 2981-2983, 1992.

12. R. Iseman, *Digital Control Systems*, Springer-Verlag, 1977.

13. M. H. Kaplan, *Modern Spacecraft Dynamics and Control*, John Wiley and Sons, 1976.

14. W. Kohn, A Declarative Theory for Rational Controllers, Proc. 27th IEEE CDC, pp. 130-136, 1988.

15. W. Kohn and T. Skillman, Hierarchical Control Systems for Autonomous Space Robots, Proc. AIAA, 382-390, 1988.

16. W. Kohn, Application of Declarative Hierarchical Methodology for the Flight Telerobotic Servicer, Boeing Document G-6630-061, Final Report of NASA-Ames Research Service Request 2072, Job Order T1988, Jan 15, 1988.

17. W. Kohn, The Rational Tree Machine: Technical Description and Mathematical Foundations, IR and D BE-499, Technical Document 905-10107-1, July 7, 1989, Boeing Computer Services.

18. W. Kohn, Rational Algebras: A Constructive Approach, IR and D BE-499, Technical Document D-905-10107-2, July 7, 1989.

19. W. Kohn, Cruise Missile Mission Planning: A Declarative Control Approach, Boeing Computer Services Technical Report, 1989.

20. W. Kohn, Declarative Multiplexed Rational Controllers, Proc. 5th IEEE Int. Symp. Intelligent Cont., pp. 794-803, 1990.

21. W. Kohn, Declarative Hierarchical Controllers, Proc. DARPA Workshop on Software Tools for Distributed Intelligent Control Systems, Domain Specific Software Initiative, pp. 141-163, Pacifica, Ca., July 17-19, 1990.

22. W. Kohn and C. Johnson, An Algebraic Approach to Formal Verification of Embedded Systems, IRD Tech. Rpt. D-180-31989-1, Boeing Computer Services, 1990.

23. W. Kohn, Advanced Architecture and Methods for Knowledge-Based Planning and Declarative Control, Boeing Computer Services Technical Document IRD BCS-021, 1990, in ISMIS 91.

24. W. Kohn and K. Carlsen, Symbolic Design and Analysis in Control, Proc. 1988 Grainger Lecture Series, U. of Illinois, pp. 40-52, 1989.

25. W. Kohn and A. Murphy, Multiple Agent Reactive Shop Control, ISMIS91.

26. W. Kohn and A. Nerode, An Autonomous Control Theory: An Overview, Proc. IEEE CACSD92, March, 1992.

27. W. Kohn and A. Nerode, Multiple Agent Autonomous Control Systems, Proc. 31st IEEE CDC (Tucson), 2956-2966.

28. W. Kohn and A. Nerode, Models for Hybrid Systems: Automata, Topologies, Controllability, Observability, this vol.

29. W. Kohn and T. Skillman, Hierarchical Control Systems for Autonomous Space Robots, Proc. AIAA Conf. on Guidance, Navigation, and Control, v. 1, pp.382-390, 1988.

30. W. Kuich and A. Salomaa, *Semirings, Automata, Languages*, Springer-Verlag, 1985.

31. J. W. S. Liu, Real Time Responsiveness in Distributed Operating Systems and Databases, Proc. Darpa Workshop on Software Tools for Intelligent Control, Domain Specific Software Initiative, pp. 185-192, Pacifica, Ca., July 17-19, 1990.

32. J. W. Lloyd, *Foundations of Logic Programming*, 2nd ed., Springer Verlag, New York, 1987.

33. M. Mesarovic and Y. Tashahara, *Theory of Hierarchical Multilevel Systems*, Academic Press, N.Y., 1970.

34. E. Mettala, Domain Specific Architectures, Proc. of Workshop on Domain Specific Software Architectures, 193-231, July 9-12, 1990, Hidden Valley, Calif.

35. A. Nerode, Modelling Intelligent Control, Proc. DARPA Workshop on Software Tools for Distributed Intelligent Control Systems, Domain Specific Software Initiative, Pacifica, Ca., July 17-19, 1990.

36. P. H. Nii, Blackboard Systems: The Blackboard Model of Problem Solving and the Evolution of the Blackboard Architecture, AI Magazine (7) 2, pp. 38-53, 1986.

37. L. E. Neustadt, *Optimization*, Princeton University Press, 1976.

38. P. Padawitz, *Computing in Horn Clause Theories*, Springer Verlag, 1988.

39. S. S. Sastry and M. Bodson, *Adaptive Control: Stability, Convergence, and Robustness*, Prentice-Hall, 1989.

40. M. Schoppers, Automatic Synthesis of Perception Driven Discrete Event Control Laws, Proc. 5th IEEE Inter. Symp. on Intelligent Control, 1990.

41. M. G. Singh, *Dynamical Hierarchical Control*, North Holland, Amsterdam, 1977.

42. T. Skillman, W. Kohn, et. al., Classes of Hierarchical Controllers and their Blackboard Implementations, J. Guidance, Control and Dynamics, (13), pp. 176-182, 1990.

43. J. Warga, *Optimal Control of Differential and Functional Equations*, Academic Press, 1972.

44. L. C. Young, *Optimal Control Theory*, Chelsea Pub. Co. N.Y, 1980.

Models for Hybrid Systems: Automata, Topologies, Controllability, Observability

Anil Nerode[1]* and Wolf Kohn[2]

[1] Mathematical Sciences Institute
Cornell University, Ithaca, New York 14850
e-mail: anil@math.cornell.edu
[2] Intermetrics Corporation
Bellevue, Washington
e-mail: wfk@minnie.bell.inmet.com

Abstract. By a "hybrid system" we mean a system of continuous plants, subject to disturbances, interacting with sequential automata in a network. "Hybrid Control" is the name we give to control of continuous plants by digital sequential control automata, that is, by control programs implemented on sequential automata. We associate, without using any approximations, sequential automata with continuous plants, and use this to bring sequential control automata and continuous plants into automata networks which themselves are sequential automata modelling hybrid systems. To be useful, the control automaton's ability to control plant trajectories in a hybrid system should be maintained under small changes in control laws, disturbance, trajectory, measurement, etc. We formalize this notion of controllability and observability for hybrid systems by continuity of system functions, including the input-output function of the control automaton, in non-Hausdorff subtopologies of the usual topologies on spaces of controls, sensor data, plants, disturbances, target sets, etc. These subtopologies arise from the limited ability of digital programs to discriminate between continuous inputs. These notions are appropriate for any knowledge or rule based system where automated deduction interacts in real time with the external world. This topological approach was announced in [46]. (What we call "controllability and observability" here is what we called "stability" in [46] in order to correspond somewhat more closely to usage in the literature [3].)

1 INTRODUCTION

"Hybrid control" is the name we give to the control of continuous plants by control programs implemented on a sequential automaton. A typical example is a closed loop system which consists of a continuous plant subject to both

---

* Research supported by ORA Corp., DARPA-US ARMY AMCCOM (Picatinny Arsenal,NJ) contract DAAA21-92-C-0013, the U. S. Army Research Office contract DAAL03-91-C-0027.

disturbance and control by a program implemented on a sequential control automaton. The control program reads sensor data, a "sensor function" of plant state sampled at discrete times, computes the next control law, and imposes it on the plant. The plant will continue to use this control law until the next such intervention. The sequence consisting of a "read the sensors" followed by a "compute and impose the next control" constitutes the control cycle. How, and when, to make these control law changes is the business of the control program. The challenge is to develop methodologies which, given a performance specification and system description, extract digital control programs which will force the plants to meet their performance specifications. W. Kohn's Declarative Control was developed to address this problem. Here we develop a foundation for declarative control and for modelling hybrid systems in general.

We can program inexpensive digital control chips which change the plant control law used every few microseconds. Plants that cannot be controlled by a single conventional control law often can be made to behave acceptably by imposing diffferent control laws on the plant in successive time intervals. This flexibility is useful for controlling many unstable high performance systems such as aerospace vehicles and structures, power grids, and automated intelligent manufacturing facilities.

There are numerous bottom-up methods for trying to identify the modes of behavior of plants, exercising different control laws in different modes ([79], [78], [31], [73], [74]). Kohn ([35], [38], [39], [40], [41], [42], [43], [45], [50]) has developed Declarative Control as a "top down" design method based on relaxed Lagrangians and automata. Recently Kohn and Nerode ([46], [47], [48]) have been developing foundations for declarative control and other hybrid theories in control and AI, along the lines of [62], [63]. This paper describes part of the needed foundations.

We believe that one should study digital control of continuous plants at the same intellectual level of sophistication as has been applied in computer science for the analysis of concurrent and distributed programs and in control theory for the analysis of single continuous control laws. This paper is an exploration of ideas and methods in hybrid systems, not a systematic development of the subject.

Some Mathematical Actors   Here is a brief description of some of the mathematical actors in this control drama. Each individual control law used for an interval is usually a standard smooth law for an off the shelf plant controller, out of a space of control laws. The actual control function $c(t)$ used is pieced together from the standard laws used in successive time intervals and is usually sectionally continuous, possibly with jump discontinuities at some of those points in time when a new control law is invoked. When the plant is governed by the usual ordinary differential equations, the trajectory $x(t)$ of the plant is usually continuous. However, at least at the times t when the script changes the control law, the plant velocity x' , that is, the direction of the vector field, often has a jump discontinuity from left value $x^{-\prime}(t)$ at the end of application of the old control law to right value $x^{+\prime}(t)$ at the beginning of application of the next control law. Other such jump discontinuities in x' can also arise from the "discontinuous right hand sides" of ordinary differential equations [22].

Two Worlds Digital programs live in a world in which inputs, outputs, and internal states are expressed using finite alphabets. The mathematical language for describing digital programs is the language of logic. The denotational semantics of the language may be state automata, or Scott continuous maps between cpo's, or process algebras, among other possibilities. Conclusions about behavior of digital programs are often drawn by reasoning essentially using logical deductions expressible in Dynamic, Hoare, Temporal, or other program logics. Methods and theorems of logic and computer science underlie verifying such properties of programs as termination or fairness or correctness.

Networks of digital programs have been studied extensively in the theory of distributed and concurrent systems. These are often described by systems of logical formulas or equations. The solutions to these systems can be thought of as state-valued functions of discrete time. These subjects are well-known to computer scientists.

In contrast, continuous devices live in a real time world in which inputs, outputs, and internal states are usually points in complete metric spaces. They are surrounded by such structures as function spaces of trajectories, or control functions, or sensor data, or disturbances. The behavior of continuous devices is usually modelled by solutions to systems of ordinary differential equations or as solutions to variational problems.

The mathematical language for describing continuous devices is conventionally the language of functional analysis. One uses theories and models and numeric and symbolic algorithms from differential equations, functional analysis, calculus of variations, dynamical systems, differential geometry, and optimization. These ideas are used to establish such features as controllability, observability and reachability. Generally these are properties of trajectories in response to control and in the presence of disturbances.

Networks of continuous devices have been studied for a long time. Such networks are well-studied by differential equations and variational calculus specialists, physicists, and control engineers, but not by most computer scientists.

Hybrid systems are systems of sequential and continuous devices interacting at discrete times. How do we reconcile the discrete time of sequential devices with the continuous time of continuous devices. How do we reconcile the finite set of states of a real time automaton with the continuum of states of a continuous device? Our description is deliberately quite general. We argue from first principles. The framework is intended to be a common substrate for all contemporary control theories on the continuous side and any of the contemporary logics and semantics of digital programs on the digital side.

First Level of Modelling Our first level of modelling is represented by Figure 1.

First, we discuss the use of non-deterministic sequential automata as control automata with input alphabet the set of sensor measurements, output alphabet the set of control laws. Thus, the control automaton will be an abstract se-

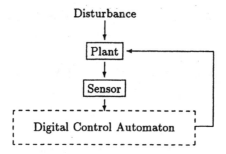

Disturbance

Plant

Sensor

Digital Control Automaton

Fig. 1.

quential automaton which operates on an infinite input alphabet of real number sensor measurements and has an infinite output alphabet of control laws. Second, we develop a sequential non-deterministic automaton model for continuous plants associated with plants given by a non-autonomous ordinary differential equation, with changes in control and disturbance introduced at times t_i, where $\Delta_i = t_{i+1} - t_i$ is variable. There are no approximations involved. Third, we model event-driven hybrid systems as non-deterministic automata networks in which a non-deterministic sequential control automaton interacts with a plant as described above at times t_i. These are the times of communications of plant and control automaton. We require that the time sequence t_i satisfiy a realizability requirement. This plays the role that fairness plays in concurrency. Fourth, we specialize this to autonomous plants where time is broken up into successive intervals of fixed length $\Delta = t_{i+1} - t_i$. Here hybrid systems are non-deterministic automaton networks with transitions at times $t_i = t_0 + i\Delta$. At the first level of modelling, by imposing more structure on the control automaton states in the fashion of predicate or temporal dynamic logic semantics, one can formulate and pursue proofs that specific control automata achieve performance specifications.

But we do not think this first level of modelling permits formulation and solution of the question of controllability and observability of hybrid systems analogous to that pursued in ordinary control theory. A cause of difficulty is that the control automata in the first level of modelling have infinite input and output alphabets, while a real time control automaton always operates based on an internal finite automaton with finite input and output alphabets. To supply a basis for a theory of controllability and observability we go on to a second level of modelling.

Second Level of Modelling Our second level of modelling models the control automaton in more detail. It is decomposed into a series of three successive units. The first unit is an analog to digital converter which converts each of an infinity of real number measurements into one of possibly finitely many symbols in the input alphabet of the "internal control automaton". (The internal control automaton, in practice, is a finite state automaton with finite input and finite output alphabets.) The second unit is the internal control automaton,

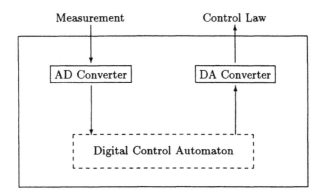

Fig. 2.

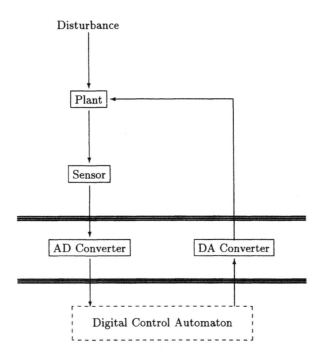

Fig. 3.

which has as input a symbolic representation of a measurement and produces as output a symbolic representation of the next control law to be imposed on the plant and how long to apply it. The third unit is a digital to analog converter which converts these output symbols representing control laws and times into the actual control laws imposed on the plant.

These three subunits respectively represent: transition from the continuous to the digital; transition from digital to digital; and, finally, transition from digital to continuous. Figure 2 represents this.

When we substitute Figure 2 into Figure 1, we get Figure 3. The digital world is represented below the lower heavy line. The analog world is represented above the upper heavy line. The interface between these two worlds consists of the converters and is represented between the heavy lines.

The Problem of Analog to Digital Conversion   This is the problem of converting each real number from a real interval I representing measurements to a letter in a finite alphabet $\{1, \ldots, k\}$, representing control laws which are outputs of a control automaton. If we think of conversion as a function $f : J \to \{1, ..., k\}$, then if the conversion is to be robust, we expect extremely close measurements from J to map into the same one of $1, \ldots, k$. That is, we expect the inverse image of any $i = 1, \ldots, k$ under f to be open. If we endow $\{1, \ldots, k\}$ with the discrete topology, which it has as a subset of the reals anyway, then f is continuous. But this is impossible unless f is constant, since the intermediate value theorem says a continuous function on an interval takes on any value between two distinct values. Measurement functions that are constant are not worth much, any more than a stopped clock. This is the root of difficulties in understanding controllability and observability for hybrid systems. We discuss this issue only at the second level of modelling, via non-Hausdorff subtopologies of the natural topologies of the spaces involved. Non-Hausdorff topologies on control and measurement spaces arise from this contrast of the possibly finite output alphabet of the internal control automaton, compared with the continuum of available measurements that have to be input to the control automaton. With this second topological level of modelling, a variety of controllability and observability definitions for hybrid systems can be investigated. This also dictates developing a topological and functional analytic automata theory, which we only hint at here.

Remark   The traditional treatment of analog to digital conversion [33], [53] rounds reals to discrete values by applying a transformation discontinuous in the standard Hausdorff topologies. We regard this as only a first approximation to a physically realistic modelling of such transductions. Such models do not impose all the mathematically and physically necessary realizability requirements. The small topologies which will be introduced later constitute a refinement with respect to understanding controllability and observability of hybrid systems using such transductions.

Discontinuity of functions with respect to small topologies can actually be detected, unlike discontinuity with respect to the usual Hausdorff topologies. Observation of such a discontinuity, which indicates that the system is violating its performance specification, can be used in real time to force a recomputation and change of the control automaton being used. This is the main adaptive feature of our multiple agent hybrid control architecture ([47], [48], [49]).

Why Allow Non-Deterministic Automata? Real time programs, including control programs, can best be modelled as denoting non-deterministic sequential automata. A digital control program runs on hardware, and started twice in the same state, may terminate differently for one of many reasons. This hardware is certainly then to be regarded as a non-deterministic automaton. Some of the reasons are listed below.

- There is a time limit on computation. The number of iterations achieved in that time limit may vary from run to run.
- Default values may sometimes have to be used because time allotted for computation has run out before termination.
- A multi-tasking system may be allocating computing time to many competing programs. These programs compete for execution time. The competitors may be different from run to run, so the portion of the run completed by time-out may be different.
- Fault tolerant systems deliberately introduce redundancy, and always use alternate execution sequences when malfunctions occur.
- Hybrid systems are inherently concurrent. Models for proving concurrent programs correct can utilize non-determinism. See Apt and Olderog [2] and Nerode, Yakhnis, and Yakhnis [66], [67].
- The fact that the set of states accessible from a given state can be empty can be used to model stuck states.
- The set of states accessible from a given state may be quite large, to model stochastic execution.
- Non-determinism can be used to reflect timing faults and other error conditions.

Why Allow Non-Deterministic Plants?

- There may be unspecified small disturbances, statistical or deterministic.
- Solutions to ordinary differential equations with given initial conditions need not be unique without a Lipschitz condition.
- Solutions to variational minimization problems with given initial conditions need not be unique in the absence of convexity assumptions.
- Ordinary differential equations are not perfect descriptions of the physical processes they model. Coefficients, boundary conditions, initial conditions, constraints will be slightly off.
- Non-linear equations do not lend themselves easily to computing error bounds for solutions resulting from inaccurate coefficients and boundary conditions.
- Striking changes of qualitative behavior may result from small changes of parameter and input values, which we cannot prevent, and can be modelled by non-determinism.
- Correctness proofs that trajectories meet performance specifications had better be robust to small changes in the numerical description of the problem.

- The theory of set valued solutions to differential equations and variational problems is available, having been extensively developed in the last twenty years. See Aubin ([6], [7], [8]).

2 NON-DETERMINISTIC AUTOMATA

Definition   Here is a traditional definition of an (input-output) automaton with one input and one output. We are given

States   A non-empty set of states S

Input   A non-empty input alphabet I

Output   A non-empty output alphabet O

Transition Function   A function assigning to each pair (i, s) consisting of an input alphabet letter i and a state s, a (possibly empty) set of states $M(i, s)$. $M(i, s)$ is called the set of immediate successor states to (i, s).

Output Function   A function O assigning to each state s an output alphabet symbol $O(s)$.

Execution Sequences   An execution sequence corresponding to input sequence $i_0, i_1, \ldots$ is a finite or infinite sequence of pairs $(i_0, s_0), (i_1, s_1), \ldots$, such that for all i, $s_{i+1} \in M(i, s_i)$. The output sequence for this execution sequence is $O(s_0), O(s_1), \ldots$.

Thus, corresponding to a single input sequence may be zero, one, or many output sequences of varying length as we vary the sequence $s_0, s_1, \ldots$ of states.

An automaton will be called sequential deterministic if $M(i, s)$ always has exactly one member.

If $\underline{i}$ is a sequence of input symbols $i_0, i_1 \ldots$, the set of execution sequences with this input sequence is the set of all sequences $\underline{s}$ of states $s_0, s_1 \ldots$ such that for all j, $s_{j+1} \in M(i_j, s_j)$. For a given input sequence, the corresponding execution sequences form a tree, whose maximal branches we call the maximal execution sequences, or plays. These may be finite or infinite. If we apply the output function O coordinatewise to an execution sequence, we get an output sequence $\underline{O}(\underline{s})$, and to the tree of execution sequences corresponds the tree of output sequences.

To introduce also automata with no input, or respectively no output, leave out all reference to the input, respectively output, alphabet. To deal with an automaton with multiple input lines with corresponding alphabets, use the Cartesian product of the input alphabets of these input lines as the input alphabet. Treat multiple output lines of automata similarly.

Sequential Deterministic Automata   These are automata in which every state has exactly one successor.

Automaton Networks   A network of input-output automata constitutes another automaton. That is, if we take some automata and connect some input lines, one on one, to some output lines with the same alphabet, we get a network. The network input lines correspond to those input lines of the component automata which are not connected to any output lines of any component automata.

The output lines of the network are similarly characterized. Those lines of the component automata which are neither input nor output lines for the network are the "internal lines" of the network.

As is standard in automata theory, the set of states of the network can be taken to be the Cartesian product of the states of the constitutent automata, the output function of the network is the coordinatewise product of the output functions of the output lines of the network. There are many algebraic tools useful for determining network behavior from the behavior of constituent automata, and for decomposing networks into constituent automata, including the Kron-Rhodes theory and Schutzenberger series [52]. Of course if the constituent automata are sequential deterministic, so is the network.

Reconciling Time Scales Non-deterministic automata are meaningful without any restriction on cardinalities of states or alphabets. We wish to use this to represent differential equations in hybrid systems in common terms with those usually used for digital control programs implemented on automata, but there is a hitch. In the usual nondeterministic sequential automaton model, as described above, the state transition functions operate in abstract non-negative integral time, state transitions take place at abstract integer times. But the state of an ordinary differential equation is a function of real numbers as time. How do we get these two time scales to interact? Since the MIT Radiation Laboratory book series shortly after World War II, it has been commonplace in control and systems theory to exposit a common generalization of machines with discrete and machines with continuous time [79]. For instance, real time can be represented by the additive group of reals, abstract integer time by a discrete additive subgroup of the reals. The most general discrete additive subgroup consists of all integer multiples of a fixed positive real Δ. So if the sequential automaton is implemented to change state at t_0 and then at regular intervals Δ, the times are simply a coset of a discrete additive subgroup. This commonality does not, however, of itself model what happens when a sequential automaton and a differential equation interact asynchronously. It also does not model what happens when the sequential automaton does not change state at uniform intervals. This is a problem that has to be faced head on by anyone modelling hybrid systems.

Real Time Automata A realizable execution sequence for a non-deterministic sequential automaton is an execution sequence together with a map of the terms of that execution sequence $(i_0, s_0), (i_1, s_1), \ldots$ to a sequence of reals $t_0, t_1, \ldots$ which is realizable in the following sense.

Definition A sequence $t_0, t_1, \ldots$ of reals is realizable if

1. $t_0 < t_1 < t_2 \ldots$
2. The $t_{i+1} - t_i$ have a positive lower bound.

There is also a notion of real time automaton with a realization of all execution sequences. There we have a map of all finite execution sequences into corresponding sequences of reals of the same length, preserving extension of sequences, with the same positive lower bound for all differences. This is useful for developing proofs of correctness, but is not needed here.

For convenience, we always start with $t_0 = 0$.

Remarks A realizable execution sequence has two properties. First, in a finite interval of time, there are only a finite number of times at which input and output symbols are produced. Second, there is a positive lower bound to transition times. These requirements are grounded in current physical theory. First, there is the quantum-mechanical limitation that state transitions are not deterministic if the time scale is small enough. Second, there is the relativistic limitation on the velocity of light which limits the speed of communication over a non-zero distance. One can develop formally the mathematics for systems violating this realizibility requirement. Philosophers and mathematicians may discuss such things, but one does not build "Zeno machines" where the n th state transition is at time $(1 - 1/2^n)$ seconds. This execution requirement must be enforced by the implementation, and has the same role here that fairness requirements have for concurrent programs, see [16], [11], [2], [55], [66].

3   NON-DETERMINISTIC PLANTS

We give a general description of continuous plants given by ordinary differential equations. But we could as well have based the later discussion on variational constraints, or difference equations, or any combination of other formulations of control problems. Differential equations merely play the role of making the discussion more concrete. What is required is the notion of a constraint on trajectories over an interval, and the notion of a control and disturbance to trajectories satisfying these constraints, expressed through spaces C of control laws c, D of disturbances d, $TRAJ$ of trajectories (solutions) X, and certain mappings between them. In variational calculus and ordinary differential equations and current control theories there is a wide variety of choices of these spaces and maps. The choice depends on what class of equations, disturbances, control, and solutions are regarded as admissible. We wish a formulation in which any one of these competing theories can be used, mix and match, but a formulation not so abstract as to obfuscate analysis, computation, and modelling.

For convenience of notation, we will look only at functions on closed intervals $[r, s]$ of the right half line $[0, \infty]$, where s is allowed to be ∞. The set C of control laws is assumed to be a set of functions each of which has domain $[0, \infty]$ with values lying in a prescribed set of control values CV. Similarly the set D of disturbances is a set of functions each of which has as domain a closed interval and whose values lie in a prescribed set of disturbance values DV.

Now let $x'(t) = f(x(t), t, a, b)$ be a vector ordinary differential equation with parameters a, b. Suppose a control function $c(t)$ and a disturbance $d(t)$ are given with domain $[0, \infty]$. Then a solution $x(t)$ on an interval $[r, s]$ is a function defined on $[r, s]$ satisfying

$$x'(t) = f(x(t), t, c(t - r), d(t - r))$$

on that interval.

Remark   We later piece together such solutions on successive intervals to get longer trajectories arising from successive intervals of disturbance and controls, in which case we might better think of the resulting trajectories as solving the integral equation

$$x(t) = \int_r^t f(x(u), u, c(u - r), d(u - r))du,$$

since we will generally get trajectories $x(t)$ which are continuous everywhere and differentiable except at isolated times. That is, there may be distinct left and right hand derivatives $x^{-\prime}(t), x^{+\prime}(t)$ at times when we change from one control law to another, or where the equation has a "discontinuous right hand side".

In specific applications the plant positions, solutions, control laws, and disturbance functions are usually subsets of specific Banach function spaces. They inherit appropriate norms, metrics, and topologies.

Let $TRAJ$ be the set of all solutions obtained as we vary the finite interval $[r, s]$ and the control and disturbance functions used on that interval. Note that we are allowing non-deterministic equations. That is, there may be many solutions X satisfying a given initial condition $X(r) = x_0$.

Closure Conditions Let F be a set of functions, each defined on a closed interval of non-negative reals, finite or infinite. Then F has a closure, PF (for piecewise F) under the first three operations below; we impose the fourth closure condition for solutions of autonomous equations.

Restriction   Any restriction of a member of PF to a smaller closed interval is also a member of PF.

Concatenation   If f_i in PF has domain $[r_i, r_{i+1}]$, $i = 1, 2$, and the f_i agree at r_2, then $f = f_1 * f_2$ is also in PF, where $f(t) = f_1(t)$ for $r_1 \le t < r_2$, $f(t) = f_2(t)$ for $r_2 \le t < r_3$.

Infinite Concatenation   If $f_i \in PF$ has domain $[r_i, r_{i+1}]$, $i = 1, 2, \ldots$, and the r_i tend to ∞, and each f_i agrees with f_{i+1} at r_{i+1}, then $f = *_\omega f_i$ is also in PF, where for $i = 1, 2, \ldots$, $r_i \le t < r_{i+1}$, $f(t) = f_i(t)$.

Translation   (For the autonomous case) If $r_1 < s_1$, $r_2 < s_2$, $r_2 < r_1$ and $s_1 - r_1 = s_2 - r_2$ and f_1 has domain $[r_1, s_1]$ and is in PF, then so is f_2, where $f_2(t) = f_1(t + (r_1 - r_2))$.

Let $PTRAJ$ be the set of piecewise solutions, which we call piecewise trajectories. Let PC be the set of piecewise controls, except that we allow concatenation of controls without requiring agreement at endpoints. We keep the left endpoint values when there is a conflict. Also PD is the set of piecewise disturbances.

Plant Automata   A control (disturbance) is one with domain $[0, \infty]$. With any plant as described above, we associate a sequential non-deterministic automaton, its plant automaton.

States   The set of states is $TRAJ$.

Input Alphabet Let the input alphabet be the set of triples (Δ, c, d) consisting of a positive real Δ, a control function c , and a disturbance function d. This is a three input line automaton with three input alphabets, the positive real times, the control functions, and the disturbance functions.

Transition Function Suppose that state X has domain $[r, s]$ and that the input is (Δ, c, d). Then state Y is an immediate successor state to X if the domain of Y is $[s, s + \Delta]$, $Y(s) = X(s)$, and Y solves

$$Y'(t) = f(Y(t), t, c(t - s), d(t - s))$$

on $[s, s + \Delta]$.

Remark A pair of controls and disturbances must be used to justify that $X \in TRAJ$. These are unconnected with the control and disturbance c, d used here in $[s, s + \Delta]$.

So the plant induces a non-deterministic automaton, with time Δ as an explicit input. This corresponds roughly to the usual differential equations reduction of non-autonomous to autonomous systems. There is also a natural output function.

Measurement Functions We usually have available for control purposes only partial information given by sensor observations. The sensors communicate only partial information about the plant trajectory. Generally, if plant sensors have previously reported sense data at time t, at the next reporting time $t + \Delta$ they report some sort of mean or average over the equation solution or plant trajectory on interval $[t, t + \Delta]$. The sensors are thus best represented as functionals on solutions to the differential equation. The usual simpler assumption is that sensors record point information, that is, are functions of current plant position at the time the measurement is communicated rather than of the whole solution over the interval $[t, t + \Delta]$.

With the automaton above, we envisage being in state X at time t, in state Y at time $t + \Delta$. We regard the sensors as represented by a "measurement function" $Meas$. In hybrid systems the intent is to use $Meas(X)$ as input to the control automaton at time $t + \Delta$.

Measurement Values M is the set of measurement values of the sensors. (This is usually a set of n-tuples of real numbers.)

Measurement Function The measurement function $Meas(X) : TRAJ \rightarrow M$, where M is the set of measurement values.

Output Function of Plant Automaton The output for plant automaton state X in $TRAJ$ is the value of the measurement function $Meas(X)$.

Control Scripts We express time and control functions in an alternate form. A pair (c, Δ), with c a control function and Δ a positive real is referred to as a finite control. A sequence of finite controls

$$(c_0, \Delta_0), (c_1, \Delta_1), \ldots$$

is called a control script provided the Δ have a positive lower bound. Setting $t_0 = 0$,

$$t_{n+1} = \Delta_0 + \Delta_1 + \ldots + \Delta_n,$$

we get that $t_0 < t_1 < t_1 < t_2 \ldots$ is a realizable sequence. This implies that either the control script is finite or the $t_i \to \infty$. Corresponding to the control script is a control function $c \in C$ defined as follows. Let $c(t) = c_i(t - t_i)$ for $t_i \leq t < t_{i+1}$. The closure conditions ensure that c is in C. But c does not contain information about the t_i.

Finally, to each control script $\{(c_i, \Delta_i,)\}$ and disturbance function $d(t)$ there corresponds a disturbance script $\{(d_i, \Delta_i)\}$, where $d_i(t) = d(t + \Delta_i)$. This divides the disturbances used in the disturbance script into parts like those in the control script. Similarly there are solution scripts each consisting of a sequence of corresponding solutions on successive intervals.

4   EVENT DRIVEN HYBRID SYSTEMS

An event-driven hybrid system consists of a plant, a control automaton, a plant sensor, and a plant disturbance, considered for realizable execution sequences. We model the control automaton by a non-deterministic automaton. We model the plant by an ordinary differential equation, and the differential equation by the corresponding plant automaton.

The plant automaton has three input lines which respectively carry positive times Δ, disturbances d, and control laws c. Its states are solutions $X \in TRAJ$, each defined on a closed interval. It has one output line, carrying measurement values M of output function $Meas(X)$. The control automaton has one input line carrying measurement values M. It has two output lines, one carrying a

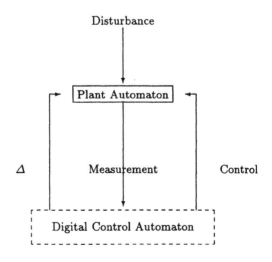

Fig. 4.

control function c, the other a positive time Δ. Here Δ is thought of as an update interval from the last transition to the current transition, and can be any positive real number. According to the realizability requirement, a used sequence of such Δ determining transition times should have a fixed positive lower bound. Technically, we are thinking of Δ as ranging over just another input alphabet, which is the positive reals. See Figure 4.

The hybrid system is thus a two automaton network consisting of the control automaton and the plant automaton. Here are the connections, or internal lines.

1. The output line of the plant automaton carrying measurement values is identified with the input line of the control automaton carrying the same.
2. The output line of the control automaton carrying control functions is identified with the input line of the plant automaton carrying the same.
3. The output line of the plant automaton carrying positive real numbers Δ is identified with the input line of the plant automaton carrying the same.

The hybrid system has one input line, and this carries the disturbance functions. There is no output line for the hybrid system.

Here is a direct description of states and state transitions for the hybrid system. Let S range over states of the control automaton, let M_c be its transition function. Let M_p be the transition function of the plant.

A state of the hybrid system is a pair (X, s), where X is a state of the plant automaton and s is a state of the control automaton. That is, the hybrid system has state space $TRAJ \times S$, where S is the set of states of the control automaton. The input alphabet is the set of all disturbance functions d. There is no output alphabet, although one can be naturally added on in several ways. The transition function $M_h(d, (X, s))$, with input d and state (X, s), of the hybrid system is defined as follows. The output function of the control automaton computes a finite control from state s , which contains a time Δ and a control function c. (The Δ tells how long c is to be employed to control the plant, during which time d is active.) Then $i = (\Delta, c, d)$ is an input symbol for the plant automaton. Also, $m = Meas(X)$ is an input symbol for the control automaton. The transition function

$M_h(d, (X, s)) = M_p(i, X) \times M_c(m, s)$.

Remark   The input alphabet for the control automaton was said to be the set M of possible sensor measurements. However, it is convenient to add an extra symbol, *Continue*. *Continue* is interpreted when received by the control automaton as meaning "no new measurements received now". Similarly we add *Continue* to the output alphabet of the control automaton. Then *Continue* is interpreted, when received by the plant, as "continue to use the same control function as before". Here is why this is a convenient convention. A consequence of imposing realizability on execution times is that the sequence of times at which the control automaton receives inputs should be realizable and the sequence of times at which the control automaton transmits an output should be realizable. In our hybrid systems model we have collapsed these two sequences into a single realizable sequence of times of mutual communication in which reception and transmission take place. *Continue* makes this collapse of receiving and transmitting non-restrictive. First, if we combine two realizable time sequences into one, the latter is also realizable. Second, what do we do if, in an application, there are times when we wish the control automaton to send a control law, but at that time we are not receiving measurements? We simply think of *Continue* as being received. What if there are times when the control automaton is receiving a measurement, but has no control law to output; then it sends *Continue*. Thus *Continue* is a convenient fiction that at all communication times, there is both an output symbol delivered from the sensor as input to the control automaton and also an output symbol delivered from the control automaton as input for the plant, and eliminates the need for two separate time sequences for transmission and reception.

Hybrid System Execution Sequences   Here is an explicit description of an execution sequence of the hybrid system above, given disturbance d and initial position x_0. A finite execution sequence of the hybrid system above is a finite sequence of triples $((c_i, \Delta_i), (d_i, \Delta_i), (X_i, s(t_i)))$, $i = 0, 1, 2, \ldots, n$ such that

1. $\{(c_i, \Delta_i)\}$ is a control script.
2. $\{(d_i, \Delta_i)\}$ is the disturbance script for d.
3. The function X_i is the successor to the plant automaton state X_{i-1} on input (Δ_i, c_i, d_i) , is defined on interval $[t_i, t_{i+1}]$ and is a solution to the differential equation on that interval with control $c_i(t - t_i)$, disturbance $d_i(t - t_i)$, and initial conditions $X_i(t_i) = X_{i-1}(t_i)$, $X_0(0) = x_0$ if the initial position is x_0.
4. The state $s(t_{i+1})$ of the control automaton at time t_{i+1} is in the set of successor states of the control automaton to the state $s(t_i)$ with input measurement value $Meas(X_i)$.

Here X_i is a solution to the differential equation on $[t_i, t_{i+1}]$ which starts with the position prescribed at the end of the previous time interval by X_{i-1} and with control and disturbance prescribed by the scripts. It is one of possibly many such solutions arising by a non-deterministic execution of the differential equation. The whole trajectory obtained is defined piecewise on intervals, that is $X(t) = X_i(t)$ for $t \in [t_i, t_{i+1}]$. Our definition of trajectory space T, by closure conditions, says X is in T.

Sequential Deterministic Plants   A plant, given as above by a differential equation and associated sets, is called sequential deterministic if for every time interval $[r, s]$, control function, disturbance function, and initial condition x_0 at r, there is exactly one solution X on that interval of the equation

$$X'(t) = f(X(t), t, c(t - r), d(t - r)).$$

Because of the reduction of event-driven hybrid systems to plant automata, the following theorem simply expresses the fact that a deterministic sequential automaton has a unique execution sequence,   where the automaton is the two automaton network which is the hybrid system. The previous 1991 proof was less perspicuous.

Theorem (Kohn-Nerode, 1991) Suppose we are given a hybrid system consisting of a sequential deterministic plant and a sequential deterministic control automaton, described as above. Suppose the greatest lower bound of the Δ in finite controls (c, Δ) in the output alphabet of the control automaton is positive. Suppose we are given, as initial conditions at time $t_0 = 0$, a disturbance function $d(t)$, an initial plant position $X_0(0)$, and an initial control automaton state s. Then there is a unique maximal hybrid system realizable execution sequence satisfying these initial conditions, and a uniquely determined plant trajectory over $[0, \infty]$.

Proof A network of two sequential deterministic automata is itself a sequential deterministic automaton, as remarked above. This network, as a deterministic sequential automaton, has a unique maximal execution sequence.

This makes it clear how to regard hybrid systems as ordinary sequential automata, provided that time intervals are regarded as inputs and that we allow automata with infinitely many symbols and states. Note that we make no assumption of constant Δ's or of infinitely many Δ's. There may be a time of last output of the control automaton, when it issues a control function used forever afterward. There is a similar theorem for general hybrid systems.

It is worthwhile to describe the operation of the system directly to see the interaction of the control automaton and the plant.

Think of the finite control (c_i, Δ_i) and the disturbance (d_i, Δ_i) as input to the plant automaton which solves its differential equation

$$X_i'(t) = f(X_i(t), t, c_i(t - t_i), d_i(t - t_i))$$

with initial condition $X_i(t_i) = X_{i-1}(t_i)$ for the new state X_i of the plant automaton, defined on domain $[t_i, t_{i+1}]$, $t_{i+1} = t_i + \Delta_i$, and delivers a new output $Meas(X_i(t_{i+1}))$ as input to the control automaton.

Remarks

1. In a sequential deterministic hybrid system for given initial conditions, the plant has a unique control script, disturbance script, solution script $\{X_i(t)\}$, control function, and trajectory $X(t)$, while the automaton has a unique state sequence $\{s(i_t)\}$ representing states at times t_i.

2. If we allow *Continue* as an output for the control automaton, and after time t_i, only *Continue* is issued, the last control $c_i(t)$ persists forever.

3. No approximation to either the control or the disturbance is involved, merely the packaging of interactions into intervals.

4. It was not assumed that the plant was autonomous. Explicit dependence on time is allowed.

5. Because we assumed for sequential deterministic hybrid systems that solutions exist everywhere and at all times, the $X_i(t)$ are differentiable except at the t_i, where the left and right hand derivatives can be distinct. The closure conditions on $PTRAJ$ assure that these piecewise differentiable continuous trajectories of the plant are in $PTRAJ$. This is the situation in both Lie semigroups and relaxed control.

We have said that the challenge is to develop methodologies which, given a performance specification and system description, extract digital control programs which will force the plants to meet their performance specifications. What is a performance specification? With any execution sequence is correlated a trajectory and a control function in the obvious way.

Performance Specifications   Suppose a hybrid system is specified as above. Define a performance specification for a hybrid system as a set of trajectories X of the plant. Say that a hybrid system satisfies a performance specification if for every admissible disturbance and initial condition, all resulting trajectories of the hybrid system in all execution sequences are in the performance specification.

Remark   For those who use logical formulas about trajectories as performance specifications, the set of trajectories satisfying the formuula is what we take to be our performance specification.

The Δ- Plant Automaton for an Autonomous Equation   We now fix a Δ for the plant automaton of an autonomous equation, so updates are at uniform times. Autonomous equations are equations of the form

$$x'(t) = f(x(t), a, b),$$

where t does not appear as an independent variable in f. We impose a fixed positive time Δ separating communications, so that $t_{i+1} - t_i = \Delta$, or $t_{i+1} = t_0 + n\Delta$. Then we speak of the Δ-plant automaton. If we drop reference to Δ, except that the set of automaton states is the set $TRAJ(\Delta)$ of all solutions X over the closed interval $[0, \Delta]$, here is what we get. This automaton will have as input alphabet the set $FC \times FD$ of pairs of control functions and disturbance functions, and one output alphabet, the set M of sensor measurements, with output function $g : TRAJ(\Delta) \to M$. The plant automaton changes state exactly at times $t(i) = t(0) + i\Delta$.

The transition function for the plant automaton assigns to each pair $(X, (c, d))$ consisting of a state $X \in TRAJ(\Delta)$ and an input $(c, d) \in FC \times FD$, the set $M(X, (c, d))$ of successor states $Y \in TRAJ(\Delta)$ such that $X(\Delta) = Y(0)$. Control scripts are simply sequences of control functions, a disturbance script is simply a sequence of disturbance functions.

Remark   If the plant is sequential deterministic, the initial position $X(0)$ determines the solution X over $[0, \Delta]$. So for sequential deterministic plants we can use PS, the usual plant states of the dynamical system, instead of $TRAJ(\Delta)$. In this case, an execution sequence of the hybrid system is a sequence of quadruples (c_i, d_i, x_i, s_i), c_i the i-th control, d_i the i-th disturbance, x_i the i-th position (at time t_i), and the state s_i the i-th control automaton state (at time t_i).

Dual Aspects   We can regard any real time sequential automaton as a continuous time device by thinking of it as maintaining its state at time $t(i)$ for all t with $t(i) \leq t < t(i + 1)$, changing state at $t(i + 1)$. This makes the hybrid system into a network of continuous time devices, because the control automaton is now regarded as a continuous time dynamical system in which the state functions have jump discontinuities. This is the same mathematical device used to regard discrete probability distributions on countable discrete sets of reals as probability distributions on all of the reals, or for relating δ-functions as measures or distributions to step functions as integrals, and leads naturally to a function space duality point of view on hybrid systems. Looking at the plant as a nondeterministic sequential automaton, and alternately replacing the control automaton by a real time dynamical system, constitute dual discrete and continuous views of the hybrid system. It is because of these dual interpretations that one can use a combination of automata theoretic methods from the Δ-automata interpretation and analytic methods from the continuous dynamical systems interpretation to investigate the behavior of hybrid systems. See Grossman and Larson, [25].

5 TOPOLOGICAL HYBRID SYSTEMS

Traditional continuous control system theories deal with the existence of, and approximation to, control laws in function spaces which produce plant trajectories which meet performance criteria, including controllability and observability. All theories of controllability and observability of continuous sytems are topologically based. In conventional control theories for continuous plants the continuity of input-output functions of a continuous network in appropriate topologies and in parameters of the system is a basic expression of controllability and observability. That is, controllability and observability should mean that for the controlled system, small changes in input and relevant parameters yields small changes in output. The topologies imposed on the relevant spaces differ from control theory to control theory and application to application. Many non-topological notions have to enter into the use of controllability and observability, such as measures, Lagrangians, Lyapounov functions, etc. But continuity in topologies plays a basic role in all treatments of controllability and observability.

Hybrid systems include continuous plants as subsystems. So it should be no surprise that topologies should play a basic role in controllability and observability of hybrid systems too. In [46] we suggested the use of auxiliary "small" topologies on the standard spaces to distinguish well-behaved from badly behaved systems extracted by Kohn's Declarative Control from the point of view of controllability and observability. What is perhaps surprising at first is that what enters is continuity with respect to "small" subtopologies of the standard topologies on the relevant spaces. These small topologies are generated by specific covers of those spaces by open sets (open covers). Generally, these are non-Hausdorff topologies satisfying no separation axioms, although each is naturally associated with a T_0-space. In many cases these topologies are literally finite topologies generated by finite open covers. These topologies usually reflect the fact that the control automaton in real time systems is guided by a finite state "internal control automaton" to be described below. A small topology imposed on a measurement space M reflects a realizability requirement that the internal control automaton has limited ability to distinguish between measurements described by real numbers. That is, this automaton reacts the same way to all input measurements which are indistinguishable by open sets in the small topology on M.

A small topology imposed on plant solution spaces $TRAJ$ reflects a realizability requirement that the goal intended for a control function imposed by an internal control automaton at the beginning of a time interval Δ, can only guarantee that the solution is in a prescribed open set in a small topology on $TRAJ$ at the end of that time interval. This limitation is a consequence of the inability of the internal control automaton to issue arbitrarily finely tuned control laws based on arbitrarily fine measurements of sense data. We observe that in case we are dealing with a sequential deterministic plant, reference to solution space $TRAJ$ is replaced here everywhere by reference to the plant state space PS. Then the goal is to use the control automaton to place the plant state at the end of the next time interval in a specified open set in a small topology on plant state space.

This is the best we can expect to do with an internal control automaton's limited ability to distinguish measurements which are used as data by the internal control automaton to issue a control function guiding the solution over the next interval of time.

To repeat, small topologies generated by finite open covers necessarily arise as soon as one tries to use a finite internal control automaton to control in which neighborhood a solution X_i will fall at the end of interval $[t_i, t_i + \Delta]$ on the basis of input information about $Meas(X_{i-1})$. Such a finite internal control automaton can be thought of as a strategy "guiding" the previously discussed control automaton. The latter had a continuous range of possible input measurements $Meas(X)$ and possible output control functions c; a finite internal control automaton has finite input and output alphabets.

Autonomous Δ-Sequential Deterministic Hybrid Systems We restrict ourselves here to autonomous sequential deterministic hybrid systems with fixed communication times at intervals of length Δ. We leave the more general case to another paper, since that case requires the language of set valued analysis. However, the required set-valued analysis is available, having been developed by analysts over the last twenty years [6], [7], [8].

We model the control automaton as a sequential composition of three units (Figure 3). The first unit is an analog to digital converter AD converting each measurement m into an input symbol u for the internal control automaton. The second unit is an internal control automaton which in state s converts input symbol u into output symbol $v = \mathcal{A}_s(u)$. The third unit is the digital to analog converter DA which converts output symbols v of the internal control automaton into control laws $c = DA(v)$ to be applied to the plant.

Control Automata From Internal Control Automata and Converters A control automaton induced by an internal control automaton and converters is a triple consisting of an:

Internal Control Automaton An input-output deterministic sequential automaton with set of input symbols U, set of output symbols V, and set S of states. This is the internal control automaton with input symbols u and output symbols v. With each state s of this automaton is associated a map $\mathcal{A}_s : U \to V$, where

$v = \mathcal{A}_s(u)$ if in states with input u, the automaton output is v.

Analog to Digital Conversion A function $AD : M \to U$. This is the "analog to digital converter", which converts a sensor measurement m into an input symbol $u = AD(m)$ for the internal control automaton.

Digital to Analog Conversion A function $DA : V \to C(\Delta)$. This is the digital to analog converter which converts an output symbol v of the internal control automaton into a control function $DA(v) = c$.

Each internal control automaton induces by composition $DA \circ \mathcal{A}_s \circ AD :$ $M \to C(\Delta)$ the state transition function for a deterministic sequential control automaton, described here state by state.

Remarks

1. We can trivially make any control automaton an internal control automaton by letting AD and DA be identity maps. But the interesting case is when U, V, S are finite, while M, $C(\Delta)$ are infinite.

2. Internal control automata are real time devices. In practice they are always implemented as finite automata. However, it is convenient to follow computer science usage and occasionally allow in the definition of internal control automaton infinitely many states and input and output alphabets. This is appropriate for PROLOG when there is an infinite Herbrand base and therefore infinitely many possible input statements and output statements. For such PROLOG programs regarded as automata, the input alphabet is the set of input statements, the output alphabet is the set of output statements. Input statements are fired by measurements, output statements fire control laws. This is a good model for Kohn's PROLOG formulation of declarative control [35] and for rule based control in general.

We assume that the spaces M, $C(\Delta)$, $TRAJ(\Delta)$, and $D(\Delta)$ come equipped with topologies of the usual sort from control theory and functional analysis making them complete metric spaces and thus topological spaces.

What choice of topologies one makes these spaces come with depends very much on the physical meaning of nearness, and on being in a space where solutions mathematically exist, and on mathematical tractability of solving equations in this domain of solutions. For instance, if the intended outputs are trajectories, we may want them close in the sup norm. If the intended outputs are functions describing the changing state of a chemical process, or the trajectory of a fighter aircraft, we may be more interested in closeness in an L_2 norm because energy may be of importance. Or we may be interested only in closeness that corresponds to finding a minimizing solution to a variational problem. Thus the idea that small changes in the input yields small changes in the output has many definitions.

For the discussion below, we need at least the following functions and their continuity. This is not an exhaustive list of definitions for development of the subject, it contains only those explicitly mentioned afterwards.

MEASUREMENT $Meas : TRAJ(\Delta) \to M$ is continuous.
EVAL Fix a $t_0 \in [0, \Delta]$, let PS be the set of plant states. Define

$$EVAL : X \in TRAJ(\Delta) \to X(t_0) \in PS.$$

Then $EVAL$ is continuous.
TRANS Fix a $d \in D(\Delta)$ and an $X \in TRAJ$. Define

$$TRANS_X : c \in C(\Delta) \to Y \in TRAJ,$$

by setting Y to be the solution to

$$Y'(t) = f(Y(t), c(t), d(t))$$

with domain $[0, \Delta]$ and initial condition $Y(0) = X(\Delta)$. Then $TRANS_X$ is continuous.

Here is our notation. Suppose the internal control automaton is in state $s = s_{i-1}$ at time t_{i-1}. The input to the analog to digital converter at time t_i is a measurement m, the input to the internal control automaton is $u = AD(m)$. In state s at time t_i, the internal control automaton outputs symbol $v = \mathcal{A}_s(u)$. This is converted finally by the digital to analog converter to $c_i = DA(v)$, the control law for the plant output at time t_i. Let U, V be the respective input and output alphabets of the internal control automaton.

5.1 Input Alphabets With The Discrete Topology

In this subsection we discuss the very simple case in which the natural input alphabet is assigned the discrete topology. On the surface, the input alphabets of finite automata look non-descript. Suppose we interpret this non-descriptness naively and put the discrete topology on the input alphabet U, that is, we declare each one element subset, and therefore every subset, of the alphabet open. If we want AD continuous, then the inverse image of these open sets must be open in M. We get the:

Discrete Input Alphabet Realizability Requirement For each input symbol $u \in U$, the associated set $\{m \in M \mid AD(m) = u\}$ is open in M.

This requirement says that for m' close enough to m in the standard topology of M, $AD(m) = u$ implies $AD(m') = u$. That is, the input measurements m, m' trigger the same input symbol u of the internal control automaton if they are close enough together in M. This ensures that arbitrarily small changes in the measurement cannot affect the input symbol u. This requirement expresses the myopia and robustness required to categorize continuous data from M continuously if the discrete topology is put on U. (This requirement is generalized later in the General Input Alphabet Realizability Requirement when the input alphabet carries a non-discrete topology.)

Difficulty In most cases we start with a natural connected space M of measurements, in which case we are in trouble because the open sets associated above with input alphabet letters are disjoint, and can exhaust a connected M only if AD is constant. So we consider the case when they do not exhaust M.

Using "*Continue*" One way to bypass the connectedness difficulty is to add to the measurement space M, whether or not it is connected, a single new element, *Continue*, getting an extended measurement space $M \cup \{Continue\}$ which is connected but is not Hausdorff. The open sets of the extended topology on $M \cup \{Continue\}$ are the open sets of M plus the one additional open set $M \cup \{Continue\}$. We can similarly extend the input alphabet U and output alphabet V of the internal control automaton and the control space $C(\Delta)$ by adding that same additional symbol *Continue* .

For those outside computer science, the origin of the use of this class of topologies in such contexts is [60], [61]. (Indeed, [61] was explicitly cited when Scott introduced his topological denotational semantics of computer programs, and the use of such topologies is now standard. Informally, the *Continue* added to M is thought of as being communicated to the DA converter at times when the m sensed for the plant is in $M - \cup_i U_i$ (is not sensed) and at times when there is no communication from sensor to input alphabet. That is, our intent is that *Continue* be the default input letter when either no measurement is observed, or a measurement m from M is observed which is not in any of the the open sets associated with input symbols u in the original alphabet U.

Define

$$AD : M \cup \{Continue\} \rightarrow U \cup \{Continue\}$$

by $AD(m) = i$ if $m \in U_i$, $AD(m) = Continue$ if m is in $(M - \cup_{u \in U} U_u) \cup \{Continue\}$. Then AD is continuous. First, any one element subset $\{i\}$ of U is open, and the inverse image under AD of any such $\{i\}$ is U_i, which is open. Thus any subset of U has an open inverse image which is a union of some of the U_i. Second, the only other open subset of $U \cup \{Continue\}$ is $U \cup \{Continue\}$, and its inverse image under AD is the whole space $M \cup \{Continue\}$.

It is natural to extend the discrete input alphabet realizability requirement to the larger alphabet $U \cup \{Continue\}$ by associating the open set $M \cup \{Continue\}$ with the new input symbol *Continue*.

Warning Under AD the inverse image of the closed set $\{Continue\}$ in $U \cup \{Continue\}$ is the closed set $M - \cup_i U_i \cup \{Continue\}$ in $M \cup \{Continue\}$. This set is closed and *not* open. (To connect with the later Generalized Input Alphabet Realizability Requirement, note that in the topology on $U \cup \{Continue\}$, the i for $i \in U$ and $\{Continue\}$ are the non-empty join irreducibles.)

Output Requirements We should also look on the output side. We most commonly use the

Deterministic Output Realizability Requirement Each output symbol v of the internal control automaton is associated with a single control law $c = DA(v)$.

Alternate to the deterministic output requirement, we might adopt an

Open Output Requirement DA is set valued. Each output symbol v of the internal control automaton is associated with an open set of control laws $DA(v)$.

The intent of the open output requirement is that when output symbol v is converted by DA, the result can be indiscriminately any control in the open set. This accords with Appendix II.

Closed Output Requirement DA is set valued. Each output symbol v of the internal control automaton is associated with a closed set $c = DA(v)$.

The intent of the closed output requirement is that when output symbol v is produced and converted to a control by DA, the result can be indiscriminately any control in the closed set. This is natural for the existence theorems of set-valued analysis ([7], [8]). But it requires set-valued analysis, which is not widely known.

All three of output requirements have uses, but in different contexts. The deterministic requirement is the simplest useful model for perfect digital to analog converters. The open requirement models non-deterministic digital to analog converters. The third is a known context for proving set-valued existence theorems for variational minima, which we need for relaxed and declarative control. See also Appendix II, where all three occur naturally. There are also other possible requirements corresponding to usage in existence theorems for stochastic differential equations.

5.2 Small Topologies

A small topology on a topological space M is the underlying set of M equipped with a subtopology T consisting of some of the open sets of M.

The most common models for measurement spaces are bounded closed regions in n-dimensional Euclidean space. Here is a standard argument from point set topology which says that any proper subtopology of such a space is necessarily not Hausdorff.

Proposition Suppose that Q is a set. Suppose that $\mathcal{U}_1$, $\mathcal{U}_2$ are (the sets of open sets of) topologies on set Q. Suppose that $\mathcal{U}_1$ is a subtopology of $\mathcal{U}_2$. If $\mathcal{U}_2$ is compact Hausdorff and $\mathcal{U}_1$ is Hausdorff, then the two topologies coincide.

Proof Otherwise, there has to be an $x \in Q$ and an open $U_2 \in \mathcal{U}_2$ such that there is no $U_1 \in \mathcal{U}_1$ such that $x \in U_1 \subseteq U_2$. Since $\mathcal{U}_1$ is Hausdorff, for each $q \in Q - U_2$, there is a pair of disjoint open sets A_q, B_q in $\mathcal{U}_1$ with $x \in A_q$, $q \in B_q$. These B_q cover $Q - U_2$. The latter is a closed subset of a compact space $\mathcal{U}_2$, hence is compact. So a finite number of B_{q_i} , $i = 1, \ldots, n$, cover $Q - U_2$. Then, contrary to hypothesis, the open (finite) intersection of the corresponding A_{q_i} is in $\mathcal{U}_1$, contains x, and is a subset of U_2.

Join Irreducibles In a lattice, an element a is called join irreducible if whenever $a = b \vee c$, then $a = b$ or $a = c$. An element a is called fully join irreducible if whenever a is the least upper bound of a (possibly infinite) family $\{a_i\}$, then $a = a_i$ for some i. Since topologies are lattices of sets, the notion of join irreducible or fully join irreducible apply to open sets in a topology.

Finite Small Topologies Suppose we are given a **finite** open cover $Z_1, \ldots, Z_n$ for M, generating a small topology T on M with a finite number of open sets. Here is further motivation for having the input alphabet associated with the non-empty join irreducibles of the lattice T. For analog to digital conversion, we visualize each open cover set Z_i as functioning independently as a "filter" Each measurement $m \in M$ passes the test represented by a "filter" Z_i in case $m \in Z_i$. The set of "filters" for which m passes the test is the set of open sets Z_i of the cover containing m. Passing the test for exactly all of these Z_i we now regard as a single event triggering one input symbol being sent to the internal control automaton by the analog to digital converter. In a finite topology, the set of all Z_i containing m has as intersection an open set. This is the smallest open set in T which contains m. This is also the smallest join irreducible containing m in T regarded as a distributive lattice. From the inclusion relations of the open sets of a finite topology, we can read off the set of join irreducibles. Since in a finite topology every open set in T is a finite union of such join irreducibles, this is the maximal "closeness" information about m that the (finite) small topology T can provide. So by using input symbols $1, 2, \ldots$ denoting the non-empty join irreducibles $U_1, U_2, \ldots$, we can provide the analog to digital converter with the most detailed information about m that the open cover and its generated topology can provide to the internal control automaton. The join irreducibles of finite topologies are very general. Up to isomorphism, any finite partially ordered set with a smallest element so arises. In summary, with a finite input alphabet, we get a finite topology T generated by the open sets associated with input letters, and can use a finite input alphabet denoting the distinct non-empty join irreducibles of T.

It is convenient to allow a larger class of topological spaces than the ones with a finite number of open sets, in such a way as to preserve the features mentioned above. In future papers this will allow mathematical modelling of controllability and observability of PROLOG control programs based on infinitely many input statements, even though the latter are not, strictly speaking, physically implementable in a strictly finite computer.

$AD-$Spaces If every point in a topological space T is contained in a smallest open set, then we call T an $AD-$space. This is a space suitable for describing an analog to digital conversion. Topological spaces with a finite number of open sets altogether are $AD-$spaces, including all topological spaces with a finite number of points. The reader can construct many others.

Proposition Suppose that U is a non-empty open set in an $AD-$space T. Then U is fully join irreducible if and only if there is an x in U such that U is the smallest open set containing x.

Proof If U is fully join irreducible, then certainly U is not a union of strictly smaller open sets. So there are $x \in U$ such that x is not in any smaller open set. So U is the smallest open set containing each such x. Conversely, suppose that open set U is the smallest open set containing x. Then if $U = \cup V_i$ with the V_i open, there exists an i such that $x \in V_i$. But U was the smallest open set containing x, so $U = V_i$, proving that U is fully join irreducible.

Essential Parts An $AD-$topological space T has a natural partition into "essential parts" obtained as follows. From every open set subtract the union of all smaller open sets. Call the remainder the "essential part" of the original open set. Then T is a disjoint union of the non-empty essential parts of open sets.

The open sets with non-empty essential parts are exactly the fully irreducible open sets. Essential parts are usually not open sets, they are the intersection of an open and a closed set.

Using AD−subtopologies we formulate a more general input alphabet realizability requirement, in a form that covers infinite, as well as finite, input alphabets for the internal control automaton.

General Input Alphabet Realizability Requirement   Let T be an AD−subtopology of the standard topology for M.

1. To every input symbol u for an internal control automaton there corresponds a non-empty fully irreducible open set U_u in T. We say that u is associated with U_u.
2. M is a (possibly infinite) union of the open sets U_u.

The subtopology T' generated by those fully join irreducible open sets of T associated with input letters may be smaller than the topology on T originally provided. It seems convenient to assume that $T = T'$, that is that the open sets associated with input letters generate the topology T.

AD-**Spaces and Analog to Digital Conversion**   Suppose the subtopology T on M satisfies the general input alphabet realizability requirement using family $\{U_u\}$ of distinct fully irreducible open sets of T, one for each input symbol u. Our intended analog to digital conversion is

$$AD : M \to U,$$

where U is the input alphabet, defined by

$$AD(m) = u,$$

where U_u is the smallest open set of the subtopology T containing m.

Example   Suppose we wish to detect whenever a gas burner is leaking and to close the gas supply off when a leak is detected. Suppose there is always a small background level of detectable gas for other reasons, whether or not there is a leak. The decision is to be based on gas concentration. Suppose the gas concentration can range over a closed interval $I = [0, p]$. We choose an a interior to that interval. Then a is interpreted as a threshold which, if exceeded, indicates leaking behavior, if fallen behind, indicates no gas leak. If the gas concentration is measured as exactly a, we regard ourselves as having insufficient information to take definite action. Here is how this is reflected by a small topology on I. Divide I into two disjoint subintervals $I_1 = [0, a)$, $I_2 = (a, p]$ separated by the one point a. Let the small topology on I have as its non-empty open sets the two intervals plus I. These three are the non-zero join irreducibles. Use these three join irreducibles as input letters for a finite control automaton. A measurement in I_1 says to the control automaton, there is no leak, be sure the gas supply is on. A measurement in I_2 says that there is indeed a leak, close the gas supply off. In the default case, the input measurement a says "continue as before". Thus measurements which fall at the division point are regarded as contributing no useful information for changing current policy.

We prefer a slightly different small topology for this problem. Expand a to a small open interval, get a covering by three sets, the left closed right open interval I_1, a left open right open interval I_3 containing a, and the left open right closed interval I_2. The small topology is then generated by these three intervals. The input alphabet consists of all the non-empty join irreducibles of the small topology. These are $I_1, I_2, I_3, I_1 \cap I_3, I_2 \cap I_3$. If the smallest join irreducible containing concentration x is one of $I_1 \cap I_3, I_2 \cap I_3, I_3$, that is if $x \in I_3$, then the gas flow should continue unchanged. If the smallest join irreducible containing x is I_1, then the gas should be on; if it is I_2, then the gas should be off.

T_0-**quotients**   A topology on set Q is usually identified with its set of open sets. The latter contains the null set $\emptyset$ and Q, and is closed under finite intersections and arbitrary unions. Let Q_{T_0} be the set of equivalence classes of elements of Q obtained by identifying points of Q if they are in the same open sets in the topology on Q. The quotient topology of Q_{T_0} declares a set of equivalence classes open in Q_{T_0} if the union of these equivalence classes is open in Q. That is,

$$U \subseteq Q_{T_0}$$

is open in Q_{T_0} iff $\cup U$ is open in Q. This is the smallest topology making the quotient map mapping each element of Q into its equivalence class continuous. A topological space is called a T_0-space if given any two points, there is an open set containing one of them but not the other. Then Q_{T_0} is always a T_0-space, and up to homeomorphism all T_0-spaces so arise. Continuous maps between topological spaces Q induce quotient continuous functions on the T_0 quotients Q_{T_0}, and this operation commutes with composition of continuous maps. Note that for AD-spaces T, the equivalence classes of the T_0-quotient can be identified with the essential parts of open sets mentioned above.

T_0-**Topologies on Input Alphabets**   Suppose that T is an $AD-$ subtopology of measurement space M. Suppose that $\{U_i\}$ is a family of distinct open sets of T with union M which are fully join irreducible in M. Suppose that the set I of subscripts of the $\{U_i\}$ is the input alphabet. This structure induces a T_0 -topology on the input alphabet I.

First, partially order the input alphabet I by declaring $i \leq j$ if $U_j \subseteq U_i$. Next, any partially ordered set $(A, \leq)$ has a topology of open sets generated by the set of all final segments (principal filters) $\{j \in A \mid i \leq j\}$ as i ranges over A. This is the topology we impose on the input alphabet.

Proposition

1. For each i in I,
$$\{j \mid U_j \subseteq U_i\}$$
   is an open subset I. These sets generate the input alphabet topology.
2. The map
$$AD : M \to I$$
   is continuous.
3. The map $AD : M \to I$ induces a homeomorphism
$$AD_{T_0} : M_{T_0} \to I.$$

Note that the requirement that distinct subscripts i yield distinct U_i is needed to verify the last assertion above. We proceed to some simple examples. These are exercises in determining the T_0- topology on the input alphabet.

Example Let M be a topological space. Let $U_2, \ldots, U_n$ be disjoint non-empty open sets in M not exhausting M. Let $U_1 = M$. So $U_1, \ldots, U_n$ is an open cover of M. The non-empty join irreducibles of the generated finite subtopology are also $U_1, \ldots, U_n$. The input alphabet is $\{1, \ldots, n\}$, where i corresponds to U_i.

1. If $m \in U_i$, and $2 \le i \le n$, then $AD(m) = i$.
2. If $m \in U_1 - (U_2 \cup \ldots \cup U_n)$, then $AD(m) = 1$.

In the partial order, the elements $2, \ldots, n$ are mutually incomparable, and all are larger than 1. The topology on the input alphabet has as its join irreducibles

$$\{2\}, \ldots, \{n\}, \{1, \ldots, n\}.$$

The open sets of the alphabet are precisely all subsets of $\{2, \ldots, n\}$, together with $\{1, \ldots, n\}$ itself. So $\{1\}$ is not open. The interpretation we usually give here of $AD(m) = 1$ is the negative one that no information is communicated, rather than the positive one that m is in

$$U_1 - (U_2 \cup \ldots \cup U_n).$$

Usually 1 is written as "bottom", $\perp$. This example corresponds to our informal discussion of *Continue* earlier.

Example Let M be a closed real interval. Let $Z_1, \ldots, Z_n$ be a finite sequence of open intervals generating a small topology on M. Suppose each interval overlaps the intervals immediately before and after, if there are such, but overlaps no others in the sequence. Suppose as well that the sequence covers M. Then the non-empty join-irreducibles of the small topology on M generated by the Z_i are

$$U_1 = Z_1, \ldots, U_n = Z_n,$$

$$U_{n+1} = Z_1 \cap Z_2, \ldots, U_{2n-1} = Z_{n-1} \cap Z_n.$$

The input alphabet is
$$\{1, \ldots, 2n - 1\},$$
where i is associated with U_i. In this example we get:

1. If $m \in Z_1 - Z_2$, then $AD(m) = 1$.
2. If $m \in Z_n - Z_{n-1}$, then $AD(m) = n$.
3. If $i = 2, \ldots, n - 1$ and $m \in Z_i - (Z_{i-1} \cup Z_{i+1})$, then $AD(m) = i$.
4. If $i = 1, \ldots, n - 1$ and $m \in Z_i \cap Z_{i+1}$, then $AD(m) = n + i$.

Equivalently, the topology on the input alphabet has as non-empty join irreducible open sets:

1. $\{1, n+1\}$
2. $\{n, 2n-1\}$
3. $\{i, n+i, n+i-1\}$ for $i = 2, \ldots, n-1$
4. $\{n+i\}$ for $i = 1, \ldots, n-1$

The one element subsets of the alphabet

$$\{1\}, \{2\}, \ldots, \{n\},$$

are not open.

Example Suppose that we extend the previous example to an infinite series of overlapping open intervals U_i, $i = 0, \pm 1, \pm 2$, which exhaust the whole line. This series forms an infinite open cover $\{U_i\}$, which generates an AD subtopology T of the topology on the line.

The $U_i \cap U_{i+1}$ and the U_i are the fully join irreducibles of T. The alphabet is

$$\{i \mid i = 0, \pm 1, \pm 2, \ldots\} \cup \{(i, i+1)\} \mid z = 0, \pm 1, \pm 2, \},$$

where i is associated with U_i and $(i, i+1)$ is associated with $U_i \cap U_{i+1}$.

Example Suppose the measurement space M consists of all the points in a closed circle C and that we cover that circle with three open circles $C_i, i = 1, 2, 3$ in general position. Then the non-empty join irreducibles of the topology generated by this cover are

$$U_1 = C_1, U_2 = C_2, U_3 = C_3, U_4 = C_1 \cap C_2,$$

$$U_5 = C_1 \cap C_3, U_6 = C_2 \cap C_3, U_7 = C_1 \cap C_2 \cap C_3.$$

Then $\{1, \ldots, 7\}$ will be the input alphabet. By construction this set will have a partial order corresponding to that of the eight element Boolean algebra generated by the C_1, C_2, C_3, without its largest element. The topology can be read off from the final segments.

Remark Instead of the language of smallest non-empty open sets containing a point, or the language of fully join irreducibles in distributive lattices [19], we could as well use the language of nerves of open coverings, simplicial complexes, and simplicial maps between nerves. The latter originates in the Čech singular cohomology theory. This apparatus allows us to represent a wide classe of continuous control automata (a term not defined here) as inverse limits of control by internal finite control automata and their DA and AD converters. Appendix II is a start.

Suppose we are given a topological hybrid system; that is, standard topologies are given on the measurement space M, control space $C(\Delta)$, solution space $TRAJ$, and position space PS, and the continuity conditions above are satisfied for each fixed disturbance d. Suppose each of these spaces is equipped with a small topology, that is, with a subtopology of its usual topology. We call this a topological hybrid system with small topologies.

Definition A topological hybrid system with small topologies is called continuous in the small if for all disturbances d

1. $Meas : TRAJ \rightarrow M$ is continuous in the small topologies on $TRAJ, M$.
2. For every state X of the plant automaton,

$$TRANS_X : C(\Delta) \rightarrow TRAJ$$

   is continuous in the small topologies on $C(\Delta)$, $TRAJ$.
3. For every state s of the internal control automaton, the map

$$DA \circ \mathcal{A}_s \circ AD$$

   mapping M to $C(\Delta)$ is continuous in the small topologies.

- The third condition gives meaning to the control automaton as a family of continuous maps, one for each initial state, just as a continuous plant is a family of continuous maps, one for each initial state.
- This is the kind of continuity required to treat controllability and observability. Finding a control automaton which not only works but is continuous in suitable small topologies is our way of approaching controllability and observability of the control automaton.
- We can also use this as a skeleton onto which to add musculature. For instance, we can ask that the functions above also be continuous in the plant states or in disturbances d, or in many other system parameters.

Remark

1. If we take the T_0-quotients of the spaces of a hybrid system with small topologies, we get a view of the hybrid system from the point of view of the finite internal control automaton, since this identifies points which the input alphabet of the control automaton cannot distinguish. If all small topologies used are finite, the range of possible control automata and their properties, relative to the fixed assignment of denotations for input and output alphabet, become subject to algebraic investigation.
2. We discuss briefly again what the choice of small topologies has to do with local control over time interval Δ. Suppose we cover $TRAJ$ with a finite cover Z_z of open sets, each of which represents a possible local goal in the following sense. We are told to place, if possible, X_{i+1} in certain of these cover sets Z_z at the end of interval $[t_i, t_{i+1}]$. Our means is to construct a suitable control automaton, which outputs control c_i at time t_i, from the finite collection of possible controls associated with output letters, which meets whatever the local goal is. A subminimal criterion for stable local behavior is to make sure there exist small topologies relative to which the system is continuous and such that all U_u are in the small topology on M and all Z_z are in the small topology on $TRAJ$.
3. Often suitable small topologies can be generated by starting with the finite topology generated by the Z_z , taking the inverse image of the topology under $Meas$, taking the inverse image of that under all the $TRANS_X$, taking the topology generated by these on $C(\Delta)$, taking its inverse image under

a proposed control automaton, and getting a subtopology of M. The first objective is to adjust the chosen internal control automaton to still meet the goals and to simultaneously make sure that the finite topology generated by the U_u is a subtopology of the small topology on M.

6 CONCLUDING REMARKS

Control-Disturbance Games The approach to extraction of digital control to meet performance specifications of the A. Nerode-A. Yakhnis control games ([66] [65]) is as follows. We examine only a simple hybrid system. Disturbance, player I, plays at the beginning of each interval, forming a disturbance sequence. The automaton receives sensor observations of the plant at that same time. The plant automaton, player II, responds with an output causing a change in control law for the plant, constructing as it goes a control sequence based on its transition table. The automaton wins the game if every trajectory in response to any disturbance is in the performance specification. A real time automaton winning the game is nothing more than a finite state winning strategy, and is a digital control program meeting the specification. Nerode, Yakhnis, and Remmel [64] are developing, and hope to implement soon, algorithms based on work of Buchi-Landweber [15], Gurevich-Harrington [28] , Yakhnis-Yakhnis [85], and McNaughton [58] for solving these games. These algorithms extract, for a wide class of games, a finite state automaton strategy which wins the game for player II, or show that there is no winning strategy for player II. This gives an approach to either find a winning strategy for the game and therefore a finite internal control automaton, or show that none exists and the problem has to be reformulated.

Logics for Hybrid Systems There is enough semantics of hybrid systems supplied above to make it possible to develop corresponding syntax and semantics for propositional and predicate modal logics of hybrid programs generalizing temporal or dynamic or process logic for hybrid systems. The hybrid state, that is the simultaneous state of the control automaton and the continuous plant, plays the role here that the state of the computer plays in modelling dynamic or temporal logic. Thus, for developing predicate dynamic logic, we need both the usual variables describing the state of register contents of the control automaton and also variables describing the state of the continuous plant. If the plant is autonomous deterministic sequential with fixed Δ, we can use the positions PS as plant states, which are usually n-tuples of reals, otherwise $TRAJ$ is the state. But we also need to refer to variables which range over measurements, controls, etc. The predicates thus have to include predicates on these variables. The formulation of the non-deterministic Δ-plant automaton given earlier is key in working this out as a generalization of dynamic logic. Such systems will be the subject of later papers.

7 APPENDICES

7.1 Appendix I : Radon and Lie Semigroup Control

Here are sketches of two control theories which fit under our model and have been employed in a wide variety of applications by Kohn using the method of relaxed Lagrangians. They are here to indicate that our formalism accomodates these approaches very directly. In a sequel we will use our formalism as as part of a mathematical foundation for Kohn's Declarative Control.

Radon Measure Valued Control A wide variety of control problems can be recast as problems in the calculus of variations. These are problems of minimizing non-negative functionals (Lagrangians) on function spaces subject to side conditions. Young [86] and Warga [81] and others introduced relaxed, or Radon measure-valued, control functions as a theoretical tool for proving the existence of solutions, and solved analytically small theoretical problems by this method. Kohn ([44], [35], [38], [39], [45], [40], [41], [42], [43]) developed techniques for computing approximations to relaxed optima by finite Radon measures. He solves the variational problem approximately with a finite Radon measure-valued control function good for a time δ in guiding the trajectory of the plant from one of finitely many regions where it was at the beginning of the interval to one of finitely many regions at the end of the interval from which the final goal is still accessible. Implementing this finite Radon measure-valued control on interval δ already involves changing the control law as many times as the Radon measure-valued control function changes value in that interval δ, so we are definitely in the domain of explicit hybrid control, which characteristically imposes many changes of control law in short time intervals.

This control function for an interval is computed as an answer substitution to a Prolog program incorporating the dynamics of the problem as well as the logical constraints and goals. If the problem is autonomous, that is, does not explicitly depend on time, and this calculation of control laws is made in advance for every possible region where the trajectory could start at the beginning of the interval, this Prolog program can be compiled to produce a finite state input-output automaton which observes where the trajectory is, and decides when to change control and what control value to change to for the next Δ time interval. The Radon measures referred to are probability measures on a space of conventional control values. The finite Radon measures are those that give non-zero positive weight α_i to only a finite number of control values c_j, with ($\sum \alpha_j = 1$).

This is often implemented as a piecewise constant control. These are also arbitrarily close to piecewise linear controls, which can be used instead. One can also use suitable splines, to get whatever degree of smoothness is required. We establish:

1. A subdivision of the basic interval δ into p successive left closed-right open subintervals I_j of relative size α_j,
2. For each such subinterval I_j the control value c_i.

We can choose a large p and set $p\Delta = \delta$, getting a small Δ. Using intervals of equal length Δ we can implement relaxed control as a control script in the sense discussed earlier in this paper.

The general strategy is to reformulate control problems as finding a control which leads to a trajectory minimizing a cost function (Lagrangian), which may be a sum of quite disparate terms. One approximates to the control for such a solution by a control script as above, and implements this approximating control sequence.

The topology used to define continuity of the plant as an input-output map from control functions to trajectories is that of L. C. Young's generalized curves [86] as reworked by Warga [81] in modern functional analysis. This is a weak topology on an appropriate dual space. Here is the rough meaning of continuity. The map from control law to solution should be such that whenever we specify finitely many continuous linear functionals (measurements) on trajectories, their values can be made to differ on trajectories simultaneously very little by assuring that the measure-valued controls differ very little.

Lie Transformation Semigroup Control We avoid discussing disturbances, which we assume are absent. One is given a family of globally defined smooth vector fields, or flows, on a manifold, which we may take here as a region in Euclidean space for discussion purposes. The idea is simply that a control function for an interval δ consists of

1. A choice of a division of δ into p consecutive subintervals of equal length Δ.
2. A choice of one vector field for each subinterval.

In each such subinterval the control consists in following the chosen vector field for that interval. That is, whichever i-th subinterval of length Δ of the δ interval you enter, follow the flow of the i-th control law throughout that i-th subinterval, switching to the $i + 1$-th control law at the end of that subinterval. This results over the δt interval in a continuous trajectory $y(t)$, but generally $y(t)$ has different left and right tangents at the end of each subinterval.

A smooth vector field is equally well represented by its infinitisimal generator, an element of the Lie algebra of all such vector fields. We assume we are given a family of physically implementable control functions in the form of infinitisimal generators of a Lie algebra. These might be for a missile, up, down, right, left, and increase burn rate. We look for an element w of the Lie algebra generated by these implementable controls which meets our goals; that is, puts the trajectory in a desired region if that trajectory is obtained by following the flow of w for the whole interval of length δt. This Lie algebra is the closure of the generators under vector addition, scalar multiplication, and Lie bracket. We have to find an implementable close approximation to w, represented by a control sequence as defined above which consists of implementable generators (and their local inverses). This implementable control sequence approximating closely to w can be computed by using Trotter's formula (the exponential series) to approximate as closely as we like to the vector sums and Lie brackets of the expression for w in terms of the implementable generators. This gives a complex control script. For some of the underlying mathematics, see Hilgert et al.[30].

Remark In Radon measure-valued (alternately, Lie transformation semi-group) control, as in all versions of control, the goals may simply not be realizable. No control sequence may be obtained because there was no measure (alternately, implementable Lie algebra element) which satisfies the requirements. The control sequence obtained may, accidentally or necessarily, not be actually implementable because of physical limitations on the plant described. It may call for faster changes in control laws than the plant can execute. A missile may be asked for a larger burn rate than the engine can produce, or a larger turn angle than the airframe can safely endure. These limitations can be built into the goal set of trajectories. It is possible that no control sequence can produce a trajectory in the goal set.

7.2 Appendix II: Automata from Covers

Why Open Performance Specifications? Performance specifications for trajectories on an interval say which trajectories are satisfactory. Performance specifications can be represented as open sets of trajectories. Consider the design of the Boeing 737 as a typical example. It was designed so that if a cup of coffee is no more than $3/4$ full and is placed anywhere in the aircraft, the cup never spills. This is a "safety requirement" supposed to hold at all times. It is definitely not an optimality criterion, but rather a "perform sufficiently well to satisfy the clients" criterion. The more stringent safety requirement that a coffee cup $9999/10000$ full never spills would be prohibitively expensive to implement, if it can be implemented at all. In engineering we never actually require an optimal solution to any problem. We only need "sufficiently close to optimal" solutions. There is always a prescribed deviation from optimality allowed. For a wide class of problems this defines the performance specification over the prescribed time interval as an open set of trajectories.

Open Specifications from Approximate Optimization Optimal control based on calculus of variations looks for a control function b over the time interval which minimizes a functional, defined on trajectories, with non-negative real values, subject to constraints. In the interests of brevity we confine our discussion to the unconstrained case. We also restrict attention to sequential deterministic plants, that is, we assume that the trajectory over an interval $[t, t + \Delta t]$ is determined by the plant state a at the beginning of the interval and the control function b used over the interval. We can then use (a, b) as a name for the resulting trajectory. Thus the functional on trajectories can be written $F(a, b)$. We usually construct $F(a, b)$ so that for a given initial state a, the smaller the value of $F(a, b)$, the better the performance of the trajectory. (In declarative control applications, F is the integral of a Lagrangian designed to capture dynamics, constraints, and system requirements). In approximate optimization we choose a positive tolerance ϵ and insist that for each a, b we choose b so that $F(a, b)/ < \epsilon + min_c F(a, c)$. With appropriate assumptions (lower semicontinuity), this defines an open set Q in $A \times B$ defined as $\{(a, b) \in A \times B \mid F(a, b) < min_c F(a, c) + \epsilon\}$ We take the performance specification from now on as an open set Q contained in $A \times B$.

Finite State Control Automata as Approximations   In declarative control [42], a continuous control is first extracted as an optimal control solving a compact convex relaxed variational problem, then this is used to extract a finite control automaton. In Nerode-Yakhnis [65] a continuous control game is first solved for a continuous control, then this is used to extract a finite state control automaton.

Is there a general topological principle behind this? How does one take a given continuous controller satisfying an open performance specification and extract a finite control automaton, with associated analog to digital and digital to analog converters and small topologies, which satisfies the same performance specification?

The fundamental idea is that we approximate to the given continuous controller by means of a finite state control automaton with associated small topologies. We confine ourselves here to the simple case when the control function f over the interval is a continuous function of the plant state at the beginning of the interval. We approximate to such a controller by a single state automaton.

Definition   A base of a topological space T is a collection of open subsets of the space, closed under finite intersection, such that every open set is a union of base sets. If bases $S(A)$, $S(B)$ are given respectively for topological spaces A, B, use Cartesian products of base elements as the base $S(A \times B)$ for $A \times B$.

Suppose that $f : A \to B$ is continuous. Suppose that A, B are compact Hausdorff spaces, so that $A \times B$ is also compact Hausdorff. Finally, suppose that Q is an open subset of $A \times B$ such that the set of ordered pairs f is a subset of Q.

We interpret these assumptions as follows.

- A consists of all plant states at the beginning of a control time interval.
- B consists of all control functions defined on the time interval that the physical controller can implement.
- f assigns to each plant state at the beginning of the interval a control function over the interval.
- Q is an open set of all pairs (plant state, control function) such that the resulting trajectory over the interval is in the performance specification.
- $f \subseteq Q$ says that the proposed control function always produces trajectories in the specification.

Since Q is open, Q is a union of a family C of open sets from the subbase for $A \times B$. Since the graph of a continuous f is closed, and $A \times B$ is compact, we conclude that f is compact. Since Q contains f and C is an open cover of f and f is compact, we conclude that C has a finite subcover C'.

Thus we have a finite cover $A'_1 \times B'_1 \cup \ldots \cup A'_r \times B'_r$ of f contained in Q.

Single State Automata from Coverings   Corresponding to any such finite open cover of f contained in the performance specification Q we extract a one-state automaton, converters, and associated small topologies.

We pinpoint this as the source of small topologies for finite control automata. Here is the construction.

Let $A_1, \ldots, A_k$ be the non-empty join irreducibles of the small topology on A generated by $A'_1, \ldots, A'_r$. Since every open set in the small topology on A is a join of join irreducibles, and every $a \in A$ is in the domain of f, for every a in A there is a a unique j such that $a \in A_j$. This A_j is the intersection of all the A'_j containing a. We assign to every non-empty join irreducible A_i of the small topology on A an open set $O(A_i) = \cup_{z \in \Gamma_i} B'_z$, where $\Gamma(i) = \{z \mid A_i \subseteq A'_z\}$. Note that $O(A_i)c$ is an open set in the small topology on B generated by the $B'_1, \ldots, B'_r$. Note that $O(A_i)$ is non-empty since there are $a \in A_i$. Since the A'_z are a cover of f, there exists a z such that $(a, f(a)) \in A'_z \times B'_z$. This defines a one state input-output automaton which maps A_i to $O(A_i)$. Its input symbols are the finitely many A_i, its output symbols must include the $O(A_i)$.

We specify that the "analog to digital converter" converts any $a \in A$ to the join irreducible A_i containing a.

We specify the "digital to analog" converter. Choose any map from output symbol $O(A_i)$ to a $b_i \in O(A_i)$. The digital to analog converter will convert $O(A_i)$ to b_i, once this map has been chosen.

In succession apply to an $a \in A$:

1. the analog to digital converter from a to that input letter A_i containing a
2. the automaton map from that input letter A_i to the output letter $O(A_i)$
3. the digital to analog converter mapping $O(A_i)$ to b_i

We get a map $g : A \to B$ from a to b_i.

Note that $g \subseteq Q$ because $(a, b_i) \in g$ implies that

$$(a, b_i) \in A_i \times O(A_i) \subseteq \cup C' \subseteq Q$$

So g also meets the performance specification.

(Open Set)-Valued Controls More generally, suppose that f is allowed to be a subset of $A \times B$, that the domain of f is A, that f is closed in the product topology on $A \times B$, and that $f \subseteq Q$, with Q open. It is natural with this formulation to relax the previously deterministic digital to analog converter to a non-deterministic one whose output for any input $O(A_i)$ is allowed to be any element $b_i \in O(A_i)$. In the previous argument, we did not use the fact that f is single valued, so the same argument yields g satisfying Q. This then gives a set-valued g whose output for input a may be indiscriminately any element of $O(A_i)$ and still satisfy Q. This fits the the open output realizability requirement introduced earlier. Any control in an open set may be issued when the output symbol is produced. All such controls have to lead to acceptable trajectories.

Remark For any topological space X, the power set $P(X)$ has a topology generated by declaring that for each open set Q in X, the power set of Q is open. Applying this to $A \times B$, we see that the finite automaton induced maps g (considered as sets of ordered pairs) are dense in the subspace of continuous functions $f : A \to B$ of $P(A \times B)$ in case A, B are both compact.

A useful way of looking at this is that we approximate to continuous maps f from A to B, each with the usual topologies. The approximations are triples (T_A, AUT, T_B), with T_A a small topology on A, T_B a small topology on B, and

AUT the finite control automaton. Thus we simultaneously approximate to the function f by a finite automaton and to the standard topologies on the domain and range of f by small topologies.

Finite State Automata from Physical Controllers We now consider a model in which the control imposed on the plant goes to a continuous physical controller which then is then used as the actuator for changing the plant state. Suppose that this continuous controller also has a compact space C of internal controller states. We model this as follows. Suppose that the continuous f is a function of *both* the continuous controller's internal state *and* the plant state at the beginning of the interval, with values in B. Then we have a continuous $f : A \times C \to B$ and an open specification $Q \subseteq A \times C \times B$ containing closed f. Apply the argument of the "one state automaton" paragraphs above, but now let $A \times C$ play the role that A played there. Use as a base of $A \times C$ a product of bases of A and C , and use a representation of Q as a union of Cartesian products of basic open sets from $A \times C$, and B. We get finite open covers for A, C, and B. As above we can extract a one state automaton satisfying the specification Q. This automaton has, as input alphabet, all pairs of non-empty join irreducibles, the first coordinate of the pair a join irreducible from the small topology generated by the finite cover of A, the second coordinate of the pair a join irreducible from the small topology on C generated by the finite cover of C. Construe the non-empty join irreducibles of the small topology on C as automaton states and the join irreducibles of the small topology on A as input letters, and use output letters as before. The join irreducibles of C constitute a "finite approximation" to the compact space C which is sufficiently accurate to be a set of states for an automaton which meets open specification Q. Further development requires too much space to be sketched further here. This idea will be elaborated elsewhere to derive Kohn's Declarative Control from Warga's Relaxed Optimal Control. There we will derive ϵ–optimal finite automata control programs from optimal continuous controls obtained by solving a relaxed variational formulation of the control problem.

More Realizability Requirements Suppose that we wish to implement a finite state control automaton induced by a finite cover. Then there are several other physical realizability requirements which have to be met by the implementation.

One requirement is that the physical sensor of plant states must be accurate enough to provide, for any input plant state, an output to the analog to digital converter which can be used to fire precisely the symbol representing the smallest open set which contains that input state, in the small topology on the plant states A. Thus the coarser the small topology on the plant states induced by the cover, the less stringent is the performance requirement imposed by the cover on the sensor and the analog to digital converter, and the cheaper it is likely to be to build them.

Another requirement is that the plant controller has to move the initial plant state at the beginning of the control interval into a final plant state at the end of the control interval which is in an open goal set of final states. The smaller this open goal set, the tighter the requirement imposed on both the the physical controller and the control applied to the controller. In applications, the open goal sets may change from control interval to control interval, and vary over a finite topology. (They do so in declarative control, where they are newly computed at the beginning of every control interval.) This can be expressed as a continuity requirement with respect to small topologies on the physical controller if the latter is regarded as an input-output device. Thus the coarser the finite topology of open goal sets of states, the less stringent is the performance requirement imposed on the physical controller and the digital to analog converter, and the cheaper it is likely to be to build them.

Acknowledgements   We wish to thank J. B. Remmel for his contribution to a precursor hybrid systems model [63], A. Yakhnis [65], [64] for substantial help, and V. S. Subrahmanian for a critical reading of an earlier draft. Appendix II was abstracted from a draft of a later paper in April 1993 in response to comments of S. Mitter and A. Benveniste. Thanks are also due to A. Yakhnis, J. N. Crossley, and M. Branicky for additional corrections. The cover approximation technique of Appendix II will be published in full later.

References

1. V. M. Alekseev, V. M. Tikhomirov, S. V. Fomin, *Optimal Control*, Consultant's Bureau, Plenum Press, 1987.

2. K.R. Apt and E-R. Olderog, *Verification of Sequential and Concurrent Programs*, Springer-Verlag, 1991.

3. M. Arbib and R. E. Kalman, *Topics in Mathematical Systems Theory*, McGraw-Hill, N. Y., 1969.

4. M. Arbib and L. Padulo, *Systems Theory*, Saunders, Philedelphia, 1974.

5. J. P. Aubin, *Convex Analysis and Optimization*, Pitman, 1982.

6. J. P. Aubin, *Differential Inclusions, Set Valued Maps, and Viability*, Springer-Verlag, 1984.

7. J. P. Aubin, *Set Valued Analysis*, Birkhauser, 1990.

8. J. P. Aubin, *Viability Theory*, Birkhauser, 1991.

9. J. P. Aubin and I. Ekeland, *Applied Non-Linear Analysis* , Wiley, 1984.

10. L.D. Berkovitz, Thirty Years of Differential Games, in *Modern Optimal Control*, Emilio O. Roxin, ed., Marcel Dekker, Inc., 1989.

11. M. Ben-Ari, *Principles of Concurrent Programming*, Prentice-Hall, 1990.

12. R. S. Boyer, M. W. Green, J. S. Moore, *The Use of a Formal Simulator to Verify a Simple Real Time Control Program*, Technical Report No. ICSA-CMP-29, Institute for Computing Science and Computer Applications, The University of Texas at Austin, SRI International, 1982.

13. E. Bradley, *Control Algorithms for Chaotic Systems*, AI Memo 1278, MIT AI Laboratory, 1991.

14. E. Bradley and F. Zhao, Phase-Space Control System Design, *IEEE CACSD92*, 1992.

15. J. R. Büchi, *The Collected Works of J. Richard Büchi* , S. MacLane, D. Siefkes, eds., Springer-Verlag, 1990.

16. K. M. Chandy and J. Misra, *An Introduction to Parallel Program Design*, Addison-Wesley, 1988.

17. N. Coleman, An Emulation-Simulation for Intelligent Controls, *Proc. of the Workshop on Software Tools for Distributed Intelligent Control Systems*, 37-52, July 17-19, 1990, Pacifica, California.

18. R. T. Dodhiawala, V. Jagoenthan, and L. S. Baum, Erasmus System Design: Performance Issues, *Proc. of the Workshop on Blackboard Implementation Issues*, AIII, Seattle, WA, July, 1987.

19. Ph. Dwinger and R. Balbes, *Distributive Lattices*, Columbia, University of Missouri Press, 1974.

20. S. Eilenberg, *Automata, Languages, and Machines* (vol. A), Academic Press, New York, 1974.

21. I. Ekeland, *Infinite Dimensional Optimization and Convexity*, University of Chicago Lecture Notes in Mathematics, University of Chicago Press, 1983.

22. A. F. Filippov, *Differential Equations with Discontinuous Right Hand Side*, Kluwer Academic Publishers, 1988.

23. A. Friedman, *Differential Games*, Wiley-Interscience, 1971.

24. G. Gierz, K. H. Hofmann, K. Keimel, J. D. Lawson, M. Mislove, D. S. Scott, *A Compendium of Continuous Lattices*, Springer-Verlag, 1980.

25. R. L. Grossman and R. G. Larson, Viewing Hybrid Systems as Products of Control Systems and Automata, *Proc. IEEE 31st CDC*, vol. 3, 2953-2955, Tucson, 1992.

26. J. Guckenheimer, A. Back, M. Myers, A Dynamical Simulation Facility for Hybrid Systems, this vol.

27. J. Guckenheimer and A. Nerode, Simulation for Hybrid and Nonlinear Control, *Proc. IEEE 31st CDC*, vol. 3, 2981-2983, 1992.

28. Y. Gurevich and L. Harrington, Trees, Automata and Games, *Proc. of the 14th Ann. ACM Symp. on Theory of Comp.*, pp. 60-65, 1982.

29. O. Hajek, *Pursuit Games*, Mathematics in Science and Engineering, vol. 120, Academic Press, New York, 1975.

30. J. Hilgert, K. H. Hofmann, J. Lawson, *Lie Groups, Convex Cones, and Semigroups*, Oxford Clarendon Press, 1988.

31. A. Isidori, *Nonlinear Control Systems, An Introduction* , Springer-Verlag, 1989.

32. R. Iseman, *Digital Control Systems*, Springer-Verlag, 1977.

33. R. Isaacs, *Differential Games*, SIAM Series in Applied Mathematics, John Wiley and Sons, Inc., 1965.

34. Y. Kesten and A. Pnueli, Timed and Hybrid Statecharts and their Textual Representation, in *Formal Techniques in Real Time and Fault Tolerant Systems*, LNCS 571, Springer-Verlag, 1992.

35. W. Kohn, A Declarative Theory for Rational Controllers, *Proc. 27th CDC*, pp. 130-136, 1988.

36. W. Kohn, *The Rational Tree Machine: Technical Description and Mathematical Foundations*, IR and DBE-499, Technical Document 905-10107-1, July 7, 1989, Boeing Computer Services.

37. W. Kohn, *Rational Algebras: A Constructive Approach*, IR and DBE-499, Technical Document D-905-10107-2, July 7, 1989.

38. W. Kohn, Hierarchical Control Systems for Autonomous Space Robots, *Proc. AIAA*, 1988.

39. W. Kohn, *Application of Declarative Hierarchical Methodology for the Flight Telerobotic Servicer*, Boeing Document G-6630-061, Final Report of NASA-Ames Research Service request 2072, Job Order T1988, Jan 15, 1988.

40. W. Kohn, *Cruise Missile Mission Planning: A Declarative Control Approach*, Boeing Computer Services technical Report, 1989.

355

41. W. Kohn, Declarative Multiplexed Rational Controllers, *Proc. 5th IEEE Int. Symp. Intelligent Cont.*, pp. 794-803, 1990.
42. W. Kohn, Declarative Hierarchical Controllers, *Proc. Darpa Workshop on Software Tools for Distributed Intelligent Control Systems, Domain Specific Software Initiative*, pp. 141-163, Pacifica, Ca., July 17-19, 1990.
43. W. Kohn, *Advanced Architecture and Methods for Knowledge-Based Planning and Declarative Control*, Boeing Computer Services Technical Document IRD BCS-021, 1990, in ISMIS 91.
44. W. Kohn and K. Carlsen, Symbolic Design and Analysis in Control, *Proc. 1988 Grainger Lecture Series*, U. of Illinois, pp. 40-52, 1989.
45. W. Kohn and A. Murphy, Multiple Agent Reactive Shop Control, *ISMIS91*.
46. W. Kohn and A. Nerode, An Autonomous Control Theory: An Overview, *Proc. IEEE CACSD92*, March, 1992.
47. W. Kohn and A. Nerode, Multiple Agent Autonomous Control Systems, *Proc. CDC92*, Dec., 1992.
48. W. Kohn and A. Nerode, Multiple Agent Autonomous Control, A Hybrid Systems Architecture, to appear in *Logical Methods: A Symposium in honor of Anil Nerode's 60th birthday*, Birkhauser, 1993.
49. W. Kohn and A. Nerode, Multiple Agent Hybrid Control Architecture, this vol.
50. W. Kohn and T. Skillman, Hierarchical Control Systems for Autonomous Space Robots, *Proc. AIAA Conf. on Guidance, Navigation, and Control*, v. 1, pp.382-390, 1988.
51. N.N. Krasovskii and A.I. Subbotin, *Game-Theoretical Control Problems*, Springer-Verlag, 1988.
52. W. Kuich and A. Salomaa, *Semirings, Automata, Languages*, Springer-Verlag, 1985.
53. B. C. Kuo, *Digital Control Systems*, Holt, Rinehart, Winston, 1980.
54. J. W. S. Liu, Real Time Responsiveness in Distributed Operating Systems and Databases, *Proc. Darpa Workshop on Software Tools for Intelligent Control, Domain Specific Software Initiative*, pp. 185-192, Pacifica, Ca., July 17-19, 1990.
55. Z. Manna and A. Pnueli, *The Temporal Logic of Reactive and Concurrent Systems*, Springer-Verlag, 1992.
56. O. Maler, Z. Manna, and A. Pnueli, From Timed to Hybrid Systems, in *Proc. Rex Workshop, Real Time in Theory and Practice*, J. W. DeBakker, C. Huizing, W. P. de Roever, G. Rozenberg, eds., LNCS 600, Springer Verlag, 1992.
57. E. Mettala, Domain Specific Architectures, *Proc. of Workshop on Domain Specific Software Architectures*, 193-231, July 9-12, 1990, Hidden Valley, Calif.
58. R. McNaughton, *Infinite Games Played on Finite Graphs*, Tech. Report 92-14, Dept. Comp. Sci., RPI, Troy, New York, May 1992.
59. M. Mesarovic and Y. Tashahara, *Theory of Hierarchical Multilevel Systems*, Academic Press, N.Y., 1970.
60. A. Nerode, General Topology and Partial Recursive Functionals, in *Summaries of Talks at the AMS Summer Insitute in Symbolic Logic*, Cornell University, 1957.
61. A. Nerode, Some Stone Spaces and Recursion Theory, *Duke Mathematics Journal*, 26, 1959, 397-406.
62. A. Nerode, Modelling Intelligent Control, *Proc. DARPA Workshop on Software Tools for Distributed Intelligent Control Systems, Domain Specific Software Initiative*, Pacifica, Ca., July 17-19, 1990.
63. A. Nerode and J.B. Remmel, A Model for Hybrid Systems, *Hybrid System Workshop Notes*, MSI, Cornell University, Ithaca, NY, June 10-12, 1991.

64. A. Nerode, J.B. Remmel and A. Yakhnis, *Playing Games on Graphs: Extracting Concurrent and Hybrid Control Programs*, in prep.

65. A. Nerode, A. Yakhnis, Modelling Hybrid Systems as Games, *Proc. CDC92*, pp.2947-2952, Dec., 1992.

66. A. Nerode, A. Yakhnis, V. Yakhnis, Concurrent Programs as Strategies in Games, in *Logic from Computer Science*, Y. Moschovakis, ed., Springer-Verlag, 1992.

67. A. Nerode, A. Yakhnis, V. Yakhnis, Distributed Concurrent Programs as Strategies in Games, to appear in *Logical Methods: A Symposium in honor of Anil Nerode's 60th birthday*, Birkhauser, 1993.

68. P. H. Nii, Blackboard Systems: The Blackboard Model of Problem Solving and the Evolution of the Blackboard Architecture, *AI Magazine*, (7) 2, pp. 38-53, 1986.

69. L. E. Neustadt, *Optimization*, Princeton University Press, 1976.

70. A. Pnueli, Application of Temporal Logic to the Specification and Verification of Reactive Systems: A Survey of Current Trends, in *Current Trends in Concurrency*, LICS 1986.

71. A. Pnueli, R. Rosner, On the Synthesis of a Reactive Module, *Proc. of 1989 POPL Conference*, January 1989.

72. L.S. Pontryagin, *On the Theory of Differential Games*, Russian Mathematical Surveys 21 (No.4), pp. 193-246, 1966.

73. M. H. Raibert, *Legged Robots that Balance*, MIT Press, 1986.

74. S. S. Sastry and M. Bodson, *Adaptive Control: Stability, Convergence, and Robustness*, Prentice-Hall, 1989.

75. M. Schoppers, Automatic Synthesis of Perception Driven Discrete Event Control Laws, *Proc. 5th IEEE Inter. Symp. on Intelligent Control*, 1990.

76. M. G. Singh, *Dynamical Hierarchical Control*, North Holland, Amsterdam, 1977.

77. T. Skillman, W. Kohn, et. al., Classes of Hierarchical Controllers and their Blackboard Implementations, *J. Guidance and Control and Dynamics*, (13), pp. 176-182, 1990.

78. J.J. E. Slotine and Weiping Li, *Applied Nonlinear Control*, Prentice Hall, 1991.

79. E. D. Sontag, *Mathematical Control Theory*, Springer-Verlag, 1990.

80. H. J. Sussman and V. Jurdjevic, Controllability of Non-linear Systems, *J. Diff. Equations*, 12 pp.95-116, 1972.

81. J. Warga, *Optimal Control of Differential and Functional Equations*, Academic Press, 1972.

82. J. Warga, Some Selected Problems of Optimal Control, in *Modern Optimal Control*, Emilio O. Roxin, ed., Marcel Dekker, Inc., 1989.

83. A. Yakhnis, *Game-Theoretic Semantics for Concurrent Programs and their Specifications*, Ph. D. Diss., Cornell University, 1990.

84. A. Yakhnis, *Hybrid Games*, Technical Report 92-38, Mathematical Sciences Institute, Cornell University, October, 1992.

85. A. Yakhnis, V. Yakhnis, Extension of Gurevich-Harrington's Restricted Memory Determinacy Theorem, *Ann. Pure and App. Logic*, 48, 277-297, 1990.

86. L. C. Young, *Optimal Control Theory*, Chelsea Pub. Co. N.Y, 1980.

87. F. Zhao, *Extracting and Representing Qualitative Behaviors of Complex Systems in Phase Space*, AI Memo 1274, MIT AI Laboratory, March, 1991.

Some Remarks About Flows in Hybrid Systems

R. L. Grossman* and R. G. Larson**

Department of Mathematics, Statistics,
& Computer Science (M/C 249)
University of Illinois at Chicago
Chicago, IL 60680, USA

Abstract. We consider hybrid systems as networks consisting of continuous input-output systems and discrete input-output automata. Some of the outputs may be connected to some of the inputs; the others server as the inputs and outputs of the hybrid system. We define a class of regular flows for such systems and make some remarks about them.

1 Introduction

In this paper, a hybrid system is a network consisting of continuous input-output systems and discrete input-output automata. Some of the outputs may be connected to some of the inputs; the others serve as the inputs and outputs of the hybrid system. We are interested in flows of hybrid systems: to completely characterize flows is too difficult a problem. Indeed, this is a generalization of the problem of characterizing the flows of dynamical systems, which is already too hard. In this paper, we first show how the characterization of flows may be reduced to an algebraic problem and then make some remarks about this problem.

The purpose of this paper is to explain these ideas simply and to give some examples. Another more technical paper is in preparation which assumes a certain amount of background material in algebra, but gives the full definitions and provides the proofs for the concepts explained here [8]. It is not hard to implement systems which simulate and analyze hybrid systems of the type described here: a proof of concept implementation is described in [4].

There are a variety of interpretations for hybrid systems currently being explored. We mention three closely connected to the point of view in this paper. An automaton may be viewed as enabling or disabling some of the continuous input-output systems on the basis of discrete input symbols [7] and [6]. In other words, the hybrid system reflects some type of generalized mode switching. Alternatively, the automaton may be viewed as selecting trajectories or collections of trajectories of the continous systems in order to satisfy performance specifications [11]. Yet another interpretation is for automaton to be used to construct

---

* This research was supported in part by NASA grant NAG2-513 and NSF grant DMS 910-1089.
** This research is supported in part by NSF grant DMS 910-1089.

control laws for continous systems [10]. In this paper, we view hybrid systems from the first point of view.

Hybrid systems have a variety of representations. The simplest are the state space representation and the input-output representation. In the state space representation, the states, inputs, and outputs of each component continuous system and automaton are described, as well as the input-output connections between the various systems. In the input-output representation, the inputs and outputs of the hybrid system as a whole are described.

In this paper, we use a different representation—the observation space representation. Roughly speaking, this may be viewed as dual to the state space representation. This representation is a very basic one: it forms the basis for the Heisenberg picture in quantum mechanics [3]; it has been used to define discrete time control systems by Sontag [13] and continuous time control systems by Bartosiewicz [1] and [2].

Using observation space representation, we define hybrid flows and regular hybrid flows. Regular hybrid flows are flows which are finite concatenations of flows of one or more continuous systems and mode switches. The important point is that at most a finite number of mode switches are involved. We first give an example of a hybrid flow which is not regular. We also give an an example of a hybrid system with the property that every point in a neighborhood of the origin can be reached using regular hybrid flows involving precisely one mode switch. Without mode switching, not every point in a nighborhood of the origin can be reached. These represent two extreme behaviors possible. It is an open problem to provide more precise characterizations of hybrid flows.

To work out the basic properties of hybrid flows from this viewpoint requires a certain amount of algebra. This is done in [8], with the results summarized here. In Section 2, we describe the observation space representation of input-output systems, automata, and hybrid systems, following [7]. In Section 3, we define hybrid flows and give the examples mentioned above.

2   Observation space representations

A basic principle is that the states X of a system can be recovered from the algebra of functions on the states $R = \text{Fun}(X)$. R is one of several *observation algebras* that can be associated with a system. This leads to the observation space representation of a system which, broadly speaking, may be thought of as dual to the state space representation. To make this precise we define both a product structure on the space of observation functions, as is usual, and a dual structure, called a coproduct. This will allow us to view hybrid systems as suitable products of discrete automata and continuous control systems. As a by product of this approach, we can obtain as special cases the approach used in [13] to study discrete time control systems and the approach used in [1] and [2] to study continous time control systems.

The basic idea is that the time evolution of a state by the dynamics may be viewed as an action on the states, and that this action corresponds to an action on the algebra R. In the case of continuous systems, this action is derivation of R; in the case of discrete state systems, this action is an endomorphism of R. We now explain this in more detail.

For continuous systems, a tangent vector to the space of states gives rise to a derivation E of the algebra R, that is, a linear map from R to itself satisfying

$$E(fg) = fE(g) + E(f)g, \qquad f, g \in R. \tag{1}$$

For discrete state systems, the action of the space of input words $w \in W$ on the states

$$X \times W \longrightarrow X, \qquad (x, w) \mapsto x' = x \cdot w$$

induces an action on the observation algebra R

$$W \times R \longrightarrow R, \qquad (w, f) \mapsto (w \cdot f)(x) = f'(x) = f(x \cdot w).$$

It is easy to see that the map

$$\sigma_w : R \longrightarrow R, \qquad \sigma_w(f) = f'$$

is an endomorphism of R, that is, a linear map from R to itself satisfying

$$\sigma(fg) = \sigma(f)\sigma(g), \qquad f, g \in R. \tag{2}$$

There is a natural generalization of these concepts which includes both of them: that is, a bialgebra H which acts on the algebra R in such a manner that

$$h(fg) = \sum_{(h)} h_{(1)}(f) h_{(2)}(g),$$

where $\Delta(h) = \sum_{(h)} h_{(1)} \otimes h_{(2)}$. We will see below how to define the coproduct Δ to obtain Equations 1 and 2.

Recall that a *coalgebra* is a vector space C with linear maps

$$\Delta : C \longrightarrow C \otimes C$$

(the coproduct) and

$$\epsilon : C \longrightarrow k$$

(the counit), satisfying conditions stating that Δ is coassociative, and that ϵ is a counit. We use the notation

$$\Delta(c) = \sum_{(c)} c_{(1)} \otimes c_{(2)},$$

to indicate that Δ sends an element c to some sum of terms of the form $c_{(1)} \otimes c_{(2)}$.

Coalgebras arise here since they are natural structures for describing actions on algebras. In particular, we will use them to describe the action on observation algebras of hybrid systems. We will require that

$$c(fg) = \Delta(c)(f \otimes g),$$

where $c \in C$ and $f, g \in R$, when C is a coalgebra acting on the algebra R. To say that c acts as a derivation is to say that c is *primitive*:

$$\Delta(c) = 1 \otimes c + c \otimes 1.$$

With this coproduct, we recover Equation 1. To say that c acts as an algebra endomorphism is to say that c is *grouplike*:

$$\Delta(c) = c \otimes c.$$

With this coproduct, we recover Equation 2. Denote the set of grouplikes of the coalgebra C by $\Gamma(C)$.

A *bialgebra* is an algebra and a coalgebra, in which the coalgebra maps are algebra homomorphisms. In the most general terms, a hybrid system in an *observation space* representation consists of a bialgebra H, a commutative algebra R with identity, and an action of H on R which satisfies

$$h(fg) = \sum_{(h)} h_{(1)}(f) h_{(2)}(g), \qquad \text{for } h \in H, \text{ and } f, g \in R.$$

That is, H acts on R with the primitives of H acting as derivations of R, and the grouplikes of H acting as endomorphisms of R. It is therefore a natural generalization of both continuous systems (modeled by derivations) and discrete state systems (modeled by discrete state transitions and the associated endomorphisms). Furthermore, as we shall see, it allows for hybrid systems exhibiting both continous and discrete behavior to built from discrete and continous components.

We now give a simple example of a hybrid system, following [7]. Continuous control systems can be viewed as special cases of this approach [5], as can discrete automata [7].

Example 1. In this example, we view a "taxicab on the streets of Manhattan" as a hybrid system with two modes: control in the first mode, corresponding to State 1, results east-west motion, but no north-south motion; control in the second mode, corresponding to State 2, results in north-south motion, but no east-west motion. See Figure 1. There are many ways of viewing this example. Of course, this system, due to its simplicity, can be modeled by a single control system in the plane consisting of a north-south vector field and an east-west vector field. The approach we describe here, on the other hand, generalizes to a large class of hybrid systems, viewed as mode switching of continous control systems, controlled by a discrete automaton. It is important to note that the automaton switches between two planar control systems, although the dynamics of each are in fact are constrained to one dimension in this case.

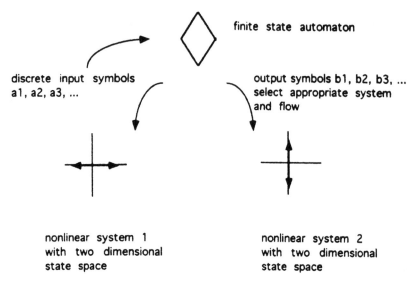

Fig. 1. The Manhattan taxicab hybrid system.

In this simple example, a finite state automaton with two states s_1 and s_2, accepts input symbols a_1, a_2, a_3, ... transits states, and outputs a symbol b_1, b_2, b_3, ... On the basis of the output symbol, a new nonlinear system and corresponding flow is selected. At the end of this flow, a new input symbol is accepted by the automaton and the cycle repeats.

We begin by defining the state space representation of the hybrid system. Consider two control systems in the space k^2 of the form

$$\dot{x}(t) = u_1(t) E_i^{(i)}(x(t)) + u_2(t) E_2^{(i)}(x(t)), \quad \text{for } i = 1, 2,$$

where

$$E_1^{(1)} = \partial/\partial X_1, \quad E_2^{(1)} = 0,$$

and

$$E_1^{(2)} = 0, \quad E_2^{(2)} = \partial/\partial X_2.$$

A two state automaton accepts an input symbol, changes states, outputs a symbol, and on the basis of the output symbol selects a nonlinear system and a corresponding flow. See Figure 1 again.

In order to define the observation space representation of the system, let $R_i = k[X_1, X_2]$, $i = 1, 2$ denote the polynomial algebra on the indicated indeterminates and let $R = R_1 \oplus R_2$. Also, let $k<\xi_1, \xi_2>$ denote the free noncommutative associative algebra on the indicated indeterminates. We specify the action of the

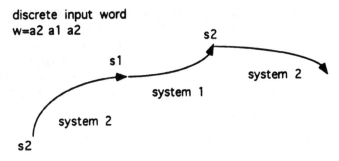

Fig. 2. Another view of the Manhattan taxicab hybrid system.

This is another view of the hybrid system illustrated in Figure 1. For this illustration, assume for simplicity that the input symbol i selects the corresponding state and nonlinear system. The flow sketched schematically above is the result of the input word $w = a_2 a_1 a_2$, which results in the state transition sequence $s_2 s_1 s_2$, which in turn selects the nonlinear systems indicated.

bialgebra $k<\xi_1, \xi_2>$ on R by specifying its actions on R_i, $i = 1, 2$:

on R_1

$$\xi_1 \text{ acts as } E_1^{(1)} = \partial/\partial X_1,$$
$$\xi_2 \text{ acts as } E_2^{(1)} = 0;$$

on R_2

$$\xi_1 \text{ acts as } E_1^{(2)} = 0,$$
$$\xi_2 \text{ acts as } E_2^{(2)} = \partial/\partial X_2.$$

Consider input symbols a_1, corresponding to east-west travel only, and a_2, corresponding to north-south travel only. Let $G = \Omega^*$ be the semigroup (that is, the input words) freely generated by the input symbols $\Omega = \{a_1, a_2\}$. The action of Ω (and thus of G) on R is given by specifying its action on R_i, $i = 1$, 2. Its action on R_1 is given as follows: let $\rho_{12} : R_1 \to R_2$ be the isomorphism sending $X_1 \in R_1$ to $X_1 \in R_2$, and $X_2 \in R_1$ to $X_2 \in R_2$. Then, for $f \in R_1$,

$$a_i \cdot f = \begin{cases} f \oplus \rho_{12}(f) & \text{if } i = 1, \\ 0 & \text{otherwise.} \end{cases}$$

Its action on R_2 is defined similarly. Intuitively, a_1 maps all states into State 1, and a_2 maps all states into State 2. The action of Ω on R is the transpose of this action. For simplicity, assume that the output symbol is given by the current state. With this assumption, the "typical" element $(u_1\xi_1 + u_2\xi_2)a_2(v_1\xi_1 + v_2\xi_2)a_1 \in k<\xi_1, \xi_2> \amalg k\Omega^*$ is to be interpreted as making a transition to State 1, flowing along $v_1 E_1^{(1)} + v_2 E_2^{(1)}$, making a transition to State 2, and then flowing along $u_1 E_1^{(2)} + u_2 E_2^{(2)}$. See Figure 2.

3 Flows

In this section, we consider flows of hybrid systems in the observation space representation. For technical reasons, we do not consider the most general type of hybrid system, but rather restrict attention to a smaller class. This restricted class includes continuous systems, automata, and products of these, as in the example in the last section.

Define a *CDP bialgebra* (Continous and Discrete Product bialgebra) to be the free product of a primitively generated bialgebra H and a semigroup algebra G. Specifically, the free product is formed in the category of augmented algebras [12]. A bialgebra is called *primitively generated* if it is generated as an algebra by its primitive elements. This is the case for bialgebras corresponding to continuous systems [5], such as $H = k<\xi_1, \xi_2>$ from Example 1. We consider a CDP bialgebra $H \amalg kG$ acting on an observation algebra R which is the direct sum of finitely many component algebras R_i which are associated with continuous systems (H, R_i). This is an immediate generalization of Example 1. The bialgebra $H \amalg kG$ acts on $R = \bigoplus_i R_i$ as follows. The bialgebra H acts on R_i as it does in the individual continuous systems (H, R_i); the semigroup G acts on the set of states $\{i\}$, and acts on the function algebra R in a fashion compatible with its action on the states. For more detail, see [7].

We discuss flows in the context of formal series. If V is a vector space, denote by V_t the space of formal power series $V[[t]] = \bigoplus_{n=0}^{\infty} Vt^n$ over V. In [9] completed tensor products of spaces of the form V_t and coalgebras of this form are discussed.

Formally, the solution to the differential equation

$$\dot{x}(t) = E(x(t)), \qquad x(0) = x^0$$

is given by

$$x(t) = e^{tE}x^0.$$

It can be shown that the fact that E is primitive implies that e^{tE} is grouplike. In the observation space representation of a continuous system, the grouplike $e^{tE} \in H_t$ is the flow corresponding to the differential equation $\dot{x}(t) = E(x(t))$. To summarize, the dynamics of a continuous systems are infinitesimally determined by derivations $E \in H$ while the flows are specified by grouplikes $K \in H_t$. One can think of K as being of the form $K = e^{tE}$.

In the observation space representation of the discrete automaton in which the alphabet of input symbols Ω acts on the state space of a finite automaton, the flows are exactly the grouplike elements of the bialgebra $(k\Omega^*)_t$, where Ω^* is the semigroup of words in the alphabet Ω. It can be shown that these elements are exactly the elements of Ω^*. To summarize, the dynamics of an automaton correspond to endomorphisms associated with input symbols, and the flows correspond to grouplike elements of $(k\Omega^*)_t$, which are words in the input symbols. The flows may be viewed as execution sequences of the automaton.

We turn to the general case now. Given any bialgebra B, an algebra of observation functions R, and a compatible action of B on R [7], the *flows* of the hybrid system are defined to be the grouplike elements of B_t. We pose the general problem:

Problem: Characterize the flows of hybrid systems.

As it stands, this is much too hard, since it includes as special cases the problems of characterizing the flows of dynamical systems, of automata, and of a large variety of systems formed from suitable products of these. In this note, we introduce a class of nicely behaved flows and make some comments about them.

In the observation space representation of the hybrid system in which the underlying bialgebra of the observation space representation is the CDP algebra $B = H \amalg kG$, the flows are the grouplikes in $(H \amalg kG)_t$. Note that the map from $H_t \amalg kG$ to $(H \amalg kG)_t$ induces a map from $\Gamma(H_t \amalg kG)$ to the flows $\Gamma((H \amalg kG)_t)$. The flows which are in the image of this map are ones which arise as a finite sequence of continuous flows (elements of $\Gamma(H_t)$) and mode switches (elements of $G = \Gamma(kG)$). We call these flows *regular flows*.

Example 2. Let $H \amalg kG$ dnote a CDP bialgebra. This example shows that not all flows are regular. Let E denote a primitive element of H, and let $g_1, g_2, \ldots$ be an infinite sequence of distinct invertible elements of G. Then

$$\prod_{n=1}^{\infty} g_n^{-1} e^{t^n E} g_n$$

is a flow in $\Gamma((H \amalg kG)_t)$ which is not in the image of $\Gamma(H_t) \amalg G$, since it cannot be expressed using only finitely many mode switches.

We observe next that the taxicab example described in Example 1 has the property that there exists a neighborhood of the origin of k^2 with the property that every point x' in the neighborhood is of the form $x' = Kx^0$, where x^0 is the origin and K is a regular flow. In other words, a neighborhood of the origin is reachable using regular flows. It is also easy to see that almost all these flows require at least one mode switch by an element of $G = \Omega^*$.

Acknowledgment: We are grateful to Moss Sweedler for his helpful suggestions in constructing a non regular flow in a hybrid system.

References

1. Z. Bartosiewicz, "Ordinary differential equations on real affine varieties," *Bulletin of the Polish Academy of Sciences: Mathematics*, Volume 35, pp. 13–18, 1987.

2. Z. Bartosiewicz, "Minimal Polynomial Realizations," *Mathematics of Control, Signals, and Systems,* Volume 1, pp. 227-237, 1988.

3. J. Glimm and A. Jaffe, *Quantum Physics: A Functional Integral Point of View,* Springer-Verlag, New York, 1981.

4. R. L. Grossman, "Path planning, persistent stores, and hybrid systems," submitted for publication.

5. R. L. Grossman and R. G. Larson, "The realization of input–output maps using bialgebras, *Forum Math.,* Volume 4, pp. 109–121, 1992.

6. R. L. Grossman and R. G. Larson, "Viewing hybrid systems as products of control systems and automata," *Proceedings of the 31st IEEE Conference on Decision and Control,* IEEE Press, 1992.

7. R. L. Grossman and R. G. Larson, "A bialgebra approach to hybrid systems," submitted for publication.

8. R. L. Grossman and R. G. Larson, "Hybrid systems as free products of bialgebras, in preparation.

9. R. L. Grossman and D. E. Radford, "Bialgebra deformations of certain universal enveloping algebras," *Communications in Pure and Applied Algebra,* to appear.

10. W. Kohn, "Declarative control theory," *AIAA GN & D Journal,* February, 1989.

11. W. Kohn and A. Nerode, "Models for hybrid sysems: Automata, topologies and stability," submitted for publication.

12. M. E. Sweedler, *Hopf algebras,* W. A. Benjamin, New York, 1969.

13. E. D. Sontag, *Polynomial Response Maps,* Lecture Notes in Control and Information Sciences, Volume 13, Springer-Verlag, New York, 1979.

Hybrid System Modeling and Autonomous Control Systems

Panos J. Antsaklis, James A. Stiver, and Michael Lemmon *

Department of Electrical Engineering
University of Notre Dame
Notre Dame, IN 46556 USA

Abstract. Hybrid control systems contain two distinct types of systems, continuous state and discrete-state, that interact with each other. Their study is essential in designing sequential supervisory controllers for continuous- state systems, and it is central in designing control systems with high degree of autonomy.

After an introduction to intelligent autonomous control and its relation to hybrid control, models for the plant, controller, and interface are introduced. The interface contains memoryless mappings between the supervisor's symbolic domain and the plant's nonsymbolic state space. The simplicity and generality afforded by the assumed interface allows us to directly confront important system theoretic issues in the design of supervisory control systems. such as determinism, quasideterminism, and the relationship of hybrid system theory to the more mature theory of logical discrete event systems.

1 Introduction

Hybrid control systems contain two distinct types of systems, continuous and discrete-state, which interact with each other. An example of such a system is the heating and cooling system of a typical home. Here the furnace and air conditioner together with the home's heat loss dynamics can be modeled as continuous-state, (continuous-time) system which is being controlled by a discrete-state system, the thermostat. Other examples include systems controlled by bang-bang control or via methods based on variable structure control.

Hybrid control systems also appear as part of *Intelligent Autonomous Control Systems*. Being able to control a continuous-state system using a discrete-state supervisory controller is a central problem in designing control systems with high degrees of autonomy. This is further discussed below.

The analysis and design of hybrid control systems requires the development of an appropriate mathematical framework. That framework must be both powerful and simple enough so it leads to manageable descriptions and efficient algorithms for such systems. Recently, attempts have been made to study hybrid control

* The partial financial support of the National Science Foundation (IRI91-09298 and MSS92-16559) is acknowledged

systems in a unified, analytical way and a number of results have been reported in the literature [Benveniste 1990] [Gollu 1989] [Grossman 1992] [Holloway 1992] [Kohn 1992] [Lemmon 1993b] [Nerode 1992] [Passino 1991b] [Peleties 1988] [Peleties 1989] [Stiver 1991a] [Stiver 1991b] [Stiver 1991c] [Stiver 1992] [Stiver 1993].

In this chapter, a novel approach to hybrid systems modeling and control is described. Descriptions of the plant to be controlled, the controller and the interface are given in Section 3. The important role of the interface is discussed at length. In Section 4, certain system theoretic questions are addressed. In particular, the concepts of determinism and quasideterminism are introduced and results are given. It is then shown how logical Discrete Event System (DES) models can be used to formulate the hybrid control problem, thus taking full advantage of existing results on DES controller design [Cassandras 1990] [Ozveren 1991] [Passino 1989a] [Passino 1989b] [Passino 1991a] [Passino 1992a] [Ramadge 1987] [Ramadge 1989] [Wonham 1987] for hybrid control systems. When the system to be controlled is changing, these fixed controllers may not be adequate to meet the control goals. In this case it is desirable to identify the plant and derive the control law on line, and this is addressed in the companion chapter in this volume titled "Event Identification and Intelligent Hybrid Control." Inductive inference methods are used to identify plant events in a computationally efficient manner.

In Section 2, after a brief introduction to Intelligent Autonomous Control, the important role hybrid control systems play in the design of Autonomous Control Systems is discussed and explained. In this way, the hybrid control problem can be seen in the appropriate setting so that its importance in the control of very complex systems may be fully understood and appreciated. Further discussion can be found in [Antsaklis 1993b]; for more information on intelligent control see [Albus 1981] [Antsaklis 1989] [Antsaklis 1991] [Antsaklis 1993b] [Antsaklis 1993a] [Antsaklis 1993c] [IEEE Computer 1989] [Passino 1993] [Saridis 1979] [Saridis 1985] [Saridis 1987] [Saridis 1989a] [Zeigler 1984].

2 On Intelligent Autonomous Control Systems

It is appropriate to first explain what is meant by the term Intelligent Autonomous Control [Antsaklis 1989] [Antsaklis 1993b]. In the design of controllers for complex dynamical systems, there are needs today that cannot be successfully addressed with the existing conventional control theory. Heuristic methods may be needed to tune the parameters of an adaptive control law. New control laws to perform novel control functions to meet new objectives should be designed while the system is in operation. Learning from past experience and planning control actions may be necessary. Failure detection and identification is needed. Such functions have been performed in the past by human operators. To increase the speed of response, to relieve the operators from mundane tasks, to protect them from hazards, a high degree of autonomy is desired. To achieve this autonomy, high level decision making techniques for reasoning under uncertainty must be utilized. These techniques, if used by humans, may be attributed to intelligence. Hence, one way to achieve high degree of autonomy is to utilize high level decision making techniques, intelligent methods, in the autonomous controller. In our view, *higher autonomy is the objective, and intelligent controllers are one way to achieve it.* The need for quantitative methods to model and analyze the dynamical behavior of such autonomous systems presents significant challenges well beyond current capabilities. It is clear that the development of autonomous controllers requires significant interdisciplinary research effort as it integrates concepts and methods from areas such as Control, Identification, Estimation, Communication Theory, Computer Science, Artificial Intelligence, and Operations Research.

Control systems have a long history. Mathematical modeling has played a central role in its development in the last century and today conventional control theory is based on firm theoretical foundations. Designing control systems with higher degrees of autonomy has been a strong driving force in the evolution of control systems for a long time. What is new today is that with the advances of computing machines we are closer to realizing highly autonomous control systems than ever before. One of course should never ignore history but learn from it. For this reason, a brief outline of conventional control system history and methods is given below.

2.1 Conventional Control - Evolution and Quest for Autonomy

The first feedback device on record was the water clock invented by the Greek Ktesibios in Alexandria Egypt around the 3rd century B.C. This was certainly a successful device as water clocks of similar design were still being made in Baghdad when the Mongols captured the city in 1258 A.D.! The first mathematical model to describe plant behavior for control purposes is attributed to J.C. Maxwell, of the Maxwell equations' fame, who in 1868 used differential equations to explain instability problems encountered with James Watt's flyball governor; the governor was introduced in the late 18th century to regulate the speed of steam engine vehicles. Control theory made significant strides in the past 120 years, with the use of frequency domain methods and Laplace transforms in the 1930s and 1940s and the development of optimal control methods and state space analysis in the 1950s and 1960s. Optimal control in the 1950s and 1960s, followed by progress in stochastic, robust and adaptive control methods in the 1960s to today, have made it possible to control more accurately significantly more complex dynamical systems than the original flyball governor.

When J.C Maxwell used mathematical modeling and methods to explain instability problems encountered with James Watt's flyball governor, he demonstrated the importance and usefulness of mathematical models and methods in understanding complex phenomena and signaled the beginning of mathematical system and control theory. It also signaled the end of the era of intuitive invention. The performance of the flyball governor was sufficient to meet the control needs of the day. As time progressed and more demands were put on the device there came a point when better and deeper understanding of the device was necessary as it started exhibiting some undesirable and unexplained behavior, in particular oscillations. This is quite typical of the situation in man made systems even today where systems based on intuitive invention rather than quantitative theory can be rather limited. To be able to control highly complex and uncertain systems we need deeper understanding of the processes involved and systematic design methods, we need quantitative models and design techniques. Such a need is quite apparent in intelligent autonomous control systems and in particular in hybrid control systems.

Conventional control design methods: Conventional control systems are designed today using mathematical models of physical systems. A mathematical model, which captures the dynamical behavior of interest, is chosen and then control design techniques are applied, aided by Computer Aided Design (CAD) packages, to design the mathematical model of an appropriate controller. The controller is then realized via hardware or software and it is used to control the physical system. The procedure may take several iterations. The mathematical model of the system must be "simple enough" so that it can be analyzed with available mathematical techniques, and "accurate enough" to describe the important aspects of the relevant dynamical behavior. It approximates the behavior of a plant in the neighborhood of an operating point.

The control methods and the underlying mathematical theory were developed to meet the ever increasing control needs of our technology. The need to achieve the demanding control specifications for increasingly complex dynamical systems has been addressed by using more complex mathematical models and by developing more sophisticated design algorithms. The use of highly complex mathematical models however, can seriously inhibit our ability to develop control algorithms. Fortunately, simpler plant models, for example linear models, can be used in the control design; this is possible because of the feedback used in control which can tolerate significant model uncertainties. Controllers can for example be designed to meet the specifications around an operating point, where the linear model is valid and then via a scheduler a controller emerges which can accomplish the control objectives over the whole operating range. This is in fact the method typically used for aircraft flight control. When the uncertainties in the plant and environment are large, the fixed feedback controllers may not be adequate, and adaptive controllers are used. Note that adaptive control in conventional control theory has a specific and rather narrow meaning. In particular it typically refers to adapting to variations in the constant coefficients in the equations describing the linear plant: these new coefficient values are identified and then used, directly or indirectly, to reassign the values of the constant coefficients in the equations describing the linear controller. Adaptive controllers provide for wider operating ranges than fixed controllers and so conventional adaptive control systems can be considered to have higher degrees of autonomy than control systems employing fixed feedback controllers. There are many cases however where conventional adaptive controllers are not adequate to meet the needs and novel methods are necessary.

2.2 Intelligent Control for High Autonomy Systems

There are cases where we need to significantly increase the operating range of control systems. We must be able to deal effectively with significant uncertainties in models of increasingly complex dynamical systems in addition to increasing the validity range of our control methods. We need to cope with significant unmodeled and unanticipated changes in the plant, in the environment and in the control objectives. This will involve the use of intelligent decision making processes to generate control actions so that a certain performance level is maintained even though there are drastic changes in the operating conditions. It is useful to keep in mind an example, the Houston Control example . It is an example that sets goals for the future and it also teaches humility as it indicates how difficult demanding and complex autonomous systems can be. Currently, if there is a problem on the space shuttle, the problem is addressed by the large number of engineers working in Houston Control, the ground station. When the problem is solved the specific detailed instructions about how to deal with the problem are sent to the shuttle. Imagine the time when we will need the tools and expertise of all Houston Control engineers aboard the space shuttle, space vehicle, for extended space travel.

In view of the above it is quite clear that in the control of systems there are requirements today that cannot be successfully addressed with the existing conventional control theory. They mainly pertain to the area of uncertainty, present because of poor models due to lack of knowledge, or due to high level models used to avoid excessive computational complexity.

The control design approach taken here is a bottom-up approach. One turns to more sophisticated controllers only if simpler ones cannot meet the required objectives. The need to use intelligent autonomous control stems from the need for an increased level of autonomous decision making abilities in achieving complex control tasks. *Note that intelligent methods are not necessary to increase the control system's autonomy. It is possible to attain higher degrees of autonomy by using methods that are not considered intelligent. It appears however that to achieve the highest degrees of autonomy, intelligent methods are necessary indeed.*

2.3 An Intelligent High Autonomy Control System Architecture For Future Space Vehicles

To illustrate the concepts and ideas involved and to provide a more concrete framework to discuss the issues, a hierarchical functional architecture of an intelligent controller that is used to attain high degrees of autonomy in future space vehicles is briefly outlined; full details can be found in [Antsaklis 1989]. This hierarchical architecture has three levels, the Execution Level, the Coordination Level, and the Management and Organization Level. The architecture exhibits certain characteristics, which have been shown in the literature to be necessary and desirable in autonomous intelligent systems.

It is important at this point to comment on the choice for a hierarchical architecture. Hierarchies offer very convenient ways to describe the operation of complex systems and deal with computational complexity issues, and they are used extensively in the modeling of intelligent autonomous control systems. Such a hierarchical approach is taken here (and in [Passino 1993]) to study intelligent autonomous and hybrid control systems.

Architecture Overview: The overall functional architecture for an autonomous controller is given by the architectural schematic of the figure below. This is a functional architecture rather than a hardware processing one; therefore, it does not specify the arrangement and duties of the hardware used to implement the functions described. Note that the processing architecture also depends on the characteristics of the current processing technology; centralized or distributed processing may be chosen for function implementation depending on available computer technology.

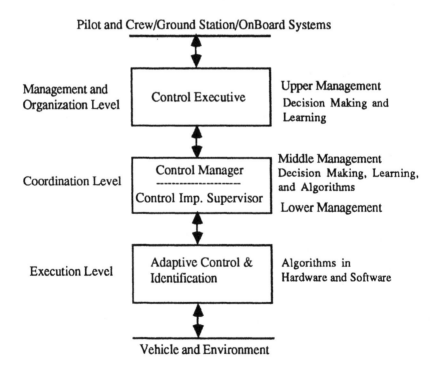

Fig. 1. Intelligent Autonomous Controller Functional Architecture

The architecture in Figure 1 has three levels; this is rather typical in the Intelligent Control literature. At the lowest level, the Execution Level, there is the interface to the vehicle and its environment via the sensors and actuators. At the highest level, the Management and Organization Level, there is the interface to the pilot and crew, ground station, or onboard systems. The middle level, called the Coordination Level, provides the link between the Execution Level and the Management Level. Note that we follow the somewhat standard viewpoint that there are three major levels in the hierarchy. It must be stressed that the system may have more or fewer than three levels. Some characteristics of the system which dictate the number of levels are the extent to which the operator can intervene in the system's operations, the degree of autonomy or level of intelligence in the various subsystems, the hierarchical characteristics of the plant. Note however that the three levels shown here in Figure 1 are applicable to most architectures of autonomous controllers, by grouping together sublevels of the architecture if necessary. As it is indicated in the figure, the lowest, Execution Level involves conventional control algorithms, while the highest, Management and Organization Level involves only higher level, intelligent, decision making methods. The Coordination Level is the level which provides the interface between the actions of the other two levels and it uses a combination of conventional and intelligent decision making methods. The sensors and actuators are implemented mainly with hardware. Software and perhaps hardware are used to implement the Execution Level. Mainly software is used for both the Coordination and Management Levels. There are multiple copies of the control functions at each level, more at the lower and fewer at the higher levels. See [Antsaklis 1989] [Antsaklis 1993b] for an extended discussion of the issues involved.

Hybrid control systems do appear in the intelligent autonomous control system framework whenever one considers the Execution level together with control functions performed in the higher Coordination and Management levels. Examples include expert systems supervising and tuning conventional controller parameters, planning systems setting the set points of local control regulators, sequential controllers deciding which from a number of conventional controllers is to be used to control a system, to mention but a few. One obtains a hybrid control system of interest whenever one considers controlling a continuous-state plant (in the Execution level) by a control algorithm that manipulates symbols, that is by a discrete-state controller (in Coordination and/or Management levels).

2.4 Quantitative Models

For highly autonomous control systems, normally the plant is so complex that it is either impossible or inappropriate to describe it with conventional mathematical system models such as differential or difference equations. Even though it might be possible to accurately describe some system with highly complex nonlinear differential equations, it may be inappropriate if this description makes subsequent analysis too difficult or too computationally complex to be useful.

The complexity of the plant model needed in design depends on both the complexity of the physical system and on how demanding the design specifications are. There is a tradeoff between model complexity and our ability to perform analysis on the system via the model. However, if the control performance specifications are not too demanding, a more abstract, higher level, model can be utilized, which will make subsequent analysis simpler. This model intentionally ignores some of the system characteristics, specifically those that need not be considered in attempting to meet the particular performance specifications. For example, a simple temperature controller could ignore almost all dynamics of the house or the office and consider only a temperature threshold model of the system to switch the furnace off or on.

The quantitative, systematic techniques for modeling, analysis, and design of control systems are of central and utmost practical importance in conventional control theory. Similar techniques for intelligent autonomous controllers do not exist. This is mainly due to the hybrid structure (nonuniform, nonhomogeneous nature) of the dynamical systems under consideration; they include both continuous-state and discrete-state systems. Modeling techniques for intelligent autonomous systems must be able to support a macroscopic view of the dynamical system, hence it is necessary to represent both numeric and symbolic information. The nonuniform components of the intelligent controller all take part in the generation of the low level control inputs to the dynamical system, therefore they all must be considered in a complete analysis. Therefore the study of modeling and control of hybrid control systems is essential in understanding highly autonomous control systems [Antsaklis 1989].

3   Hybrid Control System Modeling

The hybrid control systems considered here consist of three distinct levels; see Figure 2. The controller is a discrete-state system, a sequential machine, seen as a Discrete Event System (DES). The controller receives, manipulates and outputs events represented by symbols. The plant is a continuous-state system typically modeled by differential/difference equations and it is the system to be controlled by the discrete-state controller. The plant receives, manipulates and outputs signals represented by real variables that are typically (piecewise) continuous. The controller and the plant communicate via the interface that translates plant outputs into symbols for the controller to use, and controller output symbols into command signals for the plant input. The interface can be seen as consisting of two subsystems: the generator that senses the plant outputs and generates symbols representing plant events, and the actuator that translates the controller symbolic commands into piecewise constant plant input signals.

To develop a useful mathematical framework we keep the interface as simple as possible; this is further discussed below. The interface determines the events the controller sees and uses to decide the appropriate control action. If the plant and the interface are taken together the resulting system is a DES, called the

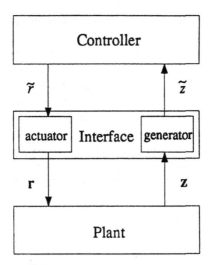

Fig. 2. Hybrid Control System

DES Plant, that the controller sees and attempts to control. Another way of expressing this is that the DES controller only sees a more abstract model of the plant; a higher level less detailed plant model than the differential / difference equation model. The complexity of this more abstract DES plant model depends on the interface. It is therefore very important to understand the issues involved in the interface design so that the appropriate DES model is simple enough so to lead to a low complexity controller. It should be noted that this lower complexity is essential for real time adaptation of hybrid control systems. All these issues pointed out here are discussed in detail later in this chapter.

It is important to identify the important concepts and develop an appropriate mathematical framework to describe hybrid control systems. Here the logical DES theory and the theory of automata are used. The aim is to take advantage as much as possible of the recent developments in the analysis and control design of DES. These include results on controllability, observability, stability of DES and algorithms for control design among others. We first present a flexible and tractable way of modeling hybrid control systems. Our goal is to develop a model which can adequately represent a wide variety of hybrid control systems, while remaining simple enough to permit analysis. We then present methods which can be used to analyze and aid in the design of hybrid control systems. These methods relate to the design of the interface which is a necessary component of a hybrid system and its particular structure reflects both the dynamics of the plant and the aims of the controller.

Below, the plant, interface and controller are described first. The assumptions made and the generality of the models are discussed. In Section 4, the DES plant model is then derived and the concepts of determinism and quasideterminism are introduced and certain results are shown. The description of the generator in the interface is discussed. Controllability of the DES plant model is studied. The selection of the interface is discussed at length and the fundamental issues are identified. Connections to Ramadge-Wonham model are shown, the difficulties involved are indicated, and some recent results are outlined. Simple examples are used throughout to illustrate and explain. Note that most of these results can be found in [Stiver 1992].

A hybrid control system, can be divided into three parts, the plant, interface, and controller as shown in Figure 2. The models we use for each of these three parts, as well as the way they interact are now described.

3.1 Plant

The system to be controlled, called the plant, is modeled as a time-invariant, continuous-time system. This part of the hybrid control system contains the entire continu-ous-time portion of the system, possibly including a continuous-time controller. Mathematically, the plant is represented by the familiar equations

$$\dot{\mathbf{x}} = f(\mathbf{x}, \mathbf{r}) \tag{1}$$
$$\mathbf{z} = g(\mathbf{x}) \tag{2}$$

where $\mathbf{x} \in \Re^n$, $\mathbf{r} \in \Re^m$, and $\mathbf{z} \in \Re^p$ are the state, input, and output vectors respectively. $f : \Re^n \times \Re^m \to \Re^n$ and $g : \Re^n \to \Re^p$ are functions. This is the common plant model used in systems and control. In our theory, developed below, it is only necessary to have a mathematical description where the state trajectories are uniquely determined by the initial state and the input signals. For the purposes of this work we assume that $\mathbf{z} = \mathbf{x}$. Note that the plant input and output are continuous-time vector valued signals. Bold face letters are used to denote vectors and vector valued signals.

3.2 Controller

The controller is a discrete event system which is modeled as a deterministic automaton. This automaton can be specified by a quintuple, $\{\tilde{S}, \tilde{Z}, \tilde{R}, \delta, \phi\}$, where $\tilde{S}$ is the (possibly infinite) set of states, $\tilde{Z}$ is the set of plant symbols, $\tilde{R}$ is the set of controller symbols, $\delta : \tilde{S} \times \tilde{Z} \to \tilde{S}$ is the state transition function, and $\phi : \tilde{S} \to \tilde{R}$ is the output function. The symbols in set $\tilde{R}$ are called controller symbols because they are generated by the controller. Likewise, the symbols in set $\tilde{Z}$ are called plant symbols and are generated by the occurrence of events in the plant. The action of the controller can be described by the equations

$$\tilde{s}[n] = \delta(\tilde{s}[n-1], \tilde{z}[n]) \tag{3}$$
$$\tilde{r}[n] = \phi(\tilde{s}[n]) \tag{4}$$

where $\tilde{s}[n] \in S, \tilde{z}[n] \in \tilde{Z}$, and $\tilde{r}[n] \in \tilde{R}$. The index n is analogous to a time index in that it specifies the order of the symbols in a sequence. The input and output signals associated with the controller are asynchronous sequences of symbols, rather than continuous-time signals. Notice that there is no delay in the controller. The state transition, from $\tilde{s}[n-1]$ to $\tilde{s}[n]$, and the controller symbol, $\tilde{r}[n]$, occur immediately when the plant symbol $\tilde{z}[n]$ occurs.

Tildes are used to indicate that the particular set or signal is made up of symbols. For example, $\tilde{Z}$ is the set of plant symbols and $z[n]$ is a sequence of plant symbols. An argument in brackets, e.g. $\tilde{z}[n]$, represents the nth symbol in the sequence $\tilde{z}$. A subscript, e.g. $\tilde{z}_i$, is used to denote a particular symbol from a set.

3.3 Interface

The controller and plant cannot communicate directly in a hybrid control system because each utilizes a different type of signal. Thus an interface is required which can convert continuous-time signals to sequences of symbols and vice versa. The interface consists of two memoryless maps, γ and α. The first map, called the actuating function or actuator, $\gamma : \tilde{R} \rightarrow \Re^m$, converts a sequence of controller symbols to a piecewise constant plant input as follows

$$\mathbf{r}(t) = \gamma(\tilde{r}[n]) \tag{5}$$

The plant input, $\mathbf{r}$, can only take on certain constant values, where each value is associated with a particular controller symbol. Thus the plant input is a piecewise constant signal which may change only when a controller symbol occurs. The second map, the plant symbol generating function or generator, $\alpha : \Re^n \rightarrow \tilde{Z}$, is a function which maps the state space of the plant to the set of plant symbols as follows

$$\tilde{z}[n] = \alpha(\mathbf{x}(t)) \tag{6}$$

It would appear from Equation 6 that, as $\mathbf{x}$ changes, $\tilde{z}$ may continuously change. That is, there could be a continuous generation of plant symbols by the interface because each state is mapped to a symbol. This is not the case because α is based upon a partition of the state space where each region of the partition is associated with one plant symbol. These regions form the equivalence classes of α. A plant symbol is generated only when the plant state, $\mathbf{x}$, moves from one of these regions to another.

3.4 Comments on the Generality of the Model

The model described above may appear at first to be too limited but this is not the case. The simplicity of this model is its strength and it does not reduce its flexibility when modeling a hybrid control system. It is tempting to add complexity to the interface, however this typically leads to additional mathematical difficulties that are not necessary. Consider first the function γ which maps controller symbols to plant inputs. Our model features only constant plant inputs, no ramps, sinusoids, or feedback strategies. The reasons for this are two fold. First, in order for the interface to generate a nonconstant signal or feedback signal it must contain components which can be more appropriately included in the continuous time plant, as is done in the model above. Second, making the interface more complex will complicate the analysis of the overall system. Keeping the function γ as a simple mapping from each controller symbol to a unique numeric value is the solution.

The interface could also be made more complex by generalizing the definition of a plant symbol. A plant symbol is defined solely by the current plant state, but this could be expanded by defining a plant symbol as being generated following the occurrence of a specific series of conditions in the plant. For example, the interface could be made capable of generating a symbol which is dependent upon the current and previous values of the state. However, doing this entails including dynamics in the interface which actually belong in the controller. The controller, as a dynamic system, is capable of using its state as a memory to keep track of previous plant symbols.

The key feature of this hybrid control system model is its simple and unambiguous nature, especially with respect to the interface. To enable analysis, hybrid control systems must be described in a consistent and complete manner. Varying the nature of the interface from system to system in an ad hoc manner, or leaving its mathematical description vague causes difficulties.

3.5 Examples

Example 1 - Thermostat/Furnace System This example will show how an actual physical system can be modeled and how the parts of the physical system correspond to the parts found in the model. The particular hybrid control system in this example consists of a typical thermostat and furnace. Assuming the thermostat is set at 70 degrees Fahrenheit, the system behaves as follows. If the room temperature falls below 70 degrees the furnace starts and remains on until the room temperature exceeds 75 degrees. At 75 degrees the furnace shuts off. For simplicity, we will assume that when the furnace is on it produces a constant amount of heat per unit time.

The plant in the thermostat/furnace hybrid control system is made up of the furnace and room. It can be modeled with the following differential equation

$$\dot{x} = .0042(T_0 - x) + 2step(r) \tag{7}$$

where the plant state, x, is the temperature of the room in degrees Fahrenheit, the input, r, is the voltage on the furnace control circuit, and T_0 is the outside temperature. The units for time are minutes. This model of the furnace is a simplification, but it is adequate for this example.

The remainder of the hybrid control system is found in the thermostat which is pictured in Figure 3. As the temperature of the room varies, the two strips of metal which form the bimetal band expand and contract at different rates thus causing the band to bend. As the band bends, it brings the steel closer to one side of the glass bulb. Inside the bulb, a magnet moves toward the nearest part of the steel and opens or closes the control circuit in the process. The bimetal band effectively partitions the state space of the plant, $\mathbf{x}$, as follows

$$\alpha(\mathbf{x}) = \begin{cases} \tilde{z}_1 \text{ if } & \mathbf{x} \le 70 \\ \tilde{z}_2 \text{ if } 70 < \mathbf{x} \le 75 \\ \tilde{z}_3 \text{ if } & \mathbf{x} > 75 \end{cases}, \tag{8}$$

where the three symbols correspond to 1) steel is moved against the left side of the bulb, 2) band is relaxed, and 3) steel is moved against the right side of the bulb.

Inside the glass bulb is a magnetic switch which is the DES controller. It has two states because the switch has two positions, on and off. The DES controller input, $\tilde{z}$, is a magnetic signal because the symbols generated by the generator are conveyed magnetically. The state transition graph of this simple controller is shown in Figure 4. The output function of the controller is essentially the following

$$\phi(\tilde{s}_1) = \tilde{r}_1 \Leftrightarrow \text{close control circuit} \tag{9}$$

$$\phi(\tilde{s}_2) = \tilde{r}_2 \Leftrightarrow \text{open control circuit} \tag{10}$$

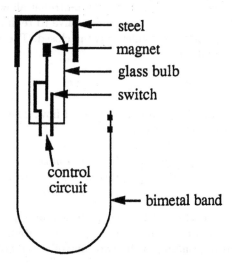

Fig. 3. Thermostat

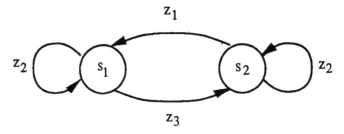

Fig. 4. Controller for Thermostat/Furnace System

The contacts on the switch which open and close the control circuit can be thought of as the actuator, although there is no logical place to separate the actuator from the DES controller. The commands from the controller to the actuator are basically a formality here because the controller and actuator are mechanically one piece. With this in mind, the actuator operates as

$$\gamma(\tilde{r}_1) = 0 \qquad (11)$$
$$\gamma(\tilde{r}_2) = 24 \qquad (12)$$

Example 2 - Surge Tank This is another example to illustrate how a simple hybrid control system can be modeled. The system consists of a surge tank which is draining through a fixed outlet valve, while the inlet valve is being controlled by a discrete event system. The controller allows the tank to drain to a minimum level and then opens the inlet valve to refill it. When the tank has reached a maximum level, the inlet valve is closed. The surge tank is modeled by a differential equation,

$$\dot{\mathbf{x}} = \mathbf{r} - \mathbf{x}^{1/2} \qquad (13)$$

where $\mathbf{x}$ is the liquid level and $\mathbf{r}$ is the inlet flow. The interface partitions the state space into three regions as follows

$$\alpha(\mathbf{x}) = \begin{cases} \tilde{z}_1 \ \textbf{if} & \mathbf{x} > max \\ \tilde{z}_2 \ \textbf{if} \ min < \mathbf{x} < max \ , \\ \tilde{z}_3 \ \textbf{if} & \mathbf{x} < min \end{cases} \qquad (14)$$

Thus when the level exceeds max, plant symbol $\tilde{z}_1$ is generated, and when the level falls below min, plant symbol $\tilde{z}_3$ is generated. The interface provides for two inputs corresponding to the two controller symbols $\tilde{r}_1$ and $\tilde{r}_2$ as follows

$$\gamma(\tilde{r}) = \begin{cases} 1 \text{ if } \tilde{r} = \tilde{r}_1 \\ 0 \text{ if } \tilde{r} = \tilde{r}_2 \end{cases}, \tag{15}$$

Since $\mathbf{r} = \gamma(\tilde{r})$, this means the inlet valve will be open following controller symbol $\tilde{r}_1$, and closed following controller symbol $\tilde{r}_2$.

The controller for the surge tank is a two state automaton which moves to state $\tilde{s}_1$ whenever $\tilde{z}_3$ is received, moves to state $\tilde{s}_2$ whenever $\tilde{z}_1$ is received and returns to the current state if $\tilde{z}_2$ is received. Furthermore $\phi(\tilde{s}_1) = \tilde{r}_1$ and $\phi(\tilde{s}_2) = \tilde{r}_2$.

4 System Theoretic Issues

4.1 The DES Plant Model

If the plant and interface of a hybrid control system are viewed as a single component, this component behaves like a discrete event system. It is advantageous to view a hybrid control system this way because it allows it to be modeled as two interacting discrete event systems which are more easily analyzed than the system in its original form. The discrete event system which models the plant and interface is called the DES Plant Model and is modeled as an automaton similar to the controller. The automaton is specified by a quintuple, $\{\tilde{P}, \tilde{Z}, \tilde{R}, \psi, \lambda\}$, where $\tilde{P}$ is the set of states, $\tilde{Z}$ and $\tilde{R}$ are the sets of plant symbols and controller symbols, $\psi : \tilde{P} \times \tilde{R} \to \tilde{P}$ is the state transition function, and $\lambda : \tilde{P} \to \tilde{Z}$ is the output function. The behavior of the DES plant is as follows

$$\tilde{p}[n+1] = \psi(\tilde{p}[n], \tilde{r}[n]) \tag{16}$$
$$\tilde{z}[n] = \lambda(\tilde{p}[n]) \tag{17}$$

where $\tilde{p}[n] \in \tilde{P}, \tilde{r}[n] \in \tilde{R}$, and $\tilde{z}[n] \in \tilde{Z}$. There are two differences between the DES plant model and the controller. First, as can be seen from Equation 16, the state transitions in the DES plant do not occur immediately when a controller symbol occurs. This is in contrast to the controller where state transitions occur immediately with the occurrence of a plant symbol. The second difference is that the automaton which models the DES plant may be non-deterministic, meaning $\tilde{p}[n+1]$ in Equation 16 is not determined exactly but rather is limited to some subset of $\tilde{P}$. The reason for these differences is that the DES plant model is a simplification of a continuous-time plant and an interface. This simplification results in a loss of information about the internal dynamics, leading to non-deterministic behavior.

The set of states, $\tilde{P}$, of the DES plant is based on the partition realized in the interface. Specifically, each state in $\tilde{P}$ corresponds to a region, in the state space of the continuous-time plant, which is equivalent under α. Thus there is a one-to-one correspondence between the set of states, $\tilde{P}$, and the set of plant symbols, $\tilde{Z}$. It is this relationship between the states of the DES plant model and the plant symbols which forms the basis for the work described in this section. It can be used to develop an expression for the state transition function, ψ. Starting with the continuous-time plant, we integrate Equation 1 to get the state after a time t, under constant input $\mathbf{r} = \gamma(\tilde{r}_k)$

$$\mathbf{x}(t) = F_k(\mathbf{x}_0, t) \tag{18}$$

Here $\mathbf{x}_0$ is the initial state, t is the elapsed time, and $\tilde{r}_k \in \tilde{R}$. $F_k(\mathbf{x}_0, t)$ is obtained by integrating $f(\mathbf{x}, \mathbf{r})$, with $\mathbf{r} = \gamma(\tilde{r}_k)$. Next we define

$$\hat{F}_k(\mathbf{x}_0) = F_k(\mathbf{x}_0, t), \tag{19}$$

where

$$\tau_0 = \inf_\tau \{\tau | \alpha(F(\mathbf{x}_0, \tau)) \neq \alpha(\mathbf{x}_0)\} \tag{20}$$

and

$$t = \tau_0 + \epsilon \tag{21}$$

for some infinitesimally small ϵ.

Equation 19 gives the state, $\mathbf{x}$, where it will cross into a new region. Now the dynamics of the DES plant model can be derived from Equations 5, 6, 19.

$$\tilde{z}[n+1] = \lambda(\psi(\tilde{p}[n], \tilde{r}[n])) \tag{22}$$

$$\tilde{z}[n+1] = \alpha(\hat{F}_k(\mathbf{x}_0)) \tag{23}$$

$$\psi(\tilde{p}[n], \tilde{r}[n]) = \lambda^{-1}(\alpha(\hat{F}_k(\mathbf{x}_0))) \tag{24}$$

where $\tilde{r}[n] = \tilde{r}_k$ and $\mathbf{x}_0 \in \{\mathbf{x} | \alpha(\mathbf{x}) = \lambda(\tilde{p}[n])\}$. As can be seen, the only uncertainty in Equation 24 is the value of $\mathbf{x}_0$. $\mathbf{x}_0$ is the state of the continuous-time plant at the time of the last plant symbol, $\tilde{z}[n]$, i.e. the time that the DES plant entered state $\tilde{p}[n]$. $\mathbf{x}_0$ is only known to within an equivalence class of α. The condition for a deterministic DES plant is that the state transition function, ψ, must be uneffected to this uncertainty.

Definition 1. A DES is *deterministic* iff for any state and any input, there is only one possible subsequent state.

The following theorem gives the conditions upon the hybrid control system such that the DES plant will be deterministic.

Theorem 2. *The DES plant will be deterministic iff given any $\tilde{p}[n] \in P$ and $\tilde{r}_k \in R$, there exists $\tilde{p}[n+1] \in P$ such that for every $\mathbf{x}_0 \in \{\mathbf{x} | \alpha(\mathbf{x}) = \lambda(\tilde{p}[n])\}$ we have $\alpha(\hat{F}_k(\mathbf{x}_0)) = \lambda(\tilde{p}[n+1])$.*

Proof: Notice that the set $\{\mathbf{x} | \alpha(\mathbf{x}) = \lambda(\tilde{p}[n])\}$ represents the set of all states, $\mathbf{x}$, in the continuous-time plant which could give rise to the state $\tilde{p}[n]$ in the DES plant. The theorem guarantees that the subsequent DES plant state, $\tilde{p}[n+1]$, is unique for a given input and thus the DES plant is deterministic.

To prove that the theorem is necessary, assume that it does not hold. There must then exist a $\tilde{p}[n] \in \tilde{P}$ and $\tilde{r}_k \in \tilde{R}$ such that no $\tilde{p}[n+1]$ exists to satisfy the condition: $\alpha(\hat{F}_k(\mathbf{x}_0)) = \lambda(\tilde{p}[n+1])$ for every $\mathbf{x}_0 \in \{\mathbf{x}|\alpha(\mathbf{x}) = \lambda(\tilde{p}[n])\}$. This is not a deterministic system because there is uncertainty in the state transition for at least one state and input. □

Theorem 2 states that the DES plant will be deterministic if all the state trajectories in the continuous-time plant, which start in the same region and are driven by the same input, move to the same subsequent region.

4.2 Double Integrator

To illustrate the DES plant model, an example of a hybrid control system containing a double integrator is given. Double integrators often arise in systems. For example, a satellite equipped with a thruster will behave as a double integrator when the thrust is considered the input and the velocity and position are the two states.

$$\dot{\mathbf{x}} = \begin{bmatrix} 0 & 1 \\ 0 & 0 \end{bmatrix} \mathbf{x} + \begin{bmatrix} 0 \\ 1 \end{bmatrix} \mathbf{r} \tag{25}$$

The general control goal in this system, which motivates the design of the interface, is to move the state of the double integrator between the four quadrants of the state-space. In the interface, the function a partitions the state space into four regions as follows,

$$\alpha(\mathbf{x}) = \begin{cases} \tilde{z}_1 & \text{if} & x_1, x_2 > 0 \\ \tilde{z}_2 & \text{if} & x_1 < 0, x_2 > 0 \\ \tilde{z}_3 & \text{if} & x_1, x_2 < 0 \\ \tilde{z}_4 & \text{if} & x_1 > 0, x_2 < 0 \end{cases}, \tag{26}$$

and the function γ provides three set points,

$$\gamma(\tilde{r}) = \begin{cases} -10 & \text{if} & \tilde{r} = \tilde{r}_1 \\ 0 & \text{if} & \tilde{r} = \tilde{r}_2 \\ 10 & \text{if} & \tilde{r} = \tilde{r}_3 \end{cases}, \tag{27}$$

So whenever the state of the double integrator enters quadrant 1, for example, the plant symbol $\tilde{z}_1$ is generated. When the controller (which is unspecified) generates controller symbol $\tilde{r}_1$, the double integrator is driven with an input of -10.

Now we know that the DES plant will have four states because there are four regions in the state space of the actual plant. By examining the various state trajectories given by Equation 28, we can find the DES plant which is shown in Figure 5. Equation 28 is obtained by integrating Equation 25 and adding $\mathbf{x}(0)$.

$$\mathbf{x} = \begin{bmatrix} 1 & t \\ 0 & 1 \end{bmatrix} \mathbf{x}_0 + \begin{bmatrix} .5t^2 \\ t \end{bmatrix} \gamma(\tilde{r}) \tag{28}$$

As can be seen in Figure 5, the DES plant is not deterministic. If we consider $\tilde{p}[n] = \tilde{p}_2$ and $\tilde{r}[n] = \tilde{r}_1$, there exists no uniquely defined $\tilde{p}[n+1]$, it could be either $\tilde{p}_1$ or $\tilde{p}_3$. This could present a problem in designing a controller for this system because it is not entirely predictable. In the following section a possible remedy for lack of determinism is presented and this example is revisited.

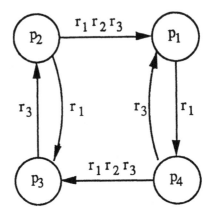

Fig. 5. DES Plant Model for Double Integrator

4.3 Partitioning and Quasideterminism

A particular problem in the design of a hybrid control system is the selection of the function α, which partitions the state-space of the plant into various regions. Since this partition is used to generate the plant symbols, it must be chosen to provide sufficient information to the controller to allow control without being so fine that it leads to an unmanageably complex system or simply degenerates the system into an essentially conventional control system.

The partition must accomplish two goals. First it must give the controller sufficient information to determine whether or not the current state is in an acceptable region. For example, in an aircraft these regions may correspond to climbing, diving, turning right, etc. Second, the partition must provide enough additional information about the state, to enable the controller to drive the plant to an acceptable region. In an aircraft, for instance, the input required to cause the plane to climb may vary depending on the current state of the plane. So to summarize, the partition must be detailed enough to answer: 1) is the current state acceptable; and 2) which input can be applied to drive the state to an acceptable region.

In a hybrid control system, the controller needs information about the plant for two reasons. First the controller must be able to assess whether the plant is operating as desired or if some new control action is needed. Second, if control action is called for, the controller needs to know which control action will achieve the desired effect. Both of these tasks require information about the plant. Consider for example a climate control system in a building. To assess the current condition, the controller needs to know whether the temperature and humidity fall within a certain range of acceptable values. If not the controller needs additional, more detailed, information about which condition is unacceptable and how much and in which direction it must be changed to reach the desired range.

To design a partition, we can start by designing a primary partition to meet the first goal mentioned above. This primary partition will identify all the desired operating regions of the plant state space, so its design will be dictated by the control goals. The final partition will represent a refinement of the primary partition which enables the controller to regulate the plant to any of the desired operating regions, thus meeting the second goal.

An obvious choice for the final partition is one which makes the DES plant deterministic and therefore guarantees that the controller will have full information about the behavior of the plant. In addition to being very hard to meet, this requirement is overly strict because the controller only needs to regulate the plant to the regions in the primary partition, not the final partition. For this reason we define quasideterminism, a weaker form of determinism. In the DES plant, the states which are in the same region of the primary partition can be grouped together, and if the DES plant is deterministic with respect to these groups, then we say it is quasideterministic. So if the DES plant is quasideterministic, then we may not be able to predict the next state exactly, but we will be able to predict its region of the primary partition and thus whether or not it is acceptable.

Definition 3. The DES plant will be quasideterministic iff given any $\tilde{p}[n] \in \tilde{P}$ and $\tilde{r}_k \in \tilde{R}$, there exists $\tilde{Q} \subset \tilde{P}$ such that for every $\mathbf{x}_0 \in \{\mathbf{x}|\alpha(\mathbf{x}) = \lambda(\tilde{p}[n])\}$ we have $\alpha_p(\hat{F}_k(\mathbf{x}_0)) = \lambda_p(\tilde{p}[n+1])$ where $\tilde{p}[n+1] \in \tilde{Q}$ and $\lambda_p(\tilde{q})$ is the same for all $\tilde{q} \in \tilde{Q}$. □

The functions α_p and λ_p are analogous to α and λ but apply to the primary partition. They are useful for comparing states but they are never implemented and their actual values are irrelevant. For example, if $\alpha_p(\mathbf{x}_1) = \alpha_p(\mathbf{x}_2)$, then $\mathbf{x}_1$ and $\mathbf{x}_2$ are in the same region of the primary partition. Or, if $\alpha_p(\mathbf{x}_1) = \lambda_p(\tilde{p}_1)$, then $\mathbf{x}_1$ is in the same region of the primary partition as $\tilde{p}_1$ in the DES plant. When used with α_p we define $\hat{F}$ as

$$\hat{F}_k(\mathbf{x}_0) = F_k(\mathbf{x}_0, t), \tag{29}$$

where

$$\tau_0 = \inf_\tau \{\tau|\alpha_p(F(\mathbf{x}_0, \tau)) \neq \alpha_p(\mathbf{x}_0)\} \tag{30}$$

and

$$t = \tau_0 + \epsilon \tag{31}$$

as before.

We would like to find the coarsest partition which meets the conditions of Definition 1 for a given primary partition. Such a partition is formed when the equivalence classes of α are as follows,

$$E[\alpha] = \inf\{E[\alpha_p], E[\alpha_p \circ \hat{F}_k]|\tilde{r}_k \in R\} \tag{32}$$

Where we use $E[\bullet]$ to denote the equivalence classes of $\bullet$. The infimum, in this case, means the coarsest partition which is at least as fine as any of the partitions in the set.

Theorem 4. *The regions described by Equation (32) form the coarsest partition which generates a quasideterministic DES plant.*

Proof: First we will prove that the partition does, in fact, lead to a quasideterministic system. For any two states, x_1 and x_2, which are in the same equivalence class of α, we apply some control $r = \gamma(\tilde{r}_k)$. The two states will subsequently enter new regions of the primary partition at $\hat{F}_k(x_1)$ and $\hat{F}_k(x_2)$ respectively. The actual regions entered are $\alpha_p(\hat{F}_k(x_1))$ and $\alpha_p(\hat{F}_k(x_2))$. Now according to Equation 32, if x_1 and x_2 are in the same equivalence class of α, then they are also in the same equivalence class of $\alpha_p \circ \hat{F}_k$. Therefore $\alpha_p(\hat{F}_k(x_1)) = \alpha_p(\hat{F}_k(x_2))$ and the system is quasideterministic.

Next we will prove that the partition is as coarse as possible. Assume there is a coarser partition which also generates a quasideterministic system. That is, there exists two states, x_3 and x_4, in the same region of the primary partition such that $\alpha(x_3) \neq \alpha(x_4)$, but $\alpha_p(\hat{F}_k(x_3)) = \alpha_p(\hat{F}_k(x_4))$ for *any* possible k. These two states would lie in the same equivalence class of $\alpha_p \circ \hat{F}_k$ for all $\tilde{r}_k \in \tilde{R}$ and therefore in the same equivalence class of $\inf\{E[\alpha_p], E[\alpha_p \circ \hat{F}_k | \tilde{r}_k \in \tilde{R}]\}$. This violates the assumption that x_3 and x_4 do not lie in the same equivalence class of α, so two such states could not exist and therefore a coarser partition can not exist. □

Quasideterminism accomplishes its goal by causing the trajectories of the various states within a given region of the final partition, under the same control, to be invariant with respect to the regions of the primary partition.

We can return now to the double integrator discussed previously and use it to illustrate quasideterminism. The state space of the double integrator had been partitioned into the four quadrants and this gave rise to the nondeterministic DES plant shown in Figure 5. Using those four regions as the primary partition, a final partition can be obtained according to Theorem 4. This partition is shown in Figure 6 and the resulting DES plant is shown in Figure 7. The final partition refined the regions in quadrants II and IV, and the DES plant is now quasideterministic (in fact it is deterministic but unfortunately that is not generally the result).

Note that the partition described in Equation 32 and discussed in Theorem 4 is not dependent upon any specific sequence of controller symbols. It is intended to yield a DES plant which is as "controllable" as possible, given the continuous-time plant and available inputs. If the specific control goals are known, it may be possible to derive a coarser partition which is still adequate. This can be done in an ad hoc fashion, for instance, by combining regions which are equivalent under the inputs which are anticipated when the plant is in those regions.

Selection of Control Action In hybrid control systems, the choice of the plant inputs which make up the range of the actuator, γ, play an important role in defining the system. At this time we have no way of systematically deriving a set of control actions which will achieve the desired control goals, either optimally or otherwise. We can assume that the control actions available are determined

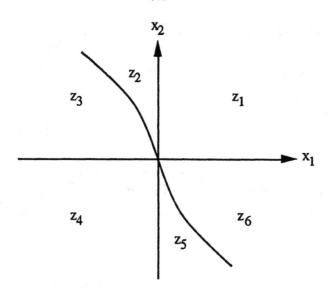

Fig. 6. State Space Partition for Double Integrator

by the plant (positions of the various valves, switches, etc.) and thus represent a constraint on the controller design.

4.4 Connections to Existing DES Control Theory

A significant amount of work has been done on the analysis and design of discrete event systems, especially the design of controllers for discrete event systems. Since the controller of a hybrid control system is a DES and we can use a DES to represent the plant in a hybrid control system, we can apply many of the theories and techniques, which were developed for DES's, to hybrid control systems. In this section, we draw on some of this work.

Stability: Several papers have been written dealing with the stability of discrete event systems, e.g. [Passino 1992a] and [Ozveren 1991]. In [Passino 1992a] the ideas of Lyapunov stability are applied to discrete event systems. These same techniques can be applied to the DES plant in a hybrid system. The states of the DES plant which are considered "desirable" are identified and a metric is defined on the remaining "undesirable" states. With a metric defined on the state space, finding a Lyapunov function will prove that the DES is stable. In the case of a hybrid control system, this interpretation of definition Lyapunov stability means the following. The state of the plant will remain within the set of states which were deemed "desirable" and if it is perturbed from this area, the state will return to it. A detailed application of these results to hybrid control systems can be found in [Stiver 1991b].

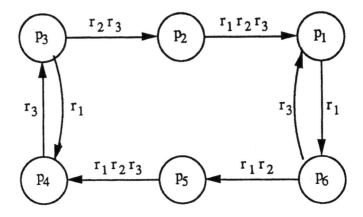

Fig. 7. DES Plant Model for Double Integrator

In [Ozveren 1991] the stability of a DES is defined as the property that the state of the DES will visit a certain subset infinitely often. This subset is analogous to the "desirable" set mention above. In a hybrid control system, this would imply that the state of the plant could leave the subset of states but would eventually return.

Controllability: Work has been done on the controllability of discrete event systems using the Ramadge-Wonham framework [Ramadge 1987] [Ramadge 1989] [Wonham 1987]. The DES models used in the Ramadge-Wonham framework differ from the models developed for hybrid control systems as described in this chapter, therefore the theorems and techniques cannot be applied directly, but must be adapted to work with a slightly different model.

The model developed by Ramadge and Wonham (henceforth RWM) features a generator and a supervisor, both DES's, which are analogous to the DES plant model and DES controller, respectively. There are, however, several differences which must be addressed first.

In the generator, the state transitions are divided into two sets, those which are controllable and those which are uncontrollable. The controllable state transitions, or symbols, can be individually enabled by a command from the supervisor, while the uncontrollable state transitions are always enabled. Also, the effect of the supervisor commands is not changed by the state of the generator. This is in contrast to our DES plant model where commands from the DES controller can enable one or more state transitions depending on the current state. The general inability to enable symbols individually and the dependence of DES controller commands upon the state of the DES plant model, are what differentiate the DES models used our work on hybrid control systems from the RWM.

The reason for the differences between the RWM and the model used for hybrid control systems is chiefly due to the fact that the RWM is suited to modeling actual discrete event systems, while the DES plant model is an abstraction of a continuous-time system. This means that a particular state of the DES plant corresponds to more than one state in the continuous-time plant.

Controllability in a DES can be characterized by the set of symbol sequences which can be made to occur in the DES plant, [Ramadge 1989]. This set is referred to as the language of the particular DES. When under control, the DES will exhibit behavior which lies in a subset of its language. A theorem has been developed to determine whether a given RWM DES can be controlled to a desired language and if not, what is the greatest portion of the desired language which can be achieved via control. With appropriate modifications this theorem can be applied to the DES plant to determine whether a given control goal is possible.

If a desired behavior (i.e. language) is not attainable for a given controlled DES, it may be possible to find a more restricted behavior which is. If so, the least restricted behavior is desirable. [Wonham 1987] provides a method for finding this behavior which is referred to as the *supremal sublanguage* of the desired language.

When a controllable language has been found for a DES plant, designing a controller is straight-forward. The controller will be another DES which produces the desired controllable language. The output from the controller enables only the symbols which are in the controller. The exact form of the above results together with their proofs are not presented here due to space limitations; they are available from the authors.

5 Concluding Remarks

This chapter has introduced a model for hybrid systems which has focused on the role of the interface between the continuous-state plant and discrete-event supervisor. An especially simple form of the interface was introduced in which symbolic events and nonsymbolic state/control vectors are related to each other via memoryless transformations. It was seen that this particular choice dichotomizes the symbolic and nonsymbolic parts of the hybrid system into two cleanly separated dynamical systems which clearly expose the relationship between plant and supervisor. With the use of the proposed interface, quasi-determinism can be used to extend controllability concepts to hybrid systems. The clear separation of symbolic and nonsymbolic domains allows the formulation of hybrid controller methodologies which are directly based on equivalent DES control methods. Finally, the acknowledgement of the different roles played by symbolic and nonsymbolic processing in hybrid systems allows the proper formulation of the hybrid system's identification problem found in the companion chapter in this volume titled "Event Identification and Intelligent Hybrid Control" [Lemmon 1993c].

The work outlined in the preceding sections is indicative of the breadth of work currently being pursued in the area of hybrid systems as a means of modeling and designing supervisory and intelligent control systems. In spite of the great strides being made in this area, there are significant issues which remain to be addressed in future work. These issues include a more rigorous examination of the traditional control concepts of controllability, observability, and stability with regard to hybrid systems. To some extent, the notions of quasi-determinism and the event identification problems are preliminary efforts to codify these extensions. Future work, however, remains before these extensions are fully understood.

References

[Acar 1990] L. Acar, U. Ozguner, "Design of Knowledge-Rich Hierarchical Controllers for Large Functional Systems", *IEEE Trans. on Systems, Man, and Cybernetics*, Vol. 20, No. 4, pp. 791-803, July/Aug. 1990.

[Albus 1981] J. Albus, et al, "Theory and Practice of Intelligent Control", *Proc. 23rd IEEE COMPCON*, pp 19-39, 1981.

[Antsaklis 1989] P. J. Antsaklis, K. M. Passino S. J. and Wang, "Towards Intelligent Autonomous Control Systems: Architecture and Fundamental Issues", *Journal of Intelligent and Robotic Systems*, Vol.1, pp. 315-342, 1989.

[Antsaklis 1990, 1992] P. J. Antsaklis, Special Issues on 'Neural Networks in Control Systems' of the IEEE Control Systems Magazine, April 1990 and April 1992.

[Antsaklis 1991] P. J. Antsaklis, K. M. Passino and S. J. Wang, "An Introduction to Autonomous Control Systems", *IEEE Control Systems Magazine*, Vol. 11, No. 4, pp.5-13, June 1991.

[Antsaklis 1993a] P. J. Antsaklis and K. M. Passino, Eds., *An Introduction to Intelligent and Autonomous Control*, 448 p., Kluwer Academic Publishers, 1993.

[Antsaklis 1993b] P. J. Antsaklis and K. M. Passino, "Introduction to Intelligent Control Systems with High Degree of Autonomy", *An Introduction to Intelligent and Autonomous Control*, P. J. Antsaklis and K. M. Passino, Eds., Chapter 1, pp. 1-26, Kluwer Academic Publishers,1993.

[Antsaklis 1993c] P. J. Antsaklis, "Neural Networks for the Intelligent Control of High Autonomy Systems", *Mathematical Studies of Neural Networks*, J.G. Taylor, Ed., Elsevier, 1993. To appear.

[Astrom 1986] K. J. Astrom, et al, "Expert Control", *Automatica*, Vol. 22, No. 3, pp. 277-286, 1986.

[Benveniste 1990] A. Benveniste, P. Le Guernic, "Hybrid Dynamical Systems and the SIGNAL Language", *IEEE Transactions on Automatic Control*, Vol. 35, No. 5, pp. 535-546, May 1990.

[Cassandras 1990] C. Cassandras, P. Ramadge, "Toward a Control Theory for Discrete Event Systems", *IEEE Control Systems Magazine*, pp. 66-68, June 1990.

[Fishwick 1991] P. Fishwick, B. Zeigler, "Creating Qualitative and Combined Models with Discrete Events", *Proceedings of The 2nd Annual Conference on AI, Simulation and Planning in High Autonomy Systems*, pp. 306-315, Cocoa Beach, FL, April 1991.

[Gollu 1989] A. Gollu, P. Varaiya, "Hybrid Dynamical Systems", *Proceedings of the 28th Conference on Decision and Control*, pp. 2708-2712, Tampa, FL, December 1989.

[Grossman 1992] R. Grossman, R. Larson, "Viewing Hybrid Systems as Products of Control Systems and Automata", *Proceedings of the 31st Conference on Decision and Control*, pp. 2953-2955, Tucson AZ, December 1992.

[Holloway 1992] L. Holloway, B. Krogh, "Properties of Behavioral Models for a Class of Hybrid Dynamical Systems", *Proceedings of the 31st Conference on Decision and Control*, pp. 3752-3757, Tucson AZ, December 1992.

[IEEE Computer 1989] Special Issue on Autonomous Intelligent Machines, IEEE Computer, Vol. 22, No. 6, June 1989.

[Isidori 1989] A. Isidori, *Nonlinear Control Systems*, 2nd Edition, Springer-Verlag, Berlin, 1989

[Kohn 1992] W. Kohn, A. Nerode, "Multiple Agent Autonomous Hybrid Control Systems", *Proceedings of the 31st Conference on Decision and Control*, pp. 2956-2966, Tucson AZ, December 1992.

[Lemmon 1992] M. D. Lemmon, "Ellipsoidal Methods for the Estimation of Sliding Mode Domains of Variable Structure Systems", *Proc. of 26th Annual Conference on Information Sciences and Systems*, Princeton N.J., March 18-20, 1992.

[Lemmon 1993a] M. D. Lemmon, "Inductive Inference of Invariant Subspaces", *Proceedings of the American Control Conference*, San Francisco, California, June 2-4, 1993.

[Lemmon 1993b] M. D. Lemmon, J. A. Stiver, P. J. Antsaklis, "Learning to Coordinate Control Policies of Hybrid Systems", *Proceedings of the American Control Conference*, San Francisco, California, June 2-4, 1993.

[Lemmon 1993c] M. D. Lemmon, J. A. Stiver, P. J. Antsaklis, "Event Identification and Intelligent Hybrid Control", This volume.

[Mendel 1968] J. Mendel and J. Zapalac, "The Application of Techniques of Artificial Intelligence to Control System Design", in *Advances in Control Systems*, C.T. Leondes, ed., Academic Press, NY, 1968.

[Mesarovic 1970] M. Mesarovic, D. Macko and Y. Takahara, *Theory of Hierarchical, Multilevel, Systems*, Academic Press, 1970.

[Narendra 1990] K. S. Narendra and K. Parthsarathy, "Identification and Control of Dynamical Systems Using Neural Networks", *IEEE Trans. Neural Networks*, Vol 1(1):4-27.

[Nerode 1992] A. Nerode, W. Kohn, "Models for Hybrid Systems: Automata, Topologies, Stability", Private Communication, November 1992.

[Ozveren 1991] C. M. Ozveren, A. S. Willsky and P. J. Antsaklis, "Stability and Stabilizabilty of Discrete Event Dynamic Systems", *Journal of the Association of Computing Machinery*, Vol 38, No 3, pp 730-752, 1991.

[Passino 1989a] K. Passino, "Analysis and Synthesis of Discrete Event Regulator Systems", Ph. D. Dissertation, Dept. of Electrical and Computer Engineering, Univ. of Notre Dame, Notre Dame, IN, April 1989.

[Passino 1989b] K. M. Passino and P. J. Antsaklis, "On the Optimal Control of Discrete Event Systems" , *Proc. of the 28th IEEE Conf. on Decision and Control*, pp. 2713-2718, Tampa, FL, Dec. 13-15, 1989.

[Passino 1991a] K. M. Passino, A. N. Michel, P. J. Antsaklis, "Lyapunov Stability of a Class of Discrete Event Systems", *Proceedings of the American Control Conference*, Boston MA, June 1991.

[Passino 1991b] K. M. Passino, U. Ozguner, 'Modeling and Analysis of Hybrid Systems: Examples", *Proc. of the 1991 IEEE Int. Symp. on Intelligent Control*, pp. 251-256, Arlington, VA, Aug. 1991.

[Passino 1992a] K. M. Passino, A. N. Michel and P. J. Antsaklis, "Ustojchivost' po Ljapunovu klassa sistem diskretnyx sobytij", *Avtomatika i Telemekhanika*, No.8, pp. 3-18, 1992. "Lyapunov Stability of a Class of Discrete Event Systems", *Journal of Automation and Remote Control*, No.8, pp. 3-18, 1992. In Russian.

[Passino 1992b] K. M. Passino and P. J. Antsaklis, "Event Rates and Aggregation in Hierarchical Discrete Event Systems", *Journal of Discrete Event Dynamic Systems*, Vol.1, No.3, pp. 271-288, January 1992.

[Passino 1993] K. M. Passino and P. J. Antsaklis, ""Modeling and Analysis of Artificially Intelligent Planning Systems", *Introduction to Intelligent and Autonomous Control*, P.J.Antsaklis and K.M.Passino, Eds., Chapter 8, pp. 191-214, Kluwer, 1993.

[Peleties 1988] P. Peleties, R. DeCarlo, "Modeling of Interacting Continuous Time and Discrete Event Systems : An Example", *Proceedings of the 26th Annual Allerton Conference on Communication, Control, and Computing*, pp. 1150-1159, Univ. of Illinois at Urbana-Champaign, October 1988.

[Peleties 1989] P. Peleties, R. DeCarlo, "A Modeling Strategy with Event Structures for Hybrid Systems", *Proceedings of the 28th Conference on Decision and Control*, pp.1308-1313, Tampa FL, December 1989.

[Peterson 1981] J. L. Peterson, *Petri Net Theory and the Modeling of Systems*, Prentice-Hall, Englewood Cliffs, NJ, 1981.

[Ramadge 1987] P. Ramadge, W. M. Wonham, "Supervisory Control of a Class of Discrete Event Processes", *SIAM Journal of Control and Optimization*, vol. 25, no. 1, pp. 206-230, Jan 1987.

[Ramadge 1989] P. Ramadge, W. M. Wonham, "The Control of Discrete Event Systems", *Proceedings of the IEEE*, Vol. 77, No. 1, pp. 81 - 98, January 1989.

[Saridis 1979] G. N. Saridis, "Toward the Realization of Intelligent Controls", *Proc. of the IEEE*, Vol. 67, No. 8, pp. 1115-1133, August 1979.

[Saridis 1983] G. N. Saridis, "Intelligent Robot Control", *IEEE Trans. on Automatic Control*, Vol. AC-28, No. 5, pp. 547-556, May 1983.

[Saridis 1985] G. N. Saridis, "Foundations of the Theory of Intelligent Controls", *Proc. IEEE Workshop on Intelligent Control*, pp 23-28, 1985.

[Saridis 1987] G. N. Saridis, "Knowledge Implementation: Structures of Intelligent Control Systems", *Proc. IEEE International Symposium on Intelligent Control*, pp. 9-17, 1987.

[Saridis 1989a] G. N. Saridis, "Analytic Formulation of the Principle of Increasing Precision with Decreasing Intelligence for Intelligent Machines", *Automatica*, Vol.25, No.3, pp. 461-467, 1989.

[Stengel 1984] R. F. Stengel, "AI Theory and Reconfigurable Flight Control Systems", Princeton University Report 1664-MAE, June 1984.

[Stiver 1991a] J. A. Stiver, "Modeling of Hybrid Control Systems Using Discrete Event System Models", M.S. Thesis, Dept. of Electrical Engineering, Univ. of Notre Dame, Notre Dame, IN, May 1991.

[Stiver 1991b] J. A. Stiver, P. J. Antsaklis, "A Novel Discrete Event System Approach to Modeling and Analysis of Hybrid Control Systems", *Control Systems Technical Report #71*, Dept. of Electrical Engineering, University of Notre Dame, Notre Dame, IN, June 1991.

[Stiver 1991c] J. A. Stiver, P. J. Antsaklis, "A Novel Discrete Event System Approach to Modeling and Analysis of Hybrid Control Systems", *Proceedings of the Twenty-Ninth Annual Allerton Conference on Communication, Control, and Computing*, University of Illinois at Urbana-Champaign, Oct. 2-4, 1991.

[Stiver 1992] J. A. Stiver, P. J. Antsaklis, "Modeling and Analysis of Hybrid Control Systems", *Proceedins of the 31st Conference on Decision and Control*, pp. 3748-3751, Tucson AZ, December 1992.

[Stiver 1993] J. A. Stiver, P. J. Antsaklis, "State Space Partitioning for Hybrid Control Systems", *Proceedings of the American Control Conference*, San Francisco, California, June 2-4, 1993.

[Turner 1984] P. R. Turner, et al, "Autonomous Systems: Architecture and Implementation", Jet Propulsion Laboratories, Report No. JPL D- 1656, August 1984.

[Valavanis 1986] K. P. Valavanis, "A Mathematical Formulation For The Analytical Design of Intelligent Machines", PhD Dissertation, Electrical and Computer Engineering Dept., Rensselaer Polytechnic Institute, Troy NY, Nov. 1986.

[Wonham 1987] W. M. Wonham, P. J. Wonham, "On the Supremal Controllable Sublanguage of a Given Language", *SIAM Journal of Control and Optimization*, vol. 25, no. 3, pp. 637-659, May 1987.

[Zeigler 1984] B. P. Zeigler, "Multifacetted Modelling and Discrete Event Simulation", Acedemic Press, NY, 1984.

[Zeigler 1987] B. P. Zeigler, 'Knowledge Representation from Newton to Minsky and Beyond", *Journal of Applied Artificial Intelligence*, 1:87-107, 1987.

[Zeigler 1989] B. P. Zeigler, "DEVS Representation of Dynamical Systems: Event Based Intelligent Control", *Proc. of the IEEE*, Vol. 77, No. 1, pp. 72-80, 1989.

Fault Accommodation in Feedback Control Systems

Mogens Blanke, Søren Bøgh Nielsen and Rikke Bille Jørgensen

Institute of Electronic Systems
Department of Control Engineering
Aalborg University, Aalborg, Denmark
E-mail : blanke@control.auc.dk

Abstract. Feedback control systems are vulnerable to faults in control loop sensors and actuators, because feedback action may cause abrupt and damaging responses when faults occur. Despite the importance of the subject, research results from analysis, design, and implementation of fault handling methods have not been widely accepted in the industry. One reason has been scarcity of realistic examples for testing of new methods against industrial systems. Another has been lack of interaction between control systems theory and computer science in this area. This paper is a contribution to help bridging these gaps.

Mathematical methods to detect and isolate faults in industrial feedback control systems are presented and analyzed, and it is shown where these methods work well and where other techniques are needed. An industrial position control loop with electro-mechanical components is used as an application example because it has several characteristics which are generic for hybrid feedback control systems. As a salient feature, the paper provides both a simple mathematical model of this system for use in the design of fault handling methods and a complete, nonlinear description for simulation and verification work. These models make it possible to make bench-mark testing of new methods including assessment of robustness against modelling uncertainty and process disturbances. These are crucial for industrial applicability but somewhat overlooked in hybrid system theory.

Key contributions are the presentation and analysis of an industrial feedback system which is suitable for bench-mark testing of new ideas in hybrid system theory, and the demonstration of state of the art in mathematical methods for fault detection and accommodation.[1]

1 Introduction

Millions of real time feedback control loops are in daily use in robotics and factory automation, but very few are protected against failure caused by faults in control loop components. Consequences, of even simple faults, may be dramatic and there are considerable incentives to enhance computerized feedback loops

---

[1] This work was funded by the Danish Research Council under contracts 16-4779 and 26-1830. This support is gratefully acknowledged.

with methods for fault detection and accommodation. A large step can be taken with mathematical methods from control engineering, but in order to arrive at efficient solutions for use in large volume, progress in the area of hybrid systems theory is mandatory.

Most automated industrial systems are hybrid by nature. They consist of continuous time processes controlled by real-time computers. Sensors on the plant provide measures of process variables, and actuators convert control signals to physical inputs to the process as visualized in figure 1.

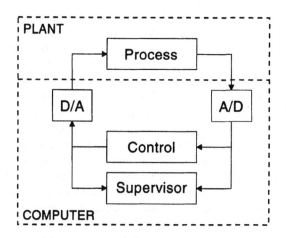

Fig. 1. Supervision of a process under control.

Feedback is established because actuator demands are calculated from the difference between a reference value and sensor measurements. Any deviation between these signals will cause an immediate reaction on the actuators when actuator demands are updated. The discrete time control algorithm makes use of both current and previous events in the plant. This makes it possible to employ, for example, prediction methods to give the control loop desired characteristics. Response time to changes in a setpoint, disturbance rejection properties, noise sensitivity, and stability properties are key attributes that are always quantified in the requirements to a particular closed loop design.

Feedback control systems are particularly sensitive to faults. A feedback sensor fault, for example, may cause a large deviation between measurement and reference. This will in most cases cause large actuator demands and eventually lead to rapid change of process state. Unacceptable excursions in process state followed by production stop, plant failure, or direct damage are experiences from actual events in industry. For these reasons, there is a need to extend feedback control systems with methods that can detect and accommodate faults. The possibilities to accomplish this depend on our ability to distinguish the conditions met in fault situations from plant behaviour in normal operation.

In normal operation, feedback control should keep a process state equal to a desired setpoint while the influence from process disturbances and measurement noise are kept minimal. This can be achieved by employing methods that estimate process states and perform optimal dynamic filtering in combination with techniques that adapt parameters in the control method to current process conditions.

In abnormal operation, when faults have occurred, the control loop should react immediately in a way that prevents a fault from developing into a malfunction of the system being controlled. This requires added functionality to well established methods in control theory.

The question of an appropriate reaction on the presence of a fault is to adapt the process, in an optimum way, to the new situation. This is referred to as fault accommodation. The goal is to obtain high *Reliability, Availability, Maintainability, and Safety.* These terms are central in the overall performance of a process plant so a definition [Lauber, 1991] in this context is suitable.

Reliability is the "probability of failure-free operation". A measure of the reliability is the mean up-time. *Availability* is the fraction of time where the system is operating compared to the intended operating time. High availability can be accomplished by redundant systems or, where this is not possible, by proper handling of faults within the system. *Maintainability* is defined as the ability to perform preventive maintenance (on-line or off-line) and repair. A good maintainability can as an example be accomplished by rapid fault identification. The *safety* aspect deals with the danger of injury to people and environmental contamination. Safety is closely related to the theory of risk analysis.

The weighing between the four goals is highly dependent of the actual application. In very high risk applications, control systems are made fully redundant to obtain a high degree of safety and availability. In normal industrial installations, however, high availability must be obtained by a fail-to-safe state design of control systems and by inclusion of methods to accommodate faults as part of the process control system.

The paper addresses fault handling from a control theory point of view. Available methods for fault detection, isolation, and accommodation are treated, and an assessment is made of the performance of various methods using an industrial position control loop as a bench-mark example. State of the art in fault handling design and implementation is discussed, and suggestions are given in the areas of supervisor design and implementation, where further theoretical progress is needed.

2 Fault Handling in Feedback Control Loops

Faults in feedback loops are in general difficult to handle. If a fault develops gradually, closed loop control will attempt to compensate for it and in this way hide the development of a malfunction. The fault may not be discovered until the control loop stops normal operation. If faults arise suddenly, the effect is amplified by the closed loop control. Production stops, process damage, or other undesired consequences may be the result. A decent system design should therefore prevent faults from developing into malfunctions.

A fail-to-safe design philosophy is needed for industrial control systems, but a methodology that supports a complete design and implementation cycle is not yet available. When fail-to-safe designs are nevertheless carried out, this is at considerable expense in the form of work by highly qualified, experienced staff. Considerable benefits could be gained if a consistent methodology existed that could support the entire design, implementation, and verification cycle of industrial system development in this area.

2.1 General Methodology

A general method for design of fault handling associated with closed loop control includes the following steps:

1. Make a Failure Modes and Effects Analysis (FMEA) related to control system components.
2. Define desired reactions to faults for each case identified by the FMEA analysis.
3. Select appropriate method for generation of residuals. This implies consideration of system architecture, available signals, and elementary models for components. Disturbance and noise characteristics should be incorporated in the design if available.
4. Select method for input-output and plant fault detection and isolation. This implies a decision on whether an event is a fault and, if this is the case, determination of which element is faulty.
5. Consider control method performance and design appropriate detectors for supervision of control effectiveness. This implies surveillance, e.g., of closed loop stability which may be affected by changes in plant parameters if controller robustness is insufficient. Design appropriate reactions.
6. Design a method for accommodation of faults according to points 2 and 5 above.
7. Implement the completed design. Separate control code from fault handling code by implementation as a supervisor structure.

2.2 Fault Analysis for Closed Loop Control Systems

Faults in a control loop can be categorized in generic types:

1. Reference value (setpoint) fault
2. Actuator element fault
3. Feedback element fault
4. Execution fault including timing fault
5. Application software, system or hardware fault in computer based controller
6. Fault in the physical plant.

Faults related to a system level include:

1. Energy supply fault or other fatal system error.

Faults in the physical plant are not a key issue in this context. The aim is to develop a methodology for fault accommodation within the control system. Many fault handling situations will require that the control system is reconfigured or, as a last resort, fails to state which is safe for the physical process. Reconfiguration of the physical process is not possible in general, and if it is, decision on such plant changes belong to a level above the individual control system.

The faults mentioned in the list are generic. The actual fault mechanisms will depend on the particular implementation. Measurements, for example, can be communicated as analog signals or they can be contained in local area network messages. The types of faults and their effects therefore need individual consideration together with the detailed architecture of the closed loop system.

Failure Modes and Effects Analysis. The initial task in fault handling is to perform a fault analysis. This includes determination of potential failure modes and an analysis of possible effects. In connection with the further work it is necessary to get a knowledge of how faults relate to desired reactions.

The basic FMEA analysis has a long history and several methods have been proposed for this purpose. The method adopted here is referred to as Matrix FMEA [Herrin, 1981]. This method offers a graphical representation of the problem and it enables backtracking from fault symptoms to a set of possible fault causes. In this approach the system is considered at a number of levels.

1. *Units,* (Sensors, Actuators)
2. *Groups,* which is a set of units
3. *Subsystems,* which is a set of groups
4. *System,* which is a set of subsystems

The basic idea in Matrix FMEA is first to determine potential failure modes at the unit and their effects in this first level of analysis. The failure effects are propagated to the second level as failure modes and the effects at that level are determined. This propagation of failure effects continues until the forth level of analysis, the system, is reached. Figure 2 illustrates the principle of the Matrix FMEA.

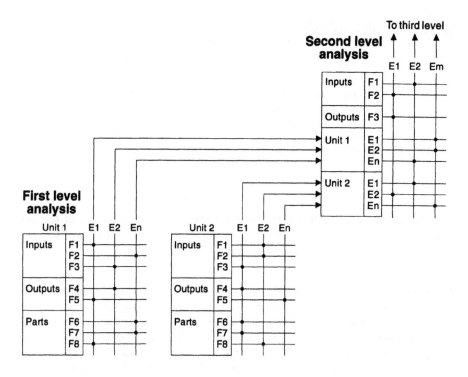

Fig. 2. The principle in FMEA analysis

When the built up process of the Matrix FMEA is completed, a reverse technique is used. This technique takes one failure effect at a higher level of analysis and tracks back through all lower levels to the contributing parts and their failure modes. The results are a listing of all contributing lower level failure effects and the units that contribute to the single failure effect at a higher level.

Each entry is finally associated with a desired remedy action which is decided on as part of the overall control system design.

Fault Consequences. The physical consequence of a fault event can often have major technical, economical, or social impact. A main classification of the consequences is done with respect to *criticality* as listed in table 1.

Hazard Classes	Consequence Description
0	No consequence
1	Decreased performance (productivity degradation, quality degradation)
2	Damage on machinery and surroundings, contamination of environment
3	Injury to people

Table 1. Classification of consequences from occurrences of faults.

This classification into *hazard classes* indicates the importance of the consequence and can be used to determine the requirements to tolerance to faults before a set of remedy actions are specified.

Detection and Isolation of Faults. Closed loop control systems can be considered to consist of four major subsystems: actuators, physical process, sensors, and control computer. A fault in physical subsystems and units may be detected as a difference between actual and expected behaviour. This can be achieved using a mathematical model. It generates a signal that shows the difference between actual and expected performance. When a method is tuned to obtain high sensitivity to certain faults and, at the same time, to be insensitive to other phenomena, the method is useful for both fault detection and isolation. Such methods are often referred to as Fault Detection and Isolation (FDI) methods.

Several FDI principles exist. Model based methods with state estimation as the main principle were described by [Gertler,1991], [Patton et al,1991], [Frank,1991]. Parameter estimation and identification techniques by [Isermann,1991], and an adaptive filtering approach by [Clark et al,1989]. Knowledge based fault detection was used by [Poucet et al,1988], and [Stengel,1991].

To apply the model based methods it is necessary that input/output signals of the monitored process are available and that dynamic characteristics are known with a reasonable degree of precision.

In the design of an FDI system, based on analytical methods, robustness properties should be considered. Robustness is defined as [Clark et. all,1989]: *The degree to which the FDI system performance is unaffected by conditions in the operating process, which turns out to be different from what they were assumed to be in the design of the FDI system.* Specific consideration need to be given to :

− parameter uncertainty
− unmodelled nonlinearities and dynamics
− disturbances and noise

When no account has been taken of the above problems the performance of the FDI system may be less than expected. The consequence is an increase in the likelihood of a missed detection, i.e., not detecting a fault when one is present. Similarly the possibility of triggering an alarm although no fault is present is increased.

In this context, reactions to faults need to be immediate, and real-time execution of an FDI method is an indispensable requirement. Artificial Intelligence (AI) based techniques have therefore not been considered.

2.3 Fault Accommodation

Autonomous reconfiguration or other action within the control system in reaction to a fault is referred to as fault accommodation. Fault accommodation is thus an implementation of the list of actions specified as part of the general method.

Fault accommodation actions. Specific actions are required to accommodate faults detected in a controlled system. Required operation can be one or more of the items listed below:

- **Change performance:**
  - Decrease performance, eg. less through-put of the controlled system.
  - Change settings in the surrounding process to decrease the requirements of the controlled system .
  - Change controller parameters.
- **Reconfigure:**
  - Use component redundancy if possible.
  - Change controller structure.
  - Replace defective sensor with signal estimator/observer. (analytical redundancy). Note, this operation may be limited in time because external disturbances may increase the estimation error.
  - If the fault is a set point error then freeze at last fault-free set point and continue control operation. Issue an alert message to operators.
- **Stop operation:**
  - Freeze controller output to a predetermined value. Zero, maximum or last fault-free value are three commonly required values - the one to be used is entirely application dependent. Finally disable the controller.
  - Fail-to-safe operation.
  - Emergency stop of physical process.

Supervisor Functionality. Implementation of the various FDI methods and methods for fault accommodation is most conveniently separated from implementation of the control method itself. The reasons are primarily software reliability due to reduced complexity of software, and enhanced testability obtained by a more modular and structured design. The tasks accomplished at the supervision level are:

- Monitoring of all input signals. The purpose is range and trend checking for signal validity verification.
- Processing of input signals and controller outputs. A set of residual generators are used for input/output faults. Other detectors are used for system errors within the control computer. The purpose is to extract features from I/O signals to enable detection of whether a fault is present or not.
- Processing of residual generator outputs in fault detectors/isolators. The purpose is to identify which fault has occurred. Information from the various residual generators may be combined at this stage.
- Determination of desired reaction based upon the particular fault.
- Reconfiguration or other desired action to accommodate the fault. Alert by message to operator level about event.

A major difficulty in the design and implementation of the supervisor function is that, due to the closed loop nature of the hybrid system, with real time execution of control code, fault detection, isolation, and accommodation must normally take place within a single sampling cycle.

2.4 State of the Art

Industrial utilization of a methodology requires a firm theoretical foundation to assure correctness of the approach, clear and rapidly applicable design procedures, and implementation tools.

When considering the general procedure for fault handling design and implementation in section 2.1, there does not at present exist a coherent solution to the entire range of difficulties.

- Fault analysis and specification of remedy actions for a system is intricate because considerable experience and knowledge about the physical system architecture is required. It seems, however, to be possible to extend existing FMEA methods to incorporate the fault handling requirements.
- Mathematical methods from control theory are proficient to detect and isolate faults that have been previously defined in a mathematical description of the physical process. Control theory has achieved very strong results when problems are well specified. Modelling of basic mathematical relations from a description of physical components at a generic level and the architecture of the process with control loop units is possible. Research prototypes exist but general tools are not available yet.
- Supervision of control method and the correct execution of a control algorithm are areas of considerable research effort. Research results are available, but have been limited to dedicated implementations.
- Design and implementation of a supervisor function for fault detection, identification, and accommodation is much more a grey area. There are not theoretical results about correctness and completeness of a fault handling supervisor available. Several questions are crucial and need answers: Is a particular set of detectors sufficient? What is the correct accommodation

action when detectors contradict each other? Are remedy actions correctly implemented?

Industrial practice is to design a control loop that assures stability and adequate robustness properties. Fault handling is either not considered, or is made after the primary control loop design. Real time systems are generally difficult to design and verify. The difficulties are much worse when the complexity from handling of faults, that occur at random, are added. The results are very high costs for design and verification.

For these reasons, new results in design and implementation of fault handling methods are believed to be important. A coherent method that supports the entire design and implementation process would be a significant research achievement and would have a major impact on industrial practice.

3 Fault Detection and Isolation

Fault detection and isolation is achieved by applying special filters, and statistical methods to the process' input and output data. The filters generate signals, referred to as residuals, which describe the difference between the state expected and the one derived from measurements. Ideally, residuals are zero in normal operation and nonzero in the case of abnormal operations. The use of model based methods for FDI requires knowledge of process dynamics both when faults are present and when they are not. In model based methods the key issue is to design residual generators such that a certain fault gives one large component of the residual. This is referred to as fault isolation. Detection is often done by comparing the residual magnitude with a preset level. When noise or stochastic changes are important, statistical methods need to be employed. The simple level detection is therefore replaced by adaptive threshold detection or statistical based detectors. [Basseville & Nikiforov,1993].

3.1 Residual Generation

A process is described by it's inputs and outputs. The most simple process model include process dynamics but exclude noise, unknown inputs and faults. The representation can be on either state space or input/output form.
Faults in the control system can be divided into:

- Input sensor faults Δu_m
- Output sensor faults Δy
- Actuator faults Δu_c

Furthermore there exists unknown inputs and plant faults Δu_d. In figure 3 these faults and unknown inputs are seen in relation to the ideal values.Only additive faults are treated.

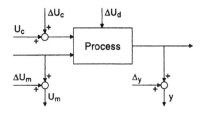

Fig. 3. Faults and unknown inputs to process.

On the state space form the process including faults and unknown inputs are modelled as:

$$x(k+1) = \Phi x(k) + \Gamma u(k) + E_1 d(k) + F_1 p(k) \tag{1}$$
$$y(k) = C x(k) + D u(k) + E_2 d(k) + F_2 q(k)$$

where

$$u(k) = [u_1(k) \ldots u_k(k)]^T \tag{2}$$
$$y(k) = [y_1(k) \ldots y_m(k)]^T$$

are input-output measurements
x are the system states
$q(k)$ are output sensor faults
$p(k)$ are actuator faults
$d(k)$ are unknown inputs and plant faults
E_1, E_2, F_1 and F_2 are system matrices for unknown inputs and faults.

On input output form:

$$y(k) = S(z)u(k) + \Delta y(k) - S(z)\Delta u(k) \tag{3}$$

where
$\Delta u = [\Delta u_m \ \Delta u_c \ \Delta u_d]^T$
Δy is output sensor fault vector.
The above equations form the basis for residual generator design.

Structured Parity Equations. Output from the parity equations are signals showing inconsistency between normal and faulty operation. In normal process operation the parity equations output approximately zero. In case of faults output will be nonzero. Fault isolation is achieved with *structured* parity equations. One element in the residual vector is unaffected by a specific fault while all the others will be affected. In that way determination of fault cause is possible.

The parity equations are designed as follows:

$$e(k) = \Delta y(k) - S(z)\Delta u(k)$$
$$= y(k) - S(z)u(k) \tag{4}$$
$$r(k) = W(z)e(k)$$

The residual vector $r(k)$ is found by multiplying a weighting filter $W(z)$ to the error $e(k)$. The filter is designed to make the j'th residual unaffected to the i'th fault. [Gertler,1991]

Diagnostic Observer. An observer is a dynamic algorithm that estimates the states of the system based on measurement of inputs and outputs. An observer has the form:

$$\hat{x}(k+1) = \Phi\hat{x}(k) + \Gamma u(k) + K[y(k) - C\hat{x}(k)]$$
$$e(k) = x(k) - \hat{x}(k) \tag{5}$$
$$\epsilon(k) = y(k) - C\hat{x}(k)$$

where
e is the state estimation error
ϵ is the output estimation error
K is the observer gain feedback

Taking the process representation of eq.1 the state estimation error and residual becomes:

$$e(k+1) = (\Phi - KC)e(k) + (E_1 - KE_2)d(k) + F_1 p(k)$$
$$-KF_2 q(k) - KDu(k) \tag{6}$$
$$r(k) = He(k) = H(zI - G)^{-1}[(E_1 - KE_2)d(k) + F_1 p(k)$$
$$-KF_2 q(k) - KDu(k)]$$

where

$$G = \Phi - KC \tag{7}$$

The K and H matrices are designed so the observer is stable and the residual has the desired properties. [Gertler,1991]

Unknown Input Observer. An unknown input observer for fault detection is used to decouple contributions from unknown inputs. It is a demand that the E_1 and E_2 matrices in eq.1 are known. Hereby the residuals will only be affected by potential faults. The general structure for the unknown input observer is:

$$z(k+1) = Fz(k) + Ju(k) + Ky(k)$$
$$\hat{x}(k) = z(k) + Hx(k) \tag{8}$$
$$e(k) = \hat{x}(k) - x(k)$$
$$r(k) = L_1 z(k) + L_2 y(k)$$

Using eq.1 the estimation error and the residual becomes:

$$e(k+1) = Fe(k) + (FT - T\Phi + KC)x(k) + (J - T\Gamma + KD)u(k)$$
$$+ KF_2 q(k) + (GE_2 - TE_1)d(k) - TF_1 p(k) \tag{9}$$
$$r(k) = L_1 e(k) + (L_1 T + L_2 C)x(k) + L_2 Du(k)$$
$$+ L_2 E_2 d(k) + L_2 F_2 q(k)$$

If the following can be made true

$$
\begin{array}{cc}
T\Phi - FT = KC\,, & T\Gamma - J = KD \\
KE_2 = TE_1\,, & L_2 E_2 = 0 \\
L_2 D = 0 & ,\ L_1 T + L_2 C = 0
\end{array} \tag{10}
$$

the residuals will only be affected by the faults $q(k)$ and $p(k)$. [Zhang],[Frank,1991]

One disadvantage using the unknown input observer for fault detection is that faults which effect the process in a similar way as the unknown input can not be detected. Such faults will be decoupled as well as the unknown input.

3.2 Detection Methods

The output from a residual generator is a time series of the vector $r(t)$, By proper design of the generator, $r(t)$ is fairly insensitive to load disturbances and model uncertainty. However, in actual applications, one will meet the difficulty that the model is not well known, disturbances do not act exactly as expected, and measurement noise is furthermore present. The residual has then no longer a unique relation with the fault vector but has stochastic elements. Detection of whether an event in $r(t)$ is caused by a fault or is an effect of random coincidence in the factors of uncertainty is therefore a key issue.

The important factor is to combine a high probability of detecting a fault with a low probability of false detections, and a low value of the time to detect. Statistical change detection theory and adaptive methods are used to make detectors work.

A basic approach is to consider the probability distribution $P_{\Theta_0}(r_i)$ of the $r(t)$ process. The detection problem is to determine whether a change has happened in the parameter vector Θ that describes the random vector $r(t)$

A change in parameters from Θ_0 to Θ_1 can be determined through investigating the properties of the log-likelihood ratio which, for a scalar parameter case, can be expressed as

$$s_i = \ln \frac{P_{\Theta_1}(r_i)}{P_{\Theta_0}(r_i)} \tag{11}$$

This simple formulation is valid for the simplest change detection problem, the change of a scalar parameter in the probability distribution of a random sequence, where both Θ_0 to Θ_1 are known.

For unknown Θ_1, we need to maximize a likelihood ratio with respect to Θ_1.

$$\Lambda = \sup_{\Theta_1} \frac{P_{\Theta_1}(r_1 \cdots r_m)}{P_{\Theta_0}(r_1 \cdots r_m)} \tag{12}$$

This is in general a nonlinear optimization problem, and the method to be applied on a particular system depends on details of the system [Basseville & Nikiforov,93].

In this context, we disregard further discussion of the detection problem and focus on the residual generation issue within FDI.

4 Bench-mark Example: Electro-mechanical Control Loop

The example chosen for bench-mark testing of various theoretical methods is an electro-mechanical control loop. In industry, such devices are used for rapid and accurate positioning of robot arms, machinery tools in manufacturing, and in a variety of energy systems. They are also used on large diesel engines for fuel injection control. Fuel injection is governed by the position of a fuel rack, which is linked to all fuel pumps on the engine. Large forces are required in this application and rapid response is needed when sudden load variations occur. A harsh environment furthermore makes the occurrence of component faults to more than a theoretical possibility. Consequences of faults may be serious because very large capital investments are involved and, when used for governing a prime mover on-board a ship, primary safety is also implicated. Our example is adopted from this application because fault handling is both crucial to the operation, and a fairly challenging task. Desired reaction onto a control loop fault in the governor is to maintain operability if possible, or eventually in a degraded mode. If continued operation is impossible, an immediate halt of the fuel rack position is required to keep steady power on the diesel engine. Close down of the diesel engine is not desirable because control of the ship is lost. [Blanke et al, 1990]. An outline of the control loop hardware is drawn in figure 4.

Main components of the position control loop are:

- A brushless synchronous DC-servo motor with built in tachogenerator, rotor-position sensor for electronic commutation and a high torque electromagnetic brake. The maximum continuous rating is 1.8 kW.
- A precision epicyclic gear with a very low backlash. Maximum output torque is 1200 Nm.
- The output shaft of the gear has a built on arm that connects to an external load via a rod. In a laboratory setup it is possible to simulate different load sequences with a maximum force of 8700 N.
- Sensors for motor current, velocity and for gear output shaft position.

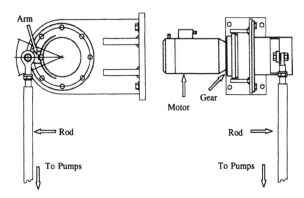

Fig. 4. Outline of Servomotor and Gear with connection rod to diesel engine fuel pumps.

The servo motor is driven by a power drive which include an analog current and velocity controller. Both have proportional and integral terms. The current generator in the power drive has adjustable peak and mean values. The power drive is able to supply a peak current which is much higher than the continuous rating of the servo motor. While this is advantageous for the acceleration performance, the peak current need to be temporarily decreased to the motors mean rating, as it will otherwise be overheated and permanently damaged. Power delivered to the motor is monitored by the power drive, and the peak current is automatically reduced when needed. The powerdrive has a number of other inherent functions to prevent damage. They include

- Temperature monitoring and automatic close down in case of excessive temperature of power switching circuits.
- Temperature monitoring of the servo motor stator winding through a build in sensor.
- End-stop position monitoring by means of switches at the gear output shaft. When an end-stop switch is sensed, the powerdrive stops delivering current in a direction that would advance the position beyond the switch.

The various safety functions are necessary to obtain overall system dependability. However, these parts are not directly active in normal operation of the control loop, and faults entering can be difficult to distinguish from the safety functions.

4.1 Servo Motor Model Overview

A block diagram of the entire system including control loops is depicted in figure 5.

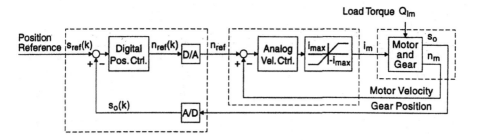

Fig. 5. Outline of position control loop for use on diesel engine

The system is divided into three subsystems to get a better overview:

1. Servo motor and gear
2. Velocity control loop
3. Position control loop

The differential equations characterizing the system, and the associated parameters, are presented below.

4.2   Basic Dynamic Equations for Servo motor with Velocity and Position Control

The servo motor can be considered as a torque device which produces a torque proportional to the average motor current. The current is controlled by the power drive and also limited to a maximum value. The current control loop within the power drive can be neglected because it's bandwidth is several times faster than the remaining system. The velocity and position of the motor are controlled with an analog and a digital controller respectively. Figure 6 shows a block diagram for the servo motor including generic faults. The velocity reference, n_{ref}, and load torque, Q_{lm}, are inputs and the motor shaft speed, n_m, and gear output position, s_o, are outputs. The following basic dynamic equations do not include the faults shown in figure 6.

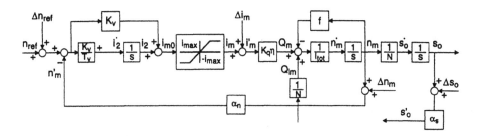

Fig. 6. Block diagram for servomotor with velocity control loop of the power drive. The speed reference is generated from a digital position controller. The figure shows the various generic faults considered

List of parameters in simple mathematical model of position loop

Variable	Range	Unit	Description
f	$19.73 \cdot 10^{-3}$	Nm/rad/s	Total friction referred to servo motor
I_{tot}	$2.53 \cdot 10^{-3}$	kgm^2	Total inertia referred to servo motor
i_{max}	30	A	Max motor current
K_q	0.54	Nm/A	Torque constant of servo motor
K_v	0.9	A/rad/s	Gain of the speed controller
N	91		Gear ratio
η	0.85		Gear efficiency
T_v	$8.8 \cdot 10^{-3}$	s	Integral time of the speed controller
α_n	1.000		Scale factor for velocity measurement
α_s	1.022		Scale factor for position measurement

The gear ratio is corrected with a factor to compensate for measurement scale errors on n_m and s_o.

List of variables in simple model of position loop

Variable	Range	Unit	Description
i_m	-30 to 30	A	Motor current. Output from power drive
n_m	-314 to 314	rad/s	Shaft speed of servo motor
n_{ref}	-314 to 314	rad/s	Shaft speed reference
Q_{lm}	-6 to 6	Nm	Load torque referred to servo motor
Q_m	-16 to 16	Nm	Torque developed by servo motor
s_o	-0.4 to 0.4	rad	Shaft angular position after gear

Velocity Controller. Input to the velocity controller is the velocity reference n_{ref} and output is the motor current i_m. The differential equations including current saturation are

$$\frac{d(i_2(t))}{dt} = \frac{K_v}{T_v}(n_{ref}(t) - n'_m(t))$$
$$i_{m0}(t) = K_v(n_{ref}(t) - n'_m(t)) + i_2(t) \qquad (13)$$
$$i_m(t) = \max(-i_{max}, \min(i_{max}, i_{m0}(t)))$$

The current saturation limits are in general important for the dynamic performance and often have to be included in a fault detector. Using observer based methods, this is straightforward. When using dynamic parity space equations, however, the nonlinear current limit does not directly fit into the general approach. In the following, we have therefore chosen to neglect the current limits in the fault detection design model.

Motor and Gear. The motor input is current i_m. The linear differential equations for the motor and gear are

$$\frac{d(n_m(t))}{dt} = \frac{1}{I_{tot}}(-fn_m(t) + K_q\eta i_m(t) + Q_{lm}(t))$$
$$\frac{d(s_o(t))}{dt} = \frac{1}{N}n_m(t) \qquad (14)$$

Discrete position controller. Inputs to the discrete position controller is a position reference, s_{ref}, and measured position s_o. Control output is the velocity reference, n_{ref}. The position control is performed digitally.

Control Loop Faults. Generic faults in electronics and other hardware for the position control loop include:

Feedback element fault: Δs_o position measurement
                             Δn_m velocity measurement
Control output fault: Δn_{ref} velocity reference
Actuator fault: Δi_m current
Control reference fault: Δs_{ref} setpoint demand to position controller

The location of these fault inputs are shown in the block diagram, figure 6.

The occurrence of a generic fault can have several reasons. In this example, we consider the following, realistic cases:

1. The wiper of a feedback potentiometer loses contact with the resistance element due to wear or dust. The fault is intermediate, and in the test case it lasts for only 0.2 second. This type of intermediate faults are very difficult to locate by service staff. The effect of this fault is an acceleration of the rod, which eventually drives the mechanism into endstop. The particular fault might cause an overspeed of the diesel engine with the shut down of the engine as the result. If this happens during manoeuvring of the ship consequences can easily become serious. This fault is a feedback element fault in s_o.

2. An end-stop switch suddenly malfunction. The reason is a broken wire or a defect in the switch element due to heavy mechanical vibration. As a result, the power drive can only deliver positive current. This fault can also be intermediate. The effect is an inability to move the rod in negative direction. This is an actuator fault in i_m.

The test sequences applied in this paper are shown in figure 7. The syste is subject to an unknown step input (load) during part of the sequence. The time of events are

Event	Start time	End time
Position fault	0.7 s	0.9 s
Current fault	2.7 s	3.0 s
Load input	1.2 s	2.3 s

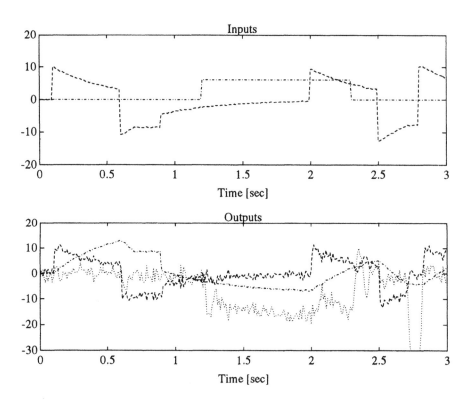

Fig. 7. Test sequencies. Velocity reference n_{ref} (–), load Q_{lm} (-.), measured current i_m (..), velocity n_m (–), and position s_o (*375, -.)

4.3 Requirements to Fault Reaction

The required reactions to faults are

1. Position measurement fault: Restructure the controller and use an estimate of the position, instead of the measured position, to calculate the control signal (second line of eq. 14). No jumps are allowed in the position, so fault accommodation must take place within the same sample cycle as detection and isolation. When the fault is detected, an alarm shall be issued to alert an operator. There shall be two possibilities for continued operation (choice is presented as part of the design):
   a) Continue with reconfigured controller until operator intervenes.
   b) Continue with reconfigured controller for 60 seconds. Thereafter velocity reference is set to zero.
2. Velocity measurement fault: Make immediate stop of servomotor by activating a stop command to the powerdrive (digital output 1), and by deactivating a magnetic brake within the motor (digital output 2). This shall be done immediately after detection. Do not wait for sampling cycle to complete.
3. Velocity reference: Reaction identical to point 2.
4. Actuator fault: Reaction identical to point 2.
5. Reference fault: Make a time-history roll back over 15 previous samples of the reference signal and use the mean of sample numbers n-10 to n-15 as the new reference signal. Restructure the controller to use this reference until the fault has been repaired.

4.4 Mathematical Bench-mark Design Model

The following variables and faults are included in the bench-mark design example

Input:	n_{ref}	Velocity reference
Outputs:	n_m	Motor velocity and
	s_o	Gear output position
Unknown input:	Q_{lm}	Load torque
Faults:	Δi_m	Motor current
	Δs_o	Position measurement

The continuous time state space description for the system can be written on the standard form

$$\dot{x}(t) = Ax(t) + Bu(t) + Ed(t) + F_1 f(t) \qquad (15)$$
$$y(t) = Cx(t) + F_2 f(t)$$

where
    $x = [i_2, n_m, s_o]^T$ is the state vector
    $u = n_{ref}$ is the controlled demand (output from position controller)
    $d = Q_{lm}$ is a vector of unknown disturbances
    $f = [\Delta i_m, \Delta s_o]^T$ is a vector comprising additive faults
    $y = [n_m', s_o']$ is a measurement vector.

With the differential equations given above, a linear state space description has the following system matrices where nonlinear control current limitation is disregarded:

$$A = \begin{bmatrix} 0 & -\frac{K_v}{T_v} & 0 \\ \frac{K_q\eta}{I_{tot}} & \frac{-f-K_vK_q\eta}{I_{tot}} & 0 \\ 0 & \frac{1}{N} & 0 \end{bmatrix} \quad B = \begin{bmatrix} \frac{K_v}{T_v} \\ \frac{K_vK_q\eta}{I_{tot}} \\ 0 \end{bmatrix} \quad E = \begin{bmatrix} 0 \\ \frac{1}{I_{tot}} \\ 0 \end{bmatrix} \tag{16}$$

$$F_1 = \begin{bmatrix} 0 & 0 \\ \frac{K_q\eta}{I_{tot}} & 0 \\ 0 & 0 \end{bmatrix} \quad C = \begin{bmatrix} 0 & 1 & 0 \\ 0 & 0 & 1 \end{bmatrix} \quad F_2 = \begin{bmatrix} 0 & 0 \\ 0 & 1 \end{bmatrix} \tag{17}$$

The continuous model is converted to discrete time by integrating the continuous state space equations over one sampling interval using the fact that the computer speed reference is constant during one sampling interval. This discretization requires the sampling time to be known and constant. The discrete representation is

$$x(k+1) = \Phi x(k) + B_m u(k) + B_d d(k) + B_f f(k) \tag{18}$$
$$y(k) = C x(k) + F_2 f(k) \tag{19}$$

Where Φ, B_m, B_d, B_f, C and F_2 are discrete system matrices. The vectors x, u, d, f, and y are discrete representations of the above defined continuous time versions. The integration to discrete time follows from:

$$\Phi(t_{k+1}, t_k) = e^{A(t_{k+1}-t_k)} \tag{20}$$
$$B_m(t_{k+1}, t_k) = \int_0^{t_{k+1}-t_k} e^{As} ds B$$

The equations for B_d and B_f are similar the last equation.

5 Performance of Model Based FDI Methods

This section considers modelbased FDI methods. FDI design is made using a simple model for the motor. Verification of the methods is made using realistic data. The faults considered are position measurement malfunction and power drive current error. The residual generators shall enable detection and isolation of the faults. The basis for design are the equations given in sections 4 for the simple model and 3 for detection algorithms, respectively. Test data for verification are generated from a complex, nonlinear model, (appendix 1).

5.1 Parity Space Approach

When applying the parity space approach the model representation must be converted to an input/output model of the form in eq. 21.

$$y(k) = S(z)u(k) \tag{21}$$

The system has one measurable input n_{ref}, one unknown input, Q_{lm}, two fault inputs, Δi_m and Δs_o and two outputs, n_m and s_o'. This gives the representation shown in eq.22.

$$\begin{bmatrix} n_m \\ s_o' \end{bmatrix} = \frac{1}{D(z)} \begin{bmatrix} S_{u11}(z) \\ S_{u21}(z) \end{bmatrix} n_{ref}$$

$$+ \frac{1}{D(z)} \left[\begin{bmatrix} S_{d11}(z) \\ S_{d21}(z) \end{bmatrix} Q_{lm} - \begin{bmatrix} S_{f11}(z) & S_{f12}(z) \\ S_{f21}(z) & S_{f22}(z) \end{bmatrix} \begin{bmatrix} \Delta i_m \\ \Delta s_o \end{bmatrix} \right] \tag{22}$$

where
$D(z)$ is the common denominator
$S_{xyz}(z)$ are polynomias in z.

The output errors becomes:

$$\begin{bmatrix} e_1 \\ e_2 \end{bmatrix} = \begin{bmatrix} n_m \\ s_o' \end{bmatrix} - \frac{1}{D(z)} \begin{bmatrix} S_{u11}(z) \\ S_{u21}(z) \end{bmatrix} n_{ref}$$

$$= \frac{1}{D(z)} \left[\begin{bmatrix} S_{d11}(z) \\ S_{d21}(z) \end{bmatrix} Q_{lm} - \begin{bmatrix} S_{f11}(z) & S_{u12}(z) \\ S_{u21}(z) & S_{u22}(z) \end{bmatrix} \begin{bmatrix} \Delta i_m \\ \Delta s_o \end{bmatrix} \right] \tag{23}$$

where the first equation is realizable because only measurable variables are included. The second is used for design of a fault isolation filter. A filter is designed to give two residuals, where the one is unaffected to Δs_o and the other to Δi_m. The unknown input, the load, is not treated which means that the residual generator will not be robust to changes in the load.

$$r(k) = W(z)e(k) = \frac{1}{D(z)} \begin{bmatrix} w_{11}(z) & w_{12}(z) \\ w_{21}(z) & w_{22}(z) \end{bmatrix} \begin{bmatrix} S_{f11}(z) & S_{u12}(z) \\ S_{u21}(z) & S_{u22}(z) \end{bmatrix} \begin{bmatrix} \Delta i_m \\ \Delta s_o \end{bmatrix} \tag{24}$$

If

$$w_{11}(z)S_{f11}(z) + w_{21}(z)S_{f21}(z) = 0 \tag{25}$$

the Δi_m contribution is eliminated and if

$$w_{12}(z)S_{f12}(z) + w_{22}(z)S_{f22}(z) = 0 \tag{26}$$

the Δs_o contribution is eliminated.

This is fulfilled by making
$w_{11}(z) = S_{f21}(z)$
$w_{12}(z) = S_{f22}(z)$
$w_{21}(z) = -S_{f11}(z)$
$w_{22}(z) = -S_{f12}(z)$

In figure 8 and 9 the errors and residuals are seen when the parity space approach is applied on the simple and complex model respectively.

With a load torque of approximately 6 Nm applied, the figures 8 and 9 shows, that it is possible to detect both the current and the position faults. The residual used to detect the current fault is sensitive to the load input. Ie. the performance of the detector for current faults depends on the size of the applied load torque. It is seen that the faults are only detectable when the system turns back to normal. The reason is that the faults enter rather slowly compared to when the system turns back to normal.

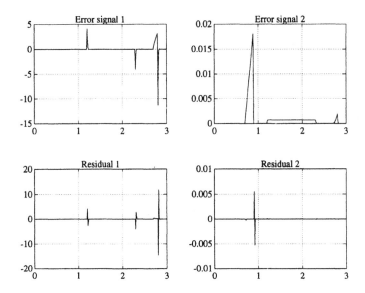

Fig. 8. Output errors and residuals generated by use of the parity space approach. Simple model.

The detector based on the parity space approach makes it possible to detect and isolate faults that gives fast process state changes. To make this detector applicable in real life it is necessary to have a very exact model of the system as it can be seen on the residuals concerning the complex model. Here there are risks for false detections due to unknown input, unmodelled dynamics and nonlinearities.

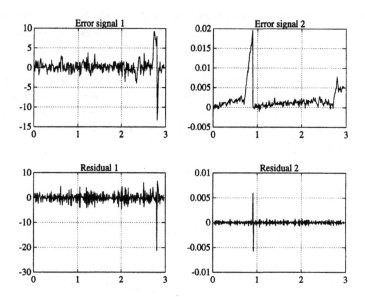

Fig. 9. Output errors and residuals generated by use of the parity space approach. Complex model.

5.2 Diagnostic Observer Approach

The diagnostic observer approach uses the state space description given in paragraph 3 and the observer has the form of eq. 27.

$$\widehat{x}(k+1) = \Phi\widehat{x}(k) + B_m u(k) + K(y(k) - \widehat{y}(k)) \tag{27}$$

The residual is given by:

$$r(k) = H(z)(y(k) - \widehat{y}(k)) \tag{28}$$

The K and H matrices are designed to make the observer stable and to give the residuals desired properties.

The state estimation error is given by:

$$
\begin{aligned}
e &= \widehat{x} - x \\
&= (zI - \Phi + KC)^{-1}(KF_2 - F_1)p - B_d d
\end{aligned}
\tag{29}
$$

When applying eq. 29 on eq. 28 the residuals become:

$$r = H(C(zI - G)^{-1})(Fp - B_d d) - F_2 p \tag{30}$$

where $F = KF_2 - F_1$ and $G = \Phi - KC$.

The right side eigenvector assignment is used, which implies that two observers are designed. Each of the two observers estimates one output. The observers are designed so that a selected columns of the fault entry matrix F is made the right side eigenvector of G. This means that for observer one (OBS1), G is designed so that the first column of F is the right side eigenvector of G and for observer two (OBS2), the same must be valid for the second column of F. Hereby OBS1 estimates the position and OBS2 estimates the velocity. Furthermore $det(I - G) = 0$ for a specified eigenvalue λ. If the above described is fulfilled it implies that

$$r = H(z^{-1}CF - F_2)p - z^{-1}CB_d d \tag{31}$$

The following definitions are given:

$$z^{-1}CF - F_2 = CF(z) = \begin{bmatrix} cf_{11}(z) & cf_{12}(z) \\ cf_{21}(z) & cf_{22}(z) \end{bmatrix}$$

$$H(z) = \begin{bmatrix} h_{11}(z) & h_{12}(z) \\ h_{21}(z) & h_{22}(z) \end{bmatrix} \tag{32}$$

H is determined to isolate the two fault contributions in the error signals. The i'th row of $H(z)$ multiplied by the j'th column of $CF(z)$ are going to be zero to eliminate the j'th error. This gives

$$h_{11}(z)cf_{12}(z) + h_{12}(z)cf_{22}(z) = 0 \tag{33}$$

to eliminate the Δs_o contribution and

$$h_{21}(z)cf_{11}(z) + h_{22}(z)cf_{21}(z) = 0 \tag{34}$$

to eliminate the Δi_m contribution.

By choosing the filter as follows eq. 33 and eq. 34 will be fulfilled

$$\begin{aligned} h_{11}(z) &= -cf_{22}(z) \wedge h_{12}(z) = cf_{12}(z) \\ h_{21(z)} &= cf_{21}(z) \quad \wedge h_{22}(z) = -cf_{11}(z) \end{aligned} \tag{35}$$

which means that residual 1 will remain unaffected by Δs_o and residual 2 unaffected by Δi_m.

The K and H matrices are:

$$K1 = \begin{bmatrix} 5.92 \cdot 10^{-2} & 1.55 \cdot 10^{-6} \\ 0.28 & 3.57 \cdot 10^{-7} \\ 2.06 & 1.00 \end{bmatrix} \quad H1 = \begin{bmatrix} (1 - 0.11z^{-1}) & (0.98z^{-1}) \\ (-3.35z^{-1}) & (0.29z^{-1}) \end{bmatrix}$$

$$K2 = \begin{bmatrix} -1.14 & 0.99 \\ -0.78 & 0.98 \\ 4.24 \cdot 10^{-5} & 0.11 \end{bmatrix} \quad H2 = \begin{bmatrix} (1 - 0.11z^{-1}) & (0.98z^{-1}) \\ (-3.35z^{-1}) & (0.29z^{-1}) \end{bmatrix} \tag{36}$$

The results can be seen in the figures 10 and 11 when the method is applied on the simple and complex models.

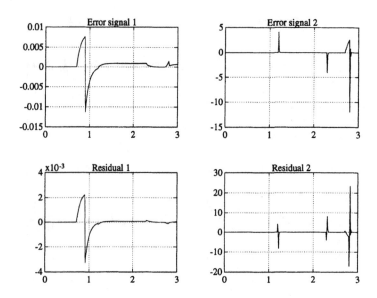

Fig. 10. Output errors and residuals generated by use of a diagnostic observer. Simple model.

The main difference between the results from the parity space approach and the diagnostic observer approach is an improved possibility in detecting the position fault. The slow entry of the fault is emphasized in the observer method.

5.3 Unknown Input Observer Approach

This approach enables elimination of influences from an unknown input. The observer is designed to attenuate contributions from the load input to the error signal. This introduces a problem in the detection of a current fault, because the load input and the current fault input enters the system at the same point. The current fault influence will be filtered away as the influences from the load. I.e. that only the position measurement fault can be detected in this case.

The unknown input observer has the form of:

$$\dot{z} = Fz + TB_mu + Ky$$
$$\hat{x} = z + Hy \qquad (37)$$
$$e = \hat{x} - x$$

The matrices F,T,K and H are designed so that the unknown input, the

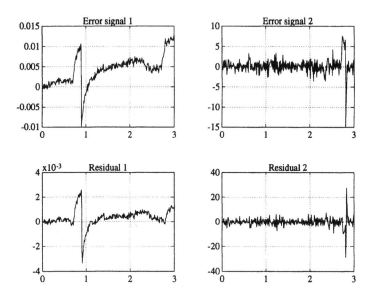

Fig. 11. Output errors and residuals generated by use of a diagnostic observer. Complex model.

load, does not affect the estimation error. The following results are obtained.

$$F = \begin{bmatrix} 1.00 & 0 & 0 \\ 0 & 0.98 & 0 \\ 0 & 0 & 0.90 \end{bmatrix} \qquad T = \begin{bmatrix} 1.00 & 2.14 & 2.46 \cdot 10^{-4} \\ 0 & 1.32 \cdot 10^{-8} & -1.15 \cdot 10^{-4} \\ 0 & -1.15 \cdot 10^{-4} & 1.00 \end{bmatrix}$$

$$K = \begin{bmatrix} -2.81 & 0 \\ -1.71 \cdot 10^{-8} & -2.30 \cdot 10^{-6} \\ 1.40 \cdot 10^{-4} & 0.10 \end{bmatrix} \quad H = \begin{bmatrix} -2.14 & -2.46 \cdot 10^{-4} \\ 1.00 & 1.15 \cdot 10^{-4} \\ 1.15 \cdot 10^{-4} & 1.32 \cdot 10^{-8} \end{bmatrix}$$

$$(38)$$

As it can be seen in figure 12 the unknown input observer does not care if changes in the load torque occur. However, as a consequence of the system structure, current faults can not be detected.

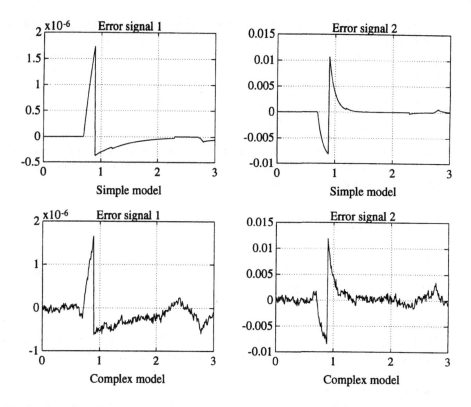

Fig. 12. Output errors generated by use of an unknown input observer. Simple and complex model.

6 Conclusions

This paper has addressed the problem of fault handling in closed loop control systems. A methodology for design of fault handling has been presented, and it has been shown how a fault accommodating subsystem implementation should be separated in a control method layer and a supervisory layer.

An assessment was made of mathematical methods for detection and isolation of faults in industrial feedback control systems. It was demonstrated that these methods work fairly well in detection and isolation of faults in sensors and actuators, and that specialized methods can be employed when disturbances and unmodelled dynamics are present. Other techniques are needed to detect and handle controller faults.

Design and implementation of the accommodation part of fault handling methods was further discussed. It was concluded that no present theory seems to provide substantial support in this part of a development. Hence, this is an area that should attract future research emphasis.

An industrial position control loop with electro-mechanical components was used as an application example. It was shown to have several characteristics which are generic for hybrid feedback control systems. The paper provided a simple mathematical model of this system for use in the design of fault handling methods and a complete, nonlinear description for simulation and verification work. These models should make it possible to make bench-mark testing of new methods in hybrid system theory.

Key contributions are believed to be the presentation and analysis of an industrial feedback system which is suitable for bench-mark testing of new ideas in hybrid system theory, and the demonstration of state of the art in mathematical methods for fault detection and accommodation.

7 References

[Bell,1989]
Trudy E. Bell *"Managing Murphy's Law: Engineering a Minimum-Risk System"* IEEE Spectrum, June 1989.

[Basseville & Nikiforov,1993]
M. Basseville and I.V. Nikiforov *"Detection of Abrupt Changes - Theory and Application "*. Prentice Hall Series in Information and System Sciences

[Benveniste et al,1987]
A. Benveniste, M. Basseville, G. Moustakides (1987) *"The Asymptotic Local Approach to Change Detection and Model Validation "*. IEEE Trans. on Automatic Control, vol AC-32, no 7 pp 583-592

[Benveniste et al,1990]
A. Benveniste, M. Metivier, P.Priouret asseville *"Adaptive Algorithms and Stochastic Approximations "*. Springer, Berlin, 1990

[Blanke & Nielsen,1990]
Mogens Blanke, Per Busk Nielsen *"The Marine Engine Governor"*. Maritime Communications and Control London, 21-23 Nov. 1990 The Institute o9f Marine Engineers, London page 11-19

[Frank, 1991]
Paul M. Frank *"Enhancement of Robustness in Observer-Based Fault Detection"*. IFAC/IMACS Symposium on "Fault Detection, Supervision and Safety for Technical Processes. - Safeprocess '91 - Vol. 1 page. 275-287.

[Gertler, 1991]
Janos Gertler *"Analytical Redundancy Methods in Fault Detection and Isolation"*. Survey and Synthesis. IFAC/IMACS Symposium on "Fault Detection, Supervision and Safety for Technical Processes. - Safeprocess '91 - Vol. 1 page.9-23.

[Herrin,1981]
Stephanie A. Herrin *"Maintainability Applications Using the Matrix FMEA Technique"*. IFAC's Transactions on Reliability, vol. R-30 No. 3, August 1981.

[Isermann,1991]
Rolf Isermann *"Fault Diagnosis of Machines via Parameter Estimation and Knowledge Processing"* IFAC/IMACS Symposium on "Fault Detection, Supervision and Safety for Technical Processes. - Safeprocess '91 - Vol. 1 page.121-133.

[Lauber,1991]
R.J.Lauber *"Aspects of Achieving Total System Availability"* IFAC/IMACS Symposium on "Fault Detection, Supervision and Safety for Technical Processes. - Safeprocess '91 - Vol. 1 page.35-41.

[McKelvei,1989]
Thomas C. McKelvey *"How to Improve the Effectiveness of Hazard & Operability Analysis"* IEEE Transactions on Reliability. Vol 37 No.7 June 1988 page 167-170

[Nielsen et al,1993]
Søren Bøgh Nielsen, Rikke Bille Jørgebsen, Mogens Blanke *"Nonlinear Characterization and Simulation of Industrial Position Control Loop"* AUC Department of Control Engineering, Report no. R-93-4012, February 1993

[Patton & Chen,1991]
R.J.Patton,J.Chen *"A Review of Parity Space Approaches to Fault Diagnosis"* IFAC/IMACS Symposium on "Fault Detection, Supervision and Safety for Technical Processes. - Safeprocess '91 - Vol. 1 page.35-41.

[Patton et al,1989]
R.J.Patton, P.Frank, R.Clark *"Fault Diagnosis in Dynamic Systems. Theory and Application"* Prentice Hall International,1989

[Poucet & Petersen ,1988]
A. Poucet, K.E. Petersen *"Knowledge Based Safety and Reliability Analysis Tools and their Application in Process Control Systems Design and Operation"* IFAC Proceedings of workshop 1989 page 83-88

[Shafaghi et al,1984]
A. Shafaghi, P.K.Andow, F.P. Lees *"Fault Tree synthesis based on Control Loop Structure"* Chemical Engineering Research and Design. Vol.62 Mar.1984

[Simulab,1991]
Simulab Users guide, January 1, 1991, The MathWorks, Inc, Natick, Massachusets

[Stengel,1991]
Robert F. Stengel *"Intelligent Failure-Tolerent Control"* IEEE Control Systems, Vol.11 No.4 1991 Page 14-23

[Ulerich et al,1988]
Nancy H. Ulerich, Gary J. Powers *"On-Line Hazard Aversion and Fault Diagnosis in Chemical Processes: The Digraph + Fault tree Methods"* IEEE Transactions on Reliability, Vol.37, No.2,June 1988, Page 171-177

[Zhang et al]
H.G. Zhang, J. Chen, H.Y.Zhang *"Decentralized Estimation using Unknown-Input Observer"*. Department of Automatic Control, Beijing University of Aeronautics & Astronautics, Beijing 100083, P. R. China.

[Åstrøm et al,1986]
K.J.Åstrøm, J.J.Anton,K.E.Årzen *"Expert Control"* Automatica, Vol. 22, No 3 page 277-286,1986

8 Appendix. 1: Complex Mathematical Model of Position Loop

A detailed model of the actuator is derived to enable analysis of the fault detection and isolation methods (FDI) designed on basis of the simple model presented in the paper. The complex model includes several nonlinear dynamic elements not included in the simple model. These constitutes unmodelled effects in the FDI design to which the FDI algorithm shall be robust. This appendix presents a brief overview of the complex model structure and nonlinear elements. Details on the mathematical equations and other results can be found in [Nielsen et al,1992]

8.1 Model Overview

The detailed model is presented in figure 13.

COMPUTER. The discrete *position controller* has input from the position reference s_{ref} and the sampled position measurement s_o . Output from the position controller is a velocity reference n_{ref} for the analog velocity control which

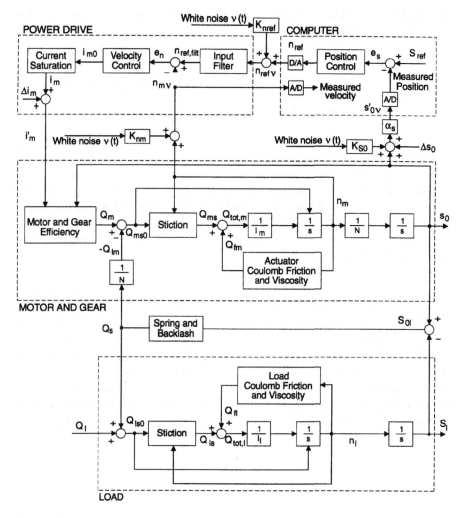

Fig. 13. Detailed model of the actuator including load dynamics and nonlinear effects.

is an inherent part of the power drive. The velocity is sampled for use in fault detection.

POWER DRIVE. The actuator motor current i_m is controlled by a proportional-integral *velocity controller*. The input to the velocity controller is a velocity reference $n_{ref,filt}$ filtered through a low pass *input filter* and the measured velocity n_m. The maximum motor current is limited by the power drive electronics and is thus subject to a *current saturation*.

MOTOR & GEAR. The motor current generates a motor torque Q_m. The motor efficiency depends on the motor shaft position such that a 20 motor torque.

The driving torque Q_{ms0} passes through a *stiction torque* function, which, for zero velocity $(n_m=0)$ compares the driving torque to the stiction torque, and only outputs the amount of driving torque that exceeds the stiction torque. When Q_{ms0} is below the stiction value the velocity remains zero.

The total acting torque $Q_{tot,m}$ accelerates the motor inertia I_m to the velocity n_m.

When the velocity is nonzero, *coulomb and viscous friction* produce damping torques which oppose the driving torque. The velocity integrates to a position through a gear with a gear ratio of N.

The actuator drives a load through a gear and a bearing. The gear has a small amount of *backlash* and the rod connecting it with the load is slightly elastic, although this is not intended. For this reason the total system is divided into the motor with gear and the load mass as two sub systems which are connected through a *spring*.

LOAD. Load dynamics can be described similarly to the motor: A load torque Q_l through a stiction function accelerates the load inertia I_l to the load velocity n_l. The load is also subject to coulomb and viscous friction when n_l is nonzero. The load position is denoted by s_l. When the spring between the actuator motor and the load is loaded $(s_{ol} \neq 0)$, the effect is a torque acting on both the actuator motor and the load. A spring torque Q_s thus adds to the driving torque. The spring torque constant was found to have different values for loading and releasing motions. Backlash is included in the spring feedback function. Measurement noise is simulated by adding discrete, random variables with gaussian distributions to inputs.

On Formal Support for Industrial-Scale Requirements Analysis

T. Anderson, R. de Lemos, J. S. Fitzgerald and A. Saeed

Department of Computing Science
University of Newcastle upon Tyne,
NE1 7RU, UK

Abstract. Drawing on practical experience in the development of dependable applications, this paper presents a number of "goals" for industrially applicable formal techniques in the specification and analysis of requirements for hybrid systems. These goals stem from domain-specific concerns such as the division between environment, plant and controller; and from the development context with its wide variety of analysis and design activities.

Motivated by some of these goals, we present a methodology, based on formal methods, for the requirements analysis of hybrid systems that are safety-critical. This methodology comprises a framework whose stages are based on levels of abstraction that follow a general structure for process control systems, a set of techniques appropriate for the issues to be analysed at each stage of the framework, and a hierarchical structure for the product of the analysis. Some aspects of the methodology are exemplified through two case studies. The extent to which this approach meets the goals espoused earlier is discussed.

Keywords: safety-critical systems, requirements analysis, formal methods, timeliness requirements, quality assessment.

1 Introduction

This paper addresses the provision of formal techniques for the analysis of requirements for safety-critical (hybrid) real-time computing systems. A particular concern in this area is the need to develop techniques which can be exploited in industrial-scale systems development.

The Dependable Computing Systems Centre at the Universities of Newcastle upon Tyne and York was established by British Aerospace, a designer and manufacturer of a full range of aircraft, space and military systems, with the intention of conducting research aimed at the development of technology which will help the sponsor and industry meet the increased dependability demands placed on computing systems in the future. As part of this work, the authors have had the opportunity to discuss both formal methods and requirements analysis with a wide range of practitioners involved in the development of a variety of systems.

The potential advantages of employing formal methods of system modelling, specification and development are well-known, and have led to their utilization in systems development being advocated as a means of improving confidence in the dependability of the finished product [14], [19]. However, a number of objections have been levelled at formal techniques concerning their applicability in the realistic industrial development of computing systems: formal techniques are hard to apply in the context of a product development, in particular there is little experience in how to manage projects employing formal methods; the techniques and notations are inaccessible to systems engineers, many of whom do not have a background in discrete mathematics; requirements change frequently, so that making the investment in formal specifications and proofs early in the development process means that modest requirements changes have substantial cost implications; specifications and proofs themselves can be quite large for realistic computing systems, and tool support for their construction is limited.

The work reported in this paper is aimed at developing requirements analysis techniques which go some way towards overcoming these objections. More specifically, the following goals are suggested for such techniques:

1. Consistency with Current Practice: While it is unrealistic to expect the exploitation of formal techniques to fit in with unmodified current industrial practice, it is reasonable to expect that the sorts of analyses currently performed in system development will be supported, e.g. Hazard and Operability Studies [12].

2. Traceability: Control of frequently changing product requirements and maintenance of the associated documentation are major problems in the industrial development of computing systems. The techniques developed here should allow design decisions to be related back to the relevant requirements, an important aspect of safety certification.

3. Exploiting Structuring: Requirements are often fixed at a highly variable level of abstraction, with details such as sensor sampling rates fixed early. Even if an early requirements specification is highly abstract, complexity soon increases in the course of a formal development by refinement. An answer to the problem of complexity is traditionally the exploitation of modularity. The main system structure can then be understood separately from the detail of its components and specifications can be re-used, along with their associated theories.

4. Notation Accessibility and Suitability: The basic concepts of mathematical modelling and analysis are familiar to many in the engineering professions. However, the introduction of discrete mathematics into the modelling of computing systems has, to some extent, been hindered by the perceived inaccessibility of the notations employed. Some graphical techniques, such as Statecharts and Statemate [2], [9], have found favour with engineers for modelling certain classes of problem (typically state-transition-based systems such as cockpit displays). It would be wise to capitalize on this experience and use notations with which engineers feel comfortable.

Different techniques are appropriate to different forms of analysis. The work reported here aims to use suitable techniques and notations at each stage in the analysis of requirements. For example, the systems under consideration in this paper are typically hybrid in that a physical process is subject to control by a computing system. The apparatus of continuous functions is appropriate to reasoning about the physical phenomena under control, while discrete mathematics forms a suitable basis for reasoning about the controller.

The rest of the paper is organised as follows. The next section describes a general structure for the class of safety-critical systems under consideration: process control systems. Section 3 the proposed methodology for the analysis of safety requirements. Our methodology has been applied to various examples drawn from a train set system [17, 3]. In this paper we illustrate the methodology by presenting its application to two examples of hybrid system: the Cat and Mouse Problem [13] in Section 4 and chemical batch processing [16] in Section 5. Finally, in Section 6 we give a preliminary evaluation of this methodology and make some concluding remarks.

2 A Structure for Safety-Critical Process Control Systems

In accordance with common practice, we partition process control systems into three different components: the *operator*, the *controller*, and the *physical process (or plant)*. The controller has two interfaces, the operator interface and the plant interface, the latter being made up of sensors and actuators. The system exists in an environment, i.e. that part of the rest of the world which may affect, or be affected by, the system.

The system is modelled by a set of variables V, called the *system variables* which consists of: a set of *physical variables* V_P which model the physical process; a set of *controller variables* V_C which model the process controller; and a set of *operator variables* V_O which model the operator. As far as the plant interface is concerned, we are only concerned with those physical variables that are being directly monitored (V_{PM}) or controlled (V_{PC}) and the controller variables that provide input to the controller (V_{CI}) or receive output from the controller (V_{CO}). Each variable (v_i) of the set of system variables ($V = \{v_1, v_2, v_3, \ldots\}$) can be represented as a function from time (T) into values for the variable from its range (R_{v_i}):

$$v_i(t): T \to R_{v_i}$$

3 Methodology for Requirements Analysis

The objective for the analysis of safety requirements is, starting from an informal description of system behaviour, to produce specifications over the controller that are sufficient to maintain the "safe behaviour" of the system. In this section we briefly present a methodology, based on formal methods, which provides a systematic way in which requirements for safety-critical systems can be analysed.

The methodology is based on the approach of separating mission requirements from safety requirements; the latter are those requirements which focus on the elimination and control of hazards, and the limitation of damage in case of an accident. The logical separation between mission and safety may be respected at the physical process level (e.g. the shut down systems of nuclear reactors), at the controller level (by implementing separate controllers) or by application of design techniques, such as "firewalls", which could prevent the logical mission controller from interfering with the logical safety-controller. However, it is *not* the intention of this approach to *force* a particular process structure on the implementation (c.f. comments on specification and implementation structuring in [5])

The proposed methodology enforces safety specifications to be formal from the initial stages of requirements analysis. An advantage of the proposed approach is that specifications can be validated and design steps verified with the assitance of formal techniques, including refinement.

3.1 Framework for Requirements Analysis

When developing safety-critical systems, a key concern is to ensure that *accidents* do not occur. The usual practice to ensure safety is to strive to identify all of the *hazards* (circumstances from which an accident might ensue) and then apply techniques to reduce the probability of these hazards occurring.

Our general framework for the analysis of safety requirements for safety-critical systems (discussed more fully in [1]) involves splitting the analysis into the stages discussed in the following sections.

Conceptual Analysis The objective of this stage is to produce an initial, informal statement of the aim and purpose of the system, including what constitutes "safety". As a product of the Conceptual Analysis we obtain the Safety Requirements, enumerating (and prohibiting) the potential *accidents*. Another activity to be performed during this stage is the identification of the *modes* of operation of the physical process; these are classes of states which group together related functions.

In the conceptual analysis, we describe system behaviour in terms of uninterpreted predicate symbols in the Predicate Event Action Model (described in Appendix A).

Safety Plant Analysis During this stage the properties of the physical process relevant to the Safety Requirements must be identified. An essential starting point is to construct a formal model of the physical process which captures its characteristics as dictated by physical laws and rules of operation. The plant analysis must also support the identification of *hazards*. As the product of the plant analysis we obtain the Safety Plant Specification which contains the *safety constraints* (usually the negation of a hazard modified to incorporate safety margins) and the *safety strategies* (ways of maintaining the safety constraints defined in terms of controllable factors). The conjunction of all of the safety constraints of a system characterizes the specified safe behaviour for that system.

At this stage, we are primarily concerned with modelling a physical process and imposing constraints on its behaviour. Therefore, we suggest that logical formalisms are appropriate and employ Timed History Logic (THL,described in Appendix B) [16] to model the continuous behaviour of the system, and the PEA Model, defined in terms of THL, for a discrete model of system behaviour.

Safety Interface Analysis The objective of this stage is to delineate the plant interface, and to specify the behaviour that must be observed at the identified interface (including properties of sensors and actuators). This stage leads to the production of the Safety Interface Specification, containing the *robust safety strategies* (strategies modified to incorporate the limitations of the sensors and actuators). Again we need a descriptive formalism. Currently we employ the same techniques as for the plant level.

Safety Controller Analysis During this stage the robust safety strategies are analysed in terms of the properties of the controller, such as the top level organization of the controller components and their failure rate. This leads to the production of the Safety Controller Specification, containing the *safety controller strategies* (refinements of a robust safety strategies incorporating some of the components of a logical architecture for the safety controller). The Safety Controller Specification is the main product of the requirements analysis, and provides the basis for subsequent development of the system.

At this stage, we analyse the structure and operations of the controller. To perform this we need an operational formalism which explicitly specifies concurrency and non-determinism, such as Statecharts [8] or PrT nets [6]. Currently we employ the latter formalism.

3.2 Safety Specification Hierarchy

The safety specifications produced during the plant, interface and controller analysis are predicates over a sequence of states that describe a possible behaviour of the system. The safety constraints and the safety strategies are defined over the physical variables (V_P), the robust safety strategies over the input and output variables (V_{CI}, V_{CO}) and the safety controller strategies over the controller variables (V_C). To organize the safety specifications we introduce the concept of a safety specification hierarchy (SSH). The top elements of an SSH are the accidents. With each accident we associate the set of hazards which could lead to that accident. To each hazard we associate the corresponding safety constraint. For each safety constraint we specify at least one safety strategy. This structure is repeated for the subsequent safety specifications: robust safety strategies and safety controller strategies.

The way in which the relationships are depicted facilitates traceability between the safety specifications. For example, during certification of the final system the SSH provides a static picture of the obligations that are specified at a particular level and how they will be discharged at a lower level.

3.3 Quality Assessment of the Safety Specifications

A SSH should be assessed both qualitatively and quantitatively. The qualitative analysis would involve establishing that a SSH is both consistent and safe. A SSH is consistent if and only if all of the safety specifications (that relate to the system in a particular mode) at each level of the hierarchy are not in conflict with each other (horizontal checks); and is safe if it can be established that safety (absence of accidents/hazards) is maintained down the hierarchy (vertical checks). The quantitative assessment should enable a risk analysis of the safety specifications. Applicable techniques include those based on Markov models, stochastic Petri nets and fault tree analysis, the application of fault tree analysis and minimal cut sets has been studied [18]. This preliminary risk analysis is useful in the determination of which strategy or combination of strategies is most suitable for the risks involved.

4 Example 1: Cat and Mouse

A sleeping cat should be awakened just in time to catch a fleeing mouse. The mouse and the cat both move along a straight line towards the mouse hole, where the mouse will be safe. They start at the same point, at a distance X from the mouse hole. The mouse immediately starts running with speed m (towards the hole). The cat (assumed to be capable of instantaneous acceleration) is awakened d time units later, and chases the mouse with speed c. Problem: design a just-in-time cat-actuator (controller). The cat and mouse are regarded as being point objects.

The system's mission requirement is to actuate the cat just in time to catch the mouse[1]. Here we show how the levels of abstraction, identified for the analysis of safety requirements, can nevertheless be applied to the analysis of mission requirements.

4.1 Conceptual Analysis

This is a self-contained system, consisting of a physical process and a controller. The physical process is composed of two physical entities: a cat and a mouse.

Modes of Operation
From the description of the problem, a high level model of the behaviour can be specified by the following modes. The analysis is performed using the PEA Model, in which modes are represented as uninterpreted states.

Idle$\langle\rangle$: the cat and the mouse are at rest, at the initial position.
Escape$\langle\rangle$: the cat is at rest and the mouse is running with speed m.
Race$\langle\rangle$: the cat is running with speed c and the mouse with speed m.
Result$\langle\rangle$: the cat has stopped or the mouse has stopped.

Rules of Operation
At this level the rules of operation specify the relationships between the modes.
*ro*1. The cat and mouse system is represented in terms of the above modes:

$$CatMouse\langle\rangle \equiv Idle\langle\rangle \prec Escape\langle\rangle \prec Race\langle\rangle \prec Result\langle\rangle$$

*ro*2. The duration of the mode *Escape*$\langle\rangle$ is d:

$$t_{\downarrow Escape} = t_{\uparrow Escape} + d$$

As a result of this stage, system behaviour is partitioned into four modes of operation, and the relationship between the modes is identified.

4.2 Plant Analysis

Here the analysis identifies the system variables that describe the behaviour of the physical process, the actions and states corresponding to the modes of operation, the rules of operation and the physical laws. The controller can observe when the mouse starts to run and can control when the cat starts to run. At this level, we use both the PEA Model and THL.

---

[1] There are no safety implications for the cat, although the mouse might see things differently.

Physical Variables

The variables of the physical process are identified by an inspection of the informal description of the modes. For example, mode *Idle*$\langle\rangle$ refers to the positions of the cat and mouse, hence the variables *Pcat* and *Pmouse* are introduced. The range of these variables are determined by identifying the initial position as zero, and specifying the distance to the end of the hole as Y. ($Y - X$ represents the length of the hole which is assumed to be non-zero.)

No.	Name	Range	Comments
p_1	Pcat	$\{x \in \mathbb{R} \mid 0 \leq x \leq X\}$	The position of the cat, on the line towards the mouse hole
p_2	Pmouse	$\{x \in \mathbb{R} \mid 0 \leq x \leq Y\}$	The position of the mouse ($Y > X$)
p_3	Scat	$\{0, c\}$	The speed of the cat
p_4	Smouse	$\{0, m\}$	The speed of the mouse

States

Next we identify the states that describe the behaviour of the cat and mouse during the modes. These states are expressed as predicates over the physical variables. Each mode is then defined in terms of states of the cat and mouse.

Cat: States

$C.rest\langle\rangle$: The cat is at rest at the initial position.

$\quad C.rest\langle\rangle(Scat = 0 \land Pcat = 0)$

$C.running\langle\rangle$: The cat is running at speed c.

$\quad C.running\langle\rangle(Scat = c)$

$C.stops\langle\rangle$: The cat has stopped at some point after the initial position.

$\quad C.stops\langle\rangle(Scat = 0 \land Pcat \neq 0)$

Mouse: States

$M.rest\langle\rangle$: The mouse is at rest at the initial position.

$\quad M.rest\langle\rangle(Smouse = 0 \land Pmouse = 0)$

$M.running\langle\rangle$: The mouse is running at speed m.

$\quad M.running\langle\rangle(Smouse = m)$

$M.stops\langle\rangle$: The mouse has stopped, at some point after the initial position.

$\quad M.stops\langle\rangle(Smouse = 0 \land Pmouse \neq 0)$

Rules of Operation

Rule of operation *ro3* expresses the relationship between the states of the cat; and rule *ro4* captures the relationship between the states of the mouse.

ro3. The behaviour of the cat is described by $Cat\langle\rangle$:

$\quad Cat\langle\rangle \equiv C.rest\langle\rangle \prec C.running\langle\rangle \prec C.stops\langle\rangle$

ro4. The behaviour of the mouse is described by $Mouse\langle\rangle$:

$\quad Mouse\langle\rangle \equiv M.rest\langle\rangle \prec M.running\langle\rangle \prec M.stops\langle\rangle$

Modes of Operation

The modes of operation of the cat and mouse system can be defined in terms of the states of the cat and mouse.

$Idle\langle\rangle \equiv (C.rest \land M.rest)\langle\rangle$

$Escape\langle\rangle \equiv (C.rest \land M.running)\langle\rangle$

$Race\langle\rangle \equiv (C.running \land M.running)\langle\rangle$

$Result\langle\rangle \equiv (C.stops \land M.stops)\langle\rangle$

Physical Laws

pl1. The distance travelled by the cat during the state $C.running\langle\rangle$ is the product of the duration of the state and c.

$$Pcat(t_{\downarrow C.running}) - Pcat(t_{\uparrow C.running}) = (t_{\downarrow C.running} - t_{\uparrow C.running}) \times c$$

pl2. The distance travelled by the mouse during the state $M.running\langle\rangle$ is the product of the duration of the state and m.

$$Pmouse(t_{\downarrow M.running}) - Pmouse(t_{\uparrow M.running}) = (t_{\downarrow M.running} - t_{\uparrow M.running}) \times m.$$

pl3. The mouse stops if and only if the mouse is not at the initial position and either the cat and mouse are at the same position or the mouse is at position Y. To capture this law, we introduce the event:

$$MouseStop = \Big|(Pmouse \neq 0 \land (Pcat = Pmouse \lor Pmouse = Y))$$

$\uparrow M.stops \Leftrightarrow MouseStop$

pl4. The cat stops if and only if the cat is not at the initial position and either the cat and mouse are at the same position or the cat is at position X. To capture this law, we introduce the event:

$$CatStop = \Big|(Pcat \neq 0 \land (Pcat = Pmouse \lor Pcat = X))$$

$\uparrow C.stops \Leftrightarrow CatStop$

Mission Aim

The cat catches the mouse just in time, that is at the instant the mouse reaches position X. This is specified by the following invariant relation which restricts the behaviour of the cat to a subset of the behaviours permitted by *pl4*. To specify the mission aim MA we introduce the event:

$$CatchAtX = \Big|(Pcat = Pmouse \land Pcat = X)$$

MA: $\uparrow C.stops \Leftrightarrow CatchAtX$

Mission Strategy

The mission strategy is based on a scheme that awakens the cat d units after the mouse, such that MA holds. This is captured by rule *msa*.

The cat enters the state $C.running\langle\rangle$ d units after the mouse enters $M.running\langle\rangle$.

$msa: t_{\uparrow C.running} = t_{\uparrow M.running} + d$

The value for d must ensure that the cat and mouse reach position X at the same time, derived by analysing the relationship between X, m and c under rule msa. From law $pl2$ we can infer the time point at which $Pmouse = X$ is:

$$t_{\uparrow M.running} + X/m$$

From rule msa and law $pl1$ we can infer the time point at which $Pcat = X$ is

$$t_{\uparrow M.running} + X/c + d$$

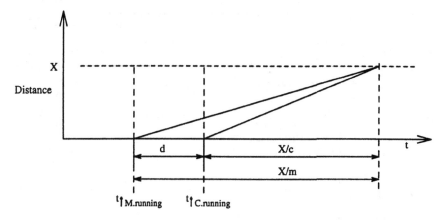

Fig. 1. Actions of the cat and mouse

Hence for the cat and mouse to reach position X at the same time, $t_{\uparrow M.running} + X/c + d = t_{\uparrow M.running} + X/m$ (See Figure 1). This can be expressed as the rule msb.

$$msb: \ d = X/m - X/c$$

$$MS = msa \wedge msb$$

Lemma 4.1. A history that satisfies rules of operation $ro3$ and $ro4$, and physical laws $pl1$ and $pl2$ must satisfy MA provided the strategy MS is satisfied[2]. The proof is shown in Figure 2, where MA is rewritten as lines 1 and 2.

As a result of this stage, the aim of the cat to just catch the mouse has been formulated (MA), and a strategy (MS) based on delaying the cat devised and verified against the stated aim.

4.3 Interface Analysis

Here the analysis identifies the variables that describe the behaviour of the sensors and actuators, and their relationship to the physical variables being monitored and controlled. These relationships are then used to develop the robust mission strategy.

---

[2] Although full proofs have been constructed for the lemmas presented in these examples, only sketches are provided here due to space limitations.

1.	$CatchAtX \Rightarrow \uparrow C.stops$		directly from $pl4$

1. $CatchAtX \Rightarrow \uparrow C.stops$ directly from $pl4$

2.   To Show $\uparrow C.stops \Rightarrow CatchAtX$

2.1. $\uparrow C.stops \Rightarrow CatStop$ from $pl4$

Case 1: $Pcat(t_{\uparrow C.stops}) = Pmouse(t_{\uparrow C.stops})$

2.1.1. $Pcat(t_{\uparrow C.stops}) = (t_{\downarrow C.running} - t_{\uparrow C.running}) \times c$ from $ro3$ and $pl1$

2.1.2. $Pmouse(t_{\uparrow C.stops}) = (t_{\downarrow M.running} - t_{\uparrow M.running}) \times m$ from $ro4$, $pl2$ and $pl3$

2.1.3. $(t_{\downarrow C.running} - t_{\uparrow C.running}) = Pcat(t_{\uparrow C.stops})/c$ $\div$ 2.1.1 by c

2.1.4. $(t_{\downarrow M.running} - t_{\uparrow M.running}) = Pcat(t_{\uparrow C.stops})/m$ from Case 1 and 2.1.2

2.1.5. $Pcat(t_{\uparrow C.stops})/m + d = Pcat(t_{\uparrow C.stops})/c$ from 2.1.3, 2.1.4 and msa

2.1.6. $Pcat(t_{\uparrow C.stops}) = X$ from msb

Case 2: $Pcat(t_{\uparrow C.stops}) = X$

2.2.1. $(t_{\downarrow C.running} - t_{\uparrow C.running}) = X/c$ from $ro3$ and $pl1$

2.2.2. $(t_{\downarrow C.running} - t_{\uparrow C.running}) = X/m + d$ from msb

2.2.3. $(t_{\downarrow C.running} - t_{\uparrow M.running}) = X/m$ from msa

2.2.4. $Pmouse(t_{\downarrow C.running}) = X$ from 2.2.3, Case 2, $pl2$ and $pl3$

2.2.5. $Pmouse(t_{\uparrow C.stops}) = X$ from 2.2.4, and $ro3$

Fig. 2. Proof of Lemma 4.1

We will suppose that the controller has access to a sensor that can detect when the mouse starts to run and an actuator that can wake the cat; this sensor and actuator are represented by the following state variables.

No.	Name	Range	Comments
p_5	Sensor	$\{rest, run\}$	The sensor detects when the mouse starts to run.
p_6	Actuator	$\{off, on\}$	The actuator is set to on to awaken the cat.

Events
Next we identify the events which describe the behaviour at the interface.
Detect: this event occurs when the state of the sensor becomes *run*.

$$Detect = \Big|(Sensor = run)$$

Awake: this event occurs when the state of the actuator becomes *on*.

$$Awake = \Big|(Actuator = on)$$

Sensor/Actuator Relations
sr1. The event *Detect* occurs if and only if the event $\uparrow M.running$ occurs.

$$Detect \Leftrightarrow \uparrow M.running$$

ar1. The event $\uparrow C.running$ occurs if and only if the event *Awake* occurs.

$$\uparrow C.running \Leftrightarrow Awake$$

The sets of monitored and controlled variables of the physical process are $V_{PM} = \{Smouse\}$ and $V_{PC} = \{Scat\}$, and the sets of input and output variables for the controller are $V_{CI} = \{Detect\}$ and $V_{CO} = \{Awake\}$.

Robust Mission Strategy
Awake must occur d units after *Detect*.

RMS: $t_{Awake} = t_{Detect} + d$

Lemma 4.2. A history that satisfies the relations $sr1$ and $ar1$ and robust mission strategy RMS must satisfy the mission strategy MS.
Proof. Rule msa follows directly from $sr1$ and $ar1$. Hence if d is chosen to satisfy rule msb the MS follows.

As a result of this stage, the plant interface has been defined as consisting of a sensor that detects when a mouse starts to run and an actuator that awakens the cat. Also the relationship that must exist between the sensor and actuator, to realize strategy MS, has been established (RMS).

4.4 Controller Analysis

The controller analysis stage aims to specify the activities to be performed by the controller to realise, at the mission controller interface, the behaviour specified by the RMS. From the complete PrT net model of the cat and mouse system, shown in Figure 3, we can observe that the modelling of the physical process, interface, and controller can be separated. The advantage of adopting this approach is that the three models can be independently developed and modified, and the behaviour of the physical process seen to correspond to the sequence of control commands issued by the controller. As can be observed, the only activity performed by the controller is to delay the awakening of the cat by d, after the mouse has been released.

The predicates of the PrT net model of the cat and mouse system are:

$C.rest$ – the cat is at rest at the initial position;
$M.rest$ – the mouse is at rest at the initial position;
$C.running$ – the cat is running with speed c;
$M.running$ – the mouse is running with speed m;
$C.stops$ – the cat has stopped;
$M.stops$ – the mouse has stopped;
$S.in$ – the mouse is running with speed m;
$A.out$ – the actuator awakes the cat;
$S.run\langle s, tstp(S.run)\rangle$ – the sensor s detects that the mouse is running;
$A.run\langle a, tstp(A.run)\rangle$ – the controller signals the actuator a for the cat to awake.

The timing constraints *transMD* and *transCD* simulate, respectively, the durations that the mouse and the cat take to run over the specified distance. The other timing constraint, *transMCS*, is the specification which is part of the Mission Controller Specification that will be used in the design of the controller.

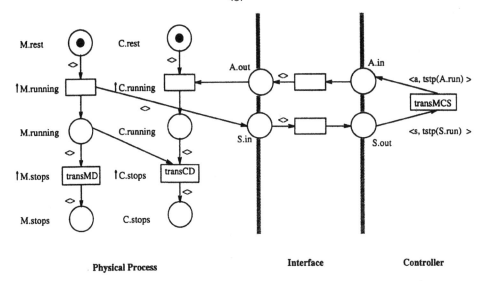

transMCS: tstp(A.run)=tstp(s.run)+d
transMD: tstp(M.stops)>tstp(M.running)+X/m
transCD: tstp(C.stops)≥tstp(C.running)+X/c

Fig. 3. PrT net model of the Cat and Mouse system

4.5 Discussion of Cat and Mouse Example

This example shows how levels of abstraction can be used to structure the requirements analysis into clearly defined stages. After an initial description of system behaviour in the PEA Model, a detailed model of the plant behaviour was "derived" from the high level model, by identification of the relevant system variables and states of the cat and mouse. A mission strategy was then formulated in terms of the plant variables, and its refinement into a specification over the sensors and actuators presented. This enabled a distinction to be maintained between the analysis of the physical laws and the properties of the sensors and actuators.

To obtain a simple and concise solution it was assumed that a perfect sensor and actuator were available. A more "realistic" analysis would be to consider the possibility of delays between the mouse starting to run and its detection by the controller and between the signal to awake being sent by the controller and the cat starting to run. The possibility of such delays would lead to the conclusion that the initial mission strategy (and mission aim) has to be modified, to take into account such limitations. For example, the mission aim could be modified to: the cat must catch the mouse near to the hole (i.e. in the range $[X - \Delta x, X]$).

5 Example 2: Chemical Batch Processing

A batch processing plant has a reaction vessel. The reaction requires a specified volume of chemical C_a and a specified volume of chemical C_b to be heated to a reaction temperature range for some specified time to produce chemical C_c. Furthermore, it is known that the product C_c is hazardous, hence the installed system must ensure that the production of C_c in the reaction vessel will not lead to an accident.

5.1 Conceptual Analysis

The system interacts with the environment by receiving reactants C_a and C_b and generating the product C_c. The physical process consists of one physical entity: a reaction vessel with three openings, O_a, O_b and O_c used to load chemicals C_a and C_b, and deliver the product C_c. From the possible accidents we consider only one:

AC_1: an explosion (Exp) in the vessel during the production of chemical C_c.

Modes of Operation
From the description of the problem a high level model of the behaviour can be specified by the following modes:

Empty$\langle\rangle$: the reaction vessel is empty.

Load$\langle\rangle$: reactant C_a is being loaded via O_a, or reactant C_b is being loaded via O_b.

Produce$\langle\rangle$: no chemicals are being loaded into the vessel and the reactants C_a and C_b are heated to the specified range to produce C_c.
Collect$\langle\rangle$: product C_c is collected and the vessel is emptied.

Rules of Operation
At this level the rules of operation specify the relationships between the modes.

ro1. The batch processing plant is represented in terms of the modes of operation:

$$\forall i \in \mathbf{N}_1 : Batch(i)\langle\rangle \equiv Empty(i)\langle\rangle \prec Load(i)\langle\rangle \prec Produce(i)\langle\rangle \prec Collect(i)\langle\rangle$$

5.2 Safety Plant Analysis

The plant analysis identifies the state variables that describe the behaviour of the physical process, the rules of operation and the physical laws. To specify the safety requirements, the hazards and safety constraints are identified and the safety strategy formulated. The controller can monitor the temperature of the vessel, and control the flow of C_a and C_b into the vessel and the flow of C_c out of the vessel.

Physical Variables
These variables are identified by examining the informal description of the modes. The type of chemical is denoted by $ch \in \mathbf{CH}$, $\mathbf{CH} = \{a, b, c\}$, the type of opening by $f \in \mathbf{F} = \{a, b, c\}$. The ranges for the temperature and volume of the chemicals in the vessel are given by the sets $\mathbf{TR} = \{x \in \mathbb{R} \mid T_l \leq x \leq T_u\}$ and $\mathbf{VR} = \{x \in \mathbb{R} \mid 0 \leq x \leq Vm\}$ and the range for the flow rate through opening O_f by $\mathbf{FR}_f = \{x \in \mathbb{R} \mid 0 \leq x \leq Fm_f\}$. The behaviour of the physical process is captured by the state variables, described in the table below.

No.	Name	Range	Comments
p_1	Temp	TR	The temperature of the contents of the vessel.
p_2	Flow	$FR_a \times FR_b \times FR_c$	The flow rates into and out of the vessel.
p_3	Vol	VR^3	The volumes of the chemicals.

Actions and States

Next we identify the actions and states that correspond to the modes of operation, these would be defined in terms of predicates over the physical variables. In the following for simplicity (and since we are primarily concerned with safety) we present only the informal definitions of the actions and states.

Empty$\langle\rangle$: In this state, the vessel is empty and no chemicals are flowing into or out of the vessel; and the temperature is below the activation value.

Load$\langle\rangle$: When this action starts the reactant C_a is being loaded via O_a or C_b is being loaded via O_b, and finishes when both reactants have been loaded.

Produce$\langle\rangle \equiv Heat\langle\rangle \prec React\langle\rangle$

Heat$\langle\rangle$: When this action starts there is no flow into or out of the vessel and the vessel contains some C_a and some C_b but no C_c; this action is finished when the temperature reaches *ActTemp* (the temperature at which the reaction between C_a and C_b is activated).

React$\langle\rangle$: When this action starts the temperature is at least *ActTemp*; this action is finished when there is no C_a or no C_b (i.e. the reaction is complete).

Collect$\langle\rangle$: When this action starts the vessel contains some C_c; this action is finished when the vessel is empty.

Hazards

A hazard analysis is performed for each action/state:

Empty, *Load*, *Heat*: An explosion cannot occur during these actions/states since the vessel does not contain any chemical C_c.

React, *Collect*: When the system is performing these actions the vessel contains some chemical C_c. Further let us suppose that the hazard analysis for both actions identifies the following hazard: an explosion may occur during the reaction or collection of C_c only if the temperature is greater than *ExpTemp*. This hazard can be expressed, in terms of the physical variables, by the system predicate:

HZ_1: $Temp > ExpTemp$

Safety Constraint

To construct the safety constraint for the reaction vessel we firstly negate the hazard:

$\neg HZ_1$: $Temp > ExpTemp$

Then we modify $\neg HZ_1$ to incorporate a safety margin on the temperature of the vessel; this is achieved by replacing *ExpTemp* by *CTemp* ($= ExpTemp - \Delta Temp$, where $\Delta Temp$ is a tolerance factor):

SC_1: $Temp \leq CTemp$

Physical Law
We introduce the following physical law over the constrained variable *Temp*.

pl1. ΔTm is the maximum rise in the temperature per second.

$$(Temp(T_1) - Temp(T_0)) < \Delta Tm \times (T_1\text{-}T_0)$$

Safety Strategy
There are two possible strategies.

SS_1: empty the contents of the vessel into a cooled vat.
SS_2: cool the chemicals while they are in the vessel.

For this case study, let us suppose that due to the structure of the chemical plant, only the first option (SS_1) is feasible.

The safety strategy SS_1 is defined by the following five rules, with respect to the notion of the vessel being drained, through a safety valve O_s:

ssa. The vessel can only start to drain if the temperature exceeds *HiActTemp*.

ssb. If the temperature exceeds *DTemp* and the vessel is not empty, the vessel must be being drained.

ssc. The vessel can only stop draining when the vessel is empty.

ssd. The vessel must stop draining before the temperature reaches *CTemp*.

sse. When the vessel is being drained there must be no flow via O_a, O_b and O_c.

The threshold *DTemp* is related to the time required to empty the vessel; the threshold *HiActTemp* is an upper bound on the expected temperature during the actions *React⟨⟩* and *Collect⟨⟩*. Obviously for the strategy to be feasible:

$$HiActTemp < DTemp < CTemp$$

The model of the plant is extended to include the flow rate out of the opening O_s into the cooled vat. This is achieved by extending the set **F** to $\{a, b, c, s\}$ and redefining *Flow* as below. (In the following *Volume* is used as an abbreviation for the sum of the volumes of C_a, C_b and C_c.)

No.	Name	Range	Comments
p_2	*Flow*	$FR_a \times FR_b \times FR_c \times FR_s$	The flow rates into and out of the vessel.

Physical Law
pl2. The vessel can be emptied via outlet O_s in a time of ΔEmp, provided there is no flow via O_a, O_b or O_c and the flow via outlet O_s is at least *Sflow(Volume)*.

$$T_1 - T_0 \geq \Delta Emp \wedge (\forall f \in \{a, b, c\}: Flow(f) = 0) \wedge Flow(s) \geq Sflow(Volume)$$
$$\Rightarrow Volume(T_1) = 0$$

The physical law $pl2$ is used to identify the following relation over the drain temperature threshold ($DTemp$).

R_1: $DTemp < CTemp - \Delta Tm.\Delta Emp$

Safety Strategy: SS$_1$
The concept of the vessel being drained is defined by the following state.

$Drain\langle\rangle$: In this state, the flow rate via O_s is at least $Sflow(Volume)$.

$Drain\langle\rangle(Flow(s) \geq Sflow(Volume))$

We then formulate the five rules of the safety strategy with respect to $Drain\langle\rangle$.

ssa. Draining can start only if the temperature is above $HiActTemp$.

$\uparrow Drain \Rightarrow Temp > HiActTemp$

ssb. If the temperature exceeds $DTemp$ and the vessel is not empty it must be draining.

$Temp > DTemp \wedge Volume \neq 0 \Rightarrow Drain$

ssc. Draining can finish only if the vessel is empty.

$\downarrow Drain \Rightarrow Volume = 0$

ssd. This rule is formalized as a bound on the duration of draining:

$Drain\langle T_0, T_1 \rangle \Rightarrow T_1 - T_0 < \frac{CTemp - DTemp}{\Delta Tm} (= \Delta Exp)$

sse. During $Drain\langle\rangle$ there is no flow via O_a, O_b and O_c.

$Drain \Rightarrow \forall f \in \{a, b, c\}: Flow(f) = 0$

Lemma 4.3. A history that satisfies physical law $pl1$ and the safety strategy SS$_1$ must satisfy the safety constraint SC$_1$.

Proof. The proof follows from the following observations. If $Temp \leq DTemp$ then SC$_1$ is satisfied, since $DTemp < CTemp$ (from relation R_1). If $Temp > DTemp$ the vessel must be being drained (from rule ssb) and must be emptied into the cooled vat before the temperature reaches $CTemp$ (from rules ssc, ssd, sse and $pl1$).

Safety Specifications

The results of the Safety Plant Analysis are summarised in the Safety Specification Hierarchy (SSH) in Figure 4. The SSH depicts the decision to simplify the safety analysis by specifying a common hazard for the actions $React\langle\rangle$ and $Collect\langle\rangle$, it also records the decision to discard the safety strategy SS$_2$.

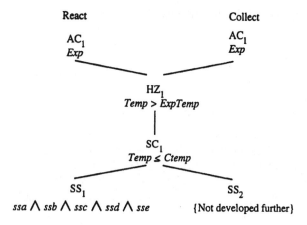

Fig. 4. Result of safety plant analysis

5.3 Safety Interface Analysis

Here the analysis identifies the variables that describe the behaviour of the sensors and actuators, and their relationship to the physical variables being monitored and controlled. These relationships are then used to develop a robust safety strategy.

The safety controller monitors the temperature via a thermometer (*Therm*). The flow rates via O_a, O_b and O_c are controlled by three locks ($Lock(l), l \in \mathbf{L}, \mathbf{L} = \{a, b, c\}$) and the flow rate via O_s by a safety valve (*Svalve*). A lock can only stop the flow, a valve can start or stop the flow. The variables that model the state of the sensors and the actuators are described below.

No.	Name	Range	Comments
p_4	*Therm*	$\{T_l, T_l + \Delta q, \ldots, T_l + Q.\Delta q\}$	The digital thermometer reading.
p_5	*Lock*	$\{on, off\}^3$	The state of the locks for O_a, O_b and O_c.
p_6	*Svalve*	$\{open, close\}$	The state of the safety valve.

Sensor/Actuator relations The relationship between the sensors/actuators and the properties of the physical process being monitored/controlled are given below.

*sr*1. If the temperature is within the range of *Therm*, the thermometer reading is always within ΔTp of the temperature.

$$T_l \leq Temp \leq T_l + Q.\Delta q \Rightarrow \mid Therm - Temp \mid \leq \Delta Tp$$

[It is assumed that all temperatures that need to be measured are within the range of *Therm*.]

*ar*1. If $Lock(l)$ is set the flow rate via O_l is zero.

$$\forall l \in \mathbf{L}: Lock(l) = on \Rightarrow Flow(l) = 0$$

*ar*2. If *Svalve* is closed the flow rate via O_s is zero.

$Svalve = close \Rightarrow Flow(s) = 0$

*ar*3. If *Svalve* is open the the flow rate via O_s is greater than *Sflow*(*Volume*) or the vessel is empty.

$Svalve = open \Rightarrow Flow(s) > Sflow(Volume) \vee Volume = 0$

*ar*4. For any two time points T_0, T_1, if the duration of $[T_0, T_1]$ is at least ΔEmp and for all time points t during $[T_0, T_1]$ the locks are on and *Svalve* is open the vessel must be empty by T_1 (c.f. physical law *pl*2).

$T_1 - T_0 \geq \Delta Emp \wedge \forall t \in [T_0, T_1]: \forall l \in L: Lock(l)(t) = on \wedge Svalve(t) = open$
$\Rightarrow Volume(T_1) = 0.$

The sets of monitored and controlled variables of the physical process are $V_{PM} = \{Temp\}$ and $V_{PC} = \{Flow\}$; the sets of input and output variables for the controller are $V_{CI} = \{Therm\}$ and $V_{CO} = \{Lock, Svalve\}$.

Robust Safety Strategy: RSS$_1$ The robust safety strategy is specified by modifying the safety strategy to tolerate the inaccuracy of the thermometer. That is, thermometer thresholds *HiActTherm* and *DTherm* are introduced for *HiActTemp* and *DTemp*:

R_2: $HiActTherm = HiActTemp + \Delta Tp$
R_3: $HiActTherm < DTherm = DTemp - \Delta Tp$

At the level of the interface, the notion of the vessel being drained via O_s is defined by the following state over *Svalve*.

OpenS$\langle\rangle$: In this state, *Svalve* is open.

$OpenS\langle\rangle(Svalve = open)$

We then define the following rules.

rssa. The safety valve can be opened only if the thermometer reading is above *HiActTherm*.

$\uparrow OpenS \Rightarrow Therm > HiActTherm$

rssb. If the thermometer reading is at least *DTherm* then the safety valve must be open.

$Therm \geq DTherm \Rightarrow OpenS$

rssc. The safety valve must be opened for a duration of at least ΔEmp.

$OpenS\langle T_0, T_1 \rangle \Rightarrow T_1 - T_0 \geq \Delta Emp$

rssd. When the safety valve is open the locks must be on.

$$OpenS \Rightarrow \forall l \in \mathbf{L} : Lock(l) = on$$

Lemma 4.4. A history that satisfies the sensor relation $sr1$, the actuator relations $ar1$ to $ar4$, relations R_1 to R_3 and RSS_1 must satisfy SS_1.

Proof. We consider each rule of the safety strategy SS_1 separately. In the proof we make use of the following assertions.

1. $\neg OpenS \Rightarrow \neg Drain$ from $ar2$ and definitions of $OpenS\langle\rangle$ and $Drain\langle\rangle$
2. $OpenS \wedge Volume \neq 0 \Rightarrow Drain$

                 from $ar4$ and definitions of $OpenS\langle\rangle$ and $Drain\langle\rangle$

ssa. $\uparrow Drain \Rightarrow Temp > HiActTemp$	from 1, rule $rssa$ and R_2
ssb. $Temp \geq DTemp \wedge Volume \neq 0 \Rightarrow Drain$	from 2, rule $rssb$ and R_3
ssc. $\downarrow Drain \Rightarrow Volume = 0$	from 2, $rssc$, $rssd$ and $ar3$
ssd. $Drain\langle T_0, T_1\rangle \Rightarrow T_1 - T_0 < \Delta Exp$	from 1, 2, $rssc$, $rssd$, $ar4$ and R_1
sse. $Drain \Rightarrow \forall f \in \{a, b, c\} : Flow(f) = 0$	from $rssd$ and $ar1$

5.4 Safety Controller Analysis

During the safety controller analysis the activities to be performed by the safety controller are specified in terms of the specifications provided by the RMS. The PrT net model of the chemical batch processing system is shown in Figure 5, and includes the models of the physical process, interface, and the controller.

The predicates of the PrT net model of the chemical batch processing are the following:

$Empty\langle Flow(a), Flow(b), Flow(c), Vol(a), Vol(b), Vol(c), Temp\rangle$ - the reaction vessel is empty;

$Load\langle Flow(a), Flow(b), Flow(c), Vol(a), Vol(b), Vol(c), Temp\rangle$ - the reactants are being loaded;

$Heat\langle Flow(a), Flow(b), Flow(c), Vol(a), Vol(b), Vol(c), Temp\rangle$ - the temperature of the reactants is being raised to the activation temperature;

$React\langle Flow(a), Flow(b), Flow(c), Vol(a), Vol(b), Vol(c), Temp\rangle$ - the chemical reaction is in progress;

$Collect\langle Flow(a), Flow(b), Flow(c), Vol(a), Vol(b), Vol(c), Temp\rangle$ - the product is collected and the vessel emptied;

$OverHeat\langle Flow(a), Flow(b), Flow(c), Flow(s)\rangle$ - the temperature of the reactor rises above a specified threshold;

$Drain\langle Flow(a), Flow(b), Flow(c), Flow(s)\rangle$ - the contents of the vessel are being drained.

$ST.in\langle Temp\rangle$ - the temperature is read from the vessel;

$ST.out\langle Therm\rangle$ - the sensor provides the temperature to the controller;

$AL.in\langle Lock(a), Lock(b), Lock(c)\rangle$ - the controller signals the actuator to lock O_a, O_b and O_c;

$AL.out\langle Flow(a), Flow(b), Flow(c)\rangle$ - there is no flow through O_a, O_b and O_c;

$AV.in\langle Svalve\rangle$ - the controller signals the actuator to open O_s;

$AV.out\langle Flow(s)\rangle$ - the actuator allows flow through O_s.

Only those controller activities related to the safety of the system are modelled. In the model we assume that only those temperatures that are higher than a predefined temperature threshold are read by the sensor. Once such a temperature is detected, the controller requests for the locks to be set, and opens the safety valve after receiving a interlocking signal from the lock actuators. The relations on $t1$ and $t2$ simulate whether the temperature of the vessel exceeds a predefined threshold.

5.5 Discussion of Chemical Batch Example

This example illustrates some of the support the proposed methodology provides for the analysis of the safety requirements. The identification of the top level actions/states enabled the hazard analysis to be focused on the actions $React\langle\rangle$ and $Collect\langle\rangle$. The model of the chemical plant enables the hazards to be directly encoded as a predicate. The SSH records the decision to discard the safety strategy SS_2, and depicts the relationship between the produced safety specifications.

This example also shows how the formal analysis can be targeted on the safety critical behaviour of the system. The analysis during the safety interface analysis and safety controller analysis was restricted to those components required to maintain the safety strategy SS_1. It should also be noted that to be effective, in an industrial context, the analysis should be performed over the whole system. Here the analysis has focused only on one accident – an explosion occurring in the reaction vessel. The possibility of an explosion in the cooled vat has not been considered, for example.

6 Preliminary Evaluation of the Methodology

Our intention in this section of the paper is to provide a preliminary evaluation of our approach in terms of the goal oriented criteria presented in Section 1.

1. Consistency with Current Practice: The basic system structure underlying our approach is taken directly from the process-control sector. Because our staged approach develops the requirements in a "top-down" fashion, the results of hazard analyses (at system or plant level) can feed directly into the appropriate stage of our requirements methodology. The notion of "modes" was introduced specifically to mirror the very common practice of structuring a group of system states to represent a set of closely linked activities, frequently referred to as modes of operation by the engineering community. The plant interface is inevitably modelled in terms of the properties and behaviour of sensors and actuators; however, requirements at this level are usually expressed in much greater detail than we have adopted (i.e. experience of practical systems indicates a considerable amount of ad hoc design being included at the requirements stage).

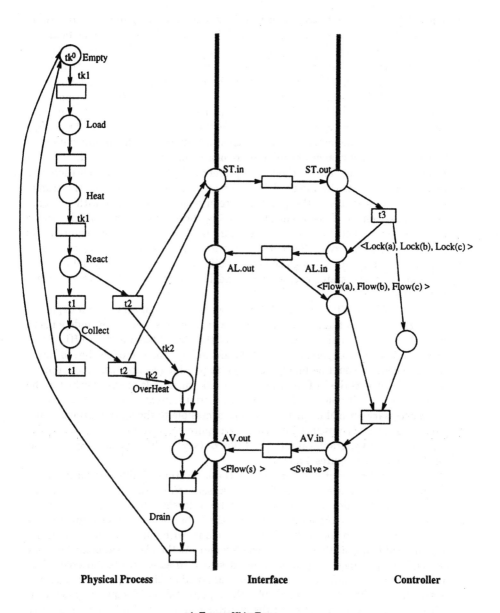

t1: Temp ≤ HiActTemp
t2: Temp > HiActTemp
t3: ∀l: Lock(l)=on
tk1: <Flow(a), Flow(b), Flow(c), Vol(a), Vol(b), Vol(c), Temp>
tk2: <Flow(a), Flow(b), Flow(c), Flow(s)>
tk⁰ = {<0,0,0,0,0,0, Temp⁰>}

Fig. 5. PrT net model of the Chemical Plant Example

2. Traceability: The key to traceability is the maintenance of a structure recording the linkages between separate elements of the results of the requirements analysis. This is embodied in the Safety Specification Hierarchy. This record provides support in tracing the consequences of change in the safety specifications. The Safety Specification Hierarchy promotes the localization of side-effects, identifying the safety specifications that must be modified, and the conditions that must be reconfirmed. Currently this localization would be performed manually, however tool support could be provided. Furthermore, the structure of the overall framework of the methodology can assist by identifying the appropriate level at which proposed changes in the requirements must be incorporated.

3. Exploit Structuring: The methodology presented here benefits from structuring in two ways. First, the range of levels of abstraction employed in the framework allows us to begin the process of requirements definition with a highly abstract statement of requirements, covering the entire system, which can be progressively refined by incorporating more detailed and specific requirements at the subsequent levels. Second, the separation of safety requirements from mission requirements – though sometimes difficult – is highly beneficial in terms of simplifying the statement, analysis and later certification of system safety properties.

4. Notation Accessibility and Suitability: Most control engineers have a degree of familiarity with mathematical models in which time is usually an independent variable, and this is also a characteristic of the PEA Model and of THL. Some engineers also have experience of state based and net based models, but for maximum benefit it would be highly desirable to provide (economical) tool support for models such as the PrT net model; currently such tools are established in the research community but much less so in industrial contexts. Tool support is even more necessary for making net formalisms accessible to the system procurer and user. It is certainly possible to envisage a requirements validation system based on a user-friendly, graphical interpretation of (say) a PrT net in a form similar to that shown in Figures 3 and 5.

A key characteristic of our approach is the utilisation of different formal techniques at different levels of the framework, selected on the basis for the anticipated characteristics of requirements at these levels. Thus, at the conceptual level, informal and more intuitive notions are to be elicited and embodied in the semi-formal notation of the PEA Model (a model with numerous virtues, including those of being small, green and wholesome). At the plant and plant interface levels, THL provides a logical notation designed to assist in the representation, refinement and verification of stipulations imposed on the temporal behaviour of a system. The PrT net model enables the representation of requirements on the high level architectural design of the controller, in terms of interaction between the controller and set of sensors and actuators.

In a future version of this paper we intend to compare a range of other approaches to requirements expression and analysis, including that emanating from the ProCos project [7]. In terms of procedures to localize faults in requirements specifications, general correctness criteria have been suggested for process control systems [10]. However, this is a technique to validate requirements specifications rather than a methodology to produce them. This discards issues such as, how the specifications are obtained and what methods are employed. The same observation could be applied to the work of Parnas & Madey [15], where a systematic way for documenting the development of computer systems is discussed. Although this approach adopts similar abstraction levels to the proposed methodology, in their work they only focus on how the specifications should be represented, but says nothing about which methods should be employed to obtain the specifications.

Acknowledgements: The authors would like to acknowledge the financial support of British Aerospace PLC (DCSC), and CAPES/ Brazil.

References

1. Tom Anderson, Rogério de Lemos, John Fitzgerald, and Amer Saeed. On Formal Support for Industrial-scale Requirements Analysis. Technical Report TR412, University of Newcastle, NE1 7RU, UK, 1992.
2. A. M. Davis. *Software Requirements: Analysis & Specification*. Prentice-Hall, 1990.
3. R. de Lemos, A. Saeed, and T. Anderson. A Train set as a Case Study for the Requirements Analysis of Safety-Critical Systems. *The Computer Journal*, 35(1):30–40, February 1992.
4. R. de Lemos, A. Saeed, and T. Anderson. Analysis of Timeliness Requirements in Safety-Critical Systems. In J. Vytopil, editor, *Proceedings of the Symposium in Formal Techniques in Real-Time and Fault-Tolerant Systems*, volume 571 of *Lecture Notes in Computer Science*, pages 171–192. Springer-Verlag, 1992.
5. J. S. Fitzgerald and C. B. Jones. Modularizing the Formal Description of a Database System. In D. Bjorner, C.A.R. Hoare, and H. Langmaack, editors, *VDM'90: VDM and Z - Formal Methods in Software Development.*, volume 428 of *Lecture Notes in Computer Science*, pages 189–210. Springer-Verlag, 1990.
6. H. Genrich. Predicate/Transition Nets. In W. Brauer, W. Reisig, and G. Rozemberg, editors, *Petri Nets: Central Models and their Properties*, volume 254 of *Lecture Notes in Computer Science*, pages 206–247. Springer-Verlag, 1987.
7. K. M. Hansen, A. P. Ravn, and H. Rischel. Specifying and Verifying Requirements of Real-Time Systems. In *Proceedings of the ACM SIGSOFT'91 Conference on Software for Critical Systems, New Orleans, Louisiana*, pages 44–54, December 1991.
8. D. Harel. Statecharts: A visual formalism for complex systems. *Science of Computer Programming*, 8:231–274, 1987.
9. D. Harel, H. Lachover, A. Naamad, and A. Pnueli. Statemate: A Working Environment for the Development of Complex reactive Systems. *IEEE Transactions on Software Engineering*, 16(4):403–414, 1990.
10. M. S. Jaffe, N. G. Leveson, M. P. E. Heimdahl, and B. E. Melhart. Software Requirements Analysis for Real-Time Process-Control Systems. *IEEE Transactions on Software Engineering*, SE-17(3):241–258, March 1991.
11. F. Jahanian and A. K. Mok. Safety Analysis of Timing Properties in Real-Time Systems. *IEEE Transactions on Software Engineering*, SE-12(9):2–13, September 1986.
12. H. G. Lawley. Hazard and Operability Studies. *Chem Eng Process*, 8(5):105–116, May 1973.
13. O. Maler, Z. Manna, and A. Pnueli. A formal approach to hybrid systems. In *Proceedings of the REX Workshop "Real-Time: Theory in Practice"*, volume 600 of *Lecture Notes in Computer Science*. Springer-Verlag, 1992.
14. L. E. Moser and P. M. Melliar-Smith. Formal Verification of Safety-Critical Systems. *Software Practice and Experience*, 20(8):799–821, August 1990.

15. D. L. Parnas and J. Madey. Functional Documentation for Computer Systems Engineering (Version 2). Technical Report CRL Report No. 237, TRIO, McMaster University, Hamilton, Ontario, 1991.

16. A. Saeed, T. Anderson, and M. Koutny. A Formal Model for Safety-Critical Computing Systems. In *SAFECOMP'90, London, UK*, pages 1–6, October 1990.

17. A. Saeed, R. de Lemos, and T. Anderson. The Role of Formal Methods in the Requirements Analysis of Safety-Critical Systems: a Train Set Example. In *Proceedings of the 21st Symposium on Fault-Tolerant Computing, Montreal, Canada*, pages 478–485, June 1991.

18. A. Saeed, R. de Lemos, and T. Anderson. An Approach to the Assessment of Requirements Specifications for Safety-Critical Systems. Technical Report No. 381, Computing Laboratory, University of Newcastle upon Tyne, 1992.

19. United Kingdom Ministry of Defence. *Interim Defence Standard 00-55: Procurement of Safety Critical Software in Defence Equipment*, 1991.

A Notation for the Predicate Event/Action Model

For the analysis of timeliness requirements we introduce a notation based on an event/action model (PEA Model) that can be employed at the different levels of abstraction. This model is based on the work of Jahanian & Mok [11], with some features that make it more applicable to requirements analysis. Specifically, the model is intended to support both discrete and dense time structures [4], has the potential of obtaining both descriptive and operational semantics, and has the ability to depict the timing analysis graphically. An overview of the model is given here.

The predicate event/action model (PEA Model) permits the analysis of the discrete behaviour of a system defined over a set of system variables. The events, actions and states are defined in terms of system predicates (i.e. predicates over the system variables).

Primitives

As our *time domain*, we adopt a structure of the form $\langle \mathbf{T}, <, 0 \rangle$, where $\mathbf{T}$ is a set of points, and $<$, a strict total order on $\mathbf{T}$ with least point 0 [4]. The PEA model consists of three basic notions: *events, states* and *actions*.

Events We use the notion of an event as a general temporal marker of no duration represented as a cut in the time line. For each event E we associate an event-occurrence function $E(t, i)$ that returns true if and only if the ith occurrence of the event occurs at time point t. For any event-occurrence function $E(t, i)$ we state the following properties: any occurrence number is marked by at most one time point (denoted as t_E^i) and the time points marking the events are ordered in accordance with occurrence numbers.

We will use $E = \big|(P)$ to denote the assertion that there exists an i such that $E(t, i)$ is true *iff* the system predicate P becomes true at time point t.

States We use the notion of a state as a means to group together system behaviour characterized by a system predicate P. For each state S we introduce a state-holding function $S(t, i)$ that returns true for time point t if and only if t is contained within the ith interval in which P holds.

State-predicate: For each state holding function S we define a state predicate $S(i)\langle t_1, t_2 \rangle$ which is true if and only if $S(t, i)$ is true for all time points t in $[t_1, t_2)$.

$$\forall i \in N_1 : \forall t_1, t_2 \in T : [S(i)\langle t_1, t_2 \rangle \Leftrightarrow \forall t \in T : (t \in [t_1, t_2) \Leftrightarrow S(t, i))]$$

Start and Finish events: With each state we associate *start* and *finish* events to mark the boundaries of the ith time the system enters the state.

$$\forall i \in N_1 : \forall t_1, t_2 \in T : \uparrow S(i, t_1) \wedge \downarrow S(i, t_2) \Leftrightarrow S(i)\langle t_1, t_2 \rangle$$

We use $S\langle\rangle(P)$ to denote the assertion that there exists an i such that $S(t, i)$ iff the system predicate P is true at time point t.

Actions We use the notion of an *action* as a basic unit of activity. For each action A we introduce an *action-execution* function $A(t, i)$ which returns true if and only if time point t is contained within the ith execution of the action.

As for states we introduce the notion of an action-predicate $A(i)\langle t_1, t_2 \rangle$ and start and finish events ($\uparrow A(t, i)$ and $\downarrow A(t, i)$) to mark the boundaries of an action.

We use $A\langle\rangle(P, Q)$ to denote the assertion that there exists an i such that $A(t, i)$ iff there exits two time points t_1 and t_2 such that P becomes true at t_1 and t_2 is the first time point after t_1 at which Q becomes true and t is in $[t_1, t_2)$.

Operators

In the following formulae $p\langle\rangle$ and $q\langle\rangle$ are action-predicates or state-predicates.
Meet:

$$\forall t_1, t_2, t_3, t_4 \in T : p\langle t_1, t_2 \rangle \prec q\langle t_3, t_4 \rangle \equiv p\langle t_1, t_2 \rangle \wedge q\langle t_3, t_4 \rangle \wedge t_2 = t_3$$

Overlap:

$$\forall t_1, t_2, t_3, t_4 \in T :$$
$$p\langle t_1, t_2 \rangle \parallel q\langle t_3, t_4 \rangle \equiv p\langle t_1, t_2 \rangle \wedge q\langle t_3, t_4 \rangle \wedge (t_1 \le t_3 \le t_2 \vee t_3 \le t_1 \le t_4)$$

Conjunction:

$$\forall t_1, t_2 \in T : (p \wedge q)\langle t_1, t_2 \rangle \equiv$$
$$\exists t_3, t_4, t_5, t_6 \in T : p\langle t_3, t_4 \rangle \wedge q\langle t_5, t_6 \rangle \wedge t_1 = max(t_3, t_5) \wedge t_2 = min(t_4, t_6)$$

Disjunction:

$$\forall t_1, t_2 \in T : (p \vee q)\langle t_1, t_2 \rangle \equiv$$
$$\exists t_3, t_4, t_5, t_6 \in T : p\langle t_3, t_4 \rangle \parallel q\langle t_5, t_6 \rangle \wedge t_1 = min(t_3, t_5) \wedge t_2 = max(t_4, t_6)$$

B   Timed History Logic

THL is a formalism based on the time domain $\langle \mathbb{R}_+, <, 0 \rangle$ and consists of three main concepts: *histories, relations* and *modes.* Here we present an overview of histories and relations; a more detailed description of the model is given elsewhere [16]. In the following we use the term interval to mean closed convex interval.

Over any interval the following functions are defined: *start point* $s(\mathbf{Int})$ – the earliest time point in $\mathbf{Int}$; *end point* $e(\mathbf{Int})$ - the latest time point in $\mathbf{Int}$; and *interval set* $SI(\mathbf{Int})$ – the set of all intervals contained within $\mathbf{Int}$. The system lifetime ($\mathbf{ST}$) is an interval which represents the operational lifetime of the system.

For a system with n state variables we have the state vector: $Sv = \langle p_1, \ldots, p_n \rangle$. The range of p_i is denoted by $\mathbf{V_{p_i}}$ and the state space by Γ. A history H of a system is a function of the form $H \colon \mathbf{ST} \to \Gamma$. The set of all "possible" histories of a system is defined as the universal history set $\Gamma\mathbf{H}$. For a history H the sequence of values taken by a state variable p_i is denoted by the function $H.p_i \colon \mathbf{ST} \to \mathbf{V_{p_i}}$.

Invariant relations are used to express relationships over the state variables which hold at every time point within $\mathbf{ST}$; these are formulated as *system predicates.*

A *system predicate* is a predicate built using n free value variables $p_1, \ldots, p_n$ of types $\mathbf{V_{p_1}}, \ldots, \mathbf{V_{p_n}}$. No other free variables may be used. A tuple of values $V = \langle x_1, \ldots, x_n \rangle$, where x_i is of type $\mathbf{V_{p_i}}$, satisfies a system predicate P if and only if substitution of each x_i for p_i within $P(x_1, \ldots, x_n)$ evaluates to true. This is denoted by: $P(V)$. We will denote $P(H.p_1(t), \ldots, H.p_n n(t))$ by $H\mathbf{sat}P@t$. A history H satisfies an invariant relation P iff: $\forall t \in \mathbf{ST} \colon H\mathbf{sat}P@t$.

History relations are used to express relationships over the state variables which hold during every interval included within $\mathbf{ST}$; these are formulated as history predicates.

A history predicate is a predicate built using two free time variables T_0, T_1 and n free function variables $p_1, \ldots, p_n$. No other free variables may be used. A history H satisfies a history predicate HP for an interval $\mathbf{Int}$ if and only if the expression resulting from substituting $s(\mathbf{Int})$ for T_0; $e(\mathbf{Int})$ for T_1 and $H.p_i$ for p_i for all i, evaluates to true. This is denoted by: $H\mathbf{sat}HP@\mathbf{Int}$. A history H satisfies an history relation HP iff: $\forall \mathbf{Int} \in SI(\mathbf{ST}) \colon H\mathbf{sat}HP@\mathbf{Int}$.

The PEA Model can be represented in THL by introducing state variables to represent the event-occurrence, action-execution and state-holding functions.

Within THL a state-holding function S has the signature $S \colon T \to \mathbb{B} \times \mathbf{N_1}$. To be consistent with the PEA Model we adopt the following convention.

$$\forall H \in \Gamma\mathbf{H} \colon \forall i \in \mathbf{N_1} \colon \forall t \in \mathbf{ST} \colon H.S(t, i) \triangleq H.S(t) = (\mathsf{true}, i)$$

Similar conventions are adopted for event-occurrence functions and action-execution functions.

The set of histories of the system are then restricted by imposing conditions over the state variables that represent functions of the PEA Model. For example a history H is well-defined for a state S characterized by a system predicate P iff:

$$\forall t \in \mathbf{ST} \colon \exists i \in \mathbf{N_1} \colon H.S(t, i) \Leftrightarrow H\mathbf{sat}P@t$$

In the analysis of a system we are only concerned with those histories that are well defined, for the identified events, states and actions.

A Formal Approach to Computer Systems Requirements Documentation

Marcin Engel, Marcin Kubica, Jan Madey[1], David Lorge Parnas[2], Anders P. Ravn[3], A. John van Schouwen[4]

[1] Institute of Informatics, Warsaw University
Banacha 2, 02-097 Warsaw, Poland
E-mail: madey@mimuw.edu.pl
[2] Communications Research Laboratory
Department of Electrical and Computer Engineering
McMaster University
Hamilton, Ontario, Canada L8S 4L7
E-mail: parnas@triose.eng.mcmaster.ca
[3] Department of Computer Science
Technical University of Denmark, Bldg. 344
DK 2800 Lyngby, Denmark
E-mail: apr@id.dth.dk
[4] Bell-Northern Research Limited
P.O. Box 3511 Station C
Ottawa, Ontario, Canada K1Y 4H7 E-mail: schouwen@bnr.ca

Abstract. This paper demonstrates how the extended duration calculus [4] can be used to support the approach to documentation of computer systems presented by in [1]. This approach uses the general concept of mathematical relations to specify properties, while the calculus of durations provides the means to reason about such specifications, and in particular, prove formally that a design implies the requirements. The presentation is based on an example originally presented in [2], and later reformulated in [3] following the approach described on [1]. In the present paper we introduce all needed relations, express them in terms of duration calculus, and formally verify software design acceptability.

1 Introduction

Software design is carried out using methods quite unlike those used by engineers in the other disciplines. It often proceeds by building first and documenting afterwards. When software documentation is produced prior to implementation, it is typically inaccurate and imprecise. The documentation uses anthropomorphic analogies and intuitive language in a way that would be considered unprofessional in other disciplines. Although the long-term benefits of precise documentation are great, developers are reluctant to do better because good documentation requires substantial effort. Documenting a software system is seen as a diversion because the end-product is not being realized.

Precise documentation has many potential benefits:

1. designers know what they are building before proceeding with further design work;
2. reviewers can check that the specifiers' intentions meet their expectations;
3. those charged with quality assurance can verify an implementation against a statement of required behaviour and can formulate validation tests without looking at source code; and,
4. maintainers can become acquainted with the system more easily, and can maintain the original design intentions when making changes.

The logical starting point for any system design should be a description of the environment in which the system is to be used and the system's required behaviour within that environment. This information is recorded in what we call a *system requirements document*. The system is modelled as a state machine that monitors and controls certain aspects of its environment, i.e. as a dynamic system. The state of the system is a total function of time. Those aspects determine the *environmental quantities of interest*. There may be physical constraints, such as feedback relations and maximum derivatives, on the environmental quantities. The computer system will be required to further restrict these relations.

This paper focuses primarily on software systems. Usually before software design begins, certain decisions will have been made about the computer(s) and peripheral devices that will be used. Specifications for these should be documented so that the software designers can refer to details about them readily. These decisions are recorded in what we call the *system design document*. It describes the interfaces to the I/O devices by presenting the relations between the I/O registers and the environmental quantities. (New environmental quantities of interest might need to Be added to represent abstract device states for some of the devices or classes thereof.) Since the software designers will need to refer to the contents of both the system requirements and system design documents, both of them can be combined into a single one called the *software requirements document*. The software designer's job is to implement a mapping from computer inputs to computer outputs such that the required relation between the environmental quantities of interest is satisfied.

In this paper, we deal with a *functional* approach[5] to documenting the systems that we wish to build as well as the devices that will exist within the system. This approach is explained in Section 3 and illustrated on examples; a more detailed discussion of it can be found in [1]. In Section 2 we introduce briefly the Water Level Monitoring System from which the examples are derived. A complete specification and discussion of this system is presented in [2], while its reformulated, shorter version is to be found in [3]. In Section 4 we first informally introduce the extended calculus of durations [4,5], and then use it to describe the requirements and verify the software's acceptability. Conclusions are presented in Section 5. The Appendix provides a more detailed description

---

[5] More precisely, we use *binary relations* rather than functions. Since, however, a binary relation $R \subseteq A \times B$ can be treated as a function $R : A \to \mathcal{P}(B)$, where $\mathcal{P}(B)$ denotes the set of all subsets of B, the term "functional approach" is well justified.

of applying the extended calculus of durations to the main example [6].

2 The Water Level Monitoring System

The Water Level Monitoring System (WLMS) is a steam generator subsystem that monitors the water level in the steam generation vessel. The system controls, among other things: a display that informs the operator of the water level; a light that notifies the operator of the fact that the water level has exceeded an upper bound; and, a switch that, when open, prevents additional water from entering the vessel.

In this paper by "WLMS" we will understand an implementation of such a system, which is described in [2]. Among the monitored environmental quantities for the WLMS is *WaterLevel*, which denotes the level of water (measured in centimetres) in the horizontally-oriented vessel measured along the vertical axis on the left side (when looking from its front to its back), 5.0 cm from the front edge. Among the controlled environmental quantities are some display windows on a CRT screen, e.g. *LevelDisplay* (denoting the current level of water) and *HighWindow* (representing a visible alarm, a light).

Typically, an independent system would control the steam generation process. It would control The rate at which water is introduced into the system and the rate at which water is boiled off. The monitoring system's purpose is twofold: it must convey information about the system's state to an operator, and it must take certain safety actions in the face of control system failure.

Further discussion of WLMS is relegated to the examples.

3 A functional approach to system documentation

In this section we will discuss the mathematical relations that should be described in the three documents mentioned in the Introduction (i.e. the system requirements, system design, and software requirements documents). Let us first explain some important terms which will be used in the sequel.

Definition 1. Let $R \subseteq X \times Y$ be a (binary) relation, then:

- The set of pairs R could also be defined by its *characteristic predicate*, $R(x, y)$, i.e.:

$$R = \{(x, y) \in X \times Y \mid R(x, y)\}.$$

- The *domain* and the *range* of R can be expressed as follows:

$$\text{domain}(R) = \{x \in X \mid R(x, y) \text{ for some } y \in Y\},$$

$$\text{range}(R) = \{y \in Y \mid R(x, y) \text{ for some } x \in X\}.$$

---

[6] D. L. Parnas participated only in the first three sections of the paper.

- The *(relational) image* of an element x from the domain of R, $R(x)$, is the following subset of the range of R:

$$R(x) = \{y \in Y \mid R(x, y)\}.$$

- If for every x from the domain of R, $R(x)$ has exactly one element — R is thus a *function* — this element is called the *value* of R at x, and we can omit the set braces; in other words, $R(x)$ will denote the value of the function R at x, rather than a singleton set.
- The notion of image can also be extended to any subset A of X, as follows:

$$R(A) = \{y \in Y \mid R(a, y) \text{ for some } a \in A\},$$

hence:

- $\text{range}(R) = R(X) = R(\text{domain}(R))$,
- $(x \notin \text{domain}(R)) \Rightarrow (R(x) = \emptyset)$, i.e. the image is empty if R is not defined at x.

3.1 The system requirements document

A critical step in documenting the requirements of a computer system treated as a "black box" is the definition of the environmental quantities of interest (those to be measured and/or controlled). They include:

- physical properties (such as temperatures),
- the readings on user-visible displays,
- administrative information (such as the number of people assigned to a given task), and even
- the wishes of a human user.

The values of these environmental quantities change with time. Hence, each of them can be expressed as a *time function* (i.e. function of time) in the way that is usual in engineering (i.e. by modelling time as a set of real numbers). The association between the physical quantities of interest and their mathematical representations must be carefully defined. (One should remember that the environmental quantities of interest can include various aspects of hardware devices — typically introduced at the system design stage.) For the rest of this paper we will assume that the system S under consideration is known and so, for simplicity, we will omit the subscript S in some of the denotations related to this system.

Definition 2.

- The set of environmental quantities is divided into two subsets:
  - *controlled* quantities, whose values the system is intended to restrict, and
  - *monitored* quantities, those that are to affect the controlled ones.
- Let p be the number of quantities to be controlled in S, and F_C a selected set of time-functions[7]. Then:

---

[7] In this and following definitions each selected set of time-functions is determined by physical characteristic of the system S.

- controlled quantities will be represented by a p-tuple called a *controlled state function*, $c = (c_1, \ldots, c_p)$, where each c_i is a member of F_C and corresponds to the i-th controlled quantity,
- the set of controlled state functions will be denoted by C_S.
- Let q be the number of quantities to be monitored in S, and F_M a selected set of time-functions. Then:
  - monitored quantities will be represented by a q-tuple called a *monitored state function*, $m = (m_1, \ldots, m_q)$, where each m_i is a member of F_M and corresponds to the i-th monitored quantity,
  - the set of monitored state functions will be denoted by M_S.

Note, that if the same quantity is to be both monitored and controlled, the corresponding time-functions must be specified to be identical (by relation NAT, see below).

Example 1. For the WLMS we have:

1. Among the monitored quantities: the level of water measured as described in Section 2 and represented by the function *WaterLevel*.
2. Among the controlled quantities:
   (a) a visible annunciation signal whose window is labelled "WATER LEVEL HIGH". This aspect is represented by the function *HighWindow*, which is described as follows:
   $$HighWindow(t) = \begin{cases} on, & \text{if "light" is on at time } t \\ off, & \text{if "light" is off at time } t \end{cases}$$
   (b) the state of a switch. The function used to represent this is named *PumpSwitch* and is described as follows:
   $$PumpSwitch(t) = \begin{cases} closed, & \text{if switch contacts are closed at time } t \\ open, & \text{if switch contacts are open at time } t \end{cases}$$
   (c) the value presented in a display labelled "WATER LEVEL". The real-valued function used to represent this is named *LevelDisplay*.

Relation NAT. Nature and previously installed systems place constraints on the values of environmental quantities. These restrictions may be documented by means of a relation, which we call NAT.

Definition 3.

- NAT $\subseteq M_S \times C_S$,
- NAT(m, c) if and only if the environmental constraints allow the controlled quantities to take on the values described by c when the values of the monitored quantities are described by m.

One should note the following:

1. domain(NAT) (respectively, range(NAT)) contains *exactly* those instances of m (respectively, c) that are allowed by the environmental constraints.
2. If there are no constraints on m and c, then NAT $= M_S \times C_S$.

3. If any values of m are not included in the domain of NAT, the system designer may assume that these values do not occur.
4. NAT is rarely a function; if it were the computer system would not be able to vary the values of the controlled quantities without affecting changes in the monitored quantities (the values of the controlled quantities would be uniquely determined by the values of the monitored ones).

Example 2. A subset of the constraints on the domain of NAT for the WLMS is:

1. $| \frac{d WaterLevel(t)}{dt} | \leq 0.375$ cm/s; and
2. 0.0 cm $\leq WaterLevel(t) \leq 30.0$ cm.

Thus, there are no values of m in the domain of NAT when the water level rises or falls more rapidly than as stated in (1), nor when the water level exceeds 0.0 cm or 30.0 cm, as stated in (2).

Relation REQ. The computer system is required to impose further constraints on the environmental quantities. The permitted values may be documented by means of a relation, which we call REQ.

Definition 4.

- REQ $\subseteq M_S \times C_S$,
- REQ(m, c) if and only if the computer system may permit the controlled quantities to take on the values described by c when the values of the monitored quantities are described by m.

One should note the following:

1. domain(REQ) contains those instances of m which are allowed by the environmental constraints.
2. range(REQ) contains only those instances of c that are considered permissible.
3. REQ is usually not a function because the application can tolerate some errors in the values of controlled quantities.

Example 3. We can illustrate REQ as follows:
In the WLMS, the value conveyed by the "WATER LEVEL" display is required to reflect water level in the vessel. In addition, the operator must be given some indication that the display device is operating properly. Let the system be powered up at time 0. Then the requirement could be stated as follows:

$$LevelDisplay(t) = \begin{cases} 0.0 & \text{if } (0 \leq t < 4) \\ \lfloor t - 4 \rfloor \times 11.1 & \text{if } (4 \leq t < 14) \\ WaterLevel(t) \pm 1\% & \text{if } (t \geq 14) \end{cases}$$

Requirements feasibility. The requirements should specify behaviour for all cases that can arise, but at the same time should not demand the impossible. Hence:

Definition 5.

- The relation REQ is called *feasible with respect to* NAT if the following two conditions hold:
  1. domain(REQ) $\supseteq$ domain(NAT),
  2. domain(REQ $\cap$ NAT) = (domain(REQ) $\cap$ domain(NAT)).
- Note that two above conditions can be reduced to:
  3. domain(REQ $\cap$ NAT) = domain(NAT)

The first condition is quite obvious; equation (2) states that for each element m that is in the domains of both REQ and NAT, both relations must share at least one common element c to which this m can be mapped. Feasibility, in the above sense, means only that nature (as described by NAT) allows at least the required behaviour (as described by REQ); it does not mean that the functions involved are computable or that an implementation is practical.

Example 4. Let us consider NAT and REQ as illustrated in Examples 2 and 3. Note that the constraints given on the domain of NAT do not suggest that NAT defines any dependence between *WaterLevel* and *LevelDisplay*. If NAT contains all pairs of values (x, y) such that x belongs to the set of values of *WaterLevel* and y belongs to the set values of *LevelDisplay*, then domain(NAT) equals domain(REQ) and REQ is feasible with respect to NAT.

3.2 The system design document

During the system design stage we use a "clear-box" approach, i.e. we identify the computers within the computer system and describe how they communicate, with emphasis on the peripheral devices. Hence, new sets of quantities are distinguished, namely input and output registers, which are also represented by time-functions. Assume a system S, and c, m, C_S, M_S, as in Section 3.1. The additional environmental quantities can be classified as follows:

Definition 6.

- The set of additional environmental quantities is divided into two subsets:
  - *inputs*, quantities that can be read by the computers in the system, associated with input registers on those computers, and
  - *outputs*, quantities whose values are set by the computers in the system, associated with output registers on those computers.
- Let u be the number of input registers on the computers in the system S, and F_I a selected set of time-functions. Then:
  - inputs will be represented by a u-tuple $i = (i_1, \ldots, i_u)$, where each i_k is a member of F_I and corresponds to the k-th input register,

- the set of all i will be denoted by I_S.
- Let v be the number of output registers on the computers in the system S, and F_O a selected set of time-functions. Then:
  - outputs will be represented by a v-tuple $o = (o_1, \ldots, o_v)$, where each o_k is a member of F_O and corresponds to the k-th output register,
  - the set of all o will be denoted by O_S.

Relation IN. The physical interpretation of the inputs and their connection with monitored quantities can be specified by a relation, which we call IN.

Definition 7.

- IN $\subseteq M_S \times I_S$,
- IN(m, i) if and only if i describes the possible values of the inputs when m describes the values of the monitored quantities.

One should note the following:

1. domain(IN) contains all possible instances of m.
2. range(IN) contains possible instances of i.
3. IN describes the behaviour of the input devices; it is a relation rather than a function because it gives some freedom to implement a measuring device with some imprecision.
4. It must be the case that domain(IN) $\supseteq$ domain(NAT), because the input device must transmit some value for every condition that can occur in nature.

Example 5. *WaterLevel* can be calculated from the measurement of differential pressure between two vertically separated points in the water vessel. The differential pressure gauge is connected to an analog-to-digital converter whose input values (to the computer) are represented by the input quantity DiffPres. That quantity is described as follows:

$$\left[(\text{DiffPres}(t) \in [1, 254]) \Rightarrow (\frac{\text{DiffPres}(t)}{255} \times 14.53\text{cm} + 12.73\text{cm}) \in I \right] \vee$$

$$[((\text{DiffPres}(t) \in \{0, 255\}) \Rightarrow (LevelDevice(t) = failed))]$$

where $I = WaterLevel(t) \pm 0.34\text{cm}$.

The differential pressure gauge is calibrated so that while the water level is between 13.0 cm and 27.0 cm, the value of DiffPres should be in the closed interval $[1, 254]$. If its value is either 0 or 255, then either the water level has drifted outside of the pressure gauge's calibrated range or the device has failed. It is to be assumed that if the value of DiffPres is 0 or 255 that the pressure gauge is not operating properly.

Relation OUT. The effects of the outputs and their connection with the controlled quantities can be specified by a relation, which we call OUT.

Definition 8.

- OUT $\subseteq O_S \times C_S$,
- OUT(o, c) if and only if c describes the possible values of the controlled quantities when o describes the values of the outputs.

One should note the following:

1. domain(OUT) contains all possible instances of o.
2. range(OUT) contains possible instances of c.
3. OUT describes the behaviour of the output devices; it is a relation rather than a function because it gives some freedom to implement an actuator with some imprecision.

Example 6. The value of PumpSwitch is affected through one bit of a discrete output register represented by the output quantity Shutdown. The effects of the relation OUT are described as follows:

$$(\text{Shutdown}(t), \text{PumpSwitch}(t)) \in \text{OUT}$$

where

$$\text{OUT} = \{(0, open), (1, closed)\}$$

3.3 The software requirements document

The software requirements are determined by the system requirements document and system design documents. As mentioned earlier, the *software requirements document* can be seen as a combination of the two documents. It would contain the relations NAT (if it does not include all combinations of m and c), REQ, IN, and OUT.

In the sequel we assume that all definitions and notational conventions from the previous sections apply. We will also assume that REQ is feasible with respect to NAT.

Relation SOF. The software implementation will yield a system with input-output behaviour that can be described by relation, which we call SOF.

Definition 9.

- SOF $\subseteq I_S \times O_S$,
- SOF(i, o) if and only if the software could produce the values described by o when the inputs are described by i.

One should note the following:

1. domain(SOF) contains all possible instances of i, i.e.

   domain(SOF) $\supseteq$ range(IN)

2. range(SOF) contains all instances of o, i.e. range(SOF) $\subseteq$ domain(OUT).
3. SOF will be a function if the software is deterministic.

Software acceptability. The resulting software must satisfy all the requirements expressed by the relations. Hence the following:

Definition 10. For the software to be *acceptable*, the relation SOF must be such, that:

   (∗) NAT$(m, c) \wedge$ IN$(m, i) \wedge$ SOF$(i, o) \wedge$ OUT$(o, c) \Rightarrow$ REQ(m, c)

   for all $m \in M_S, i \in I_S, o \in O_S, c \in C_S$

One should note the following:

1. If one or more of the predicates IN(m, i), OUT(o, c), or NAT(m, c) are false, then any software behaviour will be considered acceptable. (For example, if a given value of m is not in the domain of IN, the behaviour of acceptable software in that case is not constrained by the formula (∗) above.)
2. Using relational composition [8] the formula (∗) can be expressed in a more concise way:

   NAT $\cap$ (IN • SOF • OUT) $\subseteq$ REQ

3. If we assume that relations REQ, IN, OUT, and SOF are functions, we can use functional notation to rewrite (∗) as follows:

   $m \in$ domain(NAT) $\Rightarrow$ REQ$(m) =$ OUT(SOF(IN(m))) for all $m \in M_S$

The writers of the requirements document must describe NAT, REQ, IN, and OUT. (The implementors determine SOF and verify (∗).) A document of this type will require natural language in the description of the environmental quantities, but can otherwise be precise and mathematical. The use of natural language in the definition of the physical interpretation of mathematical quantities is unavoidable and quite usual in engineering.

4 Duration calculus and its application to WLMS

Duration calculus [4], an extension of interval temporal logic, was invented to allow reasoning about design and requirements for real-time, embedded systems.

---

[8] Given two relations $R \subseteq A \times B$, and $S \subseteq B \times C$, *relational composition* $R • S$ is defined by: $R • S = \{(a, c) \in a \times C \mid R(a, b) \wedge S(b, c)$ for some $b \in B\}$

To capture properties of piecewise continuous states the extended duration calculus (in short: EDC) is proposed in [5]. We believe it can be used in a natural way to express the relations introduced in previous sections (NAT, REQ, IN, OUT, and SOF). In this section we will formulate some aspects of the WLMS in EDC and verify formally that the formula (∗) from Section 3.3 holds. In the sequel we will consider a more general case of WLMS then the one described in [2]. Hence there will be some minor terminological differences and we will not use the particular values from [2].

4.1 Introduction to the extended calculus of duration

The extended duration calculus is a first order logic with special duration and interval operators, and with corresponding axioms. It is based on a mathematical theory of functions. The following definition summarizes the most important aspects of EDC and introduces terms needed for this paper. A more complete description of EDC is to be found in [5].

Definition 11.

- *States* are names, which are interpreted as a functions from reals (representing time) to reals.
- *State expressions* can be built from states by applying arithmetic operators (e.g: +). The result is defined pointwise (e.g. $(f + g)(t) \hat{=} f(t) + g(t)$).
- *State assertions* can be built from states and state expressions by applying property names (e.g. Continuous (.)) or relational symbols. These assertions are interpreted as operators with a value domain restricted to the subset $\{0, 1\}$ of reals (with 1 denoting the value "true").
- We assume, that any expression and assertion is finite and sectionally continuous, i.e. there can be only a finite number of discontinuities in a finite interval.
- For a state assertion P and an arbitrary observation interval (b, e), a *duration* of P, denoted $\int P$, is interpreted as follows: $\int P \hat{=} \int_b^e P(t)\, dt$. Note, that:
  - $\int P$ is the accumulated time when $P = 1$ within the observation interval,
  - $\int 1 = e - b$, i.e. it is the length of the observation interval; we will denote it in the sequel by ℓ.
- *Initial* value of a state expression f, denoted $b.f$ is interpreted as $\lim_{t \to b+} f(t)$ for the arbitrary, nonempty observation interval (b, e).
- *Final* value of a state expression f, denoted $e.f$ is interpreted as $\lim_{t \to e-} f(t)$.
- *Duration terms* are built from global variables (which are interpreted as real numbers), durations, initial and final values and n-ary operators on duration terms.
- From duration terms we can build *duration formulae*. We introduce now the most important kinds of such formulae:
  - Simple predicates can be formed in the two following ways. Any property Q which holds for any proper (non empty) interval in the mathematical theory can be lifted to a simple predicate $\lceil Q \rceil$, which is true if and

only if $(\ell > 0) \wedge Q(x)$ for all $x \in (b, e)$. Also for a given valid relation from mathematical theory $R(r_1, \ldots, r_n)$ over non-negative real numbers $r_1, \ldots, r_n$, a simple predicate is formed by substituting duration terms for $r_1, \ldots, r_n$.

- We use $\lceil \rceil$ to denote the empty interval, i.e. $\lceil \rceil \hat{=} \ell = 0$.
- We can combine duration formulae using the boolean logic operators which have here standard meaning, and the interval temporal logic operator "chop" (denoted ";"). The latter one is defined as follows. Let A and B be duration formulae. The formula $A; B$ is true for any interval which can be split into two subintervals, of which the first satisfies A and the second satisfies B.
- The modal operators $\square$ and $\Diamond$ are introduced as abbreviations of the following formulae:

$$\Diamond D \hat{=} \text{true}; D; \text{true}$$

($\Diamond D$ holds if there is a subinterval of a given interval where D holds)

$$\square D \hat{=} \neg(\Diamond \neg D)$$

(D holds in any subinterval of a given interval)

- We say that D_1 is τ-*followed* by D_2 and write it $D_1 \xrightarrow{\tau} D_2$, if: $(D_1; \ell = \tau) \Rightarrow (D_1; D_2)$

– We use the following precedence rules (with the highest priority first):

- $\int$, b., e.
- arithmetic operators
- relational operators
- $\neg, \square, \Diamond$
- ;
- $\vee, \wedge$
- $\Rightarrow$

4.2 The WLMS requirements specification

We will limit our considerations only to that part of the WLMS that is responsible for switching the pump on and off depending on the level of water (cf. Section 3.1 Example 1). The monitored quantity is represented by the function *Water-Level*, which we will abbreviate in the sequel as WL. The controlled quantity is represented by the function *PumpSwitch*, which we will abbreviate in the sequel as PS. Hence $m = $ (WL) and $c = $ (PS). We will slightly redefine the function *PumpSwitch*: its range is a subset $\{0, 1\}$ of reals (*closed* being replaced by 1, *open* by 0), i.e. :

$$\text{PS}(t) = \begin{cases} 1, \text{ if the switch contacts are closed at time } t \\ 0, \text{ if the switch contacts are open at time } t \end{cases}$$

According to the approach presented in the previous section, we should now formulate the relations NAT and REQ. Let us begin with the latter.

Relation REQ. To be able to specify the informal statement "switch the pump off if the level of water is too high", we will assume that we are given four real constants: *High*, Δ, T_1, T_2, such that $High > \Delta > 0$, $T_1 > 0, T_2 > 0$, and will formulate our requirements as follows:

1. If the level of water remains above the high mark *High* for at least time T_1, then the pump switch should be opened after the period T_1, and
2. If the level of water remains below $(High - \Delta)$ for at least time T_2, then the pump switch should be closed after the period T_2.

The above requirements can be expressed in DC as follows:

$$\mathrm{REQ}_1 \triangleq \{(\mathrm{WL}, \mathrm{PS}) \in M_S \times C_S \,|$$
$$\square(\lceil \mathrm{WL} > High \rceil \wedge (\ell > T_1) \Rightarrow \mathrm{true}; \lceil \mathrm{PS} = 0 \rceil)\}$$
$$\mathrm{REQ}_2 \triangleq \{(\mathrm{WL}, \mathrm{PS}) \in M_S \times C_S \,|$$
$$\square(\lceil \mathrm{WL} < (High - \Delta) \rceil \wedge (\ell > T_2) \Rightarrow \mathrm{true}; \lceil \mathrm{PS} = 1 \rceil)\}$$

The relation REQ is then defined:

$$\mathrm{REQ} \triangleq \mathrm{REQ}_1 \cap \mathrm{REQ}_2$$

Relation NAT. NAT should describe constraints imposed by nature. Here we should have:

$$\mathrm{NAT}_1 \triangleq \{(\mathrm{WL}, \mathrm{PS}) \in M_S \times C_S \,|\, \lceil 0 < \mathrm{WL} < \mathrm{Max} \rceil \}$$

$$\mathrm{NAT}_2 \triangleq \{(\mathrm{WL}, \mathrm{PS}) \in M_S \times C_S \,|\, \square(|\, e.\mathrm{WL} - b.\mathrm{WL} \,| \le c \cdot \ell \vee \ell = 0)\}$$

$$\mathrm{NAT}_3 \triangleq \lceil Continuous\,(\mathrm{WL}) \rceil \vee \lceil \rceil$$

$$\mathrm{NAT} \triangleq \mathrm{NAT}_1 \cap \mathrm{NAT}_2 \cap \mathrm{NAT}_3$$

The first condition states that the water level is always greater than 0, and less then a certain constant Max.

The second one bounds the pump speed by a constant c.

The third condition states that the water level is continuous.

4.3 The WLMS design specification

The relation IN and OUT will be very simple in our case. We will assume that the values of the function WL are represented by the input [9] register WL_o, and that the pump switch is actuated by the output register PS_o, i.e. $i = (\mathrm{WL}_o)$, and $o = (\mathrm{PS}_o)$.

There will be only a tolerance requirement on WL_o, which is expressed as follows:

$$\mathrm{IN} \triangleq \{(\mathrm{WL}, \mathrm{WL}_o) \in M_S \times I_S \,|\, \square(\lceil |\, \mathrm{WL} - \mathrm{WL}_o \,| < \varepsilon_1 \rceil \vee \lceil \rceil)\}$$

where ε_1 is a constant such, that $0 < \varepsilon_1 < \Delta/2$.

As far as output is concerned we will make an idealistic assumption that PS_o equals PS almost all the time:

$$\mathrm{OUT} \triangleq \{(\mathrm{PS}_o, \mathrm{PS}) \in O_S \times C_S \,|\, \square(\lceil \mathrm{PS} = \mathrm{PS}_o \rceil \vee \lceil \rceil)\}$$

---

[9] An input/output register is usually a part of a digital computer and in such a case has discrete values.

4.4 The WLMS software specification

Software specification SOF is a little bit more complicated. To describe it in terms of EDC we will need a simple model for the program. The simplest one is an automaton. We will design a three phase [10] automaton with a set of phases $\{P_0, P_1, P_2\}$. We will use a state name Phase ranging over $\{P_0, P_1, P_2\}$ to denote the phase the automaton is in. We will also use an abbreviation $\lceil P_i \rceil$ to denote that $\lceil \text{Phase} = P_i \rceil$.

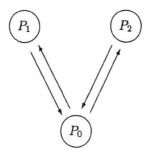

Figure 1: Three phase automaton model.

When the measured water level is above $High - \varepsilon_1$, the automaton changes its phase from P_0 to P_2. When the level is below $High - \Delta + \varepsilon_1$, the automaton changes its phase from P_0 to P_1. We know that $0 < \varepsilon_1 < \frac{\Delta}{2}$, hence $High - \Delta + \varepsilon_1 < High - \varepsilon_1$. Between these two values the automaton changes its phase to P_0. These transition properties are described by the following four conditions:

$$\text{Sof}_1 \mathrel{\hat{=}} \Box(\lceil \text{WL}_\text{o} > High - \varepsilon_1 \rceil \Rightarrow (\lceil P_0 \rceil \xrightarrow{\tau_1} \lceil P_2 \rceil))$$

$$\text{Sof}_2 \mathrel{\hat{=}} \Box(\lceil \text{WL}_\text{o} \le High - \varepsilon_1 \rceil \Rightarrow (\lceil P_2 \rceil \xrightarrow{\tau_2} \lceil P_0 \rceil; \text{true}))$$

$$\text{Sof}_3 \mathrel{\hat{=}} \Box(\lceil \text{WL}_\text{o} < High - \Delta + \varepsilon_1 \rceil \Rightarrow (\lceil P_0 \rceil \xrightarrow{\tau_3} \lceil P_1 \rceil))$$

$$\text{Sof}_4 \mathrel{\hat{=}} \Box(\lceil \text{WL}_\text{o} \ge High - \Delta + \varepsilon_1 \rceil \Rightarrow (\lceil P_1 \rceil \xrightarrow{\tau_4} \lceil P_0 \rceil; \text{true})).$$

The positive constants $\tau_1, \ldots, \tau_4$ describe transition delays. P_0 is the intermediate phase between P_1 and P_2 (cf. figure 1), what is expressed by "true" at the end of Sof_2 and Sof_4.

When the automaton is in phase P_2 the pump is turned off and when it is in phase P_1 the pump is turned on. Delays in turning the pump on and off are limited by two constants: $\delta_1 > 0$ and $\delta_2 > 0$. Behaviour of the pump is not determined when the automaton is in phase P_0. These requirements are expressed by the following two formulae:

$$\text{Sof}_5 \mathrel{\hat{=}} \Box((\lceil P_1 \rceil \wedge (\ell > \delta_1)) \Rightarrow \text{true}; \lceil \text{PS}_\text{o} = 1 \rceil)$$

$$\text{Sof}_6 \mathrel{\hat{=}} \Box((\lceil P_2 \rceil \wedge (\ell > \delta_2)) \Rightarrow \text{true}; \lceil \text{PS}_\text{o} = 0 \rceil).$$

---

[10] The word *state* is used already to denote a time function in EDC. To avoid ambiguity we will use the word *phase* to denote a state of an automaton.

Formulae $\text{Sof}_1, \ldots, \text{Sof}_4$ do not specify the behaviour of the automaton in all situations. For instance, when the system is started, the water level may be low (respectively high) and the automaton may already be in phase P_1 (respectively P_2). Hence, there are two conditions specifying that if the automaton is in state P_1 (respectively, P_2) and WL_o is less than $High - \Delta + \varepsilon_1$ (respectively, is greater than $High - \varepsilon_1$), then the automaton stays in phase P_1 (respectively, P_2):

$$\text{Sof}_7 \triangleq \Box((\lceil P_1 \rceil; \text{true} \wedge \lceil \text{WL}_\text{o} < High - \Delta + \varepsilon_1 \rceil) \Rightarrow \lceil P_1 \rceil)$$

$$\text{Sof}_8 \triangleq \Box((\lceil P_2 \rceil; \text{true} \wedge \lceil \text{WL}_\text{o} > High - \varepsilon_1 \rceil) \Rightarrow \lceil P_2 \rceil).$$

One can notice that formula Sof_7 (respectively Sof_8) forces the automaton to stay in phase P_1 (respectively P_2) as long as the water level is low (respectively high).

Due to the finite variability of states, each phase covers an interval, what is expressed by the formula:

$$\text{Sof}_9 \triangleq \Box(\lceil P_0 \rceil; \text{true} \vee \lceil P_1 \rceil; \text{true} \vee \lceil P_2 \rceil; \text{true} \vee \lceil \rceil).$$

The whole software specification is now defined as follows:

$$\text{SOF} \triangleq \{(\text{WL}_\text{o}, \text{PS}_\text{o}) \in I_S \times O_S \mid \bigwedge_{i=1,\ldots,9} \text{Sof}_i, \text{ for some Phase}\}$$

We will have to make some additional assumptions on $\tau_1, \ldots, \tau_4$, δ_1 and δ_2 constants because there is no correlation between them, and the constants T_1 and T_2. These assumptions naturally appear in a software verification.

4.5 An example of software acceptability verification

Verification of software acceptability is a proof that the software requirements, input and output specifications and the knowledge of the nature imply the system requirements [11]. The complete proof can be found in the Appendix. Here we will show the main part of it only. The system requirements consist of two conditions REQ_1 and REQ_2. Let us consider REQ_1.

$$\text{REQ}_1 = \{(\text{WL}, \text{PS}) \in M_S \times C_S \mid \Box(\lceil \text{WL} > High \rceil \wedge (\ell > T_1) \Rightarrow$$

$$\text{true}; \lceil \text{PS} = 0 \rceil)\}$$

For arbitrarily chosen pair $(\text{WL}, \text{PS}) \in \text{REQ}_1$ we have:

$$\Box(\lceil \text{WL} > High \rceil) \wedge (\ell > T_1) \Rightarrow \text{true}; \lceil \text{PS} = 0 \rceil).$$

Assume that:

$$\lceil \text{WL} > High \rceil \wedge (\ell > T_1).$$

---

[11] Since in our case the required actions (switching the pump on and off) do not depend on conditions expressed by NAT, the latter relation will not be used in the verification process.

For any WL_o satisfying $(WL, WL_o) \in IN$ we obtain:

$\lceil WL_o > High - \varepsilon_1 \rceil \wedge (\ell > T_1)$.

From Sof_9 we know that initially the automaton is in one of three phases: P_0, P_1, P_2. Assume that it is P_1.

$\lceil P_1 \rceil; true$

From Sof_4 we have:

$\lceil P_1 \rceil \overset{\tau_4}{\to} \lceil P_0 \rceil; true$.

If $T_1 > \tau_4$ then

$(\ell > \tau_4); \lceil P_0 \rceil; true; (\ell > T_1 - \tau_4) \wedge \lceil WL_o > High - \varepsilon_1 \rceil$.

From that we obtain:

$(\ell > \tau_4); (\lceil P_0 \rceil; true \wedge (\ell > T_1 - \tau_4) \wedge \lceil WL_o > High - \varepsilon_1 \rceil)$.

If $T_1 - \tau_4 > \tau_1$, then from Sof_1 we know that

$(\ell > \tau_4); ((\ell > \tau_1); (\lceil P_2 \rceil; true \wedge (\ell > T_1 - \tau_4 - \tau_1)) \wedge \lceil WL_o > High - \varepsilon_1 \rceil)$,

what can be weaken to:

$(\ell > \tau_4 + \tau_1); (\lceil P_2 \rceil; true \wedge (\ell > T_1 - \tau_4 - \tau_1) \wedge \lceil WL_o > High - \varepsilon_1 \rceil)$.

From Sof_8 we obtain:

$(\ell > \tau_4 + \tau_1); (\lceil P_2 \rceil \wedge (\ell > T_1 - \tau_4 - \tau_1) \wedge \lceil WL_o > High - \varepsilon_1 \rceil)$.

If $T_1 - \tau_4 - \tau_1 > \delta_2$ then from Sof_6 we know that

$(\ell > \tau_4 + \tau_1); true; \lceil PS_o = 0 \rceil$.

From OUT we have

true; $\lceil PS = 0 \rceil$

The successor of the implication have been derived from the predecessor under the assumption that $T_1 > \tau_1 + \tau_4 + \delta_2$.

If the automaton is initially in phase P_0 then the proof is similar to the second half of that one above. If it starts in phase P_2 then from IN and Sof_8 we obtain

$\lceil P_2 \rceil$

and from Sof_6 and OUT

true; $\lceil PS = 0 \rceil$.

The fact that WL, WL_o and PS had been chosen arbitrarily completes the proof of REQ_1.

We can similarly prove REQ_2. We will get the assumption that $T_2 > \tau_2 + \tau_3 + \delta_1$. The assumptions on constants $\tau_1, \ldots, \tau_4, \delta_1, \delta_2$ should form one more software requirement limiting these constants and leaving their concrete values up to software designers.

468

5 Conclusions

This work is a first step towards merging two distinct approaches to formal specification of real-time systems. We have shown how one can express the relations described in requirements documents in terms of extended duration calculus. This allows us to verify formally each design step, and in particular to check whether certain design decisions are correct from the requirements point of view.

We expect this first step to be followed by further investigations and by elaboration of more examples.

Acknowledgments

This work was partially supported by:

(a) In Canada:
   – Telecommunications Research Institute of Ontario (TRIO),
   – Atomic Energy Control Board (AECB),
   – Natural Sciences and Engineering Research Council of Canada (NSERC);
(b) In Europe:
   – Commission of the European Communities (CEC) under the ESPRIT programme in the field of Basic Research Action proj. no. 3104: "ProCoS: Provably Correct Systems".

References

1. Parnas, D.L., and Madey, J., "Functional Documentation for Computer Systems Engineering (Version 2)", *CRL Report 237*, Telecommunications Research Institute of Ontario (TRIO), McMaster University, Hamilton, ON, September 1991, 14pp.
2. van Schouwen, A.J., "The A-7 Requirements Model: Re-examination for Real-Time Systems and an Application to Monitoring Systems", *Technical Report 90-276*, Queen's University, C&IS, Telecommunications Research Institute of Ontario (TRIO), Kingston, ON, January 1991 (revision 3), 94pp.
3. van Schouwen, A.J., Parnas, D.L., and Madey, J., "Documentation of Requirements for Computer Systems", Proceeding of the IEEE International Symposium on Requirements Engineering, San Diego, California, January 4–6, 1993.
4. Zhou Chaochen, C.A.R. Hoare, Anders P. Ravn, "A Calculus of Durations", *Information Processing Letters 40*, 5, pp. 269 – 276, 1992.
5. Zhou Chaochen, Michael R. Hansen, Anders P. Ravn "An Extended Duration Calculus for Hybrid Real-Time Systems" in this volume.

Appendix: Software verification

We have introduced the following definitions of REQ, NAT, IN, OUT and SOF:

$$\mathrm{REQ}_1 \mathrel{\hat{=}} \{(\mathrm{WL}, \mathrm{PS}) \in M_S \times C_S \,|$$

$$\Box(\lceil \mathrm{WL} > High \rceil \wedge (\ell > T_1) \Rightarrow \mathrm{true}; \lceil \mathrm{PS} = 0 \rceil)\}$$

$REQ_2 \doteq \{(WL, PS) \in M_S \times C_S \,|\, \Box(\lceil WL < (High - \Delta)\rceil \wedge (\ell > T_2) \Rightarrow$

true; $\lceil PS = 1\rceil)\}$

$REQ \doteq REQ_1 \cap REQ_2$

$NAT_1 \doteq \{(WL, PS) \in M_S \times C_S \,|\, \lceil 0 < WL < Max\rceil\}$

$NAT_2 \doteq \{(WL, PS) \in M_S \times C_S \,|\, \Box(|\,e.WL - b.WL\,| \le c \cdot \ell \vee \ell = 0)\}$

$NAT_3 \doteq \lceil Continuous\,(WL)\rceil \vee \lceil\,\rceil$

$NAT \doteq NAT_1 \cap NAT_2 \cap NAT_3$

$IN \doteq \{(WL, WL_o) \in M_S \times I_S \,|\, \Box(\lceil|\,WL - WL_o\,| < \varepsilon_1\rceil \vee \lceil\,\rceil)\}$

$OUT \doteq \{(PS_o, PS) \in O_S \times C_S \,|\, \Box(\lceil PS = PS_o\rceil \vee \lceil\,\rceil)\}$

$Sof_1 \doteq \Box(\lceil WL_o > High - \varepsilon_1\rceil \Rightarrow (\lceil P_0\rceil \overset{\tau_1}{\dashrightarrow} \lceil P_2\rceil))$

$Sof_2 \doteq \Box(\lceil WL_o \le High - \varepsilon_1\rceil \Rightarrow (\lceil P_2\rceil \overset{\tau_3}{\dashrightarrow} \lceil P_0\rceil; true))$

$Sof_3 \doteq \Box(\lceil WL_o < High - \Delta + \varepsilon_1\rceil \Rightarrow (\lceil P_0\rceil \overset{\tau_3}{\dashrightarrow} \lceil P_1\rceil))$

$Sof_4 \doteq \Box(\lceil WL_o \ge High - \Delta + \varepsilon_1\rceil \Rightarrow (\lceil P_1\rceil \overset{\tau_4}{\dashrightarrow} \lceil P_0\rceil; true))$

$Sof_5 \doteq \Box((\lceil P_1\rceil \wedge (\ell > \delta_1)) \Rightarrow true; \lceil PS_o = 1\rceil)$

$Sof_6 \doteq \Box((\lceil P_2\rceil \wedge (\ell > \delta_2)) \Rightarrow true; \lceil PS_o = 0\rceil)$

$Sof_7 \doteq \Box((\lceil P_1\rceil; true \wedge \lceil WL_o < High - \Delta + \varepsilon_1\rceil) \Rightarrow \lceil P_1\rceil)$

$Sof_8 \doteq \Box((\lceil P_2\rceil; true \wedge \lceil WL_o > High - \varepsilon_1\rceil) \Rightarrow \lceil P_2\rceil)$

$Sof_9 \doteq \Box(\lceil P_0\rceil; true \vee \lceil P_1\rceil; true \vee \lceil P_2\rceil; true \vee \lceil\,\rceil)$

$SOF \doteq \{(WL_o, PS_o) \in I_S \times O_S \,|\, \bigwedge_{i=1,\ldots,9} Sof_i, \text{ for some Phase}\}$

We assumed that all constants in considered example are positive. We have also derived additional assumptions on constants:

$T_1 > \tau_1 + \tau_4 + \delta_2$

$T_2 > \tau_2 + \tau_3 + \delta_1$

which should be included into SOF. In particular design constants T_1, T_2 and ε_1 should be replaced by concrete numbers. Constants $\tau_1, \ldots \tau_4, \delta_1$ and δ_2 may be left up to software designers.

To prove REQ_1 we assume that for a chosen pair $(WL, PS) \in REQ_1$ and for a chosen period of time, the following holds:

$$\lceil WL > High \rceil \wedge (\ell > T_1).$$

For any WL_o such that $(WL, WL_o) \in IN$ we have:

$$\lceil WL_o > High - \varepsilon_1 \rceil \wedge (\ell > T_1).$$

From Sof_9 we know that initially the automaton is in one of three phases: P_0, P_1, P_2. Hence, we have to consider the three cases.

1° Assume that:

$$\lceil P_1 \rceil; true.$$

We have:

$$\lceil P_1 \rceil; true \wedge \lceil WL_o > High - \varepsilon_1 \rceil \wedge (\ell > T_1),$$

and from that:

$$\lceil P_1 \rceil; (\ell = \tau_4); (\ell > T_1 - \tau_4) \wedge \lceil WL_o > High - \varepsilon_1 \rceil.$$

From Sof_4 we obtain:

$$(\lceil P_1 \rceil; (\ell = \tau_4) \wedge \lceil P_1 \rceil \overset{\tau_4}{\rightarrow} \lceil P_0 \rceil; true); (\ell > T_1 - \tau_4) \wedge \lceil WL_o > High - \varepsilon_1 \rceil.$$

We can simplify it:

$$(\ell > \tau_4); \lceil P_0 \rceil; (\ell > T_1 - \tau_4) \wedge \lceil WL_o > High - \varepsilon_1 \rceil.$$

We can weaken it:

$$(\ell > \tau_4); (\lceil P_0 \rceil; (\ell > T_1 - \tau_4) \wedge \lceil WL_o > High - \varepsilon_1 \rceil).$$

From that we obtain:

$$(\ell > \tau_4); (\lceil P_0 \rceil; (\ell = \tau_1); (\ell > T_1 - \tau_4 - \tau_1) \wedge \lceil WL_o > High - \varepsilon_1 \rceil).$$

From Sof_1 we know that:

$$(\ell > \tau_4)((\lceil P_0 \rceil; (\ell = \tau_1) \wedge \lceil P_0 \rceil \overset{\tau_1}{\rightarrow} \lceil P_2 \rceil); (\ell > T_1 - \tau_4 - T_1) \wedge$$
$$\lceil WL_o > High - \varepsilon_1 \rceil),$$

hence:

$$(\ell > \tau_4); ((\ell > \tau_1); \lceil P_2 \rceil; (\ell > T_1 - \tau_4 - \tau_1) \wedge \lceil WL_o > High - \varepsilon_1 \rceil).$$

From that we obtain:

$$(\ell > \tau_4); (\ell > \tau_1); (\lceil P_2 \rceil; true \wedge (\ell > T_1 - \tau_4 - \tau_1) \wedge \lceil WL_o > High - \varepsilon_1 \rceil).$$

We can simplify it:

$$(\ell > \tau_4 + \tau_1); (\lceil P_2 \rceil; true \wedge (\ell > T_1 - \tau_4 - \tau_1) \wedge \lceil WL_o > High - \varepsilon_1 \rceil).$$

From Sof_8 we have:

$$(\ell > \tau_4 + \tau_1); (\lceil P_2 \rceil \wedge (\ell > T_1 - \tau_4 - \tau_1) \wedge \lceil WL_o > High - \varepsilon_1 \rceil).$$

From Sof_6 we know that:

$$(\ell > \tau_4 + \tau_1); true; \lceil PS_o = 0 \rceil.$$

From OUT we have:

$$(\ell > \tau_4 + \tau_1); true; \lceil PS = 0 \rceil,$$

hence:

$$true; \lceil PS = 0 \rceil.$$

Which is the desired consequence.

2° Assume that:

$$\lceil P_0 \rceil; true.$$

We have:

$$\lceil P_0 \rceil; true \wedge \lceil WL_o > High - \varepsilon_1 \rceil \wedge (\ell > T_1),$$

and from that:

$$\lceil P_0 \rceil; (\ell = \tau_1); (\ell > T_1 - \tau_1) \wedge \lceil WL_o > High - \varepsilon_1 \rceil.$$

From Sof_1 we know that:

$$(\lceil P_0 \rceil; (\ell = \tau_1) \wedge \lceil P_0 \rceil \overset{\tau_1}{\rightarrow} \lceil P_2 \rceil); (\ell > T_1 - \tau_1) \wedge \lceil WL_o > High - \varepsilon_1 \rceil.$$

We can simplify it:

$$(\ell > \tau_1); \lceil P_2 \rceil; (\ell > T_1 - \tau_1) \wedge \lceil WL_o > High - \varepsilon_1 \rceil.$$

We can simplify it:

$$(\ell > \tau_1); (\lceil P_2 \rceil; true \wedge (\ell > T_1 - \tau_1) \wedge \lceil WL_o > High - \varepsilon_1 \rceil).$$

From Sof_8 we have:

$$(\ell > \tau_1); (\lceil P_2 \rceil \wedge (\ell > T_1 - \tau_1) \wedge \lceil WL_o > High - \varepsilon_1 \rceil).$$

From Sof_6 we obtain:

$$(\ell > \tau_1); true; \lceil PS_o = 0 \rceil.$$

From OUT we have:

$$(\ell > \tau_1); true; \lceil PS = 0 \rceil,$$

hence:

$$true; \lceil PS = 0 \rceil.$$

Which is the desired consequence.

3° Assume that:

$$\lceil P_2 \rceil; true.$$

We have:

$$\lceil P_2 \rceil; true \land \lceil WL_o > High - \varepsilon_1 \rceil \land (\ell > T_1).$$

From Sof_8 we obtain:

$$\lceil P_2 \rceil \land \lceil WL_o > High - \varepsilon_1 \rceil \land (\ell > T_1).$$

From Sof_6 we know that:

$$true; \lceil PS_o = 0 \rceil.$$

From OUT we finally get:

$$true; \lceil PS = 0 \rceil$$

Which is the desired consequence.

We have considered three possible cases. The fact that we have arbitrarily chosen WL, WL_o and PS ends the proof of REQ_1.

A proof of REQ_2 is very similar to the proof of REQ_1. To prove REQ_2 we assume that for a chosen pair $(WL, PS) \in REQ_2$ and for a chosen period of time the following holds:

$$\lceil WL < High - \Delta \rceil \land (\ell > T_2).$$

For any WL_o such that $(WL, WL_o) \in IN$ we have:

$$\lceil WL_o < High - \Delta + \varepsilon_1 \rceil \land (\ell > T_2).$$

From Sof_9 we know that initially the automaton is in one of three phases: P_0, P_1, P_2. Hence, we have to consider the three cases.

1° Assume that:

$$\lceil P_2 \rceil; true.$$

We have:

$$\lceil P_2 \rceil; true \land \lceil WL_o < High - \Delta + \varepsilon_1 \rceil \land (\ell > T_2),$$

and from that:

$$\lceil P_2 \rceil; (\ell = \tau_2); (\ell > T_2 - \tau_2) \land \lceil WL_o < High - \Delta + \varepsilon_1 \rceil.$$

From Sof_2 we obtain:

$$(\lceil P_2 \rceil; (\ell = \tau_2) \land \lceil P_2 \rceil \overset{\tau_2}{\rightarrow} \lceil P_0 \rceil; true); (\ell > T_2 - \tau_2) \land$$

$$\lceil WL_o < High - \Delta + \varepsilon_1 \rceil,$$

hence:

$$(\ell > \tau_2); \lceil P_0 \rceil; (\ell > T_2 - \tau_2) \wedge \lceil \mathrm{WL_o} < High - \Delta + \varepsilon_1 \rceil.$$

We can weaken it:

$$(\ell > \tau_2); (\lceil P_0 \rceil; (\ell = \tau_3); (\ell > T_2 - \tau_2 - \tau_3) \wedge \lceil \mathrm{WL_o} < High - \Delta + \varepsilon_1 \rceil).$$

From $\mathrm{Sof_3}$ we know that:

$$(\ell > \tau_2); ((\lceil P_0 \rceil; (\ell = \tau_3) \wedge \lceil P_0 \rceil \overset{\tau_3}{\to} \lceil P_1 \rceil); (\ell > T_2 - \tau_2 - \tau_3) \wedge$$

$$\lceil \mathrm{WL_o} < High - \Delta + \varepsilon_1 \rceil),$$

hence:

$$(\ell > \tau_2); ((\ell > \tau_3); \lceil P_1 \rceil; (\ell > T_2 - \tau_2 - \tau_3) \wedge \lceil \mathrm{WL_o} < High - \Delta + \varepsilon_1 \rceil).$$

From that we obtain:

$$(\ell > \tau_2); (\ell > \tau_3); (\lceil P_1 \rceil; true \wedge (\ell > T_2 - \tau_2 - \tau_3) \wedge$$

$$\lceil \mathrm{WL_o} < High - \Delta + \varepsilon_1 \rceil).$$

We can simplify it:

$$(\ell > \tau_2 + \tau_3); (\lceil P_1 \rceil; true \wedge (\ell > T_2 - \tau_2 - \tau_3) \wedge \lceil \mathrm{WL_o} < High - \Delta + \varepsilon_1 \rceil).$$

From $\mathrm{Sof_7}$ we have:

$$(\ell > \tau_2 + \tau_3); (\lceil P_1 \rceil \wedge (\ell > T_2 - \tau_2 - \tau_3) \wedge \lceil \mathrm{WL_o} < High - \Delta + \varepsilon_1 \rceil).$$

From $\mathrm{Sof_5}$ we know that:

$$(\ell > \tau_2 + \tau_3); true; \lceil \mathrm{PS_o} = 1 \rceil.$$

From OUT we obtain:

$$(\ell > \tau_2 + \tau_3); true; \lceil \mathrm{PS} = 1 \rceil,$$

hence:

$$true; \lceil \mathrm{PS} = 1 \rceil.$$

Which is the desired consequence.

2° Assume that:

$$\lceil P_0 \rceil; true.$$

We have:

$$\lceil P_0 \rceil; true \wedge \lceil \mathrm{WL_o} < High - \Delta + \varepsilon_1 \rceil \wedge (\ell > T_2),$$

and from that:

$$\lceil P_0 \rceil; (\ell = \tau_3); (\ell > T_2 - \tau_3) \wedge \lceil \mathrm{WL_o} < High - \Delta + \varepsilon_1 \rceil.$$

From Sof_3 we know that:

$$([P_0]; (\ell = \tau_3) \wedge [P_0] \overset{\tau_3}{\to} [P_1]); (\ell > T_2 - \tau_3) \wedge [\text{WL}_o < High - \Delta + \varepsilon_1],$$

hence:

$$(\ell > \tau_3); [P_1]; (\ell > T_2 - \tau_3) \wedge [\text{WL}_o < High - \Delta + \varepsilon_1].$$

We can weaken it:

$$(\ell > \tau_3); ([P_1]; (\ell > T_2 - \tau_3) \wedge [\text{WL}_o < High - \Delta + \varepsilon_1]).$$

From that we have:

$$(\ell > \tau_3); ([P_1]; true \wedge (\ell > T_2 - \tau_3) \wedge [\text{WL}_o < High - \Delta + \varepsilon_1]).$$

From Sof_7 we obtain:

$$(\ell > \tau_3); ([P_1] \wedge (\ell > T_2 - \tau_3) \wedge [\text{WL}_o < High - \Delta + \varepsilon_1]).$$

From Sof_5 we know that:

$$(\ell > \tau_3); true; [\text{PS}_o = 1].$$

From OUT we have:

$$(\ell > \tau_3); true; [\text{PS} = 1],$$

hence:

$$true; [\text{PS} = 1].$$

Which is the desired consequence.

3° Assume that:

$$[P_1]; true.$$

We have:

$$[P_1]; true \wedge [\text{WL}_o < High - \Delta + \varepsilon_1] \wedge (\ell > T_2).$$

From Sof_7 we obtain:

$$[P_1] \wedge [\text{WL}_o < High - \Delta + \varepsilon_1] \wedge (\ell > T_2).$$

From Sof_5 we know that:

$$true; [\text{PS}_o = 1].$$

From OUT we finally get:

$$true; [\text{PS} = 1].$$

Which is the desired consequence.

The fact that we have arbitrarily chosen WL, WL_o and PS ends the proof of REQ_2. These two proofs give the full software acceptability verification.

Lecture Notes in Computer Science

For information about Vols. 1–660
please contact your bookseller or Springer-Verlag

Vol. 697: C. Courcoubetis (Ed.), Computer Aided Verification. Proceedings, 1993. IX, 504 pages. 1993.

Vol. 698: A. Voronkov (Ed.), Logic Programming and Automated Reasoning. Proceedings, 1993. XIII, 386 pages. 1993. (Subseries LNAI).

Vol. 699: G. W. Mineau, B. Moulin, J. F. Sowa (Eds.), Conceptual Graphs for Knowledge Representation. Proceedings, 1993. IX, 451 pages. 1993. (Subseries LNAI).

Vol. 700: A. Lingas, R. Karlsson, S. Carlsson (Eds.), Automata, Languages and Programming. Proceedings, 1993. XII, 697 pages. 1993.

Vol. 701: P. Atzeni (Ed.), LOGIDATA+: Deductive Databases with Complex Objects. VIII, 273 pages. 1993.

Vol. 702: E. Börger, G. Jäger, H. Kleine Büning, S. Martini, M. M. Richter (Eds.), Computer Science Logic. Proceedings, 1992. VIII, 439 pages. 1993.

Vol. 703: M. de Berg, Ray Shooting, Depth Orders and Hidden Surface Removal. X, 201 pages. 1993.

Vol. 704: F. N. Paulisch, The Design of an Extendible Graph Editor. XV, 184 pages. 1993.

Vol. 705: H. Grünbacher, R. W. Hartenstein (Eds.), Field-Programmable Gate Arrays. Proceedings, 1992. VIII, 218 pages. 1993.

Vol. 706: H. D. Rombach, V. R. Basili, R. W. Selby (Eds.), Experimental Software Engineering Issues. Proceedings, 1992. XVIII, 261 pages. 1993.

Vol. 707: O. M. Nierstrasz (Ed.), ECOOP '93 – Object-Oriented Programming. Proceedings, 1993. XI, 531 pages. 1993.

Vol. 708: C. Laugier (Ed.), Geometric Reasoning for Perception and Action. Proceedings, 1991. VIII, 281 pages. 1993.

Vol. 709: F. Dehne, J.-R. Sack, N. Santoro, S. Whitesides (Eds.), Algorithms and Data Structures. Proceedings, 1993. XII, 634 pages. 1993.

Vol. 710: Z. Ésik (Ed.), Fundamentals of Computation Theory. Proceedings, 1993. IX, 471 pages. 1993.

Vol. 711: A. M. Borzyszkowski, S. Sokołowski (Eds.), Mathematical Foundations of Computer Science 1993. Proceedings, 1993. XIII, 782 pages. 1993.

Vol. 712: P. V. Rangan (Ed.), Network and Operating System Support for Digital Audio and Video. Proceedings, 1992. X, 416 pages. 1993.

Vol. 713: G. Gottlob, A. Leitsch, D. Mundici (Eds.), Computational Logic and Proof Theory. Proceedings, 1993. XI, 348 pages. 1993.

Vol. 714: M. Bruynooghe, J. Penjam (Eds.), Programming Language Implementation and Logic Programming. Proceedings, 1993. XI, 421 pages. 1993.

Vol. 715: E. Best (Ed.), CONCUR'93. Proceedings, 1993. IX, 541 pages. 1993.

Vol. 716: A. U. Frank, I. Campari (Eds.), Spatial Information Theory. Proceedings, 1993. XI, 478 pages. 1993.

Vol. 717: I. Sommerville, M. Paul (Eds.), Software Engineering – ESEC '93. Proceedings, 1993. XII, 516 pages. 1993.

Vol. 718: J. Seberry, Y. Zheng (Eds.), Advances in Cryptology – AUSCRYPT '92. Proceedings, 1992. XIII, 543 pages. 1993.

Vol. 719: D. Chetverikov, W.G. Kropatsch (Eds.), Computer Analysis of Images and Patterns. Proceedings, 1993. XVI, 857 pages. 1993.

Vol. 720: V.Mařík, J. Lažanský, R.R. Wagner (Eds.), Database and Expert Systems Applications. Proceedings, 1993. XV, 768 pages. 1993.

Vol. 721: J. Fitch (Ed.), Design and Implementation of Symbolic Computation Systems. Proceedings, 1992. VIII, 215 pages. 1993.

Vol. 722: A. Miola (Ed.), Design and Implementation of Symbolic Computation Systems. Proceedings, 1993. XII, 384 pages. 1993.

Vol. 723: N. Aussenac, G. Boy, B. Gaines, M. Linster, J.-G. Ganascia, Y. Kodratoff (Eds.), Knowledge Acquisition for Knowledge-Based Systems. Proceedings, 1993. XIII, 446 pages. 1993. (Subseries LNAI).

Vol. 724: P. Cousot, M. Falaschi, G. Filè, A. Rauzy (Eds.), Static Analysis. Proceedings, 1993. IX, 283 pages. 1993.

Vol. 725: A. Schiper (Ed.), Distributed Algorithms. Proceedings, 1993. VIII, 325 pages. 1993.

Vol. 726: T. Lengauer (Ed.), Algorithms – ESA '93. Proceedings, 1993. IX, 419 pages. 1993

Vol. 727: M. Filgueiras, L. Damas (Eds.), Progress in Artificial Intelligence. Proceedings, 1993. X, 362 pages. 1993. (Subseries LNAI).

Vol. 728: P. Torasso (Ed.), Advances in Artificial Intelligence. Proceedings, 1993. XI, 336 pages. 1993. (Subseries LNAI).

Vol. 729: L. Donatiello, R. Nelson (Eds.), Performance Evaluation of Computer and Communication Systems. Proceedings, 1993. VIII, 675 pages. 1993.

Vol. 730: D. B. Lomet (Ed.), Foundations of Data Organization and Algorithms. Proceedings, 1993. XII, 412 pages. 1993.

Vol. 731: A. Schill (Ed.), DCE – The OSF Distributed Computing Environment. Proceedings, 1993. VIII, 285 pages. 1993.

Vol. 732: A. Bode, M. Dal Cin (Eds.), Parallel Computer Architectures. IX, 311 pages. 1993.

Vol. 733: Th. Grechenig, M. Tscheligi (Eds.), Human Computer Interaction. Proceedings, 1993. XIV, 450 pages. 1993.

Vol. 734: J. Volkert Ed.), Parallel Computation. Proceedings, 1993. VIII, 248 pages. 1993.

Vol. 735: D. Bjørner, M. Broy, I. V. Pottosin (Eds.), Formal Methods in Programming and Their Applications. Proceedings, 1993. IX, 434 pages. 1993.

Vol. 736: R. L. Grossman, A. Nerode, A. P. Ravn, H. Rischel (Eds.), Hybrid Systems. VIII, 474 pages. 1993.

Vol. 737: J. Calmet, J. A. Campbell (Eds.), Artificial Intelligence and Symbolic Mathematical Computing. Proceedings, 1992. VIII, 305 pages. 1993.

Vol. 739: H. Imai, R. L. Rivest, T. Matsumoto (Eds.), Advances in Cryptology – ASIACRYPT '91. X, 499 pages. 1993.